VOLUME ONE HUNDRED AND FIFTY

ADVANCES IN CANCER RESEARCH

Autophagy and Senescence in Cancer Therapy

VOLUME ONE HUNDRED AND FIFTY

ADVANCES IN CANCER RESEARCH

Autophagy and Senescence in Cancer Therapy

Edited by

DAVID A. GEWIRTZ
Department of Pharmacology and Toxicology and Medicine; Massey Cancer Center, Goodwin Research Laboratories, Virginia Commonwealth University, Richmond, VA, United States

PAUL B. FISHER
Department of Human and Molecular Genetics; VCU Institute of Molecular Medicine; VCU Massey Cancer Center, Virginia Commonwealth University, School of Medicine, Richmond, VA, United States

ELSEVIER

ACADEMIC PRESS
An imprint of Elsevier

Academic Press is an imprint of Elsevier
50 Hampshire Street, 5th Floor, Cambridge, MA 02139, United States
525 B Street, Suite 1650, San Diego, CA 92101, United States
The Boulevard, Langford Lane, Kidlington, Oxford OX5 1GB, United Kingdom
125 London Wall, London, EC2Y 5AS, United Kingdom

First edition 2021

ISBN: 978-0-12-824158-5
ISSN: 0065-230X

For information on all Academic Press publications
visit our website at https://www.elsevier.com/books-and-journals

Publisher: Zoe Kruze
Developmental Editor: Naiza Ermin Mendoza
Production Project Manager: James Selvam
Cover Designer: Alan Studholme

Typeset by SPi Global, India

Contents

Contributors

Corinne Abbadie
Univ Lille, CNRS, Inserm, CHU Lille, UMR9020-U1277-CANTHER-Cancer Heterogeneity Plasticity and Resistance to Therapies, Lille, France

Ahmad Alhesa
Department of Basic Medical Sciences, Faculty of Medicine, The Hashemite University, Zarqa, Jordan

Alexandru Almasan
Department of Cancer Biology, Lerner Research Institute; Department of Radiation Oncology, Taussig Cancer Institute, Cleveland Clinic; Case Comprehensive Cancer Center, Case Western Reserve University School of Medicine, Cleveland, OH, United States

Enas Alwohosh
Department of Basic Medical Sciences, Faculty of Medicine, The Hashemite University, Zarqa, Jordan

Monique Bernard
Centre de recherche du Centre hospitalier de l'Université de Montréal (CRCHUM); Institut du cancer de Montréal, Montreal, QC, Canada

Sarah Bloukh
Department of Basic Medical Sciences, Faculty of Medicine, The Hashemite University, Zarqa, Jordan

Agnieszka Bojko
Laboratory of Molecular Bases of Aging, Nencki Institute of Experimental Biology, Polish Academy of Sciences, Warsaw, Poland

Guillaume B. Cardin
Centre de recherche du Centre hospitalier de l'Université de Montréal (CRCHUM); Institut du cancer de Montréal, Montreal, QC, Canada

Liam D. Cassidy
Cancer Research UK Cambridge Research Institute, University of Cambridge, Li Ka Shing Centre, Cambridge, United Kingdom

Apostolos Christopoulos
Centre de recherche du Centre hospitalier de l'Université de Montréal (CRCHUM); Institut du cancer de Montréal; Otolaryngology-Head and Neck Surgery Service, Centre Hospitalier de l'Université de Montréal (CHUM), Montreal, QC, Canada

Swadesh K. Das
Department of Human and Molecular Genetics; VCU Institute of Molecular Medicine; VCU Massey Cancer Center, Virginia Commonwealth University, School of Medicine, Richmond, VA, United States

Magdalena Dudkowska
Laboratory of Molecular Bases of Aging, Nencki Institute of Experimental Biology, Polish Academy of Sciences, Warsaw, Poland

Luni Emdad
Department of Human and Molecular Genetics; VCU Institute of Molecular Medicine; VCU Massey Cancer Center, Virginia Commonwealth University, School of Medicine, Richmond, VA, United States

Paul B. Fisher
Department of Human and Molecular Genetics; VCU Institute of Molecular Medicine; VCU Massey Cancer Center, Virginia Commonwealth University, School of Medicine, Richmond, VA, United States

David A. Gewirtz
Department of Pharmacology and Toxicology and Medicine; Massey Cancer Center, Goodwin Research Laboratories, Virginia Commonwealth University, Richmond, VA, United States

Chanhee Kang
School of Biological Sciences, Seoul National University, Seoul, South Korea

Jaejin Kim
School of Biological Sciences, Seoul National University, Seoul, South Korea

Mi-Sung Kim
School of Biological Sciences, Seoul National University, Seoul, South Korea

Yeonghyeon Lee
School of Biological Sciences, Seoul National University, Seoul, South Korea

Nicolas Malaquin
Centre de recherche du Centre hospitalier de l'Université de Montréal (CRCHUM); Institut du cancer de Montréal, Montreal, QC, Canada

Masashi Narita
Cancer Research UK Cambridge Research Institute, University of Cambridge, Li Ka Shing Centre, Cambridge, United Kingdom

Nipa H. Patel
Department of Pharmacology and Toxicology and Medicine; Massey Cancer Center, Goodwin Research Laboratories, Virginia Commonwealth University, Richmond, VA, United States

Olivier Pluquet
Univ Lille, CNRS, Inserm, CHU Lille, UMR9020-U1277-CANTHER-Cancer Heterogeneity Plasticity and Resistance to Therapies, Lille, France

Francis Rodier
Centre de recherche du Centre hospitalier de l'Université de Montréal (CRCHUM); Institut du cancer de Montréal; Université de Montréal, Département de radiologie, radio-oncologie et médicine nucléaire, Montreal, QC, Canada

Kyeonghwan Roh
School of Biological Sciences, Seoul National University, Seoul, South Korea

Tareq Saleh
Department of Basic Medical Sciences, Faculty of Medicine, The Hashemite University, Zarqa, Jordan

Sarah Saoudaoui
Centre de recherche du Centre hospitalier de l'Université de Montréal (CRCHUM); Institut du cancer de Montréal, Montreal, QC, Canada

Arishya Sharma
Department of Cancer Biology, Lerner Research Institute, Cleveland Clinic, Cleveland, OH, United States

Ewa Sikora
Laboratory of Molecular Bases of Aging, Nencki Institute of Experimental Biology, Polish Academy of Sciences, Warsaw, Poland

Karolina Staniak
Laboratory of Molecular Bases of Aging, Nencki Institute of Experimental Biology, Polish Academy of Sciences, Warsaw, Poland

Sarmistha Talukdar
Department of Human and Molecular Genetics; VCU Institute of Molecular Medicine, Virginia Commonwealth University, School of Medicine, Richmond, VA, United States

Andrew R.J. Young
Cancer Research UK Cambridge Research Institute, University of Cambridge, Li Ka Shing Centre, Cambridge, United Kingdom

Preface

Autophagy is induced in tumor cells exposed to chemotherapy and radiation and can often be considered a "first responder." Autophagy can express a variety of functions in a tumor cell, one of the most prominent being cytoprotective, in that its inhibition is often shown to sensitize the tumors to the autophagy-initiating stimulus or stress. There is also data suggesting that this form of autophagy can confer therapeutic resistance to therapy. However, there is also evidence that in certain cases, autophagy may actually promote tumor cell death (autophagic cell death) or be a precursor to apoptotic cell death. Moreover, it has also been reported extensively on what is termed nonprotective autophagy, where autophagy inhibition by either pharmacological or genetic approaches fails to alter the therapeutic response.

Like autophagy, senescence is now recognized as one of the primary responses to chemotherapy and radiation in tumor cells that fail to undergo apoptosis or other modes of cell death. Senescence is a prolonged and durable growth arrest that has long been considered to be irreversible, which would potentially be a desirable outcome of cancer therapy. However, evidence has accumulated in recent years that at least a subset of the senescent tumor populations may be capable of escaping from senescence and recovering proliferative capacity. If this is, in fact, the case, then senescence can be envisioned as a temporary avenue of arrest whereby the tumor cells avoid signaling pathways that would otherwise promote cell death. Consequently, senescent tumor cells might also be considered a reservoir and a form of dormancy from which the cells can ultimately escape, by as yet undetermined mechanisms, thereby contributing to disease recurrence that takes place months or years after the patient has ostensibly been cured of the primary disease. While recovery in preclinical studies might reflect subpopulations of tumor cells that are in what might be a "shallow" rather than "deep" form of senescence, tumor dormancy and disease recurrence in patients suggests that reemergence of the disease can occur even from a "deep" (i.e.,) extremely prolonged state of growth arrest.

Another deleterious aspect of senescence is the associated senescence-associated secretory phenotype that is comprised of a spectrum of chemokines and cytokines that can promote tumor growth. Finally, it currently remains uncertain as to whether senescent tumor cells are recognized and cleared by the immune system or are actually immunosuppressive. It is largely for these reasons that the scientific community has recently embraced the potential use of agents that are senolytics, with the goal of eliminating the senescent tumor population in order to enhance the durability of the response to chemotherapy and radiation.

Both autophagy and senescence tend to occur in tandem, with autophagy generally considered to precede the senescence response, thereby lending credibility to the premise that these are interdependent outcomes. Furthermore, it would appear logical that autophagy might provide the pool of energy that allows the cells to enter and maintain the prolonged senescence state. However, the literature has thus far failed to establish any consistency for the "relationship" between these two outcomes. One limitation to the experimental efforts to understand whether senescence might be "autophagy-dependent" is that inhibition of the cytoprotective form of autophagy, by definition, results in tumor cell death; consequently, it becomes challenging to determine whether the disappearance of senescent tumor cells when autophagy is inhibited, either pharmacologically or genetically, is a direct result of autophagy cessation or simply an indirect consequence of the elimination of the tumor cells. Conversely, studies relating to senescence have been handicapped by the absence of specific genetic or pharmacologic inhibitors of this response.

In the case of both autophagy and senescence, a central deficiency in our understanding of their clinical implications is the dearth of information as to what extent these responses occur in patient tissues upon treatment with conventional or targeted therapies or radiation, and the degree to which senescence and/or autophagy influence tumor sensitivity or resistance to therapy.

The current special issue of Advances in Cancer provides reports by leaders in the field that address the nature and relevance of autophagy and senescence in various preclinical models and, in some cases, in clinical cancer therapy.

The chapter by Patel, Bloukh, Alwohosh, Alhesa, Saleh, and Gewirtz focuses on the induction of autophagy and senescence, individually and concomitantly, in response to anticancer therapy. Both autophagy and senescence share a number of regulators, including p53, oxidative stress and mTOR, an inhibitory effect on autophagy through ubiquitin-mediated beclin-1 degradation and inhibition of AMPK. Another common meditator of both autophagy and senescence is oxidative stress. Anticancer therapies, such as radiation and cisplatin, can result in increased ROS levels due to mitochondrial dysfunction and uncoupling, as well as genotoxic stress.

Autophagy serves to alleviate the cellular burden of the damaged proteins and mitochondria and to remove stressors, such as damaged mitochondrial DNA (mtDNA), which may stimulate senescence induction. While autophagy is induced in response to oxidative stress, prolonged autophagy impairment in the presence of oxidative stress can result in senescence induction. Furthermore, excessive ROS accumulation can further exacerbate DNA damage, result in $p53/p21^{Cip1/Mda\text{-}6/Waf1}$ activation and promote senescence. Taken together, pathways activated in response to therapy-induced damage can regulate both autophagy and senescence.

There is evidence that autophagy inhibition with HCQ can decrease survival of dormant breast cancer cells and reduce lung metastasis; however, autophagy inhibition is minimally effective once dormant cells have regained proliferative capacity, suggesting a role for autophagy in sustaining tumor dormancy in breast cancer cells. Consequently, senescence may be a potential tumor dormancy model and further studies are warranted to interrogate the contributions of autophagy in senescence-associated tumor dormancy and recovery. The chapter also discusses the use of senolytics as adjuvant cancer therapy, a proposition that is rapidly gaining interest in the field. Senolytics can be used in an effort to eliminate the persistent senescent tumor cells and interfere with their ability to recover, and thus, senolytics might be beneficial in mitigating cancer recurrence.

The chapter by J. Kim, Lee, Roh, M-S. Kim, and Kang discusses dual roles of autophagy and senescence in tumorigenesis and highlights their complex relationship, requiring consideration for autophagy- and senescence-based cancer treatment. Under normal conditions, autophagy suppresses a variety of cellular stresses, maintaining cellular homeostasis and thus limiting tumorigenesis. Once tumorigenesis is triggered, however, autophagy contributes to the stress support pathway in cancer cells by relieving metabolic stress, maintaining redox homeostasis and proteostasis, promoting therapeutic resistance to chemotherapies, modulating cancer cell survival pathways, and promoting immune evasion. The authors then catalogue several properties of the stress support network associated with autophagy addiction for cancer treatment. For example, cancer cells addicted to autophagy for immune surveillance may display reduced levels of MHC I molecules, inflammasome formation, and immune cell infiltration. They propose that autophagy inhibition will be favored under this situation to recover immune surveillance for cancer treatment. The authors also describe how senescence both promotes and inhibits tumorigenesis with a specific focus on the senescence-associated secretory phenotype (SASP).

Although senescence serves as a potent tumor-suppressive mechanism by ensuing stable growth arrest of premalignant cells, senescence can also promote nearly every aspect of tumorigenesis mainly through the SASP, including tumor growth, chemotherapy resistance, tumor stemness, metastasis, and immune suppression. Based on these associations, the authors highlight that senolytics (agents that selectively remove senescent cells) and senostatics (agents that do not kill senescent cells but modulate the SASP) could tilt the balance toward maximizing the beneficial effects of senescence while suppressing its deleterious effects for cancer treatment. Lastly, this chapter by Kim et al. notes the intertwined relationship between autophagy and senescence. Before senescence is induced, autophagy acts to maintain cellular homeostasis, suppressing senescence in a passive manner. Autophagy can also suppress senescence by directly targeting critical regulators of senescence. For example, p62-dependent selective autophagy degrades a key regulator of the SASP, GATA4, suppressing the SASP and senescence. Once senescence is triggered, however, autophagy can contribute to senescence establishment by providing amino acids to support the massive synthesis of SASP factors. Autophagy can also target negative regulators of senescence for degradation, thus actively promoting senescence. Furthermore, autophagy can dampen several senescence-associated stresses, maintaining the viability of senescent cells. Thus, their complex relationship should be carefully considered when autophagy and senescence are employed as therapeutic targets for cancer treatment. For example, the authors note that autophagy inhibition could decrease the efficacy of chemotherapies to induce senescence. In addition, as autophagy may be necessary for senescence maintenance, its inhibition after senescence inducing treatment could facilitate the escape severely damaged, potentially premalignant cells from senescence, rather accelerating, not suppressing, tumorigenesis.

The chapter by Drs. Young, Cassidy, and Narita describes the utility and insights that have been generated from genetic mouse models of aging and cancer. A particularly critical conclusion from their analysis of this literature is that elimination of senescent cells that accumulate in tissue with aging clearly has the potential to increase lifespan. Conversely, the apparent decline in autophagic capacity with age may contribute to the accumulation of senescent cells and their deleterious impact.

The chapter by Talukdar, Das, Emdad, and Fisher addresses the involvement of autophagy and senescence in stem cells. They note that autophagic-specific factors can also influence the differentiation of these multipotent SCs towards specific cell lineages, as well as impact aging-related properties, including senescence. Autophagy promotes negative characteristics of CSCs that are often unregulated in CSCs and display dual functions relative to inducing senescence and dormancy. It is currently assumed that autophagy helps to maintain the stemness of CSCs and results in the failure of anticancer therapies. Conversely, autophagy inhibition may provide a desirable outcome for CSCs thereby decreasing their resistance to anticancer therapies.

The chapter by Dudkowska, Staniak, Bojko, and Sikora focuses on the escape from senescence and the relationship of this escape to polyploidy in senescence.

Their work presents arguments for dysfunctional autophagy in an established senescence phenotype and potentially also for a role of active autophagy in the development of the senescence phenotype, during which the cell interior is rebuilt. They speculate that development of the senescence phenotype does not rely on autophagy but that senescence is frequently associated with diminished autophagic flux. They further hypothesize that autophagy is crucial for division of senescent/polyploid cancer cells, that is escape from senescence.

The chapter by Sharma and Almasan discusses DNA damage-induced autophagy and its role in regulating genomic stability and senescence in prostate and lung cancer. It describes how the dose of DNA damaging agents determines senescence *vs* cell death, which, in turn, is attributed to the amount of DNA damage and the ensuing DNA damage response. The dichotomous nature of senescence and the SASP is presented, and the various factors that regulate SASP phenotype are briefly discussed, with a focus on cGAS/STING signaling as a crucial link between DNA damage response and SASP. While both autophagy and senescence occur concurrently, and share multiple common upstream regulators of overlapping signaling mechanisms, the deterministic molecular links and mechanistic insights into regulation of senescence by autophagy remain elusive. And, finally evidence is presented that supports PTEN-loss and a mechanism of inducing cellular senescence as a failsafe process to defend against tumorigenesis.

The chapter by Pluquet and Abbadie develops the theme of the endoplasmic reticulum stress/unfolded protein response (UPR) in senescence and potentially the relationship to autophagy. A fundamental question is to understand why and how a cell acquires and maintains a senescent phenotype. Both in vitro and in vivo data support the hypothesis that the UPR (an adaptive response to counteract the endoplasmic reticulum proteostasis imbalance) plays a broader active role on the characteristic features of senescent cells. However, the effectors that link UPR to these hallmarks remain unknown. Modulating the UPR to limit senescence may in fact represent a potential therapeutic tool whenever senescence is involved. They speculate that senescent cells require an optimal UPR duration/intensity for their long-term survival. However, it is unknown whether targeting the UPR to overload the endoplasmic reticulum in senescent cells could potentially induce cell death. They hypothesized that specific pharmacological compounds—unknown to date—may act as senolytics by selectively activating the UPR arms to accelerate cell death. This would have the advantage of preventing senescence escape, and as a consequence perhaps also—in longer term—neoplastic transformation. Alternatively, inhibiting the adaptive UPR signals by several existing compounds may suppress or prevent, at least partially, some senescent characteristics. Therefore, by slowing down the onset of senescence, they could reduce the postsenescence neoplastic escape. Further data will determine which strategy may have the best potential.

The chapter by Saoudaoui, Bernard, Cardin, Malaquin, Christopoulos, and Rodier is focused on the role of mTOR in both autophagy and senescence both in models of aging- and therapy-induced senescence. mTOR is a central molecule in the promotion of both autophagy and senescence and represents a targetable signaling pathway. mTOR inhibitors such as rapamycin and rapamycin analogs are currently in various stages of clinical testing both in cancer therapeutics and aging-related pathologies. The chapter explores the possibility that the mTOR pathway could be used as a target to modify phenotypes of cell senescence that are important and relevant for affecting treatment outcome.

David A. Gewirtz, Ph.D.
Department of Pharmacology and Toxicology,
Virginia Commonwealth University,
School of Medicine, Richmond,
Virginia, 23239, United States

Paul B. Fisher, M.Ph., Ph.D., FNAI
Department of Human and Molecular Genetics,
VCU Institute of Molecular Medicine (VIMM),
VCU Massey Cancer Center,
Virginia Commonwealth University,
School of Medicine, Richmond,
Virginia 23239, United States

CHAPTER ONE

Autophagy and senescence in cancer therapy

Nipa H. Patel[a,b], Sarah Bloukh[c], Enas Alwohosh[c], Ahmad Alhesa[c], Tareq Saleh[c], and David A. Gewirtz[a,b,*]

[a]Department of Pharmacology and Toxicology and Medicine, Virginia Commonwealth University, Richmond, VA, United States
[b]Massey Cancer Center, Goodwin Research Laboratories, Virginia Commonwealth University, Richmond, VA, United States
[c]Department of Basic Medical Sciences, Faculty of Medicine, The Hashemite University, Zarqa, Jordan
*Corresponding author: e-mail address: david.gewirtz@vcuhealth.org

Contents

Abstract

Tumor cells can undergo diverse responses to cancer therapy. While apoptosis represents the most desirable outcome, tumor cells can alternatively undergo autophagy and senescence. Both autophagy and senescence have the potential to make complex contributions to tumor cell survival *via* both cell autonomous and cell non-autonomous pathways. The induction of autophagy and senescence in tumor cells, preclinically and clinically, either individually or concomitantly, has generated interest in the utilization of autophagy modulating and senolytic therapies to target autophagy and senescence,

Advances in Cancer Research, Volume 150
ISSN 0065-230X
https://doi.org/10.1016/bs.acr.2021.01.002

respectively. This chapter summarizes the current evidence for the promotion of autophagy and senescence as fundamental responses to cancer therapy and discusses the complexity of their functional contributions to cell survival and disease outcomes. We also highlight current modalities designed to exploit autophagy and senescence in efforts to improve the efficacy of cancer therapy.

Abbreviations

3-MA	3-methyladenine
5-FU	5-fluorouracil
AD	Alzheimer's disease
AMPK	AMP-activated protein kinase
ATG	autophagy regulatory proteins
ATG	autophagy-related gene
ATM	ataxia-telangiectasia mutant
BafA1	bafilomycin A1
CDK	cyclin-dependent kinase
CDKI	cyclin-dependent kinase inhibitor
CMA	chaperone-mediated autophagy
CQ	chloroquine
CTL	cytotoxic T lymphocyte
DAMP	danger-associated molecular pattern
DC	dendritic cells
DDR	DNA damage repair response
DRAM	damage-regulated autophagy modulator
Drp1	dynamin-related protein 1
DSB	DNA double-stranded breaks
EGFR	epidermal growth factor receptor
ER	estrogen receptor
GBM	glioblastoma multiforme
GSC	glioma stem cell
H3K9Me3	H3 lysine 9 tri-methylation
HCC	hepatocellular carcinoma
HCQ	hydroxychloroquine
HD	Huntington's disease
HDAC	histone deacetylase
HIF-1α	hypoxia-inducible factor-1α
HP1	heterochromatin protein 1
ICB	immune-checkpoint blockade
MAP1-LC3	microtubule-associated protein light chain 2
MHC-I	major histocompatibility complex class I
MPR	mannose-6-phosphate receptor
mtDNA	mitochondrial DNA
mTOR	mammalian target of rapamycin
MTX	mitoxantrone
NAC	*N*-acetyl-L-cysteine
NCI	National Cancer Institute
NSCLC	non-small cell lung carcinoma

OXA	oxaliplatin
PD	Parkinson's disease
PDAC	pancreatic ductal adenocarcinoma
PDGF	platelet-derived growth factor
PGE2	prostaglandin E2
RIPK 1	receptor-interacting protein kinase 1
RIPK 3	receptor-interacting protein kinase 3
ROS	reactive oxygen species
SAHA	suberoylanilide hydroxamic acid
SAHF	senescence-associated heterochromatin foci
SASP	senescence-associated secretory phenotype
SA-β-gal	senescence-associated β-galactosidase
SQSTM1	sequestosome 1
TASCC	TOR-autophagy spatial coupling compartment
TIA	therapy-induced autophagy
TIS	therapy-induced senescence
TME	tumor microenvironment
TNF	tumor necrosis factor
TP53BP1	p53 binding protein 1
TRAIL	TNF-related apoptosis-inducing ligand
ULK-1	Unc-51 like autophagy activating kinase 1
VEGF	vascular endothelial growth factor
Vps18	vacuolar protein sorting protein 18
Vps34	vacuolar protein sorting protein 34
X-gal	5-bromo-4-chloro-3-indolyl-β-D-galactopyranoside

1. Introduction

Tumor cells have the potential to exhibit a relatively wide spectrum of responses to conventional and targeted cancer therapies that will determine the outcome of treatment (Fulda, Gorman, Hori, & Samali, 2010). Classically, the ideal consequence of exposure to cancer therapy is for the tumor cells to undergo apoptosis or programmed cell death (Carneiro & El-Deiry, 2020). Apoptosis is a highly regulated cytotoxic process that is propagated through two primary pathways; the mitochondrial (intrinsic) and the death receptor (extrinsic) pathways (Fulda & Debatin, 2006). The intrinsic pathway involves permeabilization of the mitochondrial outer membrane and the release of cytochrome C, while the extrinsic pathway is induced upon the activation of death receptors including Fas, the tumor necrosis factor (TNF) receptors, and the TNF-related apoptosis-inducing ligand (TRAIL) receptors (Fulda & Debatin, 2006; Li & Yuan, 2008).

Both pathways are ultimately executed by a complex machinery resulting in the activation of caspases, which catalyze the proteolysis of cellular proteins, suborganellar structures and the endonucleolytic breakdown of chromosomal DNA, leading to cell death (Li & Yuan, 2008). Necroptosis is an alternative form of irreversible cell death, characterized by the activation of the receptor-interacting protein kinase 1 and 3 (RIPK 1 and RIPK3), rapid membrane permeabilization and release of intracellular constituents (Moriwaki, Bertin, Gough, Orlowski, & Chan, 2015; Tang et al., 2020), while pyroptosis is an inflammatory form of programmed cell death in which caspase-1 activation in response to cytokine signals mediates membrane pore formation, nuclear fragmentation and release of pro-inflammatory cytokines (Bergsbaken, Fink, & Cookson, 2009). Ferroptosis is a functionally and morphologically distinct form of iron-dependent, non-apoptotic cell death associated with oxidative stress (Li et al., 2020). Finally, mitotic catastrophe is a cell death response which is induced when cells fail to pass the mitotic checkpoints due to accumulation of defects, including—but not limited to—DNA damage, which eventually leads to cell death (Galluzzi et al., 2012). Several studies reported that induction of either necroptosis or pyroptosis can boost the immune response against cancer cells by activating macrophage engulfment of the necroptotic and pyroptotic cells (Cai et al., 2016; Krysko et al., 2017; Tang et al., 2020; Wang et al., 2018). Moreover, evidence has been generated implicating mitotic catastrophe in the inhibition of the progression of lung, colorectal and ovarian cancers (Amawi et al., 2017). Collectively, apoptosis, necrosis, necroptosis, pyroptosis and ferroptosis as well as mitotic catastrophe have all been shown to be induced in tumor cells upon exposure to cancer therapy (Kaufmann & Earnshaw, 2000; Su, Yang, Xie, Dewitt, & Chen, 2016; Wang et al., 2017; Wu et al., 2020).

Despite the extensive focus of the scientific literature on these responses for decades, it has also long been recognized that irreversible cell death is not the sole response to cancer therapy; in fact, it is likely that for any treatment modality, the tumor cells response to therapy will be heterogenous, including cell death (Fulda et al., 2010), but also autophagy and senescence (Portugal, Mansilla, & Bataller, 2009; Saleh, Bloukh, et al., 2020; Zheng, He, Kitazato, & Wang, 2019). Autophagy is a highly selective and programmed recycling process, whereby damaged DNA, proteins, and organelles are sequestered into vesicles and targeted for degradation (Murrow & Debnath, 2013). Classically, autophagy exhibits a pro-survival function in cells exposed to chemotherapy or radiation through DNA damage repair, removal of misfolded proteins and damaged organelles, scavenging of reactive oxygen species (ROS), energetic regulation and the activation of cytoprotective signals that block apoptosis (Moreau, Luo, & Rubinsztein, 2010; Ulasov, Fares, Timashev, & Lesniak, 2019). Nonetheless, current studies have demonstrated the complexity of the autophagic response in tumor cells following exposure to cancer therapeutics, wherein autophagy can exhibit cytotoxic (autophagic cell death) or cytostatic functions that conversely promote tumor cell dysfunction (Gewirtz, 2014a). Finally, autophagy can exhibit a neutral effect on survival, serving a poorly understood, nonprotective role (Gewirtz, 2020).

Cellular senescence is a state in which cells are stably growth arrested in response to stress (Sikora, Mosieniak, & Alicja Sliwinska, 2016). In nonmalignant cells, senescence can be induced in response to replicative-, oxidative-, or genotoxic stresses (Campisi, 2001). In malignant cells, senescence can be induced in response to a variety of conventional and targeted therapeutics, a process often referred to as therapy-induced senescence (TIS) (Saleh, Bloukh, et al., 2020). Senescent cells are characterized by several distinctive features including structural changes, macromolecular damage, oxidative stress, energetic dysregulation and wide genetic and epigenetic alterations (Hernandez-Segura, Nehme, & Demaria, 2018). Despite the durability of the growth arrest in senescent cells, the cells maintain metabolic activity and are capable of modulating the microenvironment via the secretion of several soluble and insoluble factors, collectively termed the senescence associated secretory phenotypes (SASPs) (Gorgoulis et al., 2019; Hernandez-Segura et al., 2018). Senescence is unique in that it uniformly commits the cells into a prolonged, durable growth arrested state (Pack, Daigh, & Meyer, 2019). In contrast, quiescence, another form of growth arrest induced by starvation or loss of adhesion, generally reflects a more transient and rapidly reversible form of growth inhibition (Pack et al., 2019). Furthermore, quiescent cells lack the characteristic morphological, biochemical and molecular alterations associated with senescence. Nevertheless, it has become evident that senescent tumor cells can evade the durable growth arrest, recover proliferative functionality, and acquire more aggressive malignant traits that may further compromise the effectiveness of therapy (Saleh, Bloukh, et al., 2020).

Evasion of apoptotic cell death is a central hallmark of cancer cells, which highlights the importance of the other non-apoptotic cell responses that potentially serve as alternative routes by which tumor cells respond to cancer treatment instead of cell death (Hanahan & Weinberg, 2011). Autophagy and senescence are the major cell stress response mechanisms that are likely to serve as avenues whereby cancer cells delay the onset of cell death or possibly develop phenotypes of resistance to therapy (Holohan, Schaeybroeck, Van Longley, & Johnston, 2013). The induction of cytotoxic/cytostatic autophagy could possibly serve as an effective approach for suppressing cancer by interfering with tumor proliferation (Gewirtz, 2009, 2014b). However, autophagy has been reported to confer resistance to therapy in a variety of tumor cell types (Sui et al., 2013), which has provided the basis for a plethora of preclinical and clinical studies that target pro-survival autophagic processes to enhance the efficacy of cancer therapeutics or to overcome acquired resistance to therapy (Kumar, Singh, & Chaudhary, 2015). Moreover, the recent notion of targeting autophagic pathways has emerged as a possible avenue for enhancing tumor killing, particularly in cancer cells that are more dependent on autophagy than normal cells, due to the protumorigenic function of autophagy (Thorburn, Thamm, & Gustafson, 2014; Zheng et al., 2019). Similarly, while senescence has long been considered as a favorable outcome of cancer therapy, recent observations suggested the possibility of a direct association of senescence with cancer relapse, tumor dormancy, development of more aggressive phenotypes and a variety of therapy-associated adverse effects (Demaria et al., 2017; Saleh, Tyutyunyk-Massey, & Gewirtz, 2019). Accordingly, clearance of senescent tumor cells using a newly emerging drug class termed senolytics has been proposed as a novel approach for the treatment of cancer (Short, Fielder, Miwa, & von Zglinicki, 2019). In this chapter, we attempt to summarize the established evidence for the induction of autophagy and senescence in response to cancer therapy, discuss their potential contributions to overall therapeutic outcomes and highlight state-of-the-art pharmacological therapies designed to target these processes for cancer treatment.

2. Therapy-induced autophagy (TIA)

2.1 Autophagy overview

The word "autophagy" is derived from the Greek words "auto" meaning self and "phagy" meaning eating to describe the cell's distinct self-degradative process designed to maintain organellar and energy turnover during injurious events (Glick, Barth, & Macleod, 2010). Accordingly, autophagy is an evolutionarily conserved intracellular catabolic process through which proteins, organelles, and pathogens are degraded for waste elimination or repurposed for the anabolic cellular needs (Feng, He, Yao, & Klionsky, 2014). Autophagy is frequently activated as part of homeostatic processes in response to cellular stresses, such as hypoxia or nutrient deprivation (Brahimi-Horn, Chiche, & Pouysségur, 2007; Maycotte & Thorburn, 2011).

Autophagy often incorporates a broader, more ubiquitous degradative process, but, in fact, there are multiple types of autophagy. The three primary types of autophagy include macroautophagy, microautophagy, and chaperone-mediated autophagy (Feng et al., 2014). In macroautophagy, components of the cytoplasm and dysfunctional organelles are sequestered into the growing phagophore, which is a *de novo* cytosolic double-membrane vesicle that engulfs cytoplasmic proteins and organelles and delivers them to the lysosome (Glick et al., 2010). Microautophagy occurs when cargo is directly taken up through invagination of the lysosomal membrane (Mijaljica, Prescott, & Devenish, 2011). While macro- and microautophagy can vary in their selectivity for the recycled cellular components, chaperone-mediated autophagy (CMA) is highly specific (Parzych & Klionsky, 2014). CMA involves the transport of unfolded proteins directly into the lysosomal membrane by heat-shock proteins, which recognize a specific consensus motif on the target protein (Feng et al., 2014; Parzych & Klionsky, 2014). For the majority of this book chapter, we will focus on macroautophagy, to which we will refer as autophagy.

During the macroautophagic process, a double-membrane vacuolar compartment is formed, the vacuoles fuse with lysosome and the degraded contents are released into the cytosol. Under normal conditions, mammalian target of rapamycin (mTOR) complex 1 negatively regulates autophagy by phosphorylating and binding to Unc-51 like autophagy activating kinase 1 (ULK-1), thus inactivating ULK-1 (Kim, Kundu, Viollet, & Guan, 2011). Under conditions of cellular stress, ULK-1 becomes dephosphorylated and dissociates from mTOR (Kim et al., 2011). Initiation of autophagy occurs through the activation of the ULK-1 complex, which triggers the nucleation of the phagophore through phosphorylation of PI3KC3, VPS34 and Beclin-1 (Dikic & Elazar, 2018; Petherick et al., 2015). While the source of the preautophagosomal structure is not fully understood, there is evidence supporting phagophore generation from the endoplasmic reticulum (Ylä-Anttila, Vihinen, Jokitalo, & Eskelinen, 2009), mitochondria (Hailey et al., 2010), Golgi apparatus (Geng, Nair, Yasumura-Yorimitsu, & Klionsky, 2010) and recycling endosomes (Puri et al., 2018).

Initial biogenesis of the autophagosome begins with the formation of the phagophore that is further elongated by the recruitment of autophagy regulatory proteins (ATG), such as ATG5 and ATG7. Microtubule-associated protein light chain 2 (MAP1-LC3) is lipidated by ATG3-mediated conjugation and incorporated into the growing phagophore, and is necessary for the closure, fusion and maturation of the autophagosome (Gammoh & Wilkinson, 2014; Lee & Lee, 2016). Cargo is sequestered into the autophagosome by SQSTM1/p62, a sequestosome that binds ubiquitinated proteins and anchors itself to LC3 located on the inner membrane of the autophagosome (Gammoh & Wilkinson, 2014). The growing phagophore extends until the two ends join and fuse together to form a double-membrane vesicle, or the autophagosome. In the final step, the mature autophagosome fuses with the lysosome to form the autolysosome. The acidic hydrolases of the lysosome then degrade the cargo within the autophagosome, which is then released into the cytoplasm for cellular repurposing (Mizushima, 2007).

The term "autophagy" was coined by Christian de Duve in the 1960s while studying lysosomes, for which he later won the Nobel Prize in Physiology or Medicine in 1974. He observed what appeared to be a double-membrane vesicle sequestering elements of the cytoplasm and other cellular organelles (Klionsky, 2007). Using electron microscopy, de Duve was able to describe the end stages of autophagy and identify autophagy morphologically. It was not until the 1990s that the mechanistic components of autophagy were demonstrated by Yoshinori Ohsumi and colleagues, who were able to identify numerous genes involved in regulating the autophagic machinery and phagophore formation by studying autophagy mutants in yeast, which led to the Nobel Prize in Physiology or Medicine in 2016. These studies provided the backbone for our current understanding of the autophagic molecular mechanism. As of this writing, 31 autophagy-related genes (*ATGs*) have been discovered, coding for various components of the autophagic machinery and essential components required for macroautophagy. Through loss-of-function studies, it has been demonstrated that autophagy is important for multiple aspects of an organism's lifespan, including maintaining cellular and tissue homeostasis, metabolism, immunity, protection against aging and early differentiation and development, as well as its role in disease states, such as cancer (Levine, Packer, & Codogno, 2015).

Autophagy has been associated with a number of pathologies, including but not limited to, cardiac disease, neurodegenerative diseases, such as Parkinson's disease (PD), Alzheimer's disease (AD) and Huntington disease (HD), and cancer (Levine & Kroemer, 2008). Specifically, autophagy has been implicated to have dual functions in cancer development and therapeutics, exhibiting both tumor-promoting and tumor-suppressive roles. Mice with *Beclin-1*$^{-/-}$ embryonic stem cells were shown to die during embryogenesis, suggesting that autophagy is essential in early development. More intriguingly, a greater number of haplosufficient *Beclin-1*$^{+/-}$ mice developed spontaneous tumors when compared to wildtype mice, suggesting that Beclin-1 may serve as a tumor-suppressor gene and that mutations in *Beclin-1*, and consequently autophagy dysregulation, could contribute to tumorigenesis (Liang et al., 1999; Takamura et al., 2011; Yue, Jin, Yang, Levine, & Heintz, 2003). With regard to the tumor-suppressive effects of autophagy, induction of autophagy prevents the accumulation of damaged organelles, protein aggregates and promotes the removal of oncogenic proteins (Elgendy, Sheridan, Brumatti, & Martin, 2011; Galluzzi et al., 2015; Mathew et al., 2009). Furthermore, autophagy induction in healthy tissue functions as a tumor-suppressive mechanism by removing dysfunctional mitochondria and ROS, which may cause DNA damage, thereby, maintaining genomic stability (Galluzzi et al., 2015; Poillet-Perez, Despouy, Delage-Mourroux, & Boyer-Guittaut, 2015). Autophagy induction also plays a role in eliciting an immune response through modulation of immunogenic cell death (ICD) by contributing to the secretion of danger-associated molecular patterns (DAMPs) (Martins et al., 2014), secretion of cytokines, as well as antigen processing (Chemali, Radtke, Desjardins, & English, 2011; Deng et al., 2014; Levine & Deretic, 2007; Levine, Mizushima, & Virgin, 2011; Li et al., 2010, 2008; Michaud et al., 2011; Xu et al., 2007). Ko et al. demonstrated that autophagy played a role in ATP release in dying cells, a DAMP that interacts with purinergic receptors on immune cells and is involved in the recruitment of immune cells to the tumor microenvironment (TME) (Ko et al., 2014). Thus, autophagy plays a role in immune infiltration in tumors and contributes to multiple aspects of immunosurveillance (Pietrocola, Bravo-San Pedro, Galluzzi, & Kroemer, 2017).

In tumor cells, depending on the extent of genomic and cellular damage, and to allow for the tolerance of stress, autophagy can promote tumor cell survival through clearance of protein aggregates and damaged organelles in response to cytotoxic therapy (Boya et al., 2005; Karantza-Wadsworth et al., 2007; Yang et al., 2011). The catabolic processes of autophagy also provide metabolic intermediates and raw materials, which feed into the intracellular anabolic processes, allowing for maintenance of cellular bioenergetics (Guo et al., 2011; Rabinowitz et al., 2010). Studies by Guo et al. demonstrated that RAS-expressing tumors had higher levels of basal autophagy, and exhibited greater tumor cell survival through mitophagy-induced clearance of dysfunctional mitochondria and maintenance of metabolically functional mitochondria (Guo et al., 2011). Autophagy can also contribute to the maintenance of low intracellular levels of ROS that contribute to activation of pro-survival pathways, such as Src and NF-κB pathways (Giannoni et al., 2008; Gough & Cotter, 2011), resulting in tumor-promotion (Harris et al., 2015; Kaminskyy, Piskunova, Zborovskaya, Tchevkina, & Zhivotovsky, 2012; Poillet-Perez et al., 2015). While autophagy is implicated in tumorigenesis to function in both tumor-promotion and tumor-suppression, autophagy is also induced in response to therapy, presumably as a central mechanism to prevent the accumulation of cellular damage (Levine, 2007; Paglin et al., 2001).

Activation of autophagy is regulated by a number of stress factors, including hypoxia, nutrient starvation, ATP/AMP levels, ROS and microbial infection (Poillet-Perez et al., 2015). Increasing ROS generation results in oxidation of lipids, proteins and DNA, ultimately resulting in global cellular damage and mitochondrial dysregulation contributing to further accumulation of oxidative stress (Poillet-Perez et al., 2015). For example, nutrient starvation-induced production of H_2O_2 results in the direct oxidation of ATG4, ultimately reducing delipidation of ATG4 and increasing autophagosomal LC3 association (Scherz-Shouval et al., 2007). Metabolic stress deregulating ATP/AMK ratios and mitochondrial function activates 5′ AMP-activated protein kinase (AMPK), which directly and indirectly modulates autophagy. AMPK inhibits mTOR, an inhibitor of autophagy, through TCS2 and raptor phosphorylation; thus, AMPK activation indirectly induces autophagy (Shaw, 2009). Furthermore, under stressful conditions, AMPK can directly phosphorylate ULK1 complexes important for the initiation of autophagy (Egan, Kim, Shaw, & Guan, 2011) and can contribute to autophagosomal maturation and lysosomal fusion (Jang et al., 2018). Oxidative stress also activates and stabilizes hypoxia-inducible factor-1α (HIF-1α), which results in its subsequent nuclear translocalization and regulates the gene transcription of essential autophagy machinery (Daskalaki, Gkikas, & Tavernarakis, 2018). Moreover, HIF-1α can regulate autophagy indirectly by altering glucose metabolism, mTOR regulation and unfolded protein responses (Daskalaki et al., 2018). HIF-1α-induced autophagy may contribute to tumor resistance to anticancer treatments, where inhibition of autophagy can restore sensitivity to therapy (Noman et al., 2011; Sun et al., 2015a). Cytotoxic therapies, acting through a multitude of mechanisms, rely largely on extended damage and impaired cellular functions to activate cell death machinery. In an effort to mitigate the cytotoxic effects of cancer therapies, autophagy is often, if not uniformly, activated to remove protein aggregates, nuclear damage and/or depolarized mitochondria (Hughson, Poon, Spowart, & Lum, 2012). Removal of these damaged moieties serves to evade activation of apoptotic cell death pathways and provides raw material for metabolic processes (Abedin, Wang, McDonnell, Lehmann, & Kelekar, 2007); however, depending on the extent of damage, excessive autophagy activation can also potentially result in cell death (Maiuri, Zalckvar, Kimchi, & Kroemer, 2007).

2.2 Autophagy in response to cancer therapy

Autophagy is often induced as a "first response" to chemotherapy and radiotherapy as the cell attempts to clear out incurred damage and evade fatal cell fate pathways (Chaachouay et al., 2011; Zhou et al., 2015). Extensive preclinical and clinical studies have demonstrated autophagy induction in response to therapy (Classen et al., 2019; Paglin et al., 2001; Rosenfeld et al., 2014). Exposure to clinically relevant doses of radiation (2 Gy) demonstrated increased LC3 and decreased SQSTM1/p62 expression based on immunohistochemical staining of 26 out of 30 NSCLC patient samples (Ren et al., 2018). In concurrence, studies by Chaachouay et al. showed that pharmacological autophagy inhibition with 3-methyladenine (3-MA) and chloroquine (CQ) radiosensitized MDA-MB-231 and HBL-100 breast cancer cells (Chaachouay et al., 2011). A screening was performed in which U20S osteosarcoma cells were exposed to 80 National Cancer Institute (NCI) anticancer drugs and the extent of apoptosis, autophagy and necrosis was measured. Of these 80 cytotoxic drugs, 59 drugs induced autophagosome formation and demonstrated autophagic flux in studies where the cells were also exposed to bafilomycin A1 (BafA1), a pharmacological inhibitor of autophagic flux (Shen et al., 2011). The remaining 21 drugs produced predominantly necrosis and apoptosis, with little autophagic flux. These latter agents were further characterized as microtubule inhibitors, which may in part be due to the requirement of microtubule function for autophagosome localization and migration to lysosomes (Köchl, Hu, Chan, & Tooze, 2006). Taken together, while there is clearly a complexity to the cellular responses activated by chemotherapy, autophagy is induced in tumor cells by many anticancer therapeutics.

2.2.1 Autophagy in response to DNA damaging therapy

A primary mechanism whereby a number of anticancer drugs, such as platinum-based compounds and topoisomerase I/II poisons, exert their cytotoxic effects is through DNA damage. DNA double-strand breaks (DSBs) activate ataxia-telangiectasia mutant (ATM), which in turn activates Chk1/2 as well as AMPK (Bordin et al., 2013). Chk1/2 phosphorylate and stabilize p53, which can modulate the transcription of key autophagy-related genes as well as DRAMs, resulting in autophagy induction (Bordin et al., 2013; Crighton et al., 2006). Furthermore, AMPK inhibits mTOR and stimulates autophagy. Autophagy also plays a role in DNA damage repair by providing metabolic precursors for ATP generation required by several DNA damage repair response (DDR) pathways, maintains dNTPs needed for DNA replication and repair and contributes to the turnover of key proteins involved in DDR and DSBs processing (Vessoni, Filippi-Chiela, Menck, & Lenz, 2013). Lin et al. demonstrated that cisplatin, a DNA cross-linking agent, induces autophagy through Beclin-1 activation in human bladder cancer (Lin et al., 2017). Similarly, Li et al. demonstrated that interference with autophagy by 3-MA or siRNA targeting ATG7, both *in vitro* and *in vivo*, sensitized colorectal cells to 5-fluorouracil (5-FU), a pyrimidine analogue and antimetabolite (Li, Hou, Faried, Tsutsumi, & Kuwano, 2010). Aydinlik et al. showed autophagy induction in response to doxorubicin, a topoisomerase II inhibitor, in triple-negative breast tumor cells; furthermore, autophagy inhibition promoted doxorubicin-induced cell death, thereby increasing drug sensitivity (Aydinlik et al., 2017).

Therapies such as radiotherapy induce excessive damage throughout the cell through direct DNA damaging capabilities and through the indirect generation of ROS (Choi et al., 2015; Noman et al., 2011; Sun, Xing, et al., 2015; Zhang et al., 2015). Both DNA damage and ROS are well characterized as inducers of autophagy. Ito et al. showed that radiation exposure in malignant glioma cell lines induced autophagy and cell cycle arrest, instead of apoptosis (Ito, Daido, Kanzawa, Kondo, & Kondo, 2005). Autophagy induction can also play a role in radioresistance. Lomonaco et al. investigated the role of autophagy in radioresistant glioma stem cells (GSCs). CD133+ GSC cells exhibited a greater extent of autophagy when exposed to γ-radiation compared to CD133- cells; moreover, exposure to BafA1 or genetic knockdown of *Beclin-1* and *ATG5* significantly reduced cell viability (Lomonaco et al., 2009). These data indicated that autophagy is cytoprotective in response to γ-radiation in CD133+ GSCs.

2.2.2 Autophagy in response to non-DNA damaging therapy

Autophagy is also induced in response to a number of chemotherapeutic modalities that are not considered to directly promote DNA damage. Moreover, as therapeutic approaches, such as targeted therapies and immunotherapy have been developed as an additional component of anticancer treatment regimen, autophagy has also been shown to be induced in response to a variety of such novel agents. For example, one of the earliest studies in the literature relating to therapy-induced cytoprotective autophagy was the finding that tamoxifen and other anti-estrogens induced protective autophagy in breast cancer (Bursch et al., 1996). Histone deacetylase (HDAC) inhibitors, such as butyrate and suberoylanilide hydroxamic acid (SAHA), were shown to induce autophagy, and more specifically, autophagic cell death in cervical cancer (Shao, Gao, Marks, & Jiang, 2004). The epidermal growth factor receptor (EGFR) represents a major target for a variety of monoclonal antibodies in the treatment of different types of cancer. EGFR activates a number of downstream pro-survival pathways, including Ras/MAPK, Jak/Stat and PI3K/AKT/mTOR signaling pathways (Henson, Chen, & Family Members, 2017). While EGFR inhibitors are often initially effective, many cancer patients become resistant to these therapies; thus, a further understanding and manipulation of autophagy may serve as a potential therapeutic strategy for the sensitization of patients to anti-EGFR therapies. In support of this approach, Li et al. demonstrated autophagy induction in response to cetuximab, an EGFR-blocking antibody, *via* inhibition of mTOR and activation of Beclin-1 in human vulvar squamous carcinoma, colorectal cancer, and NSCLC (Li & Fan, 2010). Furthermore, pharmacological inhibition of autophagy utilizing CQ and NH_4Cl, as well as genetic knockdown of *ATGs*, resulted in increased cetuximab-induced apoptosis, indicating that autophagy was cytoprotective in function (Li, Lu, Pan, & Fan, 2010). Autophagy induction was further demonstrated in response to targeted therapy in studies where hepatocellular carcinoma (HCC) cells were shown to undergo cytoprotective autophagy in response to linifanib, a vascular endothelial growth factor (VEGF) and platelet-derived growth factor (PDGF) inhibitor, through suppression of PDGFR-β and its downstream signaling pathways (Pan et al., 2014). Furthermore, genetic and pharmacological interference with autophagy sensitized the HCC cells *in vitro* and reduced tumor volume *in vivo*.

2.2.3 Non-cell autonomous effects of autophagy

Autophagy modulation may mediate both cell autonomous (direct cellular) and cell non-autonomous effects, the latter related to surrounding cells, such as immune, epithelial and other tumor cells in the TME (Fig. 1).

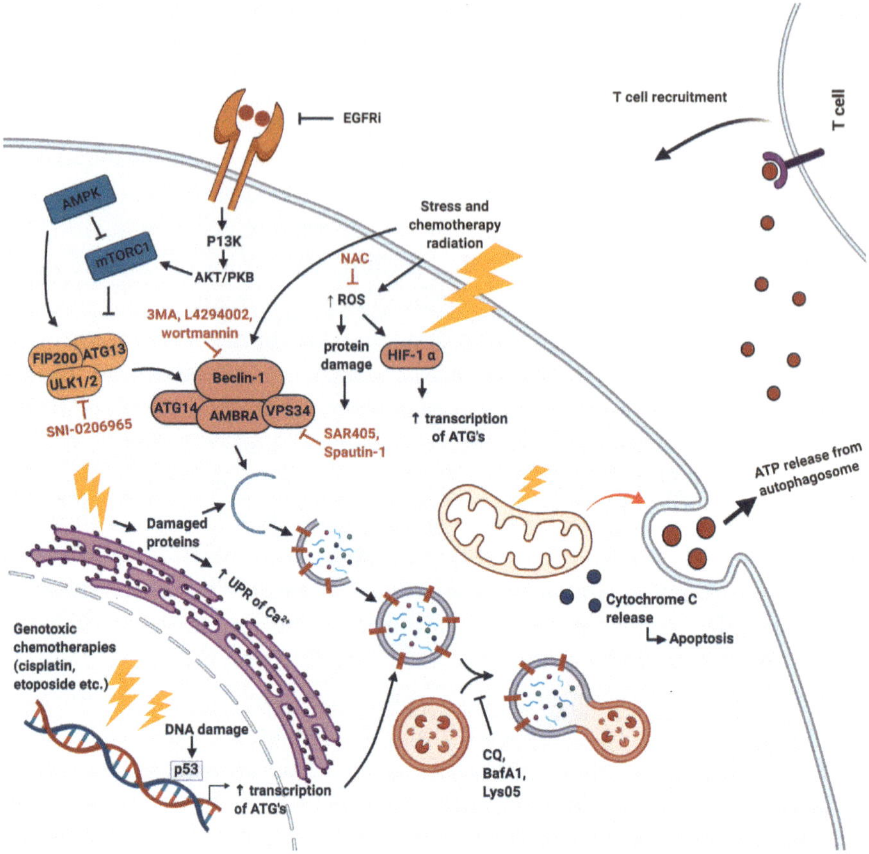

Fig. 1 Autophagy induction in response to therapy. Anticancer therapy work through multiple mechanisms of action, such as genotoxic stress, oxidative stress, and increase ER and mitochondrial damage. Genotoxic therapies, such as cisplatin and etoposide, incur DNA damage, activating DNA damage response and p53 pathways. p53 activation can promote the transcription of autophagy-related genes (ATGs), which contribute to components of the autophagic machinery. Stress, chemotherapy, and radiation can also induce protein and organelle damage, as well as the accumulation of misfolded proteins. Excessive aggregation of unfolded proteins activates the unfolded protein response, UPR, in the ER, which promotes autophagy as a means to alleviate the burden unfolded/misfolded proteins and maintain homeostasis. Aside from increased ER stress, multiple chemotherapies, and radiation promote the generation of reactive oxygen species (ROS) levels as well. Increased cellular ROS levels and oxidative stress triggers HIF-1 activation. HIF-1 can translocate into the nucleus and promote transcription of ATG genes; thus, inducing autophagy to prevent accumulation of oxidative stress. Furthermore, increasing levels of ROS oxidize surrounding proteins resulting in oxidative damage and malfunction, as well as dysregulation of mitochondrial bioenergetics, exacerbating damage, and activate intrinsic apoptotic pathways. The resulting metabolic dysfunction increases intracellular ATP, which can be shuttled into the extracellular space, bind to purogenic receptors on T cells and promote recruitment. Autophagy has been shown to play a role in ecto-ATP release in dying cells and promote immune recruitment and clearance of tumor cells. Lastly, therapies, such as EGFR inhibitors, block tyrosine kinase receptor signaling cascades promoting mTOR activation, a known inhibitor of autophagy; therefore, EGFRi block upstream receptor signaling, resulting in mTOR inhibition and Beclin1 activation. Taken together, autophagy is involved and induced in response to multiple cytotoxic therapies through a multitude of intracellular pathways as a means to prevent accumulation of damage incurred by therapy. *This figure was created with biorender.com.*

Immune evasion by tumor cells remains a major barrier to effective cancer treatment. Tumor cells can develop several mechanisms to evade host-mediated immunity, such as impaired/downregulation of antigen presentation and upregulation of immune checkpoints, which downregulate T-cell activity (Pardoll, 2012). Yamamoto et al. demonstrated that major histocompatibility complex class I (MHC-I), a molecule important in antigen presentation, was selectively degraded through autophagy-dependent mechanisms in pancreatic ductal adenocarcinoma (PDAC) cell lines (Yamamoto et al., 2020). Furthermore, in a mouse model of PDAC tumor cells expressing a doxocyclin-inducible dominant negative *ATG4B* knockdown, autophagy inhibition was shown to increase MHC-1 expression and CD8+ T cell activation, enhanced T-cell mediated killing and consequently to reduce tumor cell viability (Yamamoto et al., 2020). Reduced tumor burden and increased* tumor-infiltrating T-cells were also evident in a similar model with knockdown of *ATG7*. Given that PDAC is refractory to immune-checkpoint blockade (ICB), the authors examined whether autophagy inhibition would sensitize tumors to immune-checkpoint inhibitors. Utilizing a syngeneic mouse model with orthotopic tumors, Yamamoto et al. demonstrated a significant reduction in tumor volume and increased CD8+ T cell infiltration in the *ATG4B* knockdown mice compared to autophagy-proficient mice when exposed to dual ICB (anti-PD-1 and anti-CTLA4). Furthermore, administration of CQ significantly sensitized tumors to the ICB therapy. Finally, Liang et al. also demonstrated autophagy induction in response to high-dose interleukin-2 treatment in a model of advanced metastatic murine liver tumors (Liang et al., 2012). Addition of CQ with IL-2 therapy increased immune infiltration, reduced tumor burden and increased survival. These data suggest that autophagy inhibition may be a potential strategy for the sensitization of patients to immune-checkpoint inhibitors.

2.3 Functional forms of autophagy, contributions and outcomes

While autophagy is routinely observed in tumor cells in response to chemotherapy and radiation, the functional contribution of autophagy to the overall outcome of therapy can vary depending on a number of factors, including but not limited to, the type of therapeutic utilized, time of treatment, dose, type of cancer and cell type. Autophagy can play multiple functions in response to chemotherapy and radiation treatment; however, the functional form of autophagy can only be determined through the impact on drug or radiation sensitivity when the autophagy is inhibited pharmacologically and/or genetically. Conventional pharmacological inhibition, using agents such as CQ, BafA1 or 3-MA, or genetic silencing techniques, including siRNA or shRNA knockdown or cell-specific gene knockout of *Beclin-1* and essential *ATGs* can be utilized to determine the functional form of autophagy induced by a particular treatment (Table 1).

2.3.1 Cytoprotective autophagy

One of the best-known functions of autophagy is its conventionally *cytoprotective* function in response to cellular stresses, such as serum starvation or hypoxia. In this scenario, pharmacological and genetic inhibition of autophagy increases sensitivity to the anticancer treatment that promotes the autophagy (Gewirtz, 2013, 2014c). Increased sensitivity is often associated with the activation of cell death pathways, apparently indicating that the autophagy was acting as a *protective* cellular mechanism for evasion of cell death (Cufi et al., 2013; Mani et al., 2015). For instance, Qadir et al. demonstrated that autophagy inhibition sensitized MCF-7 breast tumor cells to tamoxifen. In these studies, sensitization was due to increased mitochondrial dysregulation and caspase 9-mediated apoptosis, indicating that inhibition of cytoprotective autophagy increased cell death (Qadir et al., 2008). Similarly, Selvakumaran et al. showed that autophagy was induced in response to oxaliplatin in multiple colon cancer cell lines under both normoxic and hypoxic conditions, and that administration of CQ significantly reduced tumor volume in response to oxaliplatin and bevacizumab in HT29-derived orthotopic mouse tumors (Selvakumaran, Amaravadi, Vasilevskaya, & O'Dwyer, 2013).

A number of miRNAs have also been identified that mediate autophagy (Sun et al., 2015b). The expression of miRNA-30a (miR-30a) is downregulated in several renal cell carcinoma (RCC) lines and patient samples and has been shown to inhibit autophagy through Beclin-1 downregulation. Knockdown of miR-30a significantly increased Beclin-1 expression and interfered with the cytotoxic effects of sorafenib in RCC cells (Zheng et al., 2015). Pharmacological autophagy inhibition through CQ, 3-MA and BafA1 reduced cell viability and increased cell death in RCC cells treated with sorafenib, indicating that autophagy was playing a cytoprotective function in response to sorafenib (Zheng et al., 2015). Collectively, these studies demonstrate the functional capacity of autophagy to protect tumor cells challenged by anticancer therapy.

Induction of cytoprotective autophagy has also been implicated in resistance to therapy and can facilitate multidrug resistance (Liang et al., 2012).

Table 1 Pharmacological autophagy inhibitors.

Autophagy inhibitor	Mechanism of inhibition	Target	Effective inhibitory concentration	References
CQ/HCQ	Accumulates within lysosome and prevents lysosomal fusion	Lysosomal fusion	5–50 μM	Mauthe et al. (2018) and Vogl et al. (2014)
BafA1	V-ATPase inhibitor, interferes with lysosomal acidification	Lysosomal fusion	1–10 nM	Boya et al. (2005) and Yoshimori, Yamamoto, Moriyama, Futai, and Tashiro (1991)
Lys05	Dimeric form of CQ, accumulates in lysosome	Lysosomal fusion	10–100 μM	McAfee et al. (2012) and Ravi and Jeffrey (2012)
3-MA	Class III PI3K inhibitor, prevents initiation	Autophagosome formation	1–10 mM	Li et al. (2009) and Wu et al. (2010)
SAR405	Vacuolar protein sorting protein 18 and 34 (Vps18 and Vps34) inhibitor	Autophagosome formation	1–10 μM	Pasquier (2015), Ronan et al. (2014), and Schlütermann et al. (2018)
Spautin-1	Ubiquitin-specific peptidase inhibitor, Vps34 inhibitor, prevents initiation	Autophagosome formation	1–20 μM	Liu et al. (2011) and Shao et al. (2014)
LY294002	PI3K inhibitor, prevents initiation	Autophagosome formation	10–100 μM	Blommaart, Krause, Schellens, Vreeling-Sindelarova, and Meijer (1997) and Feng et al. (2018)
Wortmannin	PI3K inhibitor, prevents initiation	Autophagosome formation	1–10 μM	Hansen, Olsson, and Casanova (1995)
SBI-0206965	ULK1 inhibitor, prevents initiation	Autophagosome formation	10–50 μM	Egan et al. (2015) and Qiu, Zhou, and Cao (2020)
NSC185058	ATG4B inhibitor	Autophagosome formation	1–100 μM	Akin et al. (2014)
Mdivi-1	Mitochondrial division inhibitor, inhibits fission by selective dynamin-related protein 1 (Drp1) inactivation	Mitophagy	1–10 μM	Givvimani et al. (2012) and Mizumura et al. (2014)

Utilizing epirubicin-resistant MCF-7er and SK-BR-3er cells, Sun et al. examined the role of autophagy in multidrug resistance (Sun, Lan, Gan, & Cai, 2015). MCF-7er and SK-BR-3er cells were collaterally insensitive to paclitaxel and vinorelbine compared to parental cells. Genetic silencing of *Beclin-1* and *ATG7* in MCF-7er and SK-BR-3er restored sensitivity to paclitaxel and vinorelbine. An additional factor contributing to resistance to therapy is that cells with higher basal autophagy may be permissive for tumor cells to remove therapy-induced damage more efficiently. Anthracycline-resistant MDA-MB-231-R8 and SUM159PT-R75 cells exhibited higher basal autophagic flux than parental cells (Chittaranjan et al., 2014). Genetic knockdown of *ATG5/7* and pharmacological inhibition of autophagy with CQ or hydroxychloroquine (HCQ) significantly sensitized both epirubicin-sensitive and resistant triple-negative breast cancer cell lines. Acidic lysosomes can trap weakly basic drugs, such as doxorubicin and irinotecan, thereby reducing drug efficacy and conferring resistance. Guo et al. examined the role of autophagy in doxorubicin localization as MCF-7 cells acquired resistance to increasing doses of doxorubicin (Guo, Tam, Santi, & Parissenti, 2016). Their studies demonstrated increased autophagy induction and sequestration of doxorubicin within lysosomes as MCF-7 cells developed resistance to doxorubicin treatment; furthermore, autophagy inhibition with CQ was sufficient to restore sensitivity in doxorubicin resistant MCF-7 cells.

Through modulation of ROS levels and removal of radiation-induced damage, cytoprotective autophagy can also contribute to radioresistance. Studies by Chen et al. showed low-dose ionizing radiation after high-dose radiation increased ROS levels, autophagy induction and promoted radioresistance in A549 NSCLC cells (Chen, Wu, Yuan, & Wang, 2015). Subsequent autophagy inhibition reduced cell viability of A549 cells in response to radiotherapy, suggesting that autophagy was functionally protective in this experimental model. Moreover, administration of *N*-acetyl-L-cysteine (NAC), a known free radicals scavenger, blocked autophagy and was sufficient to suppress the induced radioresistance. These studies indicate that high levels of ROS can promote cytoprotective autophagy in lung cancer cells and thereby contribute to radioresistance.

While autophagy inhibition may have beneficial outcomes, in a cell autonomous manner, the effects of autophagy inhibition on surrounding cells, or the cell non-autonomous aspects, must also be considered. Thorburn et al. demonstrated that silencing of *ATG5* and *ATG7* in MDA-MB-231 TRAIL-sensitive cells resulted in increased outgrowth of GFP-tagged TRAIL-resistant cells when compared to autophagy-proficient controls (Thorburn et al., 2017). Furthermore, autophagy inhibition through an inducible shATG12 model in a population of GL261 glioma cells sensitive to EGF inhibition promoted outgrowth of GFP-tagged EGF-resistant cells. Through these studies, Thorburn et al. also demonstrated that autophagy inhibition *in drug-sensitive cells* induced caspase activation and the subsequent stimulation of prostaglandin E_2 (PGE_2) signaling resulted in the promotion of tumor repopulation of *drug-resistant cells* (Huang et al., 2011; Thorburn et al., 2017). Taken together, autophagy inhibition may increase cell killing and sensitize drug-sensitive cells to therapy; however, this may promote the selection and growth of a resistant subpopulation, which may have detrimental therapeutic outcomes. Thus, there is a complexity to autophagic function in response to chemotherapy and radiation that must be considered prior to the adoption of autophagy inhibition therapy.

2.3.2 Cytotoxic autophagy

The generally accepted premise that autophagy is *cytoprotective* in cells in response to stress was the foundation for establishing its *cytoprotective* function in response to anticancer therapy in tumor cells. Extensive preclinical data in a number of experimental tumor models and in response to various chemotherapeutic agents does support the cytoprotective function of autophagy based on the observations outlined in the previous section. However, it should be emphasized that preclinical data has, in fact, demonstrated that autophagy can divert from this protective function and exhibit both a cytotoxic activity as well as, in certain cases, failing to modulate the tumor response to therapeutics.

Autophagy modulation may appear as a desirable therapeutic strategy; however, it is important to acknowledge the multifunctional nature of autophagy. In response to chemotherapy and radiation, cells can also undergo a *cytotoxic* form of autophagy, resulting in reduced cell viability and clonogenic survival. Cytotoxic autophagy can kill tumor cells through both apoptosis-dependent and -independent mechanisms (Cao et al., 2006; Gewirtz, 2014c; Kim et al., 2013). When autophagy expresses a cytotoxic function, autophagy inhibition *reduces* tumor cell sensitivity to therapy and *promotes* cell viability. Studies by Talarico et al. examined the effects of SI113, a selective SGK1 serine/threonine inhibitor, in combination with radiotherapy on human glioblastoma multiforme (GBM) cells (Talarico et al., 2016). Administration of SI113 significantly reduced cell viability and oxidative stress in GBM cell lines exposed to 5, 8 or 10 Gy of radiation. The authors demonstrated that SI113 induced autophagy and that addition of CQ abolished the cytotoxicity of SI113, thus indicating that the antitumor effects of SI113 were mediated through induction of cytotoxic autophagy. Kanzawa et al. demonstrated that autophagy inhibition with 3-MA and genetic silencing ameliorated temozolomide antitumor effects (Kanzawa et al., 2004). Indeed, autophagy inhibition when the autophagy is protective in nature is a therapeutic vulnerability to be exploited as a means of sensitization; however, when the autophagy is toxic in function, this can result in detrimental outcomes, such as tumor survival and drug resistance.

2.3.3 Cytostatic autophagy

Autophagy can also exhibit a *cytostatic* function in response to cancer therapeutics. Of the functional forms of autophagy, this is perhaps the least understood, although it is logical that cells exposed to stress might arrest and fail to continue to progress normally through the cell cycle. Cytostatic autophagy may also be associated with senescence, as addressed in the last section.

In the case of cytostatic autophagy, therapy-induced growth inhibition and sensitivity is mediated through autophagy, rather than apoptosis; furthermore, autophagy interference restores cellular proliferation and ameliorates antitumor therapeutic effects (Gewirtz, 2014c; Zhang et al., 2020). For instance, work by Sharma et al. demonstrated cytostatic autophagy in response to EB 1089, a vitamin D analog, in NSCLC cells when combined with radiation (Sharma et al., 2014). The authors show that administration of EB 1089 or 1,25-D_3 significantly sensitized H460 NSCLC cells to radiation. Sensitization was not associated with increased DNA damage, apoptosis, or necrosis, but rather, these cells exhibited a more pronounced growth arrest, which was characterized as a form of cytostatic autophagy. Autophagy inhibition through CQ administration, shATG5 or ShBECN1 knockdown sensitized H460 cells to radiation alone, indicative of radiation-induced cytoprotective autophagy in H460 cells; however, in cells treated with EB 1089 and radiation, autophagy inhibition reversed the sensitization induced by EB 1089. Studies by Dou et al. also demonstrated cytostatic autophagy in breast cancer cells exposed to ivermectin, an antiparasitic drug with anticancer therapeutic potential (Dou et al., 2016). The authors showed that ivermectin was sufficient to suppressed breast cancer cell proliferation and tumor burden, *in vivo*. Breast cancer cells were shown to undergo little or no significant apoptosis in response to ivermectin, but did undergo autophagy. Furthermore, autophagy inhibition was sufficient to attenuate the antiproliferative effects of ivermectin in breast cancer cells with no significant impact on the extent of apoptosis, suggesting that ivermectin's antiproliferative effects are mediated through induction of *cytostatic* autophagy (Dou et al., 2016).

2.3.4 Nonprotective autophagy

A novel function of autophagy in response to therapy is the *nonprotective* form. Preclinical studies have demonstrated that in this case, while significant autophagy may be induced, autophagy inhibition does not alter tumor cell sensitivity to chemotherapy and radiation (Eng et al., 2016; Gewirtz, 2013; Xu et al., 2018). Studies by Xu et al. demonstrated a low extent of apoptosis in response to radiotherapy in H460 NSCLC cells; however, the cells exhibited an increase in autophagy induction (Xu et al., 2018). Pharmacological and genetic autophagy interference failed to alter sensitivity to radiation exposure, suggesting autophagy was nonprotective in function. Another example of this is provided by the work of Eng et al., where pharmacological autophagy inhibition by CQ or Lys01, and genetic autophagy inhibition by deletion of *ATG7* using zinc finger nucleases did not alter sensitivity to radiation or a panel of 30 different chemotherapies in K-RAS mutant tumors *in vitro* and *in vivo* (Eng et al., 2016). These studies also demonstrated that CQ-mediated antiproliferative effects independently of autophagy, suggesting sensitization may be due to off-target effects.

2.3.5 The autophagic switch

Collectively, autophagy can play various functional roles in response to chemotherapy and radiation; moreover, the mechanism(s) determining the functional form induced in response to therapy still remains unclear. To add further complexity to these observations, a unique phenomenon whereby autophagy can "switch" between functions has also been demonstrated (Chakradeo et al., 2015; Wilson et al., 2011; Xu et al., 2018). For instance, it was demonstrated that H460 p53wt cells and H460 p53 knockout cells exhibited differential autophagic function when exposed to the same concentration of cisplatin (Xu et al., 2018). The extent of autophagy was similar in both cell lines; however, autophagy inhibition indicated a *nonprotective* function in the H460wt cells and a *cytoprotective* function in the H460 p53 KO cells. Using this set of isogenic cell lines differing only in p53 status, it was determined that cisplatin exposure induced two different functional forms of autophagy, indicating the existence of an "autophagic switch". Similarly, the "autophagic switch" was also exhibited in studies of estrogen receptor (ER) expression in breast cancer cells exposed to gemcitabine (Shen et al., 2016). Gemcitabine induced cytoprotective autophagy in ER − BCap37 breast cancer cells, but cytotoxic autophagy in ER+ Bcap37 cells. Further, genetic silencing of the ERα receptor in the ER+Bcap37 cells was sufficient to switch the cytotoxic form of autophagy to the cytoprotective form. Given ample preclinical data indicating the multifunctional nature of autophagy and the capability of autophagy to "switch" functions, it will undoubtedly be necessary to determine individual patient tumor responses to therapy prior to administration of autophagy inhibitors.

2.4 Autophagy and the immune response

Therapy-induced autophagy is an integral component of immunogenic cell death and immune modulation in response to anticancer therapy. Autophagy plays a role in the release of ATP in dying cells that act as chemotactic ligands for purogenic receptors on immune cells (Ko et al., 2014). This interaction is important in recruitment of dendritic cells (DC) to the TME and the subsequent activation of T-cells to elicit antitumor responses (Ghiringhelli et al., 2009; Michaud et al., 2014). Studies by Michaud et al. demonstrated that autophagy induction *in vitro* did not significantly alter tumor cell responses to mitoxantrone (MTX) and oxaliplatin (OXA) (Michaud et al., 2011). In dramatic contrast, in immune-competent mice, autophagy-proficient cells exhibited increased DC and T-cell infiltration compared to autophagy-deficient cells. Autophagy inhibition suppressed ATP release and subsequent immune recruitment, while increasing exogenous ATP in autophagy-deficient tumors was able to restore immune infiltration and tumor-suppressive responses (Michaud et al., 2011). These data implicate a role for autophagy in the release of ATP, a DAMP, and its function in eliciting antitumor immune responses to mitoxantrone or oxaliplatin in mouse colon carcinoma xenografts. Similarly, Ko et al. examined the immunomodulatory effects of autophagy in response to radiotherapy (Ko et al., 2014). These authors demonstrated that shRNA knockdown of *ATG5* and Beclin-1 significantly radiosensitized A549 and H460 NSCLC cells and CT26 colon carcinoma cells, *in vitro*, and in immune-deficient mice. However, autophagy depletion reduced radiosensitivity and decreased ATP release in immune-competent mice, suggesting that autophagy-induced ATP release could be critical for anticancer immune responses (Ko et al., 2014). Consistent with the studies from Michaud et al., increasing exogenous ATP in autophagy-deficient tumors was able to restore immune infiltration and radiosensitize tumors *in vivo* (Ko et al., 2014).

Autophagy modulates immune function through a number of pathways, such as pathogen degradation, antigen presentation, and processing of cytokines (Levine et al., 2011; Narita et al., 2011). Autophagy can also regulate trafficking of receptors to the cell surface, which are required for antigen presentation and complement activation (Ramakrishnan et al., 2012; Xu et al., 2007). Work by Ramakrishnan demonstrated the role of autophagy and tumor cell lysis by cytotoxic T lymphocytes (CTLs). They showed upregulation of the mannose-6-phosphate receptor (MPR) in tumor cells exposed to multiple therapies, including cisplatin, paclitaxel and doxorubicin, as well as multiple myeloma patient tumor samples (Ramakrishnan et al., 2012). Furthermore, MPR was implicated in chemotherapy-induced anti-CTL responses, reduced tumor volume, as well as cell killing of bystander tumor cells that did not express tumor antigen. Autophagy inhibition abrogated paclitaxel-induced MPR cell surface expression in melanoma cells and doxorubicin-induced MPR expression in multiple myeloma cells, as well as reducing CTL cytotoxicity (Ramakrishnan et al., 2012), implicating autophagy in the regulation of MPR expression on tumor cell surfaces and CTL cell killing. Taken together, these data indicate the immunomodulatory functions of autophagy in both innate and adaptive immune responses to chemotherapy and radiation.

2.5 Current modalities to modulate autophagy and clinical trials

Current paradigms for clinical administration of autophagy inhibitors to sensitize patients to chemotherapy and radiation are based on the premise that the autophagy induced is *cytoprotective* in function; however, the clinical utilization of autophagy inhibitors has largely produced inconsistent results (Chude, Amaravadi, Chude, & Amaravadi, 2017). Currently, the only clinically approved autophagy inhibitor is hydroxychloroquine (HCQ), an FDA-approved anti-malarial drug used for prophylactic treatment of malaria, treatment for lupus or rheumatoid arthritis (Al-Bari, 2015). HCQ acts by inhibiting lysosomal acidification, as well as preventing the fusion of the autophagosome to the lysosome in the final steps of autophagy (Mauthe et al., 2018); however, there are also off-target effects of HCQ, which still remain elusive (Lee, Mustafa, Hudes, & Kruger, 2015).

Significant preclinical data indicating the *cytoprotective* function of autophagy in response to chemotherapy and radiation (Apel, Herr, Schwarz, Rodemann, & Mayer, 2008; Chaachouay et al., 2011; Cook & Clarke, 2014) provided the foundation for the initiation of a number of clinical trials to inhibit autophagy as a strategy for sensitization of patients' tumors to therapy (Table 2). Studies published by Vogl et al. in 2014 examined the effects

Table 2 Clinical trials manipulating autophagy inhibition to sensitize cancer patients to antitumor therapy.

Type of cancer	Treatment	Results	Noted toxicities	References
Relapsed/refractory myeloma	600 mg of HCQ twice daily with a standard dose of bortezomib	Partial or minor responses in 6 (28%) of pts and stable disease in 10 (45%) of pts	Twice daily with a standard dose of bortezomib and only grade	Vogl et al. (2014)
Glioblastoma multiforme	600 mg or 800 mg HCQ combined with low-dose temozolomide and radiation therapy	No significant improvement in overall survival	At 800 mg of HCQ daily combined with temozolomide significant Grade 3 and 4 neutropenia and thrombocytopenia	Rosenfeld et al. (2014)
Metastatic prostate cancer	HCQ with docetaxel	Early termination due to a lack of improved efficacy		Identifier: NCT00786682
High Grade Glioma	Lower doses of HCQ and short course radiotherapy	When compared to radiation alone, combination did not improve pt outcome	Grade 3–5 toxicities	Brazil et al. (2018)
Metastatic pancreatic adenocarcinoma (did not respond to conventional therapy)	400 or 600 mg HCQ, twice daily as a monotherapy	10% (2 out of 20 pts) achieved 2-month PFS (poor therapeutic efficacy) and inconsistent autophagy inhibition	2 patients experienced Grade 3/4 lymphopenia and elevated alanine aminotransferase	Wolpin et al. (2014)
Advanced solid tumors and melanoma	Combined temsirolimus (rapamycin analog) (25 mg, weekly) with 600 mg, twice daily of HCQ	67% of pts. reached stable disease. Tumor biopsies demonstrated sufficient autophagy inhibition at 1200 mg HCQ	Tolerated with minor toxicities (7% anorexia, 7% fatigue and 7% nausea)	Rangwala et al. (2014)

Advanced solid tumors and melanoma	HCQ 200 to 1200 mg daily with dose-intense oral TMZ 150 mg/m (2 s) daily	Partial response in 14% of pts and stable disease in 27% of pts with metastatic melanoma was observed	At 600 mg, twice daily of HCQ, treatment (1200 mg HCQ) was tolerated with Grade 2 toxicities: fatigue (55%), anorexia (28%), nausea (48%), constipation (20%), and diarrhea (20%)	Rangwala et al. (2014)
Pancreatic Cancer	gemcitabine and nab-paclitaxel (1000 mg/m^2 and 125 mg/m^2) combined with 600 mg, twice daily of HCQ	Not powered enough to detect survivor differences, but increased immune infiltration and serum biomarker responses detected in pts. treated with HCQ		Zeh et al. (2020)
Non-small cell lung cancer	Paclitaxel with carboplatin, and bevacizumab combined with 200 or 600 mg HCQ daily	21% of pts had stable disease with addition of HCQ and median PFS was 3.7 months (3.2–5.8 months) for pts taking HCQ	Grade 3 or higher: neutropenia (35%), anemia (15%), thrombocytopenia (10%) and dehydration (10%)	Malhotra et al. (2018)
EGFR-mut NSCLC	Erlotinib (150 mg/daily) with or without HCQ (1000 mg/daily)	Combination of HCQ and Erlotinib did not improve PFS compared to Erlotinib alone	Grade 3 fatigue and nausea	Neal et al. (2014)

of HCQ on the efficacy of bortezomib, a proteasomal inhibitor, in patients with relapsed/refractory myeloma. The combination of HCQ with bortezomib demonstrated partial or minor responses in 6 (28%) patients and stable disease in 10 (45%) patients (Vogl et al., 2014). Furthermore, electron micrographs of bone marrow plasma collected from patients exhibited greater numbers of autophagic vesicles and increased misfolded protein trafficking with the combination therapy. Patients were given 600 mg of HCQ twice daily with a standard dose of bortezomib and only grade 2 gastrointestinal toxicity and cytopenia were observed. Similarly, studies by Barnard et al. examined the effects of HCQ administration in combination with doxorubicin in canine non-Hodgkin lymphoma in a Phase I clinical trial. Canine patients received HCQ orally 72 h prior to standard dosage of doxorubicin. Oral HCQ administration was well tolerated with no significant grade 3 or 4 toxicities. Combination therapy exhibited a 93% overall response rate in canines with lymphoma, with a median progression-free interval of 5 months (Barnard et al., 2014). Two separate studies by Levy et al. were able to demonstrate improved therapeutic efficacy with autophagy inhibitors in combination with vemurafenib, a $BRAF^{V600E}$ inhibitor (Levy et al., 2014; Mulcahy Levy et al., 2017). In studies examining the effects of HCQ administration in patients with $BRAF^{V600E}$-mutant brain tumors resistant to vemurafenib, patients received 500 mg CQ orally daily with continued standard dose of vemurafenib treatment. In addition to improved therapeutic efficacy (Mulcahy Levy et al., 2017), increased LC3 II accumulation was evident in peripheral white blood cells, suggesting that autophagy was sufficiently inhibited. In contrast, in a phase I/II clinical trial combining HCQ, temozolomide (an alkylating agent) and radiation performed in patients with glioblastoma multiforme (Rosenfeld et al., 2014), doses of 600 mg HCQ combined with low-dose temozolomide failed to demonstrate improved overall survival. Furthermore, when increasing doses to 800 mg of HCQ daily combined with temozolomide, significant grade 3 and 4 neutropenia and thrombocytopenia were evident. Electron microscopic images indicated increased autophagic vesicles in patients treated with radiation; however, even at the higher doses of HCQ, substantial autophagy inhibition could not be demonstrated.

In addition to the dose-limiting side effects of HCQ, another limitation likely contributing to the inconsistent clinical results is the failure of currently available pharmacological autophagy inhibitors to sufficiently interfere with autophagy in patients. In order to effectively block autophagy in patients, significantly higher doses of HCQ would be required, which may result in undesirable side effects (Chude et al., 2017).

2.6 Conclusions and future direction

Anticancer therapies can promote varied responses in the tumor cell; an abundance of preclinical and clinical data have identified autophagy as a prevalent response. Autophagy is induced in response to chemotherapy, radiotherapy, targeted therapy and immunotherapy; therefore, the contributions of autophagy to tumor cell sensitivity to treatment must be considered. A plethora of clinical trials have been launched based on the premise that autophagy induced by chemotherapy (or radiation) will exhibit a cytoprotective function; however, these clinical trials have largely failed to generate consistent or promising outcomes. While autophagy is indisputably cytoprotective in certain scenarios, preclinical data also indicates that autophagy may present in multiple functional forms; consequently, it would be necessary to determine the functional form of autophagy prior to administering autophagy inhibitors in combination with standard of care therapy in efforts to improve patient response. The precise mechanisms underlying the specific functional forms of autophagy induced in response to therapy still remains elusive and will ultimately require further investigation if autophagy inhibition is to take a place in the clinical treatment of cancer.

3. Therapy-induced senescence in cancer

3.1 Overview of senescence

Senescence is derived from the latin word "*senex*" which means "*old*." Cellular senescence is a stable form of growth arrest that develops in response to replicative-, oxidative-, genotoxic-, and therapeutic stresses. The concept of senescence was introduced in the early 1960s by Hayflick and Moorhead (1961). These investigators demonstrated that human fibroblasts grown in culture were not immortal and did not have the ability to replicate indefinitely, but rather entered a state of permanent growth arrest after a certain number of divisions (Hayflick, 1965). The point at which the cells stop dividing was later termed the "the Hayflick limit" (Shay & Wright, 2000). The definition of senescence has evolved dramatically throughout subsequent decades, and while a number of other structural, genetic and metabolic changes occur as components of the complex senescent phenotype, the stability of growth arrest remains a central element in the definition (Sharpless & Sherr, 2015).

Senescence contributes to a number of physiological processes including wound healing (Demaria et al., 2014) and embryonal development (Storer et al., 2013). Furthermore, senescence plays a pivotal role in the pathogenesis of a variety of aging-related processes and pathologies, including cancer (Campisi, 2013). Senescent cells accumulate in premalignant lesions (Collado et al., 2005), and serve as a critical tumor-suppressor mechanism against carcinogenic DNA damage and oncogene overexpression (Campisi, 2001), where transformation is prevented through the persistence of the senescent growth arrest. Paradoxically, senescent cells have also been shown to promote microenvironmental conditions favorable for cellular transformation in a cell non-autonomous fashion, through the secretion of a spectrum of soluble and insoluble factors as part of a functional process termed the senescence-associated secretory phenotype (SASP) (Davalos, Coppe, Campisi, & Desprez, 2010).

Whereas tumor cells are, by definition, immortal, they nevertheless retain the ability to undergo senescence under severe genotoxic stress, often precipitated by exposure to cancer chemotherapy and ionizing radiation, hence the terminology of therapy-induced senescence (TIS) (Saleh, Bloukh, et al., 2020). More importantly, the induction of senescence in tumor cells has been established as a fundamental cell response to a variety of cancer therapeutics in preclinical models (Schmitt, 2007) and clinical cancer (Poele et al., 2002). Early on, the induction of senescence in tumor cells was considered to be a desirable outcome of chemotherapeutic treatment, since forcing tumor cells into a prolonged growth arrest should, in theory, halt cancer progression (Ewald, Desotelle, Wilding, & Jarrard, 2010). However, both the instability of the senescent growth arrest and the ability of senescent cells to secrete microenvironment-modulating factors has led to a more in-depth reassessment of the actual contribution of senescence to the outcomes of cancer therapy (Wang, Kohli, & Demaria, 2020). Accumulating evidence indicates that not only can senescent cells regain their proliferative capacity, subsequently resulting in more aggressive tumor variants (Milanovic et al., 2018; Yang, Fang, & Chen, 2017), but senescent cells can also convert neighboring cells into more malicious phenotypes (Ewald, Desotelle, Almassi, & Jarrard, 2008; Sun et al., 2018). As a result, the concept of pharmacologically eliminating the senescent tumor cells that accumulate in response to cancer therapy using newly developing drugs with senolytic properties as a component of adjuvant cancer therapy has gained interest (Short et al., 2019). In this section, hallmarks of senescence, its stressors, its induction in response to cancer therapy, and the current modalities to target senescence in cancer treatment will be discussed.

3.2 Hallmarks and stressors

3.2.1 Hallmarks

The senescent phenotype is complex and encompasses a variety of distinct features, but with no *single* ubiquitous marker that allows for the reliable identification of senescent cells (Noren Hooten & Evans, 2017). Moreover, the expression of senescent features can differ depending on the genetic background of the cells and the senescence-inducing agent, and may be influenced by the *in vitro* and *in vivo* settings. For example, the remarkable enlargement in cell size is not as evident in animal tissues or human samples as is the case *in vitro*. Consequently, it is recommended to test for several senescence-associated markers in order to experimentally confirm the development of the senescence phenotype (Table 3) (Noren Hooten & Evans, 2017).

3.2.1.1 p53, DNA damage repair response and growth arrest

The promotion of DNA damage is often followed by activation of the DNA damage repair response (DDR) and the induction of growth arrest to allow for DNA repair to take place (Jackson & Bartek, 2009). The importance of the DDR in senescence induction was demonstrated by the attenuation of the senescent response following the knockdown of ATM, a key effector in activating DNA repair pathways, in both mice and human models (Goehe et al., 2012; Zhan, Suzuki, Aizawa, Miyagawa, & Nagai, 2010). In response to DNA damage, often precipitated by exposure to genotoxic chemotherapy or ionizing radiation, activated ATM in tumor cells results in the phosphorylation and activation of p53, which consequently leads to growth arrest (Chang et al., 1999). p53 is a fundamental regulator of senescence induction in tumor cells in response to therapy, as lack of functional p53 often (but not always) interferes with the ability of tumor cells to senesce in response to chemotherapy (Elmore et al., 2002). Additionally, patient tumor samples receiving chemotherapy where senescent cells were identified, displayed increased p53 expression (Poele et al., 2002).

The cell cycle consists of four stages gap-1 (G1), synthesis (S), gap-2 (G2) and mitosis (M), which ultimately lead to division into two identical daughter cells. In the G1 phase, the cell increases in size and duplicates its cellular content, including proteins and suborganelles. Next, in the S phase, all 23 chromosomes are replicated, resulting in double the chromosomal number (4n). This is followed by another gap phase (G2) and finally, the cell divides through the M phase, resulting in the formation of two

Table 3 Cell cycle arrest features are closely associated with senescence induction in tumor cells in response to therapy.

Test/Marker	Indication	Notes
DNA content/cell cycle analysis	Identifies the phase of senescent cell cycle arrest	Chemotherapy-induced senescent cells frequently arrest in the G2-M phase, but can be arrested at other stages of the cell cycle
Colony forming assay	Identifies the clonal survival and proliferation potential of tumor cells	Senescent tumor cells often exhibit halted clonal proliferation
BrdU incorporation assay	Identifies active DNA synthesis	Senescent cells have reduced DNA synthesis activity
CDKIs: $p16^{INK4a}$, $p21^{Cip1}$, $p15^{INK4b}$, $p27^{Kip1}$	Identify the activation of growth arrest regulators	CDKIs are frequently expressed or activated in senescent tumor cells
Phosphorylated ATM	Identifies the activation of	These markers are more successfully detected early on after cell exposure to senescence-inducing stimuli, particularly those involved in causing DNA damage γH2AX and TP53BP1 are ideally detected in colocalization reflecting the formation of DNA double-stranded breaks
p53		
H2AX phosphorylation (γH2AX)	DNA double-stranded breaks	
P53 binding protein 1 (TP53BP1)	Repair of DNA double-stranded breaks repair	
Enlarged, irregular, multinucleated	Morphology	Detected using microscopy
Condensed DNA foci	Structural chromatin changes	Detected using fluorescence microscopy or nucleic acid stains
H3K9Me3 or HP1	Epigenetic chromatin changes	Not completely specific to all types of senescence. Highly associated with oncogene-induced senescence
Lamin B1 levels	Changes in nuclear lamina	Expected to decrease in senescent cells

identical 2n cells. Cyclin-dependent kinases (CDKs) play a central role in cell cycle regulation, binding to cyclins and forming complexes which promote cell cycle progression (Martin, 2011). Conversely, cyclin-dependent kinase inhibitors (CDKIs) can drive the cell into growth arrest (Hernandez-Segura et al., 2018). $p21^{Cip1}$ and $p16^{INK4a}$, CDKIs encoded by the *CDKN1A* and *CDKN2A* genes, respectively, play a central role in enforcing the senescent growth arrest (Brenner, Stampfer, & Aldaz, 1998; Yosef et al., 2017). For example, under normal circumstances, cyclin D levels increase as the cell progresses through G1 and cyclin D binds to CDK4 and CDK6, which in turn promotes progression to the S phase. In response to stress, $p16^{INK4a}$ inhibits phosphorylation events mediated by the cyclin-CDK complex, which prevents the cell from further progression into the cell cycle, resulting in senescence (Serrano, Hannon, & Beach, 1993). $p16^{INK4a}$ expression in senescent cells is also frequently observed, including in aging mice and human cells exposed to radiation (Brenner et al., 1998; Ressler et al., 2006; Zindy, Quelle, Roussel, & Sherr, 1997), and is considered a primary marker of senescence (Brookes, Rowe, Gutierrez Del Arroyo, Bond, & Peters, 2004). Similarly, in response to chemotherapy-induced DNA damage and p53 activation, $p21^{Cip1}$ is induced and results in growth arrest at multiple stages of the cell cycle (Yosef et al., 2017), and is involved in the maintenance of the senescent growth arrest (Chen et al., 2006). Although $p16^{INK4a}$ and $p21^{Cip1}$ have an established role in cellular senescence, these regulatory proteins reflect only one of the many complex pathways by which cells senesce, and their absence does not obligatorily correspond to the cessation of senescence. However, the increased expression of different CDKIs in addition to other signs of growth arrest such as reduced DNA synthesis and markers ofDNA damage are frequently observed in senescent tumor cells (Sikora et al., 2016).

3.2.1.2 Resistance to apoptosis

Resistance to apoptosis is a primary feature of senescent cells and accounts in part for their persistence for prolonged periods *in vivo* (Kaczanowski, 2016). Senescent cells have been shown to express antiapoptotic proteins belonging to the Bcl-2 protein family (Douarre et al., 2005) as well as to downregulate the expression of apoptosis mediators such as caspase-3 (Marcotte, Lacelle, & Wang, 2004; Murata et al., 2006; Spaulding, Guo, & Effros, 1999). However, the precise mechanisms of resistance to apoptosis in senescent cells are poorly understood as some senescent cells fail to express antiapoptotic proteins and yet are capable of maintaining their viability for extended periods (Sasaki, Kumazaki, Takano, Nishiyama, & Mitsui, 2001). It is established that tumor cells exposed to anticancer therapeutics upregulate pro-survival machineries, which account for their targeting by senolytic agents, such as the Bcl-2 inhibitor, ABT-263 (Saleh, Carpenter, et al., 2020).

3.2.1.3 Enhanced lysosomal biogenesis

Senescent cells remain metabolically active and develop an increased lysosomal mass (Kurz, Decary, Hong, & Erusalimsky, 2000), a phenomenon that frequently characterizes aging cells (Carmona-Gutierrez, Hughes, Madeo, & Ruckenstuhl, 2016). This is evident by positive staining of the Senescence-Associated β-galactosidase (SA-β-gal) enzyme, reflecting the enhanced activity of lysosomes (Dimri et al., 1995). SA-β-gal is the most frequently reported marker of senescent cells and is ubiquitously observed in senescent tumor cells induced by exposure to cancer therapy (Sikora et al., 2016). *In vitro*, the expression of SA-β-gal, and morphological changes in cancer cells, usually require at least a few days (Bergsbaken et al., 2009; Fulda & Debatin, 2006; Li & Yuan, 2008; Moriwaki et al., 2015; Tang et al., 2020) following exposure to the senescence-inducing stimulus (Mikuła-Pietrasik, Niklas, Uruski, Tykarski, & Książek, 2020).

Staining for SA-β-gal staining is based on detecting the increased activity of the enzyme in higher lysosomal pH (often pH=6). Furthermore, the staining must be performed in live cells or utilizing fresh or frozen tissue due to rapid degradation of the enzyme (Ben-Porath & Weinberg, 2005; Cotarelo et al., 2016). The activity of this enzyme can be detected in tissue sections and in cultured cells using the enzyme's substrate X-gal (5-bromo-4-chloro-3-indolyl-ß-D-galactopyranoside). In the presence of the enzyme, X-gal is converted to a blue-green chromogen that can be detected by microscopy (Tominaga, Shimada, Okada, Kawamata, & Kibayashi, 2019). Unfortunately, the use of this marker alone is not always a reliable indicator of whether cells are in the senescent state due to non-specificity (Ben-Porath & Weinberg, 2005; Cotarelo et al., 2016). Furthermore, utilizing SA-β-gal for identifying senescent tumor cells in patient tumor samples is challenging since the tissue needs to be harvested without being exposed to a fixative agent, which would require the preparation of fresh frozen sections intraoperatively. Nevertheless, SA-β-gal is considered a primary marker of senescent tumor cells that can be detected *in vitro* and *in vivo* (Poele et al., 2002).

3.2.1.4 Gene expression alterations and the senescence-associated secretory phenotype

Senescent cells undergo extensive alterations in gene expression (Zhang, Pan, & Cohen, 2003), including the elevated expression of a wide variety of secreted soluble and insoluble proteins such as pro-inflammatory chemokines and cytokines (Acosta et al., 2013; Shelton, Chang, Whittier, Choi, & Funk, 1999). The senescence-associated secretory phenotype (SASP) can be a consequence of the activation of the DDR which is uniformly induced in tumor cells treated with chemotherapy or radiation (Fumagalli & d'Adda, 2009; Rodier et al., 2009). p53, a major facilitator of the senescent growth arrest, is also involved in the regulation of gene expression and SASP development during senescence (Itahana, Dimri, & Campisi, 2001). However, the SASP can be induced independently of DNA damage, as in physiological embryogenesis-related senescence (Storer et al., 2013) or during wound healing (Demaria et al., 2014). This supports the proposition that the SASP is regulated independently of cell cycle arrest and the expression of CDKIs (Coppé et al., 2011; Ferrand et al., 2015). The SASP is further regulated by stress-induced p38MAPK and increased expression of NF-κB (Freund, Patil, & Campisi, 2011). NF-κB is responsible for the increased transcription of many SASP factors in tumor cells exposed to chemotherapy (Chien et al., 2011). Lastly, mTOR appears to play a critical role in the regulation of the SASP as rapamycin promotes a robust suppression of inflammatory mediator release (Herranz et al., 2015) since mTOR inhibition prevents the translation of cytokines such as IL1α, which in turn, reduces the secretion of several other inflammatory SASP components (Laberge et al., 2015). The SASP is observed in the laboratory in tumor cells both *in vitro* an *in vivo* in response genotoxic cancer therapy (Lindgren, Stigbrand, Råberg, Riklund, & Eriksson, 2015; Taschner-mandl et al., 2015; Yu et al., 2013), as well as targeted therapies such as bevacizumab, palbociclib and aurora kinase inhibitors (Foersch et al., 2015; Hasan, Ho, Owen, & Tai, 2011; Vijayaraghavan et al., 2017; Wang et al., 2017, 2016).

3.2.1.5 Structural and morphological changes

A number of structural and morphological changes are evident in senescent cells. Senescent cells become enlarged as a result of enhanced cell mass (Bent, Gilbert, & Hemann, 2016). Cytoskeletal changes lead to the irregular neuron-like shape of senescent cells (Druelle et al., 2016). Plasma membrane composition is altered whereby the caveolae cholesterol domains make-up, which plays an important role in extracellular signaling, is modified

(Ohno-Iwashita, Shimada, Hayashi, & Inomata, 2010). Senescent cells develop increased mitochondrial fusion and disrupted mitochondrial membranes, which contribute to the high level of oxidative stress often observed in senescent cells (Korolchuk, Miwa, Carroll, & von Zglinicki, 2017). Finally, several nuclear changes take place, among which are decreased DNA content, disrupted nuclear lamina and the senescence-associated heterochromatin foci (SAHF) (Narita et al., 2003). Nuclear lamina line the inner surface of the nuclear envelope, which has a role in the regulation of gene expression. The nuclear lamina is primarily composed of Lamins (Dechat, Adam, Taimen, Shimi, & Goldman, 2010; Freund, R-MR, Demaria, & Campisi, 2012). Lamins are further classified into two types, type A and type B, based on sequence homology, expression pattern, biochemical properties and their cellular localization during mitosis. Of these, Lamin B1 loss could be potentially considered as a biomarker of significance in identifying senescent cells *in vitro* and *in vivo* (Freund et al., 2012; Hutchison, 2014). Studies have indicated that Lamin B1 expression declines in human and mouse cells undergoing senescence by diverse stimuli such as DNA damage, replicative exhaustion, or oncogene expression (Freund et al., 2012). Lamin B1 expression loss is strongly induced only either directly by the p53- or pRB-associated pathways, or through other mechanisms including ATM, the p38MAPK, nuclear factor-κB or reactive oxygen species signaling pathways (Campisi & d'Adda di Fagagna, 2007). Accordingly, Lamin B1 degradation serves as a useful complementary marker for identifying senescence.

The nucleosome is the fundamental unit of chromatin and is composed of a complex of DNA and histone proteins, H3, H4, H2A and H2B. Heterochromatin represents rich regions of condensed chromatin structure that play a role in gene expression (Filion et al., 2010). Regulation of gene expression is a complex process that is not restricted to changes in the DNA itself and can involve other mechanisms such as DNA methylation, miRNA and histone modification (Lyon et al., 2014). Cellular senescence is notably associated with alteration in the chromatin structure, usually in the form of forming heterochromatin structures designed to inhibit the expression of proliferation-related genes. These are collectively referred to as senescence-associated heterochromatic foci (SAHF). SAHF are thought to contribute to maintenance of the cell in a non-proliferating state *via* repression of the expression of proliferation-promoting genes, such as E2F and cyclin. These proteins cannot bind to the compressed and therefore inaccessible chromatin, which prevents transcription of S phase-related genes (Narita et al., 2003; Zhang, Chen, & Adams, 2007). SAHF can include several epigenetic signatures such as hypoacetylated histones, histone H3 lysine 9 tri-methylation (H3K9Me3) and the heterochromatin protein, heterochromatin protein 1 (HP1) (Aird & Zhang, 2013). The formation of SAHF structures is dependent both upon the cell type, as well as the nature of the damage inflicted and associated with the activation of the tumor-suppressor p16^{INK4a}-pRb pathway (Narita et al., 2003). SAHF represent areas of transcriptionally silent and compacted chromatin that result from the presence of repressive H3K9Me3 and absence of activating histone mark H3K4Me3 (Zhang et al., 2007). In other words, increased expression of H3K9Me3 is considered an indicator for the occurrence of cellular senescence. For example, a study indicated that SAHF formation is not only associated with the DNA damage response by several cultured cell models but that SAHF formation is associated with oncogenic stress (Kosar et al., 2011). This means that while SAHF is, of itself, not a specific feature of cellular senescence, if combined with the expression of other senescence markers, can serve as a useful indicator of TIS in tumor cells.

3.2.1.6 Senescence vs. quiescence

While senescence and quiescence both represent forms of growth arrest that are induced in tumor cells in response to stress such as exposure to therapy, one of the primary differences between the two phenotypes is the durability of this arrest (Coller, Sang, & Roberts, 2006). Quiescent cells are only transiently growth arrested in the presence of unfavorable conditions including lack of mitogenic stimulation, absence of growth factor stimulation and over-confluency and contact inhibition (Coller, 2011; Polyak et al., 1994). Once these unfavorable conditions cease to exist, the quiescent cells regain the ability to proliferate (Coller et al., 2006). In senescence, however, the growth arrest is more stable and lasts for prolonged time periods. More importantly, the senescence phenotype includes many other processes that are not observed in quiescent cells, including the distinctive irregular and enlarged cell shape, epigenetic modifications and enhanced lysosomal activity (Kuilman, Michaloglou, Mooi, & Peeper, 2010). Additionally, while senescent cells are distinguished for their active metabolic state, with increased protein biogenesis and a secretory phenotype (the SASP), this is not the case in quiescent cells that maintain minimal bioactive metabolic drive (Pereira, Gonzalez, & Dworkin, 2015). Conversely, similar to senescent cells, CDKIs such as p21^{Cip1} are detected in quiescent cells while

Table 4 Differential comparison of senescent and quiescent phenotypes.

	Quiescence	Senescence
Eliciting stimuli	Mitogenic deprivation, contact inhibition, loss of adhesion	Telomere dysfunction, oncogene overexpression, loss of tumor suppressor function, genotoxic stress, oxidative stress
Nature of growth arrest	Transient	Durable
Phase of cell cycle arrest	G_0	Variable
Associated CDKIs	$p27^{Kip1}$, $p21^{Cip1}$	$p16^{INK4a}$, $p21^{Cip1}$, $p27^{Kip1}$, $p15^{INK5b}$
Response to mitogenic stimuli	Responsive	Irrelevant
Protein synthesis	Decreased	Increased
Secretory phenotype	No	Yes
Enhanced lysosomal biogenesis	No	Yes
Heterochromatic foci formation	No	Yes

$p16^{INK4a}$ expression is more closely associated with senescent, but not quiescent, cells (Sharpless & Sherr, 2015) (Table 4).

3.2.2 Stressors

Cells can undergo senescence due to a number of different stimuli that include telomere dysfunction, due to repetitive cell division, oncogene overstimulation and cancer therapy. Pro-senescence cancer therapies include genotoxic modalities such as conventional chemotherapy and ionizing radiation as well as several targeted therapeutics (Saleh, Bloukh, et al., 2020). Telomeres are noncoding DNA segments which cap the ends of chromosomes to protect against chromosomal damage (Moyzis et al., 1988). With repetitive DNA replication, a stretch of telomere is lost, eventually leading to telomere dysfunction and consequently triggering the DDR and p53 activation (Herbig, Jobling, Chen, Chen, & Sedivy, 2004), which results in *replicative senescence*, a common phenomenon of aging cells (Lindsey, McGill, Lindsey, Green, & Cooke, 1991). Telomerase, a reverse transcriptase, prevents or overcomes telomere attrition by elongating the telomeres (Bodnar et al., 1998). Although found in 90% of cancer cells (Belair, Yeager, Lopez, & Reznikoff, 1997), most somatic cells do not express telomerase and, accordingly, ultimately undergo replicative senescence (Collins & Mitchell, 2002). In contrast to replicative senescence, *oncogene-induced senescence* is not a product of telomere shortening (Ling, Ding, & Hua, 2018). Somatic cells acquiring abnormal oncogenic drive due to K-RAS or B-RAF over-expression can undergo senescence (Michaloglou et al., 2005; Serrano, Lin, McCurrach, Beach, & Lowe, 1997), which explains, in part, the increased frequency of senescent cells in premalignant lesions (Joselow, Lynn, Terzian, & Box, 2017). Again, the induction of oncogene-induced senescence is related to the accumulation of DNA damage caused by the hyperreplication of oncogenically driven cells that results in the build up of DNA damaging ROS (Di Micco et al., 2006).

Finally, extensive evidence has been generated demonstrating the capacity of a spectrum of chemotherapeutic agents to promote senescence in various cancer cells, both *in vitro* and *in vivo* (Table 5). That is, although-tumor cells have the ability to divide indefinitely, they can enter a dormant-like period of senescence upon exposure to chemotherapy or radiation. *Therapy-induced senescence* is discussed in further detail in the next section.

3.3 Senescence in response to cancer therapy

As previously discussed, it was initially unexpected that supposedly immortal tumor cells were capable of undergoing senescence. This is in part due to their ability to express telomerase, thereby enabling the cells to divide an unlimited number of times (Bodnar et al., 1998). However, this concept was shown to be flawed after several studies demonstrated that human diploid fibroblasts were in fact capable of senescing after being exposed to non-telomeric DNA damage (Chen, Fischer, Reagan, Yan, & Ames, 1995). Wang et al. were the first to show evidence of TIS in studies where nasopharyngeal carcinoma cells were treated with cisplatin, a DNA cross-linking agent, resulting in senescent features including stable growth arrest, morphologic changes and SA-β-gal expression (Wang et al., 1998). Moreover, it was demonstrated that TIS in tumor cells is independent of telomere dysfunction (Elmore et al., 2002).

Many laboratories have since provided evidence for the capacity of numerous anticancer drugs to induce senescence in a wide range of cell lines. A report by Saleh et al. provided a comprehensive summary of the FDA-approved anticancer drugs which were shown to induce senescence (Saleh, Bloukh, et al., 2020). It is noteworthy to mention here that

Table 5 Evidence for the induction of senescence in response to cancer therapeutics.

Drug class	Drug name	Model/Cell line	References
Topoisomerase poisons/inhibitors	Doxorubicin (Adriamycin)	MCF-7, MDA-MB231, H460, A549, HCT116, HT1080, LS174T, A2780, patient breast cancer tissue samples, DU145, LNCaP, K562, rat-derived BMSCs and ADSCs, MDFs, HCA2, BJ, *in vivo* mouse model ($p16^{INK4a}$-3MR), SH-SY-5Y	Bojko, Czarnecka-Herok, Charzynska, Dabrowski, and Sikora (2019), Chang et al. (1999), Demaria et al. (2017), Elmore et al. (2002), Mosieniak et al. (2015), Poele et al. (2002), Qi et al. (2012), Schwarze, Fu, Desotelle, Kenowski, and Jarrard (2005), Su et al. (2009), Yang et al. (2012), and You, Dai, Zhang, Barding, and Raftery (2018)
	Etoposide	HepG2, U2OS, IMR-90, MEFs, BJ, B16F10, NRK-52E, follicular lymphoma 3D model	Decaup et al. (2013), Gu and Kitamura (2012), Nagano et al. (2016), Yang, Wang, Ren, Chen, and Chena (2017), and Yosef et al. (2017)
	Topotecan	MNA, STA-NB-10, CLB-Ma mouse xenograft (MYCN-amplified neuroblastoma)	Taschner-mandl et al. (2015)
Alkylating agents	Busulfan	Rat-derived BMSCs and ADSCs, WI38, U2OS, MG63, murine hematopoietic cells	Cells, Meng, Wang, Van, and Zhou (2003), Mei, Li, Quan, Liu, and Xu (2014), Probin, Wang, Bai, and Zhou (2006), Probin, Wang, and Zhou (2007), and Qi et al. (2012)
	Temozolomide	Patient derived glioma cells, GL261, LN229, *in vivo* ($p16^{INK4a}$-3MR) mouse model	Aasland et al. (2019), Demaria et al. (2017), Kipper et al. (2018), and Papait, Magrassi, Rigamonti, and Cattaneo (2009)
	Cyclophosphamide	HSC-bcl2 lymphoma	Schmitt et al. (2002)

Platinum-based	Cisplatin	A375, B16F10, B16F10 xenografts, A2780, CNE1, SKOV3, TOV-21G, HepG2, SMMC-7721, follicular lymphoma 3D model, *in vivo* mouse model ($p16^{INK4a}$-3MR)	Decaup et al. (2013), Demaria et al. (2017), Li et al. (2014), Nakayama et al. (2017), Qu et al. (2014), Sun et al. (2018), and Wang et al. (1998)
	Carboplatin	H1299, patients' lung tumor samples	Roberson, Kussick, Vallieres, Chen, and Wu (2005)
	Oxaliplatin	PROb, CT26, HepG2, SMMC-7721, patients' colorectal tumor samples	Bojko et al. (2019) and Seignez et al. (2014)
Antimetabolites	Methotrexate	C85, MCF-7, rat-derived BMSCs and ADSCs, A549, SH-SY-5Y, HCT116, MDA-MB-231	Dabrowska, Skoneczny, and Rode (2011), Dabrowska, Skoneczny, Uram, and Rode (2019), Hattangadi et al. (2004), and Qi et al. (2012)
	Pemetrexed	H1650, A549, H2228, H292, H226 and H1650, A549 xenografts	Ge et al. (2018) and Tanino, Tsubata, Harashima, Harada, and Isobe (2018)
	Azacitidine	U2OS, MCF-7, TPC-1, KKU100, HuCCA1, RMCCA1, DU145, LNCaP	Moolmuang, Singhirunnusorn, and Ruchirawat (2016), Putri, Widodo, Sakamoto, Kaul, and Wadhwa (2017), Schwarze et al. (2005), and Wu et al. (2016)
Microtubule inhibitors/poisons	Paclitaxel	Human mesenchymal stem cells, MCF-7, MEFs, A549, SH-SY-5Y, HCT116, MDA-MB-231	Bojko et al. (2019), Khongkow et al. (2016), and McKenna, Traganos, Zhao, and Darzynkiewicz (2012)
	Docetaxel	DU145, LNCaP, PTEN null prostate tumors	Di Mitri et al. (2014) and Schwarze et al. (2005)
	Vinblastine	Patient derived glioma cells	Kipper et al. (2018)

Continued

Table 5 Evidence for the induction of senescence in response to cancer therapeutics.—cont'd

Drug class	Drug name	Model/Cell line	References
Hormonal therapy	Tamoxifen	MCF-7, HCT116	Lee, Kang, and Bae (2014) and Tuttle et al. (2012)
	Fulvestrant	MCF-7, T-47D	Dolfi et al. (2014) and Wu, Pepowski, Takahashi, and Kron (2019)
	Androgen Deprivation	LNCaP, LAPC4, ICR castrated mice, LuCaP, xenografts patient samples	Barakat, Zhang, Barberi, Denmeade, and Friedman (2015), Burton et al. (2013), Ewald et al. (2016), and Kawata, Kamiakito, Nakaya, and Komatsubara (2017)
Kinase inhibitors	Dasatinib	H1666, Cal12T, A549, H661	Peng et al. (2016) and Sen et al. (2013)
	Trametinib	DMBC11, DMBC12, DMBC21, DMBC28, DMBC17, H2030, H460, A549, MSK-LX68 patient-derived xenografts, H1944, H358, H441, H2009, HCC441	Hartman, Sztiller-Sikorska, and Czyz (2019), Ruscetti et al. (2018), and Wang et al. (2018)
	Vemurafenib	DMBC11, DMBC12, DMBC21, DMBC28, DMBC17, MM034, MM070, MM074, SKMEL-28, MM050, SK-MEL-28, Mel2a, M19-Mel, UACC-62, UACC-257, and FM88, M14, Malme 3M, Mel2a, SK-MEL-mouse xenografts	Haferkamp et al. (2013), Hartman et al. (2019), and Krayem et al. (2018)
mTOR inhibitors	Rapamycin (Sirolimus)	SMMC-7721, HUVECs	Bu et al. (2008) and Ota et al. (2009)

Monoclonal antibodies	Rituximab	EHEB, RC-K8, and SD-1, follicular lymphoma 3D model	Däbritz et al. (2016) and Decaup et al. (2013)
	Obinutuzumab	Follicular lymphoma 3D model	Decaup et al. (2013)
	Bevacizumab	MIP101, RKO, SW620, SW480, MIP101 mouse xenografts	Hasan et al. (2011)
CDK 4/6 inhibitors	Palbociclib	U87MG, U138MG, Hs683, H4, A172, LN18, LN229, CCF-STTG1, T98G, DBTRG-05MG, DKMG, GAMG, SNB19, AM38, NMC-G1, KG-1-C, U87MG and GBM39 xenograft, HEK293, HeLa, U2OS, LS8817, LS141, LS0082, 1205Lu, 983B, 983BR, B16-F1, B16-F10, NL212, NL216, TRIA, SK-MEL-103, NCI-H226, Huh7, SAOS-2, UT-SCC-42B, AGS, MKN-45, MCF-7, skHep1, Huh7 mouse xenografts, lung sections of Cdk4-deficient mice, MOUSE-derived sarcoma cells/tissues	Anders et al. (2011), Bollard et al. (2017), Guan et al. (2017), Kovatcheva et al. (2015), Michaud et al. (2010), Miettinen et al. (2018), Muñoz-Espín et al. (2018), Perez, Muñoz-Galván, Jiménez-García, Marín, and Carnero (2015), Puyol et al. (2010), Valenzuela, Vargas, Martinez, Bravo, and Brown (2017), and Yoshida, Lee, and Diehl (2016)
	Abemaciclib	MCF-7	Torres-Guzmán et al. (2017)
	Ribociclib	Hey1	Iyengar et al. (2018)
PARP inhibitors	Olaparib	HCT116, OV1369 (R2), OV90, OV4453, OV1946, MDA-MB-231	Alotaibi et al. (2016) and Fleury et al. (2019)
	Niraparib	HCT116	Alotaibi et al. (2016)
	Rucaparib	PC3, LNCaP, DU145, VCaP	Chatterjee et al. (2013)
Proteasome inhibitors	Bortezomib	U87, T98	Morelli et al. (2017)

exposing various cell lines to radiation (Jones et al., 2005; Nardella, Clohessy, Alimonti, & Pandolfi, 2011) or non-DNA damaging agents such as the CDK4/6 inhibitor palbociclib (Anders et al., 2011; Bollard et al., 2017; Guan et al., 2017; Kovatcheva et al., 2015; Michaud et al., 2010; Miettinen et al., 2018; Muñoz-Espín et al., 2018; Perez et al., 2015; Puyol et al., 2010; Valenzuela et al., 2017; Yoshida et al., 2016) has also been established to induce senescence in tumor cells.

Doxorubicin (Adriamycin) belongs to the anthracycline family and is widely used to treat a broad spectrum of malignancies such as breast, lung, and bladder cancers and lymphoma (Yang, Kemp, & Henikoff, 2015; Zhao, Woodle, & Mixson, 2018). Doxorubicin has several distinct mechanisms of action to promote tumor cell injury including DNA intercalation, topoisomerase II inhibition and induction of oxidative stress (Yang et al., 2015; Zhao et al., 2018). Several reports have confirmed that doxorubicin is a potent trigger of therapy-induced senescence in a wide variety of tumor cell types and doxorubicin can now be considered as a prototypical senescence inducer (Ewald et al., 2010; Inao et al., 2019; Mikuła-Pietrasik et al., 2020).

Cyclophosphamide is an oxazaphosphinan agent widely used in the treatment of patients with multiple forms of cancer such as breast and ovarian cancer and bone and soft tissue sarcomas (Emadi, Jones, & Brodsky, 2009). Cyclophosphamide is a DNA-alkylating agent that is converted to aldophosphamide by the P450 microsomal system which then breaks it down to the alkylating metabolite, phosphoramide mustard, in target cells (Ahmed & Hombal, 1984; Emadi et al., 2009). The active metabolite forms interstrand cross linkages in the DNA, which leads to inhibition of DNA replication and ultimately induction of cell death (Emadi et al., 2009). It has been established that cyclophosphamide can also trigger senescence in breast cancer cells both *in vitro* and *in vivo* (Ewald et al., 2010; Poele et al., 2002).

Docetaxel is a cytotoxic chemotherapeutic drug belonging to the taxane family (Pazdur, Kudelka, Kavanagh, Cohen, & Raber, 1993) that is currently used as a monotherapy or in combination with other agents to treat multiple forms of cancers including breast, prostate, lung, and ovarian cancer (Pazdur et al., 1993). Docetaxel has a unique mechanism of action as an inhibitor of microtubule polymerization that disrupts cell division, leading to cell death (Lyseng-Williamson & Fenton, 2005; Pazdur et al., 1993). Moreover, several studies have been reported that docetaxel induces senescence in prostate cancer models (Di Mitri et al., 2014; Ewald et al., 2008; Schwarze et al., 2005).

TIS is also induced in tumor cells by a variety of other drugs. For example, topoisomerase I and II poisons such as camptothecin and etoposide, respectively, have been shown to induce cellular senescence in several different tumor cell lines (Elmore et al., 2002; Saleh et al., 2019; Zhang et al., 2014). In addition, ionizing radiation, which is frequently used in the treatment of multiple malignancies, has also been shown to induce TIS in several tumor cell lines, further suggesting that senescence is a primary response to anticancer therapy (Alotaibi et al., 2016). Several other drugs, such as the CDK 4/6 inhibitor, palbociclib, which is not traditionally associated with DNA damage, have been shown to induce senescence in breast cancer cell lines (Muñoz-Espín et al., 2018). Palbociclib is an established component of the systemic treatment of ER positive breast cancer. Finally, hormonal deprivation precipitated by, for instance, androgen receptor antagonists, has been shown to induce senescence in prostate cancer cell lines, indicating that senescence is not strictly limited to genotoxic or oxidative stress in tumor cells (Pernicová et al., 2011). Limited evidence is available as to whether antiestrogen modulators activate senescence in breast tumor cells.

While TIS has been extensively established in many preclinical models, the induction, and the potential role of senescence in tumor cells of cancer patients receiving anticancer treatment is not yet well researched. For example, the seminal report by Poele et al. showed that senescence is detected in approximately 40% of tumor samples collected from 36 patients with breast cancer receiving neoadjuvant therapy (Poele et al., 2002). This study also showed that evidence of senescence was detected only in 10% of tumor samples derived from patients that did not receive chemotherapy (Poele et al., 2002). However, this study utilized archived, frozen tumor samples where the activity of SA-β-gal might be altered. Moreover, the induction of senescence was investigated in response to only a single regimen of chemotherapy. Another report by Roberson et al. presented evidence of senescence induction in tumor samples derived from lung cancer patients receiving platinum-based/taxane therapy (Roberson et al., 2005). However, the latter report included results from only three patients. This small number of subjects included in these very few studies in the literature represents a significant limitation to understanding the extent of senescence induction *in vivo* as a primary response to anticancer therapy. Also, these studies only focused on patients receiving DNA damaging chemotherapy, without establishing the same observation in patients receiving radiotherapy or targeted therapy. Furthermore, these studies relied primarily on the detection of the SA-β-gal staining. While SA-β-gal is the classical marker of senescence, its sole use as a proof of senescence induction has been received with multiple concerns including non-specificity (Yang & Hu, 2005). Finally, while these reports provided suggestive evidence for senescence induction in preclinical and clinical cancer, a more comprehensive analysis of the induction of senescence in response to cancer treatment in the clinic would clearly be of significant benefit to the field.

3.4 Therapy-induced senescence: Beneficial or harmful?

The induction of senescence in tumor cells is likely to have an *initial* positive therapeutic impact by the enforcement of a stable growth arrest in the tumor cells. However, due to extensive evidence that has now been generated demonstrating that (some) senescent tumor cells can escape this growth arrest, and emerge to form viable colonies *in vitro* and proliferating tumors *in vivo* (Saleh, Bloukh, et al., 2020), it appears to be the case that TIS also can be considered as a strategy whereby the tumor cells actually evade the potential cytotoxicity of chemotherapy or radiation therapy. Furthermore, therapy-induced senescence could reflect one form of tumor dormancy, from which the cells' reproductive capacity could eventually be recovered (Saleh, Tyutyunyk-Massey, & Gewirtz, 2019). It should be emphasized that escape from senescence in this context does not equate to the tumor cells directly reverting back to their pre-senescent state. Rather, it may reflect the development of cells with a more malignant state, through multiple phenotypic and genotypic alterations (Milanovic, Fan, et al., 2018; Yang, Fang, & Chen, 2017). Furthermore, the "reversibility" from the senescent state in not necessarily a universal phenomenon applicable to all types of senescence or all cells within a senescent tumor population. Instead, it is highly likely to represent an infrequent event that takes place only in specific cells under select biologic conditions, and that may predominate in senescent tumor cells that possess heterogenous genetic backgrounds.

Multiple laboratories have provided evidence in support of the premise that senescence need not obligatorily reflect a permanent form of growth arrest. For example, H1299 non-small cell lung cancer (NSCLC) cells that were induced into senescence upon exposure to several chemotherapeutic drugs such as camptothecin, etoposide and cisplatin demonstrated typical senescent features including morphology, increased granularity, and SA-β-gal expression. Three to 4 weeks after drug removal, a very small subpopulation displayed proliferative capacity (Roberson et al., 2005). Studies from our own laboratory have demonstrated that therapy-induced senescent lung and breast tumor cells are capable of regaining proliferative capacity following a period of growth arrest (Saleh, Tyutyunyk-Massey, Murray, et al., 2019). To rigorously confirm that recovery in this experimental system could occur from senescent cells, SA-β-gal positive cells were isolated and enriched using flow cytometry and the senescence-sorted population was monitored over time to detect restoration of proliferative capacity, the ability to develop spontaneous mitotic events and to form viable tumors *in vivo*. Under all conditions, a subpopulation of enriched senescent tumor cells were able to overcome the stable growth arrest and proliferate (Saleh, Tyutyunyk-Massey, Murray, et al., 2019). Escape from senescence was observed using the same approach in colon cancer cells induced to senescence by ionizing radiation or, more robustly, ionizing radiation combined with PARP inhibitors (Alotaibi et al., 2016).

Several mechanisms have been proposed to explain the ability of senescent tumor cells to escape the senescent growth arrest including mutational deficiency of p53 or CDKIs, polyploidy and genetic reprogramming that favors the evolution of escaping variants (Milanovic, Fan, et al., 2018; Puig et al., 2008). Of those, the vast genetic and epigenetic modulations that occur during TIS may account for the selection of stem cell-like cells that slip through the senescent growth arrest (Milanovic, Fan, et al., 2018). Moreover, these senescence escapers assume more malignant qualities that lead to the formation of more aggressive tumors *in vivo* (Milanovic, Yu, & Schmitt, 2018). This recent evidence indicates that TIS might only be of short-term benefit due to the reversibility of the growth arrest in a subpopulation of tumor cells, but also, that TIS can give rise to more ominous forms of tumor cells that can complicate therapy outcomes.

In addition to the ability of senescent tumor cells to revert to a proliferative state, these cells also can secrete several factors that modulate the microenvironment. The effect of the SASP on non-malignant cells has been shown to have a protumorigenic potential, but further, the accumulation of therapy-induced non-malignant senescent cells have other deleterious, SASP-mediated effects, in several organ system dysfunction of the heart, kidneys, bone marrow and bone (Demaria et al., 2017). Nevertheless, the influence of the tumor cell SASP on the surrounding microenvironment has not been fully elucidated (Saleh et al., 2018). Recent evidence suggested that chemotherapy-induced senescent cells are capable of evading recognition by the immune system through the secretion of matrix metalloproteases which are major components of the SASP, suggesting that senescent tumor cells might acquire tools to escape immunosurveillance (Pereira et al., 2019).

Conversely, senescent cells have been shown to be identified and eliminated by several components of the immune system, suggesting that probably only a small variant subpopulation is capable of escaping immune recognition (Kale et al., 2020). Collectively, the ability of senescent tumor cells to escape growth inhibition and the deleterious adverse effects mediated by the SASP support the contention that *TIS is likely to be a largely unfavorable outcome of cancer treatment*.

3.5 Senolytics in the potential treatment of cancer

Recent reports have demonstrated that the selective clearance of accumulating senescent cells from aging animals delays the onset of, or interferes with the progression of several aging-associated processes and mitigates some of the established features of aging (Baker et al., 2011; Xu et al., 2015). The cornerstone of these studies was based on a transgenic mouse model of a $p16^{INK4a}$ suicide gene which allowed for the selective removal of senescent cells upon the use of an inert antiviral drug (Zhu et al., 2015). These efforts paved the way for the development of a novel drug class, where many of its drug members were identified using large-scale drug screening approaches, termed *senolytic agents* (Zhu et al., 2015). The primary senolytic agents are established drugs that have been repurposed due to their novel ability to target and eliminate senescent cells. The majority of investigations were focused on the ability of senolytics to modify features of progeroid models with highly promising results in disorders such as Alzheimer's, lung fibrosis and cardiac dysfunction (Lehmann et al., 2017; Walaszczyk et al., 2019; Zhang et al., 2019).

Based on the promising potential of senolytics to eliminate senescent cells effectively, and the recent shift in the paradigm that identified TIS as an unfavorable outcome of therapy, the use of senolytics has been proposed as a component of adjuvant cancer therapy (Short et al., 2019). This approach is based on the premise that clearance of senescent tumor cells will definitely reduce the tumor cell load, since these cells are often persistent and can possibly evade immune clearance; additionally, pre-emptive removal of senescent tumor cells could reduce the risk of reversion to a proliferative state and/or generation of more malignant variants. Several reports have now demonstrated the efficacy of the most frequently utilized senolytics in inducing selective killing in senescent tumor cell models. Of those, the established Bcl-2/Bcl-xl inhibitor ABT263 (navitoclax), which was first shown to eliminate senescent cells from animals exposed to total body radiation, resulting in rejuvenation of their hematopoietic functions (Chang et al., 2016; Yosef et al., 2016), has demonstrated similar senolytic ability in tumor models (Saleh, Carpenter, et al., 2020). ABT-263 was able to eliminate chemotherapy- and radiotherapy-induced senescent lung and breast cancer cells both *in vitro* and *in vivo* (Saleh, Carpenter, et al., 2020). In addition to ABT-263, several other senolytic agents have recently been identified as senescence-clearing therapies in cancer models including the HDAC inhibitor panobinostat (Samaraweera, Adomako, Rodriguez-Gabin, & McDaid, 2017), and several cardiac glycosides (Fuhrmann-Stroissnigg et al., 2017). On the other hand, senolytics such as the combination of dasatinib + quercetin (D + Q) have failed to demonstrate a robust senolytic activity in tumor cell models as opposed to previous results in aging models (Kovacovicova et al., 2018).

While drugs such as ABT-263 are usually administered together with the chemotherapy (Tolcher et al., 2015), we have recently proposed a novel therapeutic approach based on the senolytic potential of ABT263, in that once the tumor cells have entered into an established state of senescence, ABT263 could be used as a "clearing" agent to eliminate the residual senescent tumor cells in order to interfere with disease recurrence. Given the sensitivity of senescent cells to senolytics, drugs such as ABT-263 are also likely to be most effective following the use of chemotherapeutic agents that either promote primarily senescence or that leave a residual surviving senescent cell population. Each of these compounds is approved for use in investigational clinical trials either alone or in concert with traditional chemotherapies (Slingerland, Guchelaar, & Gelderblom, 2014; Zagouri et al., 2013). However, the most effective scheduling to maximize their clinical benefits against solid tumors remains to be determined. It is possible that drug efficacy would be maximized by treating patients when their tumors have achieved peak levels of senescence (either as a primary response or in residual cells). This would reduce the total exposure to senolytic agents, possibly allowing the use of effective lower doses, and for extended periods of time.

3.6 Conclusions and future directions

Senescence is a fundamental component of tumor cell response to therapy. This observation has been largely based on *in vitro*, and to a lesser extent, *in vivo* evidence. The contribution of senescence to the overall outcome of cancer therapy is yet to be determined; however, recent data suggest that the accumulation of senescent tumor cells can be associated with resistance to therapy, tumor dormancy, cancer recurrence and a spectrum of therapy-related untoward effects. Accordingly, the elimination of senescent tumor cells using senolytic drugs has been proposed and shown to be effective in reducing senescent tumor cell burden. Nevertheless, further studies are required to obtain a better understanding of the contribution of senescence to cancer outcomes and the potential benefit from using senolytic agents as part of cancer treatment. First, a better understanding of *in vivo* senescence and senescence in patients undergoing chemotherapy is needed; this requires high-throughput, reliable testing approaches that identify multiple senescence-associated markers. Second, the development of experimental models that directly address the involvement of senescence in tumor dormancy should be encouraged. Lastly, the toxicity of certain senolytic agents that have proven effective in the clearance of senescent tumor cells argues for the development of less toxic, more selective senolytics that can target harmful senescent cells but spare essential physiological functions that might be impaired when senescence cannot be targeted with a high degree of specificity in the tumor cell.

4. Autophagy and senescence: The connection

The previous sections focused on the induction of autophagy and senescence, individually, in response to anticancer therapy; however, both of these processes can be induced concomitantly in tumor cells in response to therapy. Given the coexistence of autophagy and senescence in response to therapy and overlapping mechanistic triggers, it is necessary to gain further insights into the relationship between the two processes.

Both autophagy and senescence share a number of regulators, including p53, oxidative stress and mTOR (Azad, Chen, & Gibson, 2009; Mrakovcic, Fröhlich, Mrakovcic, & Fröhlich, 2018; Nam et al., 2013; Tai et al., 2017). Therapy-induced damage activates p53, an essential tumor suppressor and stress sensor that can modulate the activity of both senescence and autophagy. Depending on the cellular localization of p53, nuclear p53 can activate autophagy through the transcriptional regulation of key ATG proteins or directly stimulate autophagy through DRAMs (Cui et al., 2016). In contrast, cytoplasmic p53 mediates an inhibitory effect on autophagy through ubiquitin-mediated Beclin-1 degradation and inhibition of AMPK (Tasdemir et al., 2008; Tripathi, Ash, & Shaha, 2014). Similarly, as discussed in depth earlier, therapy-induced DNA damage activates p53, resulting in $p21^{Cip1}$ induction and cellular growth arrest

(Chen et al., 2006; Yosef et al., 2017). Another common meditator of both autophagy and senescence is oxidative stress. Anticancer therapies, such as radiation and cisplatin, can result in increased ROS levels due to mitochondrial dysfunction and genotoxic stress. Autophagy serves to alleviate the cellular burden of the damaged proteins and mitochondria and to remove stressors, such as damaged mitochondrial DNA (mtDNA), which may stimulate senescence induction (Han et al., 2016; Kang, Lee, Kim, Choi, & Park, 2011; Rubinsztein, Mariño, & Kroemer, 2011). While autophagy is induced in response to oxidative stress, prolonged autophagy impairment in the presence of oxidative stress can result in senescence induction (Kwon et al., 2017; Tai et al., 2017). Furthermore, excessive ROS accumulation can further exacerbate DNA damage, result in p53/ $p21^{Cip1}$ activation and promotion of senescence (Fitzgerald et al., 2015; Luo, Yang, Schulte, Wargovich, & Wang, 2013; Macip et al., 2002; Passos et al., 2010). Taken together, pathways activated in response to therapy-induced damage can regulate both autophagy and senescence (Fig. 2).

While autophagy and senescence are induced in parallel in response to anticancer therapy, whether the two processes are interdependent still remains elusive. Seminal work by Narita and Young identified a specialized compartment, termed TOR-autophagy spatial coupling compartment (TASCC), in which products from the cells catabolic processes and autophagy degradation could provide raw materials for protein synthesis of certain SASP factors, including IL-6 and IL-8 (Narita et al., 2011; Young, Narita, & Narita, 2011). These studies suggest that autophagy may be an essential component of SASP processing and could contribute to the senescent phenotype. Alternatively, previous studies by our laboratory demonstrated that pharmacological and genetic inhibition of autophagy did not alter senescence induction in lung, breast and colorectal cell lines exposed to etoposide, doxorubicin and radiation, respectively (Saleh et al., 2020). All of these therapies induced nonprotective autophagy and a transient senescent growth arrest followed by proliferative recovery; moreover, autophagy interference did not further alter the proliferative capacity of these cells. In contrast, Vijayaraghavan et al. examined the effects of autophagy inhibition in breast cancer cells exposed to hormonal therapy in combination with CDK 4/6 inhibitors (Vijayaraghavan et al., 2017). These authors demonstrated that autophagy inhibition significantly reduced cell viability and tumor burden of breast tumors exposed to palbociclib, as well as the combination of palbociclib and letrazole; furthermore, this sensitization resulting from administration of an autophagy inhibitor in

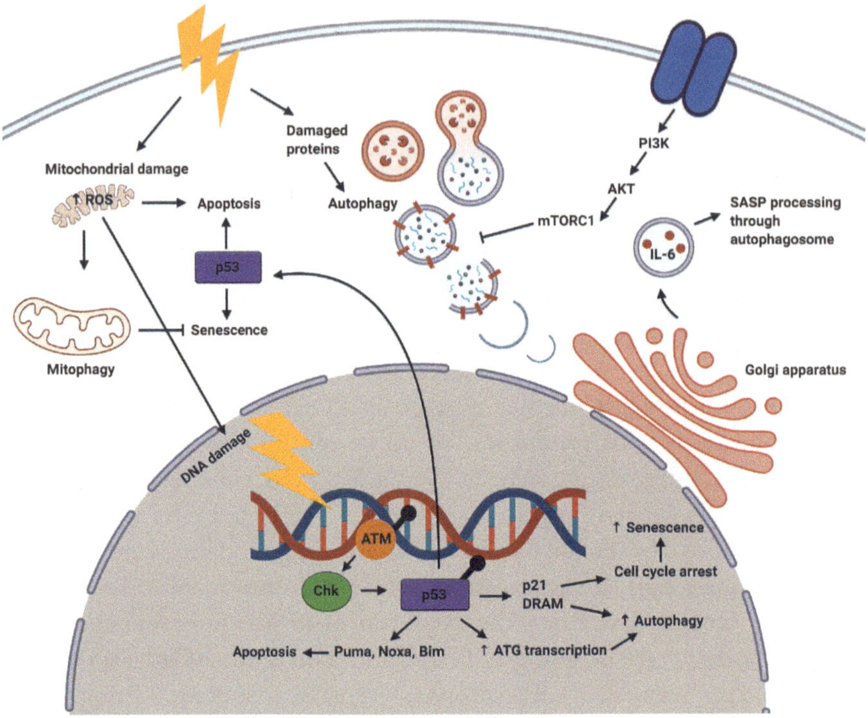

Fig. 2 Interconnectivity of cellular senescent and autophagic pathways. Given that both autophagy and senescence are two pathways activated by cellular stress and damage, it is plausible and well-exhibited in preclinical models that the two process often occur in parallel in response to chemotherapy and radiation. Both autophagy and senescence converge on a number of molecular signaling pathways and share multiple regulators, such as p53, mTOR and ROS. Therapies inducing DNA damage, such as etoposide, cisplatin or radiotherapy, is recognized by ATM and results in its subsequent phosphorylation. Phosphorylated ATM activates DDR pathways, as well as initiates signaling cascades for p53 phosphorylation and activation. Activated p53 regulates a number of responses, including p21 induction resulting in cellular growth arrest and senescence, as well as transcription of proapoptotic proteins. Furthermore, subcellular localization of p53 can also modulate autophagy. Nuclear p53 and DRAM/p53 signaling axis leads to increased transcription of autophagic machinery; whereas, cytoplasmic p53 can block autophagy induction and promote apoptosis and growth arrest. Anticancer therapies increasing intracellular ROS levels and inducing mitochondrial dysfunction can result in upregulation of mitophagy. Clearance of dysfunctional and aged mitochondria can prevent senescence, which occurs due to damage accumulation and oxidative stress. Moreover, elevated ROS levels can exacerbate DNA damage activating cellular growth arrest and apoptotic pathways. Lastly, autophagy may contribute to the senescent phenotype through processing of SASP factors, such as IL-6 and IL-8, via a specialized compartment, TASCC, in which products from the cells catabolic autophagic processes provide raw material for the anabolic protein synthesis processes. Collectively, while both autophagy and senescence are induced in parallel in response to anticancer therapeutics, whether the two processes are interdependent still remains elusive and requires further inquiry. *This figure was created with biorender.com.*

combination with palbociclib was mediated through the synergistic induction of senescence. Similarly, Vera-Ramirez et al. showed autophagy inhibition with HCQ decreased survival of dormant breast cancer cells and reduced lung metastasis; however, HCQ administration was minimally effective once dormant cells had regained proliferative capacity, suggesting a role for autophagy in sustaining tumor dormancy in breast cancer cells (Vera-Ramirez, Vodnala, Nini, Hunter, & Green, 2018). While the extent of senescence was not assessed in these studies, senescence may be a potential tumor dormancy model and further studies are warranted to interrogate the contributions of autophagy in senescence-associated tumor dormancy and recovery (Saleh, Tyutyunyk-Massey, & Gewirtz, 2019).

Lastly, several agents that modulate autophagy-related processes have been shown to exhibit senolytic effect, despite the fact that autophagy is not uniformly essential for the survival of senescent tumor cells, particularly, when autophagy is nonprotective in function (Saleh, Tyutyunyk-Massey, et al., 2020). For example, Was et al. demonstrated the use of BafA1 is capable of delaying the ability of therapy-induced senescent tumor cells to regain proliferative recovery (Was et al., 2017). However, the long-term consequences of single-exposure of senescent tumor cells to BafA1 led to increased proliferation *in vitro*, and enhanced tumor formation *in vivo*, suggesting a more complex contribution of autophagy to the survival of senescent tumor cells (Was et al., 2017). A seminal report by Wakita et al. showed that the senolytic potential of ARV825, a BET degrader agent, is highly dependent on the promotion of autophagy, and that autophagy inhibition, both pharmacologically and genetically, interfered with the elimination of senescent cells (Wakita et al., 2020). Furthermore, the macrolide antibiotics, azithromycin and roxithromycin, have been shown to selectively kill senescent fibroblasts, a mechanism that was also associated with robust induction of aerobic glycolysis and autophagy (Ozsvari, Nuttall, Sotgia, & Lisanti, 2018). Overall, the utilization of autophagy modulators as senolytic agents requires further investigation, and must be based on better understanding of the connection between senescence and autophagy in cancer.

Acknowledgments

Research relating to therapy-induced senescence in Dr David A. Gewirtz's laboratory is supported by Grant # RO1 CA-239706 from the National Institutes of Health/National Cancer Institute. Research relating to therapy-induced autophagy is supported by the Office of the Assistant Secretary of Defense for Health Affairs through Breast Cancer

Research Program Grant # W81XWH-19-1-0490. Nipa Patel is supported by the National Institutes of Health grant # 5T32DA007027. Core facilities are supported by Massey Cancer Support Grant # P30 CA016059. Dr Tareq Saleh is supported, in part, by the Deanship of Scientific Research at The Hashemite University, Jordan.

References

Aasland, D., Gotzinger, L., Hauck, L., Berte, N., Meyer, J., Effenberger, M., et al. (2019). Temozolomide induces senescence and repression of DNA repair pathways in glioblastoma cells via activation of ATR–Chk1, p21, and NF-kB. *Cancer Research*, *79*(1), 99–113.

Abedin, M. J., Wang, D., McDonnell, M. A., Lehmann, U., & Kelekar, A. (2007). Autophagy delays apoptotic death in breast cancer cells following DNA damage. *Cell Death and Differentiation*, *14*(3), 500–510.

Acosta, J. C., Banito, A., Wuestefeld, T., Georgilis, A., Janich, P., Morton, J. P., et al. (2013). A complex secretory program orchestrated by the inflammasome controls paracrine senescence. *Nature Cell Biology*, *15*(8), 978–990.

Ahmed, A. R., & Hombal, S. M. (1984). Cyclophosphamide (Cytoxan): A review on relevant pharmacology and clinical uses. *Journal of the American Academy of Dermatology*, *11*, 1115–1126.

Aird, K. M., & Zhang, R. (2013). Detection of senescence-associated heterochromatin foci (SAHF). *Methods in Molecular Biology*, *965*, 185–196.

Akin, D., Wang, S. K., Habibzadegah-Tari, P., Law, B., Ostrov, D., Li, M., et al. (2014). A novel ATG4B antagonist inhibits autophagy and has a negative impact on osteosarcoma tumors. *Autophagy*, *10*(11), 2021–2035.

Al-Bari, M. A. A. (2015). Chloroquine analogues in drug discovery: New directions of uses, mechanisms of actions and toxic manifestations from malaria to multifarious diseases. *The Journal of Antimicrobial Chemotherapy*, *70*(6), 1608–1621.

Alotaibi, M., Sharma, K., Saleh, T., Povirk, L. F., Hendrickson, E. A., & Gewirtz, D. A. (2016). Radiosensitization by PARP inhibition in DNA repair proficient and deficient tumor cells: Proliferative recovery in senescent cells. *Radiation Research*, *185*(3), 229–245.

Amawi, H., Karthikeyan, C., Pathak, R., Hussein, N., Robey, R., Ashby, C. R., et al. (2017). Thienopyrimidine derivatives exert their anticancer efficacy via apoptosis induction, oxidative stress and mitotic catastrophe. *European Journal of Medicinal Chemistry*, *138*, 1053–1065.

Anders, L., Ke, N., Hydbring, P., Choi, Y. J., Widlund, H. R., Chick, J. M., et al. (2011). A systematic screen for CDK4/6 substrates links FOXM1 phosphorylation to senescence suppression in cancer cells. *Cancer Cell*, *20*(5), 620–634.

Apel, A., Herr, I., Schwarz, H., Rodemann, H. P., & Mayer, A. (2008). Blocked autophagy sensitizes resistant carcinoma cells to radiation therapy. *Cancer Research*, *68*(5), 1485–1494.

Aydinlik, S., Erkisa, M., Cevatemre, B., Sarimahmut, M., Dere, E., Ari, F., et al. (2017). Enhanced cytotoxic activity of doxorubicin through the inhibition of autophagy in triple negative breast cancer cell line. *Biochimica et Biophysica Acta - General Subjects*, *1861*(2), 49–57.

Azad, M. B., Chen, Y., & Gibson, S. B. (2009). Regulation of autophagy by reactive oxygen species (ROS): Implications for cancer progression and treatment. *Antioxidants & Redox Signaling*, *11*(4), 777–790.

Baker, D. J., Wijshake, T., Tchkonia, T., Lebrasseur, N. K., Childs, B. G., Sluis, B., et al. (2011). Clearance of p16Ink4a-positive senescent cells delays ageing-associated disorders. *Nature*, *479*(7372), 232–236.

Barakat, D. J., Zhang, J., Barberi, T., Denmeade, S. R., & Friedman, A. D. (2015). CCAAT/Enhancer binding protein β controls androgen-deprivation-induced senescence in prostate cancer cells. *Oncogene*, *34*(48), 5912–5922.

Barnard, R. A., Wittenburg, L. A., Amaravadi, R. K., Gustafson, D. L., Thorburn, A., & Thamm, D. H. (2014). Phase I clinical trial and pharmacodynamic evaluation of combination hydroxychloroquine and doxorubicin treatment in pet dogs treated for spontaneously occurring lymphoma. *Autophagy*, *10*(8), 1415–1425.

Belair, C. D., Yeager, T. R., Lopez, P. M., & Reznikoff, C. A. (1997). Telomerase activity: A biomarker of cell proliferation, not malignant transformation. *Proceedings of the National Academy of Sciences of the United States of America*, *94*(25), 13677–13682.

Ben-Porath, I., & Weinberg, R. A. (2005). The signals and pathways activating cellular senescence. *The International Journal of Biochemistry & Cell Biology*, *37*(5), 961–976.

Bent, E. H., Gilbert, L. A., & Hemann, M. T. (2016). A senescence secretory switch mediated by PI3K/AKT/mTOR activation controls chemoprotective endothelial secretory responses. *Genes & Development*, *30*(16), 1811–1821.

Bergsbaken, T., Fink, S. L., & Cookson, B. T. (2009). Pyroptosis: Host cell death and inflammation. *Nature Reviews. Microbiology*, 7(2), 99–109.

Blommaart, E. F. C., Krause, U., Schellens, J. P. M., Vreeling-Sindelarova, H., & Meijer, A. J. (1997). The phosphatidylinositol 3-kinase inhibitors wortmannin and LY294002 inhibit autophagy in isolated rat hepatocytes. *European Journal of Biochemistry*, *243*(1–2), 240–246.

Bodnar, A. G., Ouellette, M., Frolkis, M., Holt, S. E., Chiu, C. P., Morin, G. B., et al. (1998). Extension of life-span by introduction of telomerase into normal human cells. *Science*, *279*(5349), 349–352.

Bojko, A., Czarnecka-Herok, J., Charzynska, A., Dabrowski, M., & Sikora, E. (2019). Diversity of the senescence phenotype of cancer cells treated with chemotherapeutic agents. *Cell*, *8*(12), 1501.

Bollard, J., Miguela, V., Ruiz De Galarreta, M., Venkatesh, A., Bian, C. B., Roberto, M. P., et al. (2017). Palbociclib (PD-0332991), a selective CDK4/6 inhibitor, restricts tumour growth in preclinical models of hepatocellular carcinoma. *Gut*, *66*(7), 1286–1296.

Bordin, D. L., Lima, M., Lenz, G., Saffi, J., Meira, L. B., Mesange, P., et al. (2013). DNA alkylation damage and autophagy induction. *Mutation Research*, *753*(2), 91–99.

Boya, P., Gonzalez-Polo, R.-A., Casares, N., Perfettini, J.-L., Dessen, P., Larochette, N., et al. (2005). Inhibition of macroautophagy triggers apoptosis. *Molecular and Cellular Biology*, *25*(3), 1025–1040.

Brahimi-Horn, M. C., Chiche, J., & Pouysségur, J. (2007). Hypoxia and cancer. *Journal of Molecular Medicine*, *85*(12), 1301–1307.

Brazil, L., Swampillai, A., Mak, K., Hackshaw, A., Edwards, D., Mesiri, P., et al. (2018). P01.072 Hydroxychloroquine and short course radiotherapy for elderly patients with glioma: A randomised study. *Neuro-Oncology*, *2*(1), vdaa046.

Brenner, A. J., Stampfer, M. R., & Aldaz, C. M. (1998). Increased p16 expression with first senescence arrest in human mammary epithelial cells and extended growth capacity with p16 inactivation. *Oncogene*, *17*(2), 199–205.

Brookes, S., Rowe, J., Gutierrez Del Arroyo, A., Bond, J., & Peters, G. (2004). Contribution of p16INK4a to replicative senescence of human fibroblasts. *Experimental Cell Research*, *298*(2), 549–559.

Bu, X., Le, C., Jia, F., Guo, X., Zhang, L., Zhang, B., et al. (2008). Synergistic effect of mTOR inhibitor rapamycin and fluorouracil in inducing apoptosis and cell senescence in hepatocarcinoma cells. *Cancer Biology & Therapy*, 7(3), 392–396.

Bursch, W., Ellinger, A., Kienzl, H., Török, L., Pandey, S., Sikorska, M., et al. (1996). Active cell death induced by the anti-estrogens tamoxifen and ICI 164 384 in human mammary carcinoma cells (MCF-7) in culture: The role of autophagy. *Carcinogenesis*, *17*(8), 1595–1607.

Burton, D. G. A., Giribaldi, M. G., Munoz, A., Halvorsen, K., Patel, A., Jorda, M., et al. (2013). Androgen deprivation-induced senescence promotes outgrowth of androgen-refractory prostate cancer cells. *PLoS One*, *8*(6), e68003.

Cai, Z., Zhang, A., Choksi, S., Li, W., Li, T., Zhang, X., et al. (2016). Activation of cell-surface proteases promotes necroptosis, inflammation and cell migration. *Nature Publishing Group*, *26*(8), 886–900.

Campisi, J. (2001). Cellular senescence as a tumor-suppressor mechanism. *Trends in Cell Biology*, *11*(11), 27–31.

Campisi, J. (2013). Aging, cellular senescence, and cancer. *Annual Review of Physiology*, *75*, 685–705.

Campisi, J., & d'Adda di Fagagna, F. (2007). Cellular senescence: When bad things happen to good cells. *Nature Reviews. Molecular Cell Biology*, *8*(9), 729–740.

Cao, C., Subhawong, T., Albert, J. M., Kim, K. W., Geng, L., Sekhar, K. R., et al. (2006). Inhibition of mammalian target of rapamycin or apoptotic pathway induces autophagy and radiosensitizes PTEN null prostate cancer cells. *Cancer Research*, *66*(20), 10040–10047.

Carmona-Gutierrez, D., Hughes, A. L., Madeo, F., & Ruckenstuhl, C. (2016). The crucial impact of lysosomes in aging and longevity. *Ageing Research Reviews*, *32*, 2–12.

Carneiro, B. A., & El-Deiry, W. S. (2020). Targeting apoptosis in cancer therapy. *Nature Reviews. Clinical Oncology*, *17*(7), 395–417.

Cells, M. H., Meng, A., Wang, Y., Van, Z. G., & Zhou, D. (2003). Ionizing radiation and busulfan induce premature senescence in murine bone marrow hematopoietic cells. *Cancer Research*, *63*(17), 5414–5419.

Chaachouay, H., Ohneseit, P., Toulany, M., Kehlbach, R., Multhoff, G., & Rodemann, H. P. (2011). Molecular radiobiology autophagy contributes to resistance of tumor cells to ionizing radiation. *Radiotherapy and Oncology*, *99*(3), 287–292.

Chakradeo, S., Sharma, K., Alhaddad, A., Bakhshwin, D., Le, N., Harada, H., et al. (2015). Yet another function of p53—The switch that determines whether radiation-induced autophagy will be cytoprotective or nonprotective: Implications for autophagy inhibition as a therapeutic strategy. *Molecular Pharmacology*, *87*(5), 803–814.

Chang, J., Wang, Y., Shao, L., Laberge, R., Demaria, M., Campisi, J., et al. (2016). Clearance of senescent cells by ABT263 rejuvenates aged hematopoietic stem cells in mice. *Nature Medicine*, *22*(1), 78–83.

Chang, B. D., Xuan, Y., Broude, E. V., Zhu, H., Schott, B., Fang, J., et al. (1999). Role of p53 and p21(waf1/cip1) in senescence-like terminal proliferation arrest induced in human tumor cells by chemotherapeutic drugs. *Oncogene*, *18*(34), 4808–4818.

Chatterjee, P., Choudhary, G. S., Sharma, A., Singh, K., Heston, W. D., Ciezki, J., et al. (2013). PARP inhibition sensitizes to low dose-rate radiation TMPRSS2-ERG fusion gene-expressing and PTEN-deficient prostate cancer cells. *PLoS One*, *8*(4), 1–12.

Chemali, M., Radtke, K., Desjardins, M., & English, L. (2011). Alternative pathways for MHC class I presentation: A new function for autophagy. *Cellular and Molecular Life Sciences*, *68*(9), 1533–1541.

Chen, Q., Fischer, A., Reagan, J. D., Yan, L. J., & Ames, B. N. (1995). Oxidative DNA damage and senescence of human diploid fibroblast cells. *Proceedings of the National Academy of Sciences of the United States of America*, *92*(10), 4337–4341.

Chen, J., Huang, X., Halicka, D., Brodsky, S., Avram, A., Eskander, J., et al. (2006). Contribution of p16 INK4a and p21 CIP1 pathways to induction of premature senescence of human endothelial cells: Permissive role of p53. *American Journal of Physiology. Heart and Circulatory Physiology*, *290*(4), H1575–H1586.

Chen, N., Wu, L., Yuan, H., & Wang, J. (2015). ROS/Autophagy/Nrf2 pathway mediated low-dose radiation induced radio-resistance in human lung adenocarcinoma A549 cell. *International Journal of Biological Sciences*, *11*(7), 833–844.

Chien, Y., Scuoppo, C., Wang, X., Fang, X., Balgley, B., Bolden, J. E., et al. (2011). Control of the senescence-associated secretory phenotype by NF-kB promotes senescence and enhances chemosensitivity. *Genes & Development*, *25*(20), 2125–2136.

Chittaranjan, S., Bortnik, S., Dragowska, W. H., Xu, J., Abeysundara, N., Leung, A., et al. (2014). Autophagy inhibition augments the anticancer effects of epirubicin treatment in anthracycline-sensitive and -resistant triple-negative breast cancer. *Clinical Cancer Research*, *20*(12), 3159–3173.

Choi, Y.-M., Kim, H.-K., Shim, W., Anwar, M. A., Kwon, J.-W., Kwon, H.-K., et al. (2015). Mechanism of cisplatin-induced cytotoxicity is correlated to impaired metabolism due to mitochondrial ROS generation. *PLoS One*, *10*(8), e0135083. Mukhopadhyay P., ed.

Chude, C. I., Amaravadi, R. K., Chude, C. I., & Amaravadi, R. K. (2017). Targeting autophagy in cancer: Update on clinical trials and novel inhibitors. *International Journal of Molecular Sciences*, *18*(6), 1279.

Classen, F., Kranz, P., Riffkin, H., Pompsch, M., Wolf, A., Göpelt, K., et al. (2019). Autophagy induced by ionizing radiation promotes cell death over survival in human colorectal cancer cells. *Experimental Cell Research*, *374*(1), 29–37.

Collado, M., Gil, J., Efeyan, A., Guerra, C., Schuhmacher, A. J., Barradas, M., et al. (2005). Tumour biology: Senescence in premalignant tumours. *Nature*, *436*(7051), 642.

Coller, H. A. (2011). The essence of quiescence. *Science*, *334*(6059), 1074–1075.

Coller, H. A., Sang, L., & Roberts, J. M. (2006). A new description of cellular quiescence. *PLoS Biology*, *4*(3), 329–349.

Collins, K., & Mitchell, J. R. (2002). Telomerase in the human organism. *Oncogene*, *21*(4), 564–579.

Cook, K. L., & Clarke, R. (2014). Estrogen receptor-α signaling and localization regulates autophagy and unfolded protein response activation in ER+ breast cancer. *Receptors Clinical Investigation*, *1*(6), e316.

Coppé, J., Rodier, F., Patil, C. K., Freund, A., Desprez, P., & Tumor Suppressor, C. J. (2011). Aging biomarker p16INK4a induces cellular senescence without the associated inflammatory secretory phenotype. *The Journal of Biological Chemistry*, *286*(42), 36396–36403.

Cotarelo, C. L., Schad, A., Kirkpatrick, C. J., Sleeman, J. P., Springer, E., Schmidt, M., et al. (2016). Detection of cellular senescence within human invasive breast carcinomas distinguishes different breast tumor subtypes. *Oncotarget*, 7(46), 74846–748597.

Crighton, D., Wilkinson, S., O'Prey, J., Syed, N., Smith, P., Harrison, P. R., et al. (2006). DRAM, a p53-induced modulator of autophagy, is critical for apoptosis. *Cell*, *126*(1), 121–134.

Cufí, S., Vazquez-Martin, A., Oliveras-Ferraros, C., Corominas-Faja, B., Cuyàs, E., López-Bonet, E., et al. (2013). The anti-malarial chloroquine overcomes primary resistance and restores sensitivity to Trastuzumab in HER2-positive breast cancer. *Scientific Reports*, *3*(1), 2469.

Cui, L., Song, Z., Liang, B., Jia, L., Ma, S., & Liu, X. (2016). Radiation induces autophagic cell death via the p53/DRAM signaling pathway in breast cancer cells. *Oncology Reports*, *35*(6), 3639–3647.

Däbritz, J. H. M., Yu, Y., Milanovic, M., Schönlein, M., Rosenfeldt, M. T., Dörr, J. R., et al. (2016). CD20-targeting immunotherapy promotes cellular senescence in B-cell lymphoma. *Molecular Cancer Therapeutics*, *15*(5), 1074–1081.

Dabrowska, M., Skoneczny, M., & Rode, W. (2011). Functional gene expression profile underlying methotrexate-induced senescence in human colon cancer cells. *Tumor Biology*, *32*(5), 965–976.

Dabrowska, M., Skoneczny, M., Uram, L., & Rode, W. (2019). Methotrexate-induced senescence of human colon cancer cells depends on p53 acetylation, but not genomic aberrations. *Anti-Cancer Drugs*, *30*(4), 374–382.

Daskalaki, I., Gkikas, I., & Tavernarakis, N. (2018). Hypoxia and selective autophagy in cancer development and therapy. *Frontiers in Cell and Development Biology*, *6*, 104.

Davalos, A. R., Coppe, J. P., Campisi, J., & Desprez, P. Y. (2010). Senescent cells as a source of inflammatory factors for tumor progression. *Cancer Metastasis Reviews*, *29*(2), 273–283.

Decaup, E., Jean, C., Laurent, C., Gravelle, P., Fruchon, S., Capilla, F., et al. (2013). Anti-tumor activity of obinutuzumab and rituximab in a follicular lymphoma 3D model. *Blood Cancer Journal*, *3*(8), e131.

Dechat, T., Adam, S. A., Taimen, P., Shimi, T., & Goldman, R. D. (2010). Nuclear lamins. *Cold Spring Harbor Perspectives in Biology*, *2*(11), a000547.

Demaria, M., Leary, M. N. O., Chang, J., Shao, L., Liu, S., Alimirah, F., et al. (2017). Cellular senescence promotes adverse effects of chemotherapy and cancer relapse. *Cancer Discovery*, 7(2), 165–177.

Demaria, M., Ohtani, N., Youssef, S. A., Rodier, F., Toussaint, W., Mitchell, J. R., et al. (2014). An essential role for senescent cells in optimal wound healing through secretion of PDGF-AA. *Developmental Cell*, *31*(6), 722–733.

Deng, L., Liang, H., Xu, M., Yang, X., Burnette, B., Arina, A., et al. (2014). STING-dependent cytosolic DNA sensing promotes radiation-induced type I interferon-dependent antitumor immunity in immunogenic tumors. *Immunity*, *41*(5), 843–852.

Di Micco, R., Fumagalli, M., Cicalese, A., Piccinin, S., Gasparini, P., Luise, C., et al. (2006). Oncogene-induced senescence is a DNA damage response triggered by DNA hyper-replication. *Nature*, *444*(7119), 638–642.

Di Mitri, D., Toso, A., Chen, J. J., Sarti, M., Pinton, S., Jost, T. R., et al. (2014). Tumour-infiltrating Gr-1 + myeloid cells antagonize senescence in cancer. *Nature*, *515*(7525), 134–137.

Dikic, I., & Elazar, Z. (2018). Mechanism and medical implications of mammalian autophagy. *Nature Reviews. Molecular Cell Biology*, *19*(6), 349–364.

Dimri, G. P., Lee, X., Basile, G., Acosta, M., Scott, G., Roskelley, C., et al. (1995). A biomarker that identifies senescent human cells in culture and in aging skin in vivo. *Proceedings of the National Academy of Sciences*, *92*(20), 9363–9367.

Dolfi, S. C., Jäger, A. V., Medina, D. J., Haffty, B. G., Yang, J. M., & Hirshfield, K. M. (2014). Fulvestrant treatment alters MDM2 protein turnover and sensitivity of human breast carcinoma cells to chemotherapeutic drugs. *Cancer Letters*, *350*(1–2), 52–60.

Dou, Q., Chen, H. N., Wang, K., Yuan, K., Lei, Y., Li, K., et al. (2016). Ivermectin induces cytostatic autophagy by blocking the PAK1/Akt axis in breast cancer. *Cancer Research*, *76*(15), 4457–4469.

Douarre, C., Gomez, D., Morjani, H., Zahm, J. M., O'Donohue, M. F., Eddabra, L., et al. (2005). Overexpression of Bcl-2 is associated with apoptotic resistance to the G-quadruplex ligand 12459 but is not sufficient to confer resistance to long-term senescence. *Nucleic Acids Research*, *33*(7), 2192–2203.

Druelle, C., Drullion, C., Deslé, J., Martin, N., Saas, L., Cormenier, J., et al. (2016). ATF6a regulates morphological changes associated with senescence in human fibroblasts. *Oncotarget*, 7(42), 67699–67715.

Egan, D. F., Chun, M. G. H., Vamos, M., Zou, H., Rong, J., Miller, C. J., et al. (2015). Small molecule inhibition of the autophagy kinase ULK1 and identification of ULK1 substrates. *Molecular Cell*, *59*(2), 285–297.

Egan, D., Kim, J., Shaw, R. J., & Guan, K.-L. (2011). The autophagy initiating kinase ULK1 is regulated via opposing phosphorylation by AMPK and mTOR. *Autophagy*, 7(6), 643–644.

Elgendy, M., Sheridan, C., Brumatti, G., & Martin, S. J. (2011). Oncogenic Ras-induced expression of noxa and beclin-1 promotes autophagic cell death and limits clonogenic survival. *Molecular Cell*, *42*(1), 23–35.

Elmore, L. W., Rehder, C. W., Di, X., McChesney, P. A., Jackson-cook, C. K., Gewirtz, D. A., et al. (2002). Adriamycin-induced senescence in breast tumor cells involves functional p53 and telomere dysfunction. *The Journal of Biological Chemistry*, *277*(38), 35509–35515.

Emadi, A., Jones, R. J., & Brodsky, R. A. (2009). Cyclophosphamide and cancer: Golden anniversary. *Nature Reviews. Clinical Oncology*, *6*(11), 638–647.

Eng, C. H., Wang, Z., Tkach, D., Toral-Barza, L., Ugwonali, S., Liu, S., et al. (2016). Macroautophagy is dispensable for growth of KRAS mutant tumors and chloroquine efficacy. *Proceedings of the National Academy of Sciences*, *113*(1), 182–187.

Ewald, J. A., Desotelle, J. A., Almassi, N., & Jarrard, D. F. (2008). Drug-induced senescence bystander proliferation in prostate cancer cells in vitro and in vivo. *British Journal of Cancer*, *98*(7), 1244–1249.

Ewald, J. A., Desotelle, J. A., Wilding, G., & Jarrard, D. F. (2010). Therapy-induced senescence in cancer. *Journal of the National Cancer Institute*, *102*(20), 1536–1546.

Ewald, J. A., Joshua, A. D., Church, D. R., Yang, B., Hyang, W., Laurila, T. A., et al. (2016). Androgen deprivation induces senescence characteristics in prostate cancer cells in vitro and in vivo. *The Prostate*, *73*(4), 337–345.

Feng, Y., Gao, Y., Wang, D., Xu, Z., Sun, W., & Ren, P. (2018). Autophagy inhibitor (LY294002) and 5-fluorouracil (5-FU) combination-based nanoliposome for enhanced efficacy against esophageal squamous cell carcinoma. *Nanoscale Research Letters*, *13*(1), 325.

Feng, Y., He, D., Yao, Z., & Klionsky, D. J. (2014). The machinery of macroautophagy. *Cell Research*, *24*(1), 24–41.

Ferrand, M., Kirsh, O., Griveau, A., Vindrieux, D., Martin, N., Defossez, P. A., et al. (2015). Screening of a kinase library reveals novel pro-senescence kinases and their common NF-κB-dependent transcriptional program. *Aging (Albany NY)*, *7*(11), 986–999.

Filion, G. J., van Bemmel, J. G., Braunschweig, U., Talhout, W., Kind, J., Ward, L. D., et al. (2010). Systematic protein location mapping reveals five principal chromatin types in Drosophila cells. *Cell*, *143*(2), 212–224.

Fitzgerald, A. L., Osman, A. A., Xie, T.-X., Patel, A., Skinner, H., Sandulache, V., et al. (2015). Reactive oxygen species and p21Waf1/Cip1 are both essential for p53-mediated senescence of head and neck cancer cells. *Cell Death & Disease*, *6*(3), e1678.

Fleury, H., Malaquin, N., Tu, V., Gilbert, S., Martinez, A., Olivier, M. A., et al. (2019). Exploiting interconnected synthetic lethal interactions between PARP inhibition and cancer cell reversible senescence. *Nature Communications*, *10*(1), 2556.

Foersch, S., Sperka, T., Lindner, C., Taut, A., Rudolph, K. L., Breier, G., et al. (2015). VEGFR2 signaling prevents colorectal cancer cell senescence to promote tumorigenesis in mice with colitis. *Gastroenterology*, *149*(1), 177–189.

Freund, A., Patil, C. K., & Campisi, J. (2011). p38MAPK is a novel DNA damage response-independent regulator of the senescence-associated secretory phenotype. *The EMBO Journal*, *30*(8), 1536–1548.

Freund, A., R-MR, L., Demaria, M., & Campisi, J. (2012). Lamin B1 loss is a senescence-associated biomarker. *Molecular Biology of the Cell*, *23*(11), 2066–2075.

Fuhrmann-Stroissnigg, H., Ling, Y. Y., Zhao, J., McGowan, S. J., Zhu, Y., Brooks, R. W., et al. (2017). Identification of HSP90 inhibitors as a novel class of senolytics. *Nature Communications*, *8*(1), 422.

Fulda, S., & Debatin, K. M. (2006). Extrinsic versus intrinsic apoptosis pathways in anticancer chemotherapy. *Oncogene*, *25*(34), 4798–4811.

Fulda, S., Gorman, A. M., Hori, O., & Samali, A. (2010). Cellular stress responses: Cell survival and cell death. *International Journal of Cell Biology*, *2010*, 214074.

Fumagalli, M., & d'Adda, d. F. F. (2009). SASPense and DDRama in cancer and ageing. *Nature Cell Biology*, *11*(8), 921–923. Nature Publishing Group.

Galluzzi, L., Pietrocola, F., Bravo-San Pedro, J. M., Amaravadi, R. K., Baehrecke, E. H., Cecconi, F., et al. (2015). Autophagy in malignant transformation and cancer progression. *The EMBO Journal*, *34*(7), 856–880.

Galluzzi, L., Vitale, I., Abrams, J. M., Alnemri, E. S., Baehrecke, E. H., Blagosklonny, M. V., et al. (2012). Molecular definitions of cell death subroutines: Recommendations of the Nomenclature Committee on Cell Death 2012. *Cell Death and Differentiation*, *19*(1), 107–120.

Gammoh, N., & Wilkinson, S. (2014). Autophagy in cancer biology and therapy. *Frontiers of Biology*, *9*(1), 35–50.

Ge, H., Ke, J., Xu, N., Li, H., Gong, J., Li, X., et al. (2018). Dexamethasone alleviates pemetrexed-induced senescence in non-small-cell lung cancer. *Food and Chemical Toxicology*, *119*, 86–97.

Geng, J., Nair, U., Yasumura-Yorimitsu, K., & Klionsky, D. J. (2010). Post-Golgi Sec proteins are required for autophagy in Saccharomyces cerevisiae. *Molecular Biology of the Cell*, *21*(13), 2257–2269.

Gewirtz, D. A. (2009). Autophagy, senescence and tumor dormancy in cancer therapy. *Autophagy*, *5*(8), 1232–1234.

Gewirtz, D. A. (2013). Cytoprotective and nonprotective autophagy in cancer therapy. *Autophagy*, *9*(9), 1263–1265.

Gewirtz, D. A. (2014a). The four faces of autophagy: Implications for cancer therapy. *Cancer Research*, *74*(3), 647–651.

Gewirtz, D. A. (2014b). Autophagy and senescence in cancer therapy. *Journal of Cellular Physiology*, *229*(1), 6–9.

Gewirtz, D. A. (2020). The switch between protective and nonprotective autophagy; Implications for autophagy inhibition as a therapeutic strategy in cancer. *Biology (Basel)*, *9*(12), 1–6.

Ghiringhelli, F., Apetoh, L., Tesniere, A., Aymeric, L., Ma, Y., Ortiz, C., et al. (2009). Activation of the NLRP3 inflammasome in dendritic cells induces IL-1β-dependent adaptive immunity against tumors. *Nature Medicine*, *15*(10), 1170–1178.

Giannoni, E., Buricchi, F., Grimaldi, G., Parri, M., Cialdai, F., Taddei, M. L., et al. (2008). Redox regulation of anoikis: Reactive oxygen species as essential mediators of cell survival. *Cell Death and Differentiation*, *15*(5), 867–878.

Givvimani, S., Munjal, C., Tyagi, N., Sen, U., Metreveli, N., & Tyagi, S. C. (2012). Mitochondrial division/mitophagy inhibitor (Mdivi) ameliorates pressure overload induced heart failure. *PLoS One*, 7(3), e32388. Tjwa M, ed.

Glick, D., Barth, S., & Macleod, K. F. (2010). Autophagy: Cellular and molecular mechanisms. *The Journal of Pathology*, *221*(1), 3–12.

Goehe, R. W., Di, X., Sharma, K., Bristol, M. L., Henderson, S. C., Valerie, K., et al. (2012). The autophagy-senescence connection in chemotherapy: Must tumor cells (self) eat before they sleep? *The Journal of Pharmacology and Experimental Therapeutics*, *343*(3), 763–778.

Gorgoulis, V., Adams, P. D., Alimonti, A., Bennett, D. C., Bischof, O., Bishop, C., et al. (2019). Cellular senescence: Defining a path forward. *Cell*, *179*(4), 813–827.

Gough, D. R., & Cotter, T. G. (2011). Hydrogen peroxide: A Jekyll and Hyde signalling molecule. *Cell Death & Disease*, *2*(10), e213.

Gu, L., & Kitamura, M. (2012). Sensitive detection and monitoring of senescence-associated secretory phenotype by SASP-RAP assay. *PLoS One*, 7(12), 1–9.

Guan, X., LaPak, K. M., Hennessey, R. C., Yu, C. Y., Shakya, R., Zhang, J., et al. (2017). Stromal senescence by prolonged CDK4/6 inhibition potentiates tumor growth. *Molecular Cancer Research*, *15*(3), 237–249.

Guo, J. Y., Chen, H.-Y., Mathew, R., Fan, J., Strohecker, A. M., Karsli-Uzunbas, G., et al. (2011). Activated Ras requires autophagy to maintain oxidative metabolism and tumorigenesis. *Genes & Development*, *25*(5), 460–470.

Guo, B., Tam, A., Santi, S. A., & Parissenti, A. M. (2016). Role of autophagy and lysosomal drug sequestration in acquired resistance to doxorubicin in MCF-7 cells. *BMC Cancer, 16*(1), 762.

Haferkamp, S., Borst, A., Adam, C., Becker, T. M., Motschenbacher, S., Windhövel, S., et al. (2013). Vemurafenib induces senescence features in melanoma cells. *The Journal of Investigative Dermatology, 133*(6), 1601–1609.

Hailey, D. W., Rambold, A. S., Satpute-Krishnan, P., Mitra, K., Sougrat, R., Kim, P. K., et al. (2010). Mitochondria supply membranes for autophagosome biogenesis during starvation. *Cell, 141*(4), 656–667.

Han, X., Tai, H., Wang, X., Wang, Z., Zhou, J., Wei, X., et al. (2016). AMPK activation protects cells from oxidative stress-induced senescence via autophagic flux restoration and intracellular NAD(+) elevation. *Aging Cell, 15*(3), 416–427.

Hanahan, D., & Weinberg, R. A. (2011). Hallmarks of cancer: The next generation. *Cell, 144*(5), 646–674.

Hansen, S. H., Olsson, A., & Casanova, J. E. (1995). Wortmannin, an inhibitor of phosphoinositide 3-kinase, inhibits transcytosis in polarized epithelial cells. *The Journal of Biological Chemistry, 270*(47), 28425–28432.

Harris, I. S., Treloar, A. E., Inoue, S., Sasaki, M., Gorrini, C., Lee, K. C., et al. (2015). Glutathione and thioredoxin antioxidant pathways synergize to drive cancer initiation and progression. *Cancer Cell, 27*(2), 211–222.

Hartman, M. L., Sztiller-Sikorska, M., & Czyz, M. (2019). Whole-exome sequencing reveals novel genetic variants associated with diverse phenotypes of melanoma cells. *Molecular Carcinogenesis, 58*(4), 588–602.

Hasan, M. R., Ho, S. H. Y., Owen, D. A., & Tai, I. T. (2011). Inhibition of VEGF induces cellular senescence in colorectal cancer cells. *International Journal of Cancer, 129*(9), 2115–2123.

Hattangadi, D. K., DeMasters, G. A., Walker, T. D., Jones, K. R., Di, X., Newsham, I. F., et al. (2004). Influence of p53 and caspase 3 activity on cell death and senescence in response to methotrexate in the breast tumor cell. *Biochemical Pharmacology, 68*(9), 1699–1708.

Hayflick, L. (1965). The limited in vitro lifetime of human diploid cell strains. *Experimental Cell Research, 636*, 614–636.

Hayflick, L., & Moorhead, P. S. (1961). The serial cultivation of human diploid cell strains. *Experimental Cell Research, 25*, 585–621.

Henson, E., Chen, Y., & Family Members, G. S. E. G. F. R. (2017). Regulation of autophagy is at a crossroads of cell survival and death in cancer. *Cancers (Basel), 9*(4), 27.

Herbig, U., Jobling, W. A., Chen, B. P. C., Chen, D. J., & Sedivy, J. M. (2004). Telomere shortening triggers senescence of human cells through a pathway involving ATM, p53, and p21CIP1, but not p16INK4a. *Molecular Cell, 14*(4), 501–513.

Hernandez-Segura, A., Nehme, J., & Demaria, M. (2018). Hallmarks of cellular senescence. *Trends in Cell Biology, 28*(6), 436–453.

Herranz, N., Gallage, S., Mellone, M., Wuestefeld, T., Klotz, S., Hanley, C. J., et al. (2015). mTOR regulates MAPKAPK2 translation to control the senescence-associated secretory phenotype. *Nature Cell Biology, 17*(9), 1205–1217.

Holohan, C., Schaeybroeck, S., Van Longley, D. B., & Johnston, P. G. (2013). Cancer drug resistance: An evolving paradigm. *Nature Publishing Group, 13*(10), 714–726.

Huang, Q., Li, F., Liu, X., Li, W., Shi, W., Liu, F.-F., et al. (2011). Caspase 3-mediated stimulation of tumor cell repopulation during cancer radiotherapy. *Nature Medicine, 17*(7), 860–866.

Hughson, L. R. K., Poon, V. I., Spowart, J. E., & Lum, J. J. (2012). Implications of therapy-induced selective autophagy on tumor metabolism and survival. *International Journal of Cell Biology, 2012*, 1–11.

Hutchison, C. J. (2014). B-type lamins in health and disease. *Seminars in Cell & Developmental Biology*, *29*(100), 158–163.

Inao, T., Kotani, H., Iida, Y., Kartika, I. D., Okimoto, T., Tanino, R., et al. (2019). Different sensitivities of senescent breast cancer cells to immune cell-mediated cytotoxicity. *Cancer Science*, *110*(9), 2690–2699.

Itahana, K., Dimri, G., & Campisi, J. (2001). Regulation of cellular senescence by p53. *European Journal of Biochemistry*, *268*(10), 2784–2791.

Ito, H., Daido, S., Kanzawa, T., Kondo, S., & Kondo, Y. (2005). Radiation-induced autophagy is associated with LC3 and its inhibition sensitizes malignant glioma cells. *International Journal of Oncology*, *26*(5), 1401–1410.

Iyengar, M., Hayer, P. O., Cole, A., Sebastian, T., Yang, K., Coffman, L., et al. (2018). CDK4/6 inhibition as maintenance and combination therapy for high grade serous ovarian cancer. *Oncotarget*, *9*(21), 15658–15672.

Jackson, S. P., & Bartek, J. (2009). The DNA-damage response in human biology and disease. *Nature*, *461*(7267), 1071–1078.

Jang, M., Park, R., Kim, H., Namkoong, S., Jo, D., Huh, Y. H., et al. (2018). AMPK contributes to autophagosome maturation and lysosomal fusion. *Scientific Reports*, *8*(1), 12637.

Jones, K. R., Elmore, L. W., Jackson-Cook, C., Demasters, G., Povirk, L. F., Holt, S. E., et al. (2005). p53-Dependent accelerated senescence induced by ionizing radiation in breast tumour cells. *International Journal of Radiation Biology*, *81*(6), 445–458.

Joselow, A., Lynn, D., Terzian, T., & Box, N. F. (2017). Senescence-like phenotypes in human nevi. *Methods in Molecular Biology*, *1534*, 175–184.

Kaczanowski, S. (2016). Apoptosis: Its origin, history, maintenance and the medical implications for cancer and aging. *Physical Biology*, *13*(3), 031001.

Kale, A., Sharma, A., Stolzing, A., Stolzing, A., Desprez, P. Y., Desprez, P. Y., et al. (2020). Role of immune cells in the removal of deleterious senescent cells [Internet]. In *Vol. 17. Immunity and ageing* (pp. 1–9). BioMed Central Ltd. [cited 2020 Dec 3]. Available from https://doi.org/10.1186/s12979-020-00187-9.

Kaminskyy, V. O., Piskunova, T., Zborovskaya, I. B., Tchevkina, E. M., & Zhivotovsky, B. (2012). Suppression of basal autophagy reduces lung cancer cell proliferation and enhances caspase-dependent and -independent apoptosis by stimulating ROS formation. *Autophagy*, *8*(7), 1032–1044.

Kang, H. T., Lee, K. B., Kim, S. Y., Choi, H. R., & Park, S. C. (2011). Autophagy impairment induces premature senescence in primary human fibroblasts. *PLoS One*, *6*(8), e23367.

Kanzawa, T., Germano, I. M., Komata, T., Ito, H., Kondo, Y., & Kondo, S. (2004). Role of autophagy in temozolomide-induced cytotoxicity for malignant glioma cells. *Cell Death and Differentiation*, *11*(4), 448–457.

Karantza-Wadsworth, V., Patel, S., Kravchuk, O., Chen, G., Mathew, R., Jin, S., et al. (2007). Autophagy mitigates metabolic stress and genome damage in mammary tumorigenesis. *Genes & Development*, *21*(13), 1621–1635.

Kaufmann, S. H., & Earnshaw, W. C. (2000). Induction of apoptosis by cancer chemotherapy. *Experimental Cell Research*, *256*(1), 42–49.

Kawata, H., Kamiakito, T., Nakaya, T., & Komatsubara, M. (2017). Stimulation of cellular senescent processes, including secretory phenotypes and anti-oxidant responses, after androgen deprivation therapy in human prostate cancer. *The Journal of Steroid Biochemistry and Molecular Biology*, *165*(Pt. B), 219–227.

Khongkow, P., Gomes, A. R., Gong, C., Man, E. P. S., Tsang, J. W. H., Zhao, F., et al. (2016). Paclitaxel targets FOXM1 to regulate KIF20A in mitotic catastrophe and breast cancer paclitaxel resistance. *Oncogene*, *35*(8), 990–1002.

Kim, E. J., Jeong, J.-H., Bae, S., Kang, S., Kim, C. H., & Lim, Y.-B. (2013). mTOR inhibitors radiosensitize PTEN-deficient non-small-cell lung cancer cells harboring an EGFR activating mutation by inducing autophagy. *Journal of Cellular Biochemistry*, *114*(6), 1248–1256.

Kim, J., Kundu, M., Viollet, B., & Guan, K.-L. (2011). AMPK and mTOR regulate autophagy through direct phosphorylation of Ulk1. *Nature Cell Biology*, *13*(2), 132–141.

Kipper, F. C., Silva, A. O., Marc, A. L., Confortin, G., Junqueira, A. V., Neto, E. P., et al. (2018). Vinblastine and antihelmintic mebendazole potentiate temozolomide in resistant gliomas. *Investigational New Drugs*, *36*(2), 323–331.

Klionsky, D. J. (2007). Autophagy: From phenomenology to molecular understanding in less than a decade. *Nature Reviews. Molecular Cell Biology*, *8*(11), 931–937.

Ko, A., Kanehisa, A., Martins, I., Senovilla, L., Chargari, C., Dugue, D., et al. (2014). Autophagy inhibition radiosensitizes in vitro, yet reduces radioresponses in vivo due to deficient immunogenic signalling. *Cell Death and Differentiation*, *21*(1), 92–99.

Köchl, R., Hu, X. W., Chan, E. Y. W., & Tooze, S. A. (2006). Microtubules facilitate autophagosome formation and fusion of autophagosomes with endosomes. *Traffic*, 7(2), 129–145.

Korolchuk, V. I., Miwa, S., Carroll, B., & von Zglinicki, T. (2017). Mitochondria in cell senescence: Is mitophagy the weakest link? *eBioMedicine*, *21*, 7–13.

Kosar, M., Bartkova, J., Hubackova, S., Hodny, Z., Lukas, J., & Bartek, J. (2011). Senescence-associated heterochromatin foci are dispensable for cellular senescence, occur in a cell type- and insult-dependent manner, and follow expression of p16ink4a. *Cell Cycle*, *10*(3), 457–468.

Kovacovicova, K., Skolnaja, M., Heinmaa, M., Mistrik, M., Pata, P., Pata, I., et al. (2018). Senolytic cocktail dasatinib + quercetin (D + Q) does not enhance the efficacy of senescence-inducing chemotherapy in liver cancer. *Frontiers in Oncology*, *8*, 459.

Kovatcheva, M., Liu, D. D., Dickson, M. A., Klein, M. E., Connor, O., Wilder, F. O., et al. (2015). MDM2 turnover and expression of ATRX determine the choice between quiescence and senescence in response to CDK4 inhibition. *Oncotarget*, *6*(10), 8226–8243.

Krayem, M., Najem, A., Journe, F., Morandini, R., Sales, F., Awada, A., et al. (2018). Acquired resistance to BRAFi reverses senescence-like phenotype in mutant BRAF melanoma. *Oncotarget*, *9*(61), 31888–31903.

Krysko, O., Løve, T., Kagan, V. E., Herde, K. D., Bachert, C., Leybaert, L., et al. (2017). Necroptotic cell death in anti-cancer therapy. *Immunological Reviews*, *280*(1), 207–219.

Kuilman, T., Michaloglou, C., Mooi, W. J., & Peeper, D. S. (2010). The essence of senescence. *Genes & Development*, *24*(22), 2463–2479.

Kumar, A., Singh, U. K., & Chaudhary, A. (2015). Targeting autophagy to overcome drug resistance in cancer therapy. *Future Medicinal Chemistry*, 7(12), 1535–1542.

Kurz, D. J., Decary, S., Hong, Y., & Erusalimsky, J. D. (2000). Senescence-associated (beta)-galactosidase reflects an increase in lysosomal mass during replicative ageing of human endothelial cells. *Journal of Cell Science*, *113*(20), 3613–3622.

Kwon, Y., Kim, J. W., Jeoung, J. A., Kim, M., Kang, C., & Autophagy, I. (2017). Pro-senescence when seen in close-up, but anti-senescence in long-shot. *Molecules and Cells*, *40*(9), 607–612.

Laberge, R., Sun, Y., Orjalo, A. V., Patil, C. K., Freund, A., Zhou, L., et al. (2015). MTOR regulates the pro-tumorigenic senescence-associated secretory phenotype by promoting IL1A translation. *Nature Cell Biology*, *17*(8), 1049–1061.

Lee, Y. H., Kang, B. S., & Bae, Y. S. (2014). Premature senescence in human breast cancer and colon cancer cells by tamoxifen-mediated reactive oxygen species generation. *Life Sciences*, *97*(2), 116–122.

Lee, Y.-K., & Lee, J.-A. (2016). Role of the mammalian ATG8/LC3 family in autophagy: Differential and compensatory roles in the spatiotemporal regulation of autophagy. *BMB Reports*, *49*(8), 424–430.

Lee, H.-O., Mustafa, A., Hudes, G. R., & Kruger, W. D. (2015). Hydroxychloroquine destabilizes phospho-S6 in human renal carcinoma cells. *PLoS One*, *10*(7), e0131464.

Lehmann, M., Korfei, M., Mutze, K., Klee, S., Skronska-Wasek, W., Alsafadi, H. N., et al. (2017). Senolytic drugs target alveolar epithelial cell function and attenuate experimental lung fibrosis ex vivo. *The European Respiratory Journal, 50*(2), 1602367.

Levine, B. (2007). Autophagy and cancer. *Nature, 446*(7137), 745–747.

Levine, B., & Deretic, V. (2007). Unveiling the roles of autophagy in innate and adaptive immunity. *Nature Reviews. Immunology*, 7(10), 767–777.

Levine, B., & Kroemer, G. (2008). Autophagy in the pathogenesis of disease. *Cell, 132*(1), 27–42.

Levine, B., Mizushima, N., & Virgin, H. W. (2011). Autophagy in immunity and inflammation. *Nature, 469*(7330), 323–335.

Levine, B., Packer, M., & Codogno, P. (2015). Development of autophagy inducers in clinical medicine. *The Journal of Clinical Investigation, 125*(1), 14–24.

Levy, J. M. M., Thompson, J. C., Griesinger, A. M., Amani, V., Donson, A. M., Birks, D. K., et al. (2014). Autophagy inhibition improves chemosensitivity in BRAFV600E brain tumors. *Cancer Discovery, 4*(7), 773–780.

Li, J., Cao, F., Liang, Y. H., Jian, H. Z., Tao, L. Z., Mao, N., et al. (2020). Ferroptosis: Past, present and future. *Cell Death & Disease, 11*(2), 88.

Li, X., & Fan, Z. (2010). The epidermal growth factor receptor antibody cetuximab induces autophagy in cancer cells by downregulating HIF-1alpha and Bcl-2 and activating the beclin 1/hVps34 complex. *Cancer Research, 70*(14), 5942–5952.

Li, J., Hou, N., Faried, A., Tsutsumi, S., & Kuwano, H. (2010). Inhibition of autophagy augments 5-fluorouracil chemotherapy in human colon cancer in vitro and in vivo model. *European Journal of Cancer, 46*(10), 1900–1909.

Li, J., Hou, N., Faried, A., Tsutsumi, S., Takeuchi, T., & Kuwano, H. (2009). Inhibition of autophagy by 3-MA enhances the effect of 5-FU-induced apoptosis in colon cancer cells. *Annals of Surgical Oncology, 16*(3), 761–771.

Li, B., Lei, Z., Lichty, B. D., Li, D., Zhang, G.-M., Feng, Z.-H., et al. (2010). Autophagy facilitates major histocompatibility complex class I expression induced by IFN-γ in B16 melanoma cells. *Cancer Immunology, Immunotherapy, 59*(2), 313–321.

Li, X., Lu, Y., Pan, T., & Fan, Z. (2010). Roles of autophagy in cetuximab-mediated cancer therapy against EGFR. *Autophagy, 6*(8), 1066–1077.

Li, W., Wang, W., Dong, H., Li, Y., Li, L. I., Han, L., et al. (2014). Cisplatin-induced senescence in ovarian cancer cells is mediated by GRP78. *Oncology Reports, 31*(6), 2525–2534.

Li, Y., Wang, L.-X., Yang, G., Hao, F., Urba, W. J., & Hu, H.-M. (2008). Efficient cross-presentation depends on autophagy in tumor cells. *Cancer Research, 68*(17), 6889–6895.

Li, J., & Yuan, J. (2008). Caspases in apoptosis and beyond. *Oncogene, 27*(48), 6194–6206.

Liang, X., De Vera, M. E., Buchser, W. J., de Vivar Chavez, A. R., Loughran, P., Stolz, D. B., et al. (2012). Inhibiting systemic autophagy during interleukin 2 immunotherapy promotes long-term tumor regression. *Cancer Research, 72*(11), 2791–2801.

Liang, X. H., Jackson, S., Seaman, M., Brown, K., Kempkes, B., Hibshoosh, H., et al. (1999). Induction of autophagy and inhibition of tumorigenesis by beclin 1. *Nature, 402*(6762), 672–676.

Liang, B., Kong, D., Liu, Y., Liang, N., He, M., Ma, S., et al. (2012). Autophagy inhibition plays the synergetic killing roles with radiation in the multi-drug resistant SKVCR ovarian cancer cells. *Radiation Oncology*, 7(1), 213.

Lin, J.-F., Lin, Y.-C., Tsai, T.-F., Chen, H.-E., Chou, K.-Y., & TI-S, H. (2017). Cisplatin induces protective autophagy through activation of BECN1 in human bladder cancer cells. *Drug Design, Development and Therapy, 11*, 1517–1533.

Lindgren, T., Stigbrand, T., Råberg, A., Riklund, K., & Eriksson, D. (2015). Genome wide expression analysis of radiation- induced DNA damage responses in isogenic HCT116 p53+/+ and HCT116 p53−/− colorectal carcinoma cell lines. *International Journal of Radiation Biology, 91*(1), 99–111.

Lindsey, J., McGill, N. I., Lindsey, L. A., Green, D. K., & Cooke, H. J. (1991). In vivo loss of telomeric repeats with age in humans. *Mutation Research, 256*(1), 45–48.
Ling, L. X., Ding, J., & Hua, M. L. (2018). Oncogene-induced senescence: A double edged sword in cancer. *Acta Pharmacologica Sinica, 39*(10), 1553–1558.
Liu, J., Xia, H., Kim, M., Xu, L., Li, Y., Zhang, L., et al. (2011). Beclin1 controls the levels of p53 by regulating the deubiquitination activity of USP10 and USP13. *Cell, 147*(1), 223–234.
Lomonaco, S. L., Finniss, S., Xiang, C., DeCarvalho, A., Umansky, F., Kalkanis, S. N., et al. (2009). The induction of autophagy by γ-radiation contributes to the radioresistance of glioma stem cells. *International Journal of Cancer, 125*(3), 717–722.
Luo, H., Yang, A., Schulte, B. A., Wargovich, M. J., & Wang, G. Y. (2013). Resveratrol induces premature senescence in lung cancer cells via ROS-mediated DNA damage. *PLoS One, 8*(3), e60065.
Lyon, D., Elmore, L., Aboalela, N., Merrill-Schools, J., McCain, N., Starkweather, A., et al. (2014). Potential epigenetic mechanism(s) associated with the persistence of psychoneurological symptoms in women receiving chemotherapy for breast cancer: A hypothesis. *Biological Research for Nursing, 16*(2), 160–174.
Lyseng-Williamson, K. A., & Fenton, C. (2005). Docetaxel: A review of its use in metastatic breast cancer. *Drugs, 65*(17), 2513–2531.
Macip, S., Igarashi, M., Fang, L., Chen, A., Pan, Z., Lee, S. W., et al. (2002). Inhibition of p21-mediated ROS accumulation can rescue p21-induced senescence. *The EMBO Journal, 21*(9), 2180–2188.
Maiuri, M. C., Zalckvar, E., Kimchi, A., & Kroemer, G. (2007). Self-eating and self-killing: Crosstalk between autophagy and apoptosis. *Nature Reviews. Molecular Cell Biology, 8*(9), 741–752.
Malhotra, J., Jabbour, S., Orlick, M., Riedlinger, G., Joshi, S., Guo, J. Y., et al. (2018). Modulation of autophagy with hydroxychloroquine in patients with advanced non-small cell lung cancer (NSCLC): A phase Ib study. *Journal of Clinical Oncology, 36*, e21138.
Mani, J., Vallo, S., Rakel, S., Antonietti, P., Gessler, F., Blaheta, R., et al. (2015). Chemoresistance is associated with increased cytoprotective autophagy and diminished apoptosis in bladder cancer cells treated with the BH3 mimetic (−)-Gossypol (AT-101). *BMC Cancer, 15*(1), 224.
Marcotte, R., Lacelle, C., & Wang, E. (2004). Senescent fibroblasts resist apoptosis by down-regulating caspase-3. *Mechanisms of Ageing and Development, 125*(10–11), 777–783.
Martin, S. (2011). Deconstructing the cell cycle. *Nature Reviews Molecular Cell Biology, 12*, 689. Nature Publishing Group.
Martins, I., Wang, Y., Michaud, M., Ma, Y., Sukkurwala, A. Q., Shen, S., et al. (2014). Molecular mechanisms of ATP secretion during immunogenic cell death. *Cell Death and Differentiation, 21*(1), 79–91.
Mathew, R., Karp, C. M., Beaudoin, B., Vuong, N., Chen, G., Chen, H.-Y., et al. (2009). Autophagy suppresses tumorigenesis through elimination of p62. *Cell, 137*(6), 1062–1075.
Mauthe, M., Orhon, I., Rocchi, C., Zhou, X., Luhr, M., Hijlkema, K.-J., et al. (2018). Chloroquine inhibits autophagic flux by decreasing autophagosome-lysosome fusion. *Autophagy, 14*(8), 1435–1455.
Maycotte, P., & Thorburn, A. (2011). Autophagy and cancer therapy. *Cancer Biology & Therapy, 11*(2), 127–137.
McAfee, Q., Zhang, Z., Samanta, A., Levi, S. M., Ma, X.-H., Piao, S., et al. (2012). Autophagy inhibitor Lys05 has single-agent antitumor activity and reproduces the phenotype of a genetic autophagy deficiency. *Proceedings of the National Academy of Sciences of the United States of America, 109*(21), 8253–8258.
McKenna, E., Traganos, F., Zhao, H., & Darzynkiewicz, Z. (2012). Persistent DNA damage caused by low levels of mitomycin C induces irreversible cell senescence. *Cell Cycle, 11*(16), 3132–3140.

Mei, Q., Li, F., Quan, H., Liu, Y., & Xu, H. (2014). Busulfan inhibits growth of human osteosarcoma through miR-200 family microRNAs in vitro and in vivo. *Cancer Science, 105*(7), 755–762.

Michaloglou, C., Vredeveld, L. C. W., Soengas, M. S., Denoyelle, C., Kuilman, T., Van Der Horst, C. M. A. M., et al. (2005). BRAFE600-associated senescence-like cell cycle arrest of human naevi. *Nature, 436*(7051), 720–724.

Michaud, M., Martins, I., Sukkurwala, A. Q., Adjemian, S., Ma, Y., Pellegatti, P., et al. (2011). Autophagy-dependent anticancer immune responses induced by chemotherapeutic agents in mice. *Science, 334*(6062), 1573–1578.

Michaud, K., Solomon, D. A., Oermann, E., Kim, J., Zhong, W., Prados, M. D., et al. (2010). Pharmacologic inhibition of cyclin-dependent kinases 4 and 6 arrests the growth of glioblastoma multiforme intracranial xenografts. *Cancer Research, 70*(8), 3228–3239.

Michaud, M., Xie, X., Bravo-San Pedro, J. M., Zitvogel, L., White, E., & Kroemer, G. (2014). An autophagy-dependent anticancer immune response determines the efficacy of melanoma chemotherapy. *Oncoimmunology, 3*(7), e944047.

Miettinen, T. P., Peltier, J., Härtlova, A., Gierliński, M., Jansen, V. M., Trost, M., et al. (2018). Thermal proteome profiling of breast cancer cells reveals proteasomal activation by CDK 4/6 inhibitor palbociclib. *The EMBO Journal, 37*(10), 1–19.

Mijaljica, D., Prescott, M., & Devenish, R. J. (2011). Microautophagy in mammalian cells: Revisiting a 40-year-old conundrum. *Autophagy, 7*(7), 673–682.

Mikuła-Pietrasik, J., Niklas, A., Uruski, P., Tykarski, A., & Książek, K. (2020). Mechanisms and significance of therapy-induced and spontaneous senescence of cancer cells. *Cellular and Molecular Life Sciences, 77*(2), 213–229.

Milanovic, M., Fan, D. N. Y., Belenki, D., Däbritz, J. H. M., Zhao, Z., Yu, Y., et al. (2018). Senescence-associated reprogramming promotes cancer stemness. *Nature, 553*(7686), 96–100.

Milanovic, M., Yu, Y., & Schmitt, C. A. (2018). The senescence–stemness alliance—A cancer-hijacked regeneration principle. *Trends in Cell Biology, 28*(12), 1049–1061.

Mizumura, K., Cloonan, S. M., Nakahira, K., Bhashyam, A. R., Cervo, M., Kitada, T., et al. (2014). Mitophagy-dependent necroptosis contributes to the pathogenesis of COPD. *The Journal of Clinical Investigation, 124*(9), 3987–4003.

Mizushima, N. (2007). Autophagy: Process and function. *Genes & Development, 21*(22), 2861–2873.

Moolmuang, B., Singhirunnusorn, P., & Ruchirawat, M. (2016). Effects of 5-aza-2′-deoxycytidine, bromodeoxyuridine, interferons and hydrogen peroxide on cellular senescence in cholangiocarcinoma cells. *Asian Pacific Journal of Cancer Prevention, 17*(3), 957–963.

Moreau, K., Luo, S., & Rubinsztein, D. C. (2010). Cytoprotective roles for autophagy. *Current Opinion in Cell Biology, 22*(2), 206–211.

Morelli, M. B., Amantini, C., Nabissi, M., Cardinali, C., Santoni, M., Bernardini, G., et al. (2017). Axitinib induces senescence-associated cell death and necrosis in glioma cell lines: The proteasome inhibitor, bortezomib, potentiates axitinib-induced cytotoxicity in a p21(Waf/Cip1) dependent manner. *Oncotarget, 8*(2), 3380–3395.

Moriwaki, K., Bertin, J., Gough, P. J., Orlowski, G. M., & Chan, F. K. M. (2015). Differential roles of RIPK1 and RIPK3 in TNF-induced necroptosis and chemotherapeutic agent-induced cell death. *Cell Death & Disease, 6*(2), 1–11.

Mosieniak, G., Sliwinska, M. A., Alster, O., Strzeszewska, A., Sunderland, P., Piechota, M., et al. (2015). Polyploidy formation in doxorubicin-treated cancer cells can favor escape from senescence. *Neoplasia, 17*(12), 882–893.

Moyzis, R. K., Buckingham, J. M., Cram, L. S., Dani, M., Deaven, L. L., Jones, M. D., et al. (1988). A highly conserved repetitive DNA sequence, (TTAGGG)n, present at the telomeres of human chromosomes. *Proceedings of the National Academy of Sciences of the United States of America, 85*(18), 6622–6626.

Mrakovcic, M., Fröhlich, L. F., Mrakovcic, M., & Fröhlich, L. F. (2018). p53-mediated molecular control of autophagy in tumor cells. *Biomolecules*, *8*(2), 14.
Mulcahy Levy, J. M., Zahedi, S., Griesinger, A. M., Morin, A., Davies, K. D., Aisner, D. L., et al. (2017). Autophagy inhibition overcomes multiple mechanisms of resistance to BRAF inhibition in brain tumors. *eLife*, *6*, e19671.
Muñoz-Espín, D., Rovira, M., Galiana, I., Giménez, C., Lozano-Torres, B., Paez-Ribes, M., et al. (2018). A versatile drug delivery system targeting senescent cells. *EMBO Molecular Medicine*, *10*(9), e9355.
Murata, Y., Wakoh, T., Uekawa, N., Sugimoto, M., Asai, A., Miyazaki, T., et al. (2006). Death-associated protein 3 regulates cellular senescence through oxidative stress response. *FEBS Letters*, *580*(26), 6093–6099.
Murrow, L., & Debnath, J. (2013). Autophagy as a stress-response and quality-control mechanism: implications for cell injury and human disease. *Annual Review of Pathology*, *8*, 105–137.
Nagano, T., Nakano, M., Nakashima, A., Onishi, K., Yamao, S., Enari, M., et al. (2016). Identification of cellular senescence-specific genes by comparative transcriptomics. *Scientific Reports*, *22*(6), 1–13.
Nakayama, K., Rahman, M., Rahman, M. T., Nakamura, K., Sato, E., Katagiri, H., et al. (2017). Nucleus accumbens-1/GADD45GIP1 axis mediates cisplatin resistance through cellular senescence in ovarian cancer. *Oncology Letters*, *13*(6), 4713–4719.
Nam, H. Y., Han, M. W., Chang, H. W., Lee, Y. S., Lee, M., Lee, H. J., et al. (2013). Radioresistant cancer cells can be conditioned to enter senescence by mTOR inhibition. *Cancer Research*, *73*(14), 4267–4277.
Nardella, C., Clohessy, J. G., Alimonti, A., & Pandolfi, P. P. (2011). Pro-senescence therapy for cancer treatment. *Nature Reviews. Cancer*, *11*(7), 503–511.
Narita, M., Nun, S., Heard, E., Narita, M., Lin, A. W., Hearn, S. A., et al. (2003). Rb-mediated heterochromatin formation and silencing of E2F target genes during cellular senescence. *Cell*, *113*(6), 703–716.
Narita, M., Young, A. R. J., Arakawa, S., Samarajiwa, S. A., Nakashima, T., Yoshida, S., et al. (2011). Spatial coupling of mTOR and autophagy augments secretory phenotypes. *Science*, *332*(6032), 966–970.
Neal, J. W., Wakelee, H. A., Feliciano, J. L., Goldberg, S. B., Morgensztern, D., Das, M. S., et al. (2014). A multicenter randomized phase II trial of erlotinib with and without hydroxychloroquine (HCQ) in TKI-naive patients (pts) with epidermal growth factor receptor (EGFR) mutant advanced non-small cell lung cancer (NSCLC). *Journal of Clinical Oncology*, *32*, 8088.
Noman, M. Z., Janji, B., Kaminska, B., Van Moer, K., Pierson, S., Przanowski, P., et al. (2011). Blocking hypoxia-induced autophagy in tumors restores cytotoxic T-cell activity and promotes regression. *Cancer Research*, *71*(18), 5976–5986.
Noren Hooten, N., & Evans, M. K. (2017). Techniques to induce and quantify cellular senescence. *Journal of Visualized Experiments*, *123*, 1–14.
Ohno-Iwashita, Y., Shimada, Y., Hayashi, M., & Inomata, M. (2010). Plasma membrane microdomains in aging and disease. *Geriatrics & Gerontology International*, *10*, S41–S52.
Ota, H., Eto, M., Ako, J., Ogawa, S., Iijima, K., Akishita, M., et al. (2009). Sirolimus and everolimus induce endothelial cellular senescence via sirtuin 1 down-regulation: Therapeutic implication of cilostazol after drug-eluting stent implantation. *Journal of the American College of Cardiology*, *53*(24), 2298–2305.
Ozsvari, B., Nuttall, J. R., Sotgia, F., & Lisanti, M. P. (2018). Azithromycin and Roxithromycin define a new family of " senolytic " drugs that target senescent human fibroblasts. *Aging (Albany NY)*, *10*(11), 3294–3307.
Pack, L. R., Daigh, L. H., & Meyer, T. (2019). Putting the brakes on the cell cycle: Mechanisms of cellular growth arrest. *Current Opinion in Cell Biology*, *60*, 106–113.

Paglin, S., Hollister, T., Delohery, T., Hackett, N., Mcmahill, M., Sphicas, E., et al. (2001). A novel response of cancer cells to radiation involves autophagy and formation of acidic vesicles. *Cancer Research*, *61*(2), 439–444.

Pan, H., Wang, Z., Jiang, L., Sui, X., You, L., Shou, J., et al. (2014). Autophagy inhibition sensitizes hepatocellular carcinoma to the multikinase inhibitor linifanib. *Scientific Reports*, *4*(1), 6683.

Papait, R., Magrassi, L., Rigamonti, D., & Cattaneo, E. (2009). Temozolomide and carmustine cause large-scale heterochromatin reorganization in glioma cells. *Biochemical and Biophysical Research Communications*, *379*(2), 434–439.

Pardoll, D. M. (2012). The blockade of immune checkpoints in cancer immunotherapy. *Nature Reviews. Cancer*, *12*(4), 252–264.

Parzych, K. R., & Klionsky, D. J. (2014). An overview of autophagy: Morphology, mechanism, and regulation. *Antioxidants & Redox Signaling*, *20*(3), 460–473.

Pasquier, B. (2015). SAR405, a PIK3C3/Vps34 inhibitor that prevents autophagy and synergizes with MTOR inhibition in tumor cells. *Autophagy*, *11*(4), 725–726.

Passos, J. F., Nelson, G., Wang, C., Richter, T., Simillion, C., Proctor, C. J., et al. (2010). Feedback between p21 and reactive oxygen production is necessary for cell senescence. *Molecular Systems Biology*, *6*(1), 347.

Pazdur, R., Kudelka, A. P., Kavanagh, J. J., Cohen, P. R., & Raber, M. N. (1993). The taxoids: Paclitaxel (Taxol®) and docetaxel (Taxotere®). *Cancer Treatment Reviews*, *19*(4), 351–386.

Peng, S., Sen, B., Mazumdar, T., Byers, L. A., Diao, L., Wang, J., et al. (2016). Dasatinib induces DNA damage and activates DNA repair pathways leading to senescence in non-small cell lung cancer cell lines with kinase-inactivating BRAF mutations. *Oncotarget*, 7(1), 565–579.

Pereira, B. I., Devine, O. P., Vukmanovic-Stejic, M., Chambers, E. S., Subramanian, P., Patel, N., et al. (2019). Senescent cells evade immune clearance via HLA-E-mediated NK and CD8+ T cell inhibition. *Nature Communications*, *10*(1), 2387.

Pereira, S. F. F., Gonzalez, R. L., & Dworkin, J. (2015). Protein synthesis during cellular quiescence is inhibited by phosphorylation of a translational elongation factor. *Proceedings of the National Academy of Sciences of the United States of America*, *112*(25), 3274–3281.

Perez, M., Muñoz-Galván, S., Jiménez-García, M. P., Marín, J. J., & Carnero, A. (2015). Efficacy of CDK4 inhibition against sarcomas depends on their levels of CDK4 and p16ink4 mRNA. *Oncotarget*, *6*(38), 40557–40574.

Pernicová, Z., Slabáková, E., Kharaishvili, G., Bouchal, J., Kral, M., Kunicka, Z., et al. (2011). Androgen depletion induces senescence in prostate cancer cells through. *Neoplasia*, *13*(6), 526–536.

Petherick, K. J., Conway, O. J. L., Mpamhanga, C., Osborne, S. A., Kamal, A., Saxty, B., et al. (2015). Pharmacological inhibition of ULK1 kinase blocks mammalian target of rapamycin (mTOR)-dependent autophagy. *The Journal of Biological Chemistry*, *290*(18), 11376–11383.

Pietrocola, F., Bravo-San Pedro, J. M., Galluzzi, L., & Kroemer, G. (2017). Autophagy in natural and therapy-driven anticancer immunosurveillance. *Autophagy*, *13*(12), 2163–2170.

Poele, R. H., Okorokov, A. L., Jardine, L., Cummings, J., Joel, S. P., te Poele, R. H., et al. (2002). DNA damage is able to induce senescence in tumor cells in vitro and in vivo. *Cancer Research*, *62*(6), 1876–1883.

Poillet-Perez, L., Despouy, G., Delage-Mourroux, R., & Boyer-Guittaut, M. (2015). Interplay between ROS and autophagy in cancer cells, from tumor initiation to cancer therapy. *Redox Biology*, *4*, 184–192.

Polyak, K., Kato, J. Y., Solomon, M. J., Sherr, C. J., Massague, J., Roberts, J. M., et al. (1994). p27(Kip1), a cyclin-Cdk inhibitor, links transforming growth factor-β and contact inhibition to cell cycle arrest. *Genes & Development*, *8*(1), 9–22.
Portugal, J., Mansilla, S., & Bataller, M. (2009). Mechanisms of drug-induced mitotic catastrophe in cancer cells. *Current Pharmaceutical Design*, *16*(1), 69–78.
Probin, V., Wang, Y., Bai, A., & Zhou, D. (2006). Busulfan selectively induces cellular senescence but not apoptosis in WI38 fibroblasts via a p53-independent but extracellular signal-regulated kinase-p38 mitogen-activated protein kinase-dependent mechanism. *The Journal of Pharmacology and Experimental Therapeutics*, *319*(2), 551–560.
Probin, V., Wang, Y., & Zhou, D. (2007). Busulfan-induced senescence is dependent on ROS production upstream of the MAPK pathway. *Free Radical Biology & Medicine*, *42*(12), 1858–1865.
Puig, P.-E. E., Guilly, M.-N. N., Bouchot, A. A., Droin, N., Cathelin, D., Bouyer, F., et al. (2008). Tumor cells can escape DNA-damaging cisplatin through DNA endoreduplication and reversible polyploidy. *Cell Biology International*, *32*(9), 1031–1043.
Puri, C., Vicinanza, M., Ashkenazi, A., Gratian, M. J., Zhang, Q., Bento, C. F., et al. (2018). The RAB11A-positive compartment is a primary platform for autophagosome assembly mediated by WIPI2 recognition of PI3P-RAB11A. *Developmental Cell*, *45*(1), 114–131.e8.
Putri, J. F., Widodo, N., Sakamoto, K., Kaul, S. C., & Wadhwa, R. (2017). Induction of senescence in cancer cells by 5′-Aza-2′-deoxycytidine: Bioinformatics and experimental insights to its targets. *Computational Biology and Chemistry*, *70*, 49–55.
Puyol, M., Martín, A., Dubus, P., Mulero, F., Pizcueta, P., Khan, G., et al. (2010). A synthetic lethal interaction between K-Ras oncogenes and Cdk4 unveils a therapeutic strategy for non-small cell lung carcinoma. *Cancer Cell*, *18*(1), 63–73.
Qadir, M. A., Kwok, B., Dragowska, W. H., To, K. H., Le, D., Bally, M. B., et al. (2008). Macroautophagy inhibition sensitizes tamoxifen-resistant breast cancer cells and enhances mitochondrial depolarization. *Breast Cancer Research and Treatment*, *112*(3), 389–403.
Qi, Z., Zhang, Y., Liu, L., Guo, X., Qin, J., & Cui, G. (2012). Mesenchymal stem cells derived from different origins have unique sensitivities to different chemotherapeutic agents. *Cell Biology International*, *36*(9), 857–862.
Qiu, L., Zhou, G., & Cao, S. (2020). Targeted inhibition of ULK1 enhances daunorubicin sensitivity in acute myeloid leukemia. *Life Sciences*, *243*, 117234.
Qu, K., Lin, T., Wang, Z., Liu, S., Chang, H., Xu, X., et al. (2014). Reactive oxygen species generation is essential for cisplatin induced accelerated senescence in hepatocellular carcinoma. *Frontiers of Medicine in China*, *8*(2), 227–235.
Rabinowitz, J. D., White, E., Sukkurwala, A. Q., Adjemian, S., Ma, Y., Pellegatti, P., et al. (2010). Autophagy and metabolism. *Science*, *330*(6009), 1344–1348.
Ramakrishnan, R., Huang, C., Cho, H.-I., Lloyd, M., Johnson, J., Ren, X., et al. (2012). Autophagy induced by conventional chemotherapy mediates tumor cell sensitivity to immunotherapy. *Cancer Research*, *72*(21), 5483–5493.
Rangwala, R., Chang, Y. C., Hu, J., Algazy, K. M., Evans, T. L., Fecher, L. A., et al. (2014). Combined MTOR and autophagy inhibition: phase I trial of hydroxychloroquine and temsirolimus in patients with advanced solid tumors and melanoma. *Autophagy*, *10*(8), 1391–1402.
Rangwala, R., Leone, R., Chang, Y. C., Fecher, L. A., Schuchter, L. M., Kramer, A., et al. (2014). Phase I trial of hydroxychloroquine with dose-intense temozolomide in patients with advanced solid tumors and melanoma. *Autophagy*, *10*(8), 1369–1379.
Ravi, A. K., & Jeffrey, W. D. (2012). Lys05 a new lysosomal autophagy inhibitor. *Autophagy*, (September), 1383–1384.

Ren, J., Liu, T., Han, Y., Wang, Q., Chen, Y., Li, G., et al. (2018). GSK-3β inhibits autophagy and enhances radiosensitivity in non-small cell lung cancer. *Diagnostic Pathology*, *13*(1), 33.

Ressler, S., Bartkova, J., Niederegger, H., Bartek, J., Scharffetter-Kochanek, K., Jansen-Dürr, P., et al. (2006). p16INK4A is a robust in vivo biomarker of cellular aging in human skin. *Aging Cell*, *5*(5), 379–389.

Roberson, R. S., Kussick, S. J., Vallieres, E., Chen, S. Y. J., & Wu, D. Y. (2005). Escape from therapy-induced accelerated cellular senescence in p53-null lung cancer cells and in human lung cancers. *Cancer Research*, *65*(7), 2795–2803.

Rodier, F., Coppé, J., Patil, C. K., Hoeijmakers, W. A. M., Muñoz, D. P., Raza, S. R., et al. (2009). Persistent DNA damage signalling triggers senescence- associated inflammatory cytokine secretion. *Nature Cell Biology*, *11*(8), 973–979.

Ronan, B., Flamand, O., Vescovi, L., Dureuil, C., Durand, L., Fassy, F., et al. (2014). A highly potent and selective Vps34 inhibitor alters vesicle trafficking and autophagyalters vesicle trafficking and autophagy. *Nature Chemical Biology*, *10*(12), 1013–1019.

Rosenfeld, M. R., Ye, X., Supko, J. G., Desideri, S., Grossman, S. A., Brem, S., et al. (2014). A phase I/II trial of hydroxychloroquine in conjunction with radiation therapy and concurrent and adjuvant temozolomide in patients with newly diagnosed glioblastoma multiforme. *Autophagy*, *10*(8), 1359–1368.

Rubinsztein, D. C., Mariño, G., & Kroemer, G. (2011). Autophagy and aging. *Cell*, *146*(5), 682–695.

Ruscetti, M., Leibold, J., Bott, M. J., Fennell, M., Kulick, A., Salgado, N. R., et al. (2018). NK cell-mediated cytotoxicity contributes to tumor control by a cytostatic drug combination. *Science*, *362*(6421), 1416–1422.

Saleh, T., Bloukh, S., Carpenter, V. J., Alwohoush, E., Bakeer, J., Darwish, S., et al. (2020). Therapy-induced senescence: An "old" friend becomes the enemy. *Cancers (Basel)*, *12*(4), 822.

Saleh, T., Carpenter, V., Tyutyunyk-Massey, L., Murray, G., Leverson, J., Souers, A., et al. (2020). Clearance of therapy-induced senescent tumor cells by the senolytic ABT-263 via interference with BCL-X L-BAX Interaction. *Molecular Oncology*, *14*(10), 2504–2519.

Saleh, T., Tyutynuk-Massey, L., Cudjoe, E. K. E. K., Idowu, M. O. M. O., Landry, J. W. J. W., Gewirtz, D. A. D. A., et al. (2018). Non-cell autonomous effects of the senescence-associated secretory phenotype in cancer therapy. *Frontiers in Oncology*, *8*(164), 1–14.

Saleh, T., Tyutyunyk-Massey, L., & Gewirtz, D. A. D. A. (2019). Tumor cell escape from therapy-induced senescence as a model of disease recurrence after dormancy. *Cancer Research*, *79*(6), 1044–1046.

Saleh, T., Tyutyunyk-Massey, L., Murray, G. F., Alotaibi, M. R., Kawale, A. S., Elsayed, Z., et al. (2019). Tumor cell escape from therapy-induced senescence. *Biochemical Pharmacology*, *162*, 202–212.

Saleh, T., Tyutyunyk-Massey, L., Patel, N. H., Cudjoe, K. E., Alotaibi, M., & Gewirtz, D. A. (2020). Studies of non-protective autophagy provide evidence that recovery from therapy-induced senescence is independent of early autophagy. *International Journal of Molecular Sciences*, *21*(4), 1427.

Samaraweera, L., Adomako, A., Rodriguez-Gabin, A., & McDaid, H. M. (2017). A novel indication for panobinostat as a senolytic drug in NSCLC and HNSCC. *Scientific Reports*, *7*(1), 1–11.

Sasaki, M., Kumazaki, T., Takano, H., Nishiyama, M., & Mitsui, Y. (2001). Senescent cells are resistant to death despite low Bcl-2 level. *Mechanisms of Ageing and Development*, *122*(15), 1695–1706.

Scherz-Shouval, R., Shvets, E., Fass, E., Shorer, H., Gil, L., & Elazar, Z. (2007). Reactive oxygen species are essential for autophagy and specifically regulate the activity of Atg4. *The EMBO Journal, 26*(7), 1749–1760.

Schlütermann, D., Skowron, M. A., Berleth, N., Böhler, P., Deitersen, J., Stuhldreier, F., et al. (2018). Targeting urothelial carcinoma cells by combining cisplatin with a specific inhibitor of the autophagy-inducing class III PtdIns3K complex. *Urologic Oncology, 36*(4), 160.e1–160.e13.

Schmitt, C. A. (2007). Cellular senescence and cancer treatment. *Biochimica et Biophysica Acta. Reviews on Cancer, 1775*(1), 5–20.

Schmitt, C. A., Fridman, J. S., Yang, M., Lee, S., Baranov, E., Hoffman, R. M., et al. (2002). A senescence program controlled by p53 and p16INK4acontributes to the outcome of cancer therapy. *Cell, 109*(3), 335–346.

Schwarze, S. R., Fu, V. X., Desotelle, J. A., Kenowski, M. L., & Jarrard, D. F. (2005). The identification of senescence-specific genes during the induction of senescence in prostate cancer cells. *Neoplasia*, 7(9), 816–823.

Seignez, C., Martin, A., Rollet, C. E., Racoeur, C., Scagliarini, A., Jeannin, J. F., et al. (2014). Senescence of tumor cells induced by oxaliplatin increases the efficiency of a lipid A immunotherapy via the recruitment of neutrophils. *Oncotarget, 5*(22), 11442–11451.

Selvakumaran, M., Amaravadi, R. K., Vasilevskaya, I. A., & O'Dwyer, P. J. (2013). Autophagy inhibition sensitizes colon cancer cells to antiangiogenic and cytotoxic therapy. *Clinical Cancer Research, 19*(11), 2995–3007.

Sen, B., Peng, S., Tang, X., Erickson, H. S., Galindo, H., Mazumdar, T., et al. (2013). Kinase impaired BRAF mutations confer lung cancer sensitivity to dasatinib. *Science Translational Medicine, 4*(136), 1–21.

Serrano, M., Hannon, G. J., & Beach, D. (1993). A new regulatory motif in cell-cycle control causing specific inhibition of cyclin D/CDK4. *Nature, 366*(6456), 704–707.

Serrano, M., Lin, A. W., McCurrach, M. E., Beach, D., & Lowe, S. W. (1997). Oncogenic ras provokes premature cell senescence associated with accumulation of p53 and p16 (INK4a). *Cell, 88*(5), 593–602.

Shao, Y., Gao, Z., Marks, P. A., & Jiang, X. (2004). Apoptotic and autophagic cell death induced by histone deacetylase inhibitors. *Proceedings of the National Academy of Sciences of the United States of America, 101*(52), 18030–18035.

Shao, S., Li, S., Qin, Y., Wang, X., Yang, Y., Bai, H., et al. (2014). Spautin-1, a novel autophagy inhibitor, enhances imatinib-induced apoptosis in chronic myeloid leukemia. *International Journal of Oncology, 44*(5), 1661–1668.

Sharma, K., Goehe, R. W., Di, X., Hicks, M. A., Torti, S. V., Torti, F. M., et al. (2014). A novel cytostatic form of autophagy in sensitization of non-small cell lung cancer cells to radiation by vitamin D and the vitamin D analog, EB 1089. *Autophagy, 10*(12), 2346–2361.

Sharpless, N. E., & Sherr, C. J. (2015). Forging a signature of in vivo senescence. *Nature Reviews. Cancer, 15*(7), 397–408.

Shaw, R. J. (2009). LKB1 and AMP-activated protein kinase control of mTOR signalling and growth. *Acta Physiologica (Oxford, England), 196*(1), 65–80.

Shay, J. W., & Wright, W. E. (2000). Hayflick, his limit, and cellular ageing. *Nature Reviews. Molecular Cell Biology, 1*(1), 72–76.

Shelton, D. N., Chang, E., Whittier, P. S., Choi, D., & Funk, W. D. (1999). Microarray analysis of replicative senescence. *Current Biology, 9*(17), 939–945.

Shen, P., Chen, M., He, M., Chen, L., Song, Y., Xiao, P., et al. (2016). Inhibition of ERα/ERK/P62 cascades induces "autophagic switch" in the estrogen receptor-positive breast cancer cells exposed to gemcitabine. *Oncotarget*, 7(30), 48501–48516.

Shen, S., Kepp, O., Michaud, M., Martins, I., Minoux, H., Métivier, D., et al. (2011). Association and dissociation of autophagy, apoptosis and necrosis by systematic chemical study. *Oncogene, 30*(45), 4544–4556.

Short, S., Fielder, E., Miwa, S., & von Zglinicki, T. (2019). Senolytics and senostatics as adjuvant tumour therapy. *eBioMedicine*, *41*, 683–692.

Sikora, E., Mosieniak, G., & Alicja Sliwinska, M. (2016). Morphological and functional characteristic of senescent cancer cells. *Current Drug Targets*, *17*(4), 377–387.

Slingerland, M., Guchelaar, H.-J., & Gelderblom, H. (2014). Histone deacetylase inhibitors: An overview of the clinical studies in solid tumors. *Anti-Cancer Drugs*, *25*(2), 140–149.

Spaulding, C., Guo, W., & Effros, R. B. (1999). Resistance to apoptosis in human CD8+ T cells that reach replicative senescence after multiple rounds of antigen-specific proliferation. *Experimental Gerontology*, *34*(5), 633–644.

Storer, M., Mas, A., Robert-Moreno, A., Pecoraro, M., Ortells, M. C., Di Giacomo, V., et al. (2013). Senescence is a developmental mechanism that contributes to embryonic growth and patterning. *Cell*, *155*(5), 1119–1130.

Su, Z., Yang, Z., Xie, L., Dewitt, J. P., & Chen, Y. (2016). Cancer therapy in the necroptosis era. *Cell Death and Differentiation*, *23*(5), 748–756.

Su, D., Zhu, S., Han, X., Feng, Y., Huang, H., Ren, G., et al. (2009). BMP4-Smad signaling pathway mediates adriamycin-induced premature senescence in lung cancer cells. *The Journal of Biological Chemistry*, *284*(18), 12153–12164.

Sui, X., Chen, R., Wang, Z., Huang, Z., Kong, N., Zhang, M., et al. (2013). Autophagy and chemotherapy resistance: A promising therapeutic target for cancer treatment. *Cell Death & Disease*, *4*, e838.

Sun, W. L., Lan, D., Gan, T. Q., & Cai, Z. W. (2015). Autophagy facilitates multidrug resistance development through inhibition of apoptosis in breast cancer cells. *Neoplasma*, *62*(2), 199–208.

Sun, X., Shi, B., Zheng, H., Min, L., Yang, J., Li, X., et al. (2018). Senescence-associated secretory factors induced by cisplatin in melanoma cells promote non-senescent melanoma cell growth through activation of the ERK1/2-RSK1 pathway. *Cell Death & Disease*, *9*(3), 260.

Sun, Y., Xing, X., Liu, Q., Wang, Z., Xin, Y., Zhang, P., et al. (2015). Hypoxia-induced autophagy reduces radiosensitivity by the HIF-1α/miR-210/Bcl-2 pathway in colon cancer cells. *International Journal of Oncology*, *46*(2), 750–756.

Tai, H., Wang, Z., Gong, H., Han, X., Zhou, J., Wang, X., et al. (2017). Autophagy impairment with lysosomal and mitochondrial dysfunction is an important characteristic of oxidative stress-induced senescence. *Autophagy*, *13*(1), 99–113.

Takamura, A., Komatsu, M., Hara, T., Sakamoto, A., Kishi, C., Waguri, S., et al. (2011). Autophagy-deficient mice develop multiple liver tumors. *Genes & Development*, *25*(8), 795–800.

Talarico, C., Dattilo, V., D'Antona, L., Barone, A., Amodio, N., Belviso, S., et al. (2016). SI113, a SGK1 inhibitor, potentiates the effects of radiotherapy, modulates the response to oxidative stress and induces cytotoxic autophagy in human glioblastoma multiforme cells. *Oncotarget*, *7*(13), 15868–15884.

Tang, R., Xu, J., Zhang, B., Liu, J., Liang, C., Hua, J., et al. (2020). Ferroptosis, necroptosis, and pyroptosis in anticancer immunity. *Journal of Hematology & Oncology*, *7*, 1–18.

Tanino, R., Tsubata, Y., Harashima, N., Harada, M., & Isobe, T. (2018). Novel drug-resistance mechanisms of pemetrexed-treated nonsmall cell lung cancer. *Oncotarget*, *9*(24), 16807–16821.

Taschner-mandl, S., Schwarz, M., Blaha, J., Kauer, M., Kromp, F., Frank, N., et al. (2015). Metronomic topotecan impedes tumor growth of MYCN-amplified neuroblastoma cells in vitro and in vivo by therapy induced senescence. *Oncotarget*, *7*(3), 3571–3586.

Tasdemir, E., Maiuri, M. C., Galluzzi, L., Vitale, I., Djavaheri-Mergny, M., D'Amelio, M., et al. (2008). Regulation of autophagy by cytoplasmic p53. *Nature Cell Biology*, *10*(6), 676–687.

Thorburn, J., Staskiewicz, L., Goodall, M. L., Dimberg, L., Frankel, A. E., Ford, H. L., et al. (2017). Non-cell-autonomous effects of autophagy inhibition in tumor cells promote growth of drug-resistant cells. *Molecular Pharmacology*, *91*(1), 58–64.

Thorburn, A., Thamm, D. H., & Gustafson, D. L. (2014). Autophagy and cancer therapy. *Molecular Pharmacology*, *85*(6), 830–838.

Tolcher, A. W., Lorusso, P., Arzt, J., Busman, T. A., Lian, G., Rudersdorf, N. S., et al. (2015). Safety, efficacy, and pharmacokinetics of navitoclax (ABT-263) in combination with erlotinib in patients with advanced solid tumors. *Cancer Chemotherapy and Pharmacology*, *76*(5), 1025–1032.

Tominaga, T., Shimada, R., Okada, Y., Kawamata, T., & Kibayashi, K. (2019). Senescence-associated-β-galactosidase staining following traumatic brain injury in the mouse cerebrum. *PLoS One*, *14*(3), e0213673.

Torres-Guzmán, R., Calsina, B., Hermoso, A., Baquero, C., Alvarez, B., Amat, J., et al. (2017). Preclinical characterization of abemaciclib in hormone receptor positive breast cancer. *Oncotarget*, *8*(41), 69493–69507.

Tripathi, R., Ash, D., & Shaha, C. (2014). Beclin-1-p53 interaction is crucial for cell fate determination in embryonal carcinoma cells. *Journal of Cellular and Molecular Medicine*, *18*(11), 2275–2286.

Tuttle, R., Miller, K. E., Maiorano, J. N., Termuhlen, P. M., Gao, Y., & Berberich, S. J. (2012). Novel senescence associated gene, YPEL3, is repressed by estrogen in ER+ mammary tumor cells and required for tamoxifen-induced cellular senescence. *International Journal of Cancer*, *130*(10), 2291–22997.

Ulasov, I., Fares, J., Timashev, P., & Lesniak, M. S. (2019). Editing cytoprotective autophagy in glioma: An unfulfilled potential for therapy. *Trends in Molecular Medicine*, *26*(3), 252–262.

Valenzuela, C. A., Vargas, L., Martinez, V., Bravo, S., & Brown, N. E. (2017). Palbociclib-induced autophagy and senescence in gastric cancer cells. *Experimental Cell Research*, *360*(2), 390–396.

Vera-Ramirez, L., Vodnala, S. K., Nini, R., Hunter, K. W., & Green, J. E. (2018). Autophagy promotes the survival of dormant breast cancer cells and metastatic tumour recurrence. *Nature Communications*, *9*(1), 1944.

Vessoni, A. T., Filippi-Chiela, E. C., Menck, C. F., & Lenz, G. (2013). Autophagy and genomic integrity. *Cell Death and Differentiation*, *20*(11), 1444–1454.

Vijayaraghavan, S., Karakas, C., Doostan, I., Chen, X., Bui, T., Yi, M., et al. (2017). CDK4/6 and autophagy inhibitors synergistically induce senescence in Rb positive cytoplasmic cyclin e negative cancers. *Nature Communications*, *8*, 1–17.

Vogl, D. T., Stadtmauer, E. A., Tan, K.-S., Heitjan, D. F., Davis, L. E., Pontiggia, L., et al. (2014). Combined autophagy and proteasome inhibition: A phase 1 trial of hydroxychloroquine and bortezomib in patients with relapsed/refractory myeloma. *Autophagy*, *10*(8), 1380–1390.

Wakita, M., Takahashi, A., Sano, O., Loo, T. M., Imai, Y., Narukawa, M., et al. (2020). A BET family protein degrader provokes senolysis by targeting NHEJ and autophagy in senescent cells. *Nature Communications*, *11*(1), 1935.

Walaszczyk, A., Dookun, E., Redgrave, R., Tual-Chalot, S., Victorelli, S., Spyridopoulos, I., et al. (2019). Pharmacological clearance of senescent cells improves survival and recovery in aged mice following acute myocardial infarction. *Aging Cell*, *18*(3), e12945.

Wang, Y., Gao, W., Shi, X., Ding, J., Liu, W., He, H., et al. (2017). Chemotherapy drugs induce pyroptosis through caspase-3 cleavage of a gasdermin. *Nature*, *547*(7661), 99–103.

Wang, B., Kohli, J., & Demaria, M. (2020). Senescent cells in cancer therapy: Friends or foes? *Trends in Cancer*, *6*(10), 838–857.

Wang, L., Leite de Oliveira, R., Wang, C., Fernandes Neto, J. M., Mainardi, S., Evers, B., et al. (2017). High-throughput functional genetic and compound screens identify targets for senescence induction in cancer. *Cell Reports*, *21*(3), 773–783.

Wang, Y., Li, N., Jiang, W., Deng, W., Ye, R., Xu, C., et al. (2018). Mutant LKB1 confers enhanced radiosensitization in combination with trametinib in KRAS-mutant non-small cell lung cancer. *Clinical Cancer Research*, *24*(22), 5744–5756.

Wang, F., Liu, W., Ning, J., Wang, J., Lang, Y., Jin, X., et al. (2018). Simvastatin suppresses proliferation and migration in non-small cell lung cancer via pyroptosis. *International Journal of Biological Sciences*, *14*(4), 406–417.

Wang, L. X., Wang, J. D., Chen, J. J., Long, B., Liu, L. L., Tu, X. X., et al. (2016). Aurora A kinase inhibitor AKI603 induces cellular senescence in chronic myeloid leukemia cells harboring T315I mutation. *Scientific Reports*, *6*(35533), 1–13.

Wang, X., Wong, S. C. H. H., Pan, J., Cells, C., Tsao, S. W., Fung, K. H. Y. Y., et al. (1998). Evidence of cisplatin-induced senescent-like growth arrest in nasopharyngeal carcinoma cells. *Cancer Research*, *58*(22), 5019–5022.

Was, H., Barszcz, K., Czarnecka, J., Kowalczyk, A., Uzarowska, E., Koza, P., et al. (2017). Bafilomycin A1 triggers proliferative potential of senescent cancer cells in vitro and in NOD/SCID mice. *Oncotarget*, *8*(6), 9303–9322.

Wilson, E. N., Bristol, M. L., Di, X., Maltese, W. A., Koterba, K., Beckman, M. J., et al. (2011). A switch between cytoprotective and cytotoxic autophagy in the radio-sensitization of breast tumor cells by chloroquine and vitamin D. *Hormones & Cancer*, *2*(5), 272–285.

Wolpin, B. M., Rubinson, D. A., Wang, X., Chan, J. A., Cleary, J. M., Enzinger, P. C., et al. (2014). Phase II and pharmacodynamic study of autophagy inhibition using hydroxychloroquine in patients with metastatic pancreatic adenocarcinoma. *The Oncologist*, *19*(6), 637–638.

Wu, D., Pepowski, B., Takahashi, S., & Kron, S. J. (2019). A cmap-enabled gene expression signature-matching approach identifies small-molecule inducers of accelerated cell senescence. *BMC Genomics*, *20*(1), 1–12.

Wu, Y.-T., Tan, H.-L., Shui, G., Bauvy, C., Huang, Q., Wenk, M. R., et al. (2010). Dual role of 3-methyladenine in modulation of autophagy via different temporal patterns of inhibition on class I and III phosphoinositide 3-kinase. *The Journal of Biological Chemistry*, *285*(14), 10850–10861.

Wu, Y., Yu, C., Luo, M., Cen, C., Qiu, J., Zhang, S., et al. (2020). Ferroptosis in cancer treatment: Another way to Rome. *Frontiers in Oncology*, *10*, 571127.

Wu, W., Zhang, L., Lin, J., Huang, H., Shi, B., Lin, X., et al. (2016). Hypermethylation of the HIC1 promoter and aberrant expression of HIC1/SIRT1 contribute to the development of thyroid papillary carcinoma. *Oncotarget*, 7(51), 84416–84427.

Xu, Y., Jagannath, C., Liu, X.-D., Sharafkhaneh, A., Kolodziejska, K. E., & Eissa, N. T. (2007). Toll-like receptor 4 is a sensor for autophagy associated with innate immunity. *Immunity*, *27*(1), 135–144.

Xu, M., Palmer, A. K., Ding, H., Weivoda, M. M., Pirtskhalava, T., White, T. A., et al. (2015). Targeting senescent cells enhances adipogenesis and metabolic function in old age. *eLife*, *19*(4), e12997.

Xu, J., Patel, N. H., Saleh, T., Cudjoe, E. K., Alotaibi, M., Wu, Y., et al. (2018). Differential radiation sensitivity in p53 wild-type and p53-deficient tumor cells associated with senescence but not apoptosis or (nonprotective) autophagy. *Radiation Research*, *190*(5), 538–557.

Yamamoto, K., Venida, A., Yano, J., Biancur, D. E., Kakiuchi, M., Gupta, S., et al. (2020). Autophagy promotes immune evasion of pancreatic cancer by degrading MHC-I. *Nature*, *581*(7806), 100–105.

Yang, L., Fang, J., & Chen, J. (2017). Tumor cell senescence response produces aggressive variants. *Cell Death & Disease*, *3*, 17049.

Yang, N., & Hu, M. (2005). The limitations and validities of senescence associated-b--galactosidase activity as an aging marker for human foreskin fibroblast Hs68 cells. *Experimental Gerontology, 40*(10), 813–819.

Yang, F., Kemp, C. J., & Henikoff, S. (2015). Anthracyclines induce double-strand DNA breaks at active gene promoters. *Mutation Research, Fundamental and Molecular Mechanisms of Mutagenesis, 773*, 9–15.

Yang, M. Y., Lin, P. M., Liu, Y. C., Hsiao, H. H., Yang, W. C., Hsu, J. F., et al. (2012). Induction of cellular senescence by doxorubicin is associated with upregulated miR-375 and induction of autophagy in K562 cells. *PLoS One*, 7(5), e37205.

Yang, S., Wang, X., Contino, G., Liesa, M., Sahin, E., Ying, H., et al. (2011). Pancreatic cancers require autophagy for tumor growth. *Genes & Development, 25*(7), 717–729.

Yang, H., Wang, H., Ren, U., Chen, Q., & Chena, Z. J. (2017). CGAS is essential for cellular senescence. *Proceedings of the National Academy of Sciences of the United States of America, 114*(23), E4612–E4620.

Ylä-Anttila, P., Vihinen, H., Jokitalo, E., & Eskelinen, E.-L. (2009). 3D tomography reveals connections between the phagophore and endoplasmic reticulum. *Autophagy, 5*(8), 1180–1185.

Yosef, R., Pilpel, N., Papismadov, N., Gal, H., Ovadya, Y., Vadai, E., et al. (2017). p21 maintains senescent cell viability under persistent DNA damage response by restraining JNK and caspase signaling. *The EMBO Journal, 36*(15), 2280–2295.

Yosef, R., Pilpel, N., Tokarsky-amiel, R., Biran, A., Ovadya, Y., Cohen, S., et al. (2016). Directed elimination of senescent cells by inhibition of BCL-W and BCL-XL. *Nature Communications*, 7, 1–11.

Yoshida, A., Lee, E. K., & Diehl, J. A. (2016). Induction of therapeutic senescence in vemurafenib-resistant melanoma by extended inhibition of CDK4/6. *Cancer Research, 76*(10), 2990–3003.

Yoshimori, T., Yamamoto, A., Moriyama, Y., Futai, M., & Tashiro, Y. (1991). Bafilomycin A1, a specific inhibitor of vacuolar-type H(+)-ATPase, inhibits acidification and protein degradation in lysosomes of cultured cells. *The Journal of Biological Chemistry, 266*(26), 17707–17712.

You, R., Dai, J., Zhang, P., Barding, G. A., & Raftery, D. (2018). Dynamic metabolic response to adriamycin-induced senescence in breast cancer cells. *Metabolites, 8*(4), E95.

Young, A. R. J., Narita, M., & Narita, M. (2011). Spatio-temporal association between mTOR and autophagy during cellular senescence. *Autophagy*, 7(11), 1387–1388.

Yu, Y., Yang, P., Chuah, Q., Huang, Y., Peng, C., Lee, Y., et al. (2013). Radiation-induced senescence in securin-deficient cancer cells promotes cell invasion involving the IL-6/STAT3 and PDGF-BB/PDGFR pathways. *Scientific Reports, 3*, 1–11.

Yue, Z., Jin, S., Yang, C., Levine, A. J., & Heintz, N. (2003). Beclin 1, an autophagy gene essential for early embryonic development, is a haploinsufficient tumor suppressor. *Proceedings of the National Academy of Sciences, 100*(25), 15077–15082.

Zagouri, F., Sergentanis, T. N., Chrysikos, D., Papadimitriou, C. A., Dimopoulos, M. A., & Psaltopoulou, T. (2013). Hsp90 inhibitors in breast cancer: A systematic review. *Breast, 22*(5), 569–578.

Zeh, H. J., Bahary, N., Boone, B. A., Singhi, A. D., Miller-Ocuin, J. L., Normolle, D. P., et al. (2020). A randomized phase II preoperative study of autophagy inhibition with high-dose hydroxychloroquine and gemcitabine/nab-paclitaxel in pancreatic cancer patients. *Clinical Cancer Research, 26*(13), 3126–3134.

Zhan, H., Suzuki, T., Aizawa, K., Miyagawa, K., & Nagai, R. (2010). Ataxia telangiectasia mutated (ATM)-mediated DNA damage response in oxidative stress-induced vascular endothelial cell senescence. *The Journal of Biological Chemistry, 285*(38), 29662–29670.

Zhang, R., Chen, W., & Adams, P. D. (2007). Molecular dissection of formation of senescence-associated heterochromatin foci. *Molecular and Cellular Biology*, *27*(6), 2343–2358.

Zhang, P., Kishimoto, Y., Grammatikakis, I., Gottimukkala, K., Cutler, R. G., Zhang, S., et al. (2019). Senolytic therapy alleviates Aβ-associated oligodendrocyte progenitor cell senescence and cognitive deficits in an Alzheimer's disease model. *Nature Neuroscience*, *22*(5), 719–728.

Zhang, H., Pan, K., & Cohen, S. N. (2003). Senescence-specific gene expression fingerprints reveal cell-type-dependent physical clustering of up-regulated chromosomal loci. *Proceedings of the National Academy of Sciences of the United States of America*, *100*(6), 3251–3256.

Zhang, M., Qiu, Q., Li, Z., Sachdeva, M., Min, H., Cardona, D. M., et al. (2015). HIF-1 alpha regulates the response of primary sarcomas to radiation therapy through a cell autonomous mechanism. *Radiation Research*, *183*(6), 594.

Zhang, Q., Wang, X., Cao, S., Sun, Y., He, X., Jiang, B., et al. (2020). Berberine represses human gastric cancer cell growth in vitro and in vivo by inducing cytostatic autophagy via inhibition of MAPK/mTOR/p70S6K and Akt signaling pathways. *Biomedicine & Pharmacotherapy*, *128*, 110245.

Zhang, J., Zhang, S., Song, J., Sun, K., Zong, C., Zhao, Q., et al. (2014). Autophagy inhibition switches low-dose camptothecin-induced premature senescence to apoptosis in human colorectal cancer cells. *Biochemical Pharmacology*, *90*(3), 265–275.

Zhao, N., Woodle, M. C., & Mixson, A. J. (2018). Advances in delivery systems for doxorubicin. *Journal of Nanoscience and Nanotechnology*, *9*(5), 519.

Zheng, K., He, Z., Kitazato, K., & Wang, Y. (2019). Selective autophagy regulates cell cycle in cancer therapy. *Theranostics*, *9*(1), 104–125.

Zheng, B., Zhu, H., Gu, D., Pan, X., Qian, L., Xue, B., et al. (2015). MiRNA-30a-mediated autophagy inhibition sensitizes renal cell carcinoma cells to sorafenib. *Biochemical and Biophysical Research Communications*, *459*(2), 234–239.

Zhou, Y., Sun, K., Ma, Y., Yang, H., Zhang, Y., Kong, X., et al. (2015). Autophagy inhibits chemotherapy-induced apoptosis through downregulating Bad and Bim in hepatocellular carcinoma cells. *Scientific Reports*, *4*(1), 5382.

Zhu, Y., Tchkonia, T., Pirtskhalava, T., Gower, A. C., Ding, H., Giorgadze, N., et al. (2015). The Achilles' heel of senescent cells: From transcriptome to senolytic drugs. *Aging Cell*, *14*(4), 644–658.

Zindy, F., Quelle, D. E., Roussel, M. F., & Sherr, C. J. (1997). Expression of the p16(INK4a) tumor suppressor versus other INK4 family members during mouse development and aging. *Oncogene*, *15*(2), 203–211.

CHAPTER TWO

Targeting the stress support network regulated by autophagy and senescence for cancer treatment

Jaejin Kim[†], Yeonghyeon Lee[†], Kyeonghwan Roh, Mi-Sung Kim, and Chanhee Kang*

School of Biological Sciences, Seoul National University, Seoul, South Korea
*Corresponding author: e-mail address: chanhee.kang@snu.ac.kr

Contents

[†] These authors contributed equally.

Advances in Cancer Research, Volume 150
ISSN 0065-230X
https://doi.org/10.1016/bs.acr.2021.01.003

Abstract

Autophagy and cellular senescence are two potent tumor suppressive mechanisms activated by various cellular stresses, including the expression of activated oncogenes. However, emerging evidence has also indicated their pro-tumorigenic activities, strengthening the case for the complexity of tumorigenesis. More specifically, tumorigenesis is a systemic process emanating from the combined accumulation of changes in the tumor support pathways, many of which cannot cause cancer on their own but might still provide excellent therapeutic targets for cancer treatment. In this review, we discuss the dual roles of autophagy and senescence during tumorigenesis, with a specific focus on the stress support networks in cancer cells modulated by these processes. A deeper understanding of such context-dependent roles may help to enhance the effectiveness of cancer therapies targeting autophagy and senescence, while limiting their potential side effects. This will steer and accelerate the pace of research and drug development for cancer treatment.

1. Introduction

Tumorigenesis is considered a multistep process that requires normal cells to progressively acquire several genetic changes, circumventing multiple barriers to tumor formation and progression (Hanahan & Weinberg, 2000, 2011). Cancer cells that have evaded such barriers display a specific set of hallmarks, which can be categorized into three types according to their characteristics: (1) growth-related hallmarks (sustaining proliferative signaling, evading growth suppressors, enabling replicative immortality); (2) survival-related hallmarks (resisting cell death, avoiding immune destruction); and (3) support-related hallmarks (inducing angiogenesis, activating invasion and metastasis, genome instability and mutation, deregulating cellular energetics, and tumor-promoting inflammation). Numerous biological processes have been identified to modulate these hallmarks, yet only a few can affect all types of cancer hallmarks. Macroautophagy (hereafter "autophagy") and cellular senescence (hereafter "senescence") are two rare representative processes that can exert such pleiotropic effects in a context-dependent manner during tumorigenesis (Coppe, Desprez, Krtolica, & Campisi, 2010; Doria, Gatto, & Punzi, 2013; Lee & Schmitt, 2019; Levine & Kroemer, 2019; White, 2015).

Autophagy is a major lysosomal degradation pathway that is essential for cellular and organismal homeostasis (Doria et al., 2013; Levine & Kroemer, 2019; Mizushima & Komatsu, 2011). Autophagy can randomly degrade a portion of the cytoplasm (general autophagy) as well as selectively target several cellular macromolecules such as damaged organelles, protein aggregates, and invading pathogens (selective autophagy) (Anding & Baehrecke, 2017; Johansen & Lamark, 2020; Kirkin, 2020). Recent studies also suggest that selective autophagy can degrade an individual protein to modulate specific biological processes, including stress responses (Belaid, Ndiaye, Klionsky, Hofman, & Mograbi, 2013; Kang & Elledge, 2016; Kang et al., 2015; Sanchez-Martin & Komatsu, 2020). Autophagy is intricately modulated by multiple regulators, including autophagy-related genes (ATG), and their mutations have been linked to an increased risk of developing several types of cancer. For example, mice carrying inactivated alleles of BECN1 (Beclin-1, a mammalian homolog of Atg6) display increased susceptibility to cancer (Levine & Klionsky, 2004), suggesting a role for autophagy in tumor suppression under basal conditions. However, under stressful conditions such as poor vascularization and chemotherapy, autophagy is induced in cancer cells and protects cells from apoptosis. These observations have informed the concept that autophagy may play dual roles in tumorigenesis, depending on the timing of autophagy activation and the type of autophagy/its target substrates (White, 2015).

Senescence is a cellular defense mechanism that prevents premalignant cells from further proliferation by ensuing stable cell cycle arrest, acting as a key barrier to tumorigenesis (Coppe et al., 2010; Gorgoulis et al., 2019; He & Sharpless, 2017; Herranz & Gil, 2018; Kuilman, Michaloglou, Mooi, & Peeper, 2010; Munoz-Espin & Serrano, 2014). Consistently, senescent cells are frequently observed to accumulate in premalignant lesions, yet are rarely found in the fully developed cancer region. However, this view only represents the surface of the complex relationship between senescence and tumorigenesis, as in the case of autophagy. Senescent cells are metabolically active and extensively secrete many pro-inflammatory cytokines, chemokines, growth factors, and proteases that can affect neighboring cells and the tissue microenvironment. This feature is collectively termed the senescence-associated secretory phenotype (SASP) (Coppe et al., 2008). The SASP is a built-in checkpoint for the removal of senescent cells by the immune system to coordinate senescence-associated biological processes in a systemic manner (e.g., tissue repair (Demaria et al., 2014; Krizhanovsky et al., 2008)). However, when this process is not correctly executed, the SASP can promote nearly every aspect of tumorigenesis: cytokines and chemokines for immunosuppressive chronic inflammation, growth factors for tumor growth, and proteases for cancer metastasis

(Coppe et al., 2010; Gonzalez-Meljem, Apps, Fraser, & Martinez-Barbera, 2018). Thus, senescence may toggle the balance of anti- and pro-tumorigenic mechanisms in a context-dependent manner. Targeting senescence for cancer treatment therefore needs to be carefully tailored.

In addition to the pleiotropic effects of autophagy and senescence in tumorigenesis, autophagy and senescence have been shown to be intertwined, which should also be considered when designing therapeutic strategies to target autophagy and senescence for cancer treatment (Gewirtz, 2013; Kang & Elledge, 2016). In this review, we discuss the context-dependent roles of autophagy and senescence in tumorigenesis, including their established functions in cell growth, cell survival, and stress homeostasis. Furthermore, we catalog cellular and molecular markers that can be exploited to guide a customized therapeutic strategy to enhance the effectiveness of therapies targeting autophagy and senescence for treating cancer.

2. Overview of autophagy and senescence

Autophagy and senescence are sophisticatedly regulated biological processes, and understanding how they are regulated took decades to unravel. Now, we have a near complete picture of the core regulatory pathways for autophagy and senescence, including transcriptional regulation, epigenetic modulation, post-translational modification, and regulated protein stability. In addition, their effector programs have been rapidly expanding, such as cellular and tissue homeostasis, development, stem cell maintenance, cell growth control, metabolism, immune modulation, tissue regeneration, development, and aging (Chan & Narita, 2019; DesMarais & Costa, 2019; Gorgoulis et al., 2019; He & Sharpless, 2017; Herranz & Gil, 2018; Higuchi-Sanabria, Frankino, Paul, Tronnes, & Dillin, 2018; Leidal, Levine, & Debnath, 2018; Levine & Kroemer, 2019).

2.1 General autophagy and selective autophagy

Autophagy is a complex yet tightly orchestrated process, which is initiated by the engulfment of a portion of the cytoplasm with a double membrane vesicle called the autophagosome. The autophagosome subsequently fuses with a lysosome to degrade and recycle its engulfed contents. The core autophagy machinery was originally characterized from yeast genetic screens, identifying over 30 autophagy-related genes (ATGs) that mainly control autophagosome formation. This regulatory mechanism has been vastly expanded, as different types of autophagy that act under specific conditions have been added to the autophagy regulatory network (Levine & Klionsky, 2004; Levine & Kroemer, 2019).

Starvation is the best-studied inducer of autophagy, mainly acting through several nutrient and energy sensing pathways, such as the mammalian target of rapamycin (mTOR) and AMP-activated protein kinase (AMPK) pathways. In addition, many cellular stresses implicated in cancer have also been shown to stimulate autophagy, which has multifaceted roles in tumorigenesis. For example, oxidative stress, the DNA damage response (DDR), the unfolded protein response (UPR), hypoxia, and inflammation can potently activate autophagy through various signaling transducers (Kroemer, Marino, & Levine, 2010). In most cases, such autophagy protects cells from damage-induced pathological responses, including transformation and cell death. However, as we discuss below, these protective roles of autophagy can also be operational in cancer cells, maintaining homeostasis and paradoxically promoting tumorigenesis.

Autophagy can be largely divided into two types according to its cargo selectivity: general and selective autophagy (Johansen & Lamark, 2020; Kirkin, 2020). General (or bulk) autophagy degrades a portion of the cytoplasm in a random manner upon induction. The best-known role for general autophagy is to provide building blocks and energy for supporting metabolic homeostasis, and thus cell survival, in response to nutrient starvation. Basal levels of general autophagy are also essential for protein homeostasis (proteostasis), especially in either slowly proliferating cells (e.g., stem cells) or fully differentiated/non-dividing cells (e.g., neurons and senescent cells). Suppressing basal autophagy is deleterious to these cells, either disrupting their functions or sensitizing them to cell death (Levine & Kroemer, 2019; Murrow & Debnath, 2013). Diverse components of the cell can also be selectively targeted for degradation by autophagy on demand. Numerous studies have reported that selective autophagy degrades various cellular organelles, including mitochondria, endoplasmic reticulum (ER), peroxisomes, ribosomes, and lipid droplets, as well as protein aggregates, invading pathogens, and individual proteins (Johansen & Lamark, 2020; Kirkin, 2020). The cargo selection for selective autophagy is mainly mediated by its direct or indirect binding to the ATG8 family of proteins (MAP1LC3s or GABARAPSs) that are conjugated to the autophagosomal membrane. The LC3-interacting region or LIR motif plays a key role in these types of recognition, as the cargo itself contains the motif (direct recognition) or several autophagy receptors, such as p62, NBR1, NDP52, and OPTN. These receptors carry the LIR motif to bridge between the cargo and the ATG8 family of proteins (indirect recognition). Most cargoes recognized through autophagy receptors need to be labeled with ubiquitin (Ji & Kwon, 2017; Kirkin, 2020; Pohl & Dikic, 2019), and how such modification is modulated for selective autophagy is an active area of research.

Selective autophagy has enormously expanded the physiological role of autophagy beyond metabolic homeostasis during nutrient stress. Mitophagy, for example, selectively degrades damaged mitochondria, preventing the accumulation of reactive oxygen species (ROS) and thus maintaining redox homeostasis (Youle & Narendra, 2011). Upon proteotoxic stress, such as the massive synthesis of proteins frequently observed in cancer cells, ER-phagy targets a portion of the endoplasmic reticulum (ER) network to maintain proteostasis (Chino & Mizushima, 2020). Other homeostatic functions of selective autophagy include the DDR, inflammatory signaling, hypoxia, and immune modulation, many of which are implicated in tumorigenesis (Levine & Kroemer, 2019).

2.2 Cellular senescence and its intrinsic and extrinsic features

Senescence is characterized by several intrinsic and extrinsic features that can crosstalk with each other, even across individual cells. It is generally assumed that its intrinsic features contribute to the development of the senescent state and support its extrinsic features, which in turn exert systemic effects on neighboring cells, the tissue microenvironment, and even the immune system of the host (Gorgoulis et al., 2019; He & Sharpless, 2017; Herranz & Gil, 2018; Kuilman et al., 2010).

The best-characterized intrinsic feature of senescence is stable growth arrest, in which two tumor suppressive pathways, the p53-p21 and p16-Rb pathways, play a major role. Upon senescence induction, p53 is stabilized to induce the CDK inhibitor p21, which immediately halts the cell cycle of damaged cells. Subsequently, if the damage is irreparable, this cell cycle arrest is potentiated by increased expression of p16, another CDK inhibitor involved in senescence. This in turn induces the retinoblastoma (Rb)-mediated senescence-associated heterochromatic foci that stably repress E2F target genes, contributing to stable growth arrest in senescent cells (Kuilman et al., 2010). Remodeling of the nuclear lamina structure in senescent cells contributes to the relaxation of preexisting heterochromatin regions and thus increases transcriptional noise during senescence. The ER-Golgi network is massively reorganized in senescent cells, which might be necessary for the SASP, a prominent extrinsic feature of senescent cells, as discussed below (Hernandez-Segura et al., 2017). Although it is not well characterized, another noticeable feature of senescence is senescent hypertrophy, which increases the volume of the cell with a flattened morphology. Currently, the functional consequence of senescent hypertrophy remains elusive, but it might be related to the expansion of the ER-Golgi network to support the massive secretory phenotype in senescent cells (Hernandez-Segura, Nehme, & Demaria, 2018).

The notable extrinsic feature of senescence is the SASP, which involves robust secretion of pro-inflammatory cytokines, chemokines, proteases, and growth factors, as well as other non-proteinous molecules including exosomes (Basisty et al., 2020; Coppe et al., 2010). The composition of the SASP can be highly varied depending on the cell type undergoing senescence, tissue of origin, and the inducer of senescence. In addition, many SASP factors are not exclusive to senescence but can have well-characterized roles in other biological contexts. For example, the inflammatory arm of the SASP, such as IL-1A and IL-6, is well known to be activated during acute inflammation. The key difference between the SASP and acute inflammation is the kinetics of each response. The SASP is a delayed response that takes several days to be fully developed upon senescence induction. The SASP is also very stable through the positive feed-forward activation loop, and thus can be maintained until senescent cells are physically removed by the immune system, including macrophages, natural killer cells (NK), and T lymphocytes (Bataller & Brenner, 2005; Brighton et al., 2017; Egashira et al., 2017; Kang et al., 2011; Lujambio et al., 2013; Mevorach et al., 2010; Munoz et al., 2019; Pereira et al., 2019). As the SASP consists of a wide-range of factors, it is not surprising to find that the SASP has pleiotropic effects on the tissue microenvironment. The SASP can affect senescence reinforcement in both autocrine and paracrine manners (Acosta et al., 2008, 2013; Coppe et al., 2008), proliferation, migration, and stemness of neighboring cells (Mosteiro et al., 2016; Ritschka et al., 2017), angiogenesis (Ruscetti et al., 2020), and immune modulation, such as macrophage differentiation and recruitment of innate and adaptive immune cells (He & Sharpless, 2017; Herranz & Gil, 2018). All of these features are closely associated with tumorigenesis. Currently, functional characterization of each SASP factor has gained attention, as this may provide a therapeutic option to preserve the beneficial role of senescence while decreasing its deleterious effect; this will be discussed below in the case of tumorigenesis. Although recent studies indicated a role for senescence-associated exosomes as a part of senescent secretome (Basisty et al., 2020; Takasugi et al., 2017), other non-proteinous, intracellular signaling molecules (e.g., metabolites) are yet to be discovered.

During the last decade, great advances have been made in understanding how the SASP is regulated at the transcriptional and post-transcriptional levels. Several transcriptional factors and epigenetic regulators, including nuclear factor-κB (NF-κB) (Chien et al., 2011), CCAAT/enhancer-binding protein-β (C/EBP-β) (Acosta et al., 2008), GATA4 (Kang et al., 2015; Mazzucco et al., 2017), NOTCH1 (Hoare et al., 2016), HMGB2 (Aird et al., 2016), and BRD4 (Tasdemir et al., 2016), have been identified to modulate the SASP at the transcriptional level. In addition, cytoplasmic chromatin fragments (CCFs) that pinch off from intact nuclei of senescent cells and LINE-1 elements reactivated during senescence stimulate the cGAS-STING pathway to activate the interferon 1 (IFN-1) response and thus maintain the SASP (De Cecco et al., 2019; Dou et al., 2015, 2017). At the post-transcriptional level, the mTOR pathway plays a central role in modulating the SASP. The TOR-autophagy spatial coupling compartment (TASCC), where the autolysosome and mTOR accumulate, facilitates the massive synthesis of secretory proteins by supplying amino acids to stimulate mTOR activity (Young, Narita, & Narita, 2011). The mTOR pathway preferentially stimulates translation of specific SASP factors and modulates the mRNA stability of SASP factors (Herranz et al., 2015; Laberge et al., 2015). Interestingly, it has been shown that senescent cells maintain a higher level of autophagy activity under this enhanced mTOR activity, which generally acts as a potent suppressor of autophagy (Dunlop & Tee, 2014). How such an uncoupling between autophagy and the mTOR pathway occurs currently remains unknown; considering the current therapeutic interest in modulating the activities of autophagy and the mTOR pathway for aging treatment, this will be an important topic in future research. This set of discoveries provides a molecular basis for developing therapeutic strategies to modulate the SASP, recently termed senostatics or senomorphics (Gorgoulis et al., 2019; Kang, 2019; Kirkland & Tchkonia, 2017; van Deursen, 2019), which will increase the repertoire of senescence-based therapies in a wide range of age-related diseases, including cancer.

3. Autophagy in tumors

3.1 Autophagy and tumor suppression

3.1.1 Metabolic stress and autophagy

It has been well established that cancer cells significantly rewire cellular metabolism to match the demand of their high growth rate, which includes increased levels of glycolysis, glutaminolysis, amino acid and lipid metabolism, and macromolecule biosynthesis (Ward & Thompson, 2012). Interestingly, several genetic mutations in metabolic enzymes are reported to be associated with tumorigenesis including succinate dehydrogenase (SDH), fumarate hydratase (FH), and isocitrate dehydrogenase 1/2 (IDH1/2) (Morin, Letouze, Gimenez-Roqueplo, & Favier, 2014). Although a single mutation in metabolic enzymes does not recapitulate all metabolic changes in cancer cells, these findings strongly suggest that metabolic stress can promote tumorigenesis in certain cases. For example, nutrient deprivation increases ROS production and causes an imbalance in the thiol redox state, resulting in oxidative stress that can promote tumorigenesis (Filomeni, De Zio, & Cecconi, 2015). Thus, autophagy may protect cells from tumorigenesis by alleviating metabolic stress. However, this might be coupled with other homeostatic functions of autophagy, such as mitochondrial homeostasis (mitophagy) or lipid droplet mobilization (lipophagy).

3.1.2 Oxidative stress and autophagy

Oxidative stress can damage cellular macromolecules such as DNA, RNA, proteins, and lipids, which create pro-tumorigenic environments by introducing oncogenic mutations and activating pro-survival pathways, such as the NF-κB pathway. Oxidative stress induces autophagy through multiple mechanisms, including mTOR, protein kinase R (PKR)-like endoplasmic reticulum kinase (PERK), c-Jun N-terminal kinase (JNK), and p53. Upon activation, autophagy can relieve oxidative stress at multiple levels (Filomeni et al., 2015). As mitochondria are the main source of cellular oxidative stress, autophagy suppresses oxidative stress by removing damaged mitochondria through mitophagy (Galluzzi et al., 2017). Consistently, mice deficient in the key regulators of autophagy, such as Beclin-1, ATG5, and ATG7, harbor abnormal mitochondria associated with increased oxidative damage (Takamura et al., 2011). In addition, autophagy can remove protein aggregates caused by oxidative stress, maintaining proteostasis during oxidative stress (Filomeni et al., 2015). Lastly, autophagy can promote the DDR to repair DNA lesions caused by oxidative stress, maintaining genome integrity and suppressing oncogenic mutations (Gomes, Menck, & Leandro, 2017).

Autophagy is also involved in modulating cellular antioxidant programs to protect cells from oxidative stress. Upon oxidative stress, the autophagy receptor p62 (also known as SQSTM1) binds to kelch-like ECH associated protein 1 (KEAP1), an E3 ligase adaptor for NF-E2-related factor 2 (NRF2), and either sequesters KEAP1 from NRF2 or targets KEAP1 for autophagic degradation (Komatsu et al., 2010). NRF2 then translocates to the nucleus and triggers the expression of antioxidant genes and detoxifying enzymes, enhancing redox homeostasis and cell survival. Collectively, it is plausible to assume that autophagy deficiency causes the oxidative stress that drives tumorigenesis. Indeed, autophagy deficiency is reported to promote hepatocellular adenoma/carcinoma with elevated levels of oxidative stress and damaged mitochondria in a manner dependent on the p62-KEAP1-NRF2 axis (Inami et al., 2011). Deletion of either p62 or NRF2 largely reverses the tumorigenic phenotypes observed in autophagy- deficient mice, suggesting that a dysregulated antioxidant program of the NRF2 pathway plays a major role (Ni et al., 2014; Takamura et al., 2011). Indeed, NRF2 also positively regulates genes involved in cancer metabolism, including the pentose phosphate pathway, purine-nucleotide synthesis, and glutaminolysis, redirecting glucose and glutamine into anabolic pathways (Mitsuishi et al., 2012).

3.1.3 The DDR and autophagy

Genome instability is one of the key hallmarks of cancers. Genetic mutations in genes related to DNA damage repair evoke so-called mutator phenotypes, which are predisposed to multiple types of cancer. It has been well established that multiple genotoxic stresses activate autophagy mostly, but not exclusively, in a DDR-dependent manner (Eliopoulos, Havaki, & Gorgoulis, 2016). The DNA damage response is also a potent inducer of senescence, suggesting a role for autophagy as an effector program of senescence (Hernandez-Segura et al., 2018). A major sensor kinase of double-strand breaks, ataxia-telangiectasia mutated (ATM), activates AMP-activated protein kinase (AMPK), which in turn suppresses mTOR by activating tuberous sclerosis complex 2 (TSC2) and relieves its inhibitory action on autophagy. In addition, ATM directly activates Unc-51 like autophagy activating kinase 1 (ULK1), which modulates early steps of autophagosome formation. ATM-checkpoint kinase 2 (CHK2) or ataxia telangiectasia and Rad3-related protein (ATR)-checkpoint kinase 1 (CHK1) pathways, a key DDR pathway, activates p53, which transcriptionally induces genes involved in autophagy, including ATG2, ATG4, ATG7, ATG10, and ULK1 (Kenzelmann Broz et al., 2013). Additionally, p53 upregulates Sestrin 1/2, which modulates the AMPK-mTOR-autophagy axis (Budanov & Karin, 2008). Sestrin 1/2 also inhibits mTOR via GAP activity toward Rags (GATOR) 1-GATOR 2 in an amino acid leucine-dependent manner, activating autophagy (Wolfson et al., 2016).

Upon activation, autophagy promotes DNA damage repair by enhancing a wide range of DNA repair pathways, including homologous recombination (HR) and nucleotide exchange repair (NER) (Gillespie & Ryan, 2016; Hewitt et al., 2016; Qiang et al., 2016; Wang et al., 2016). Autophagy degrades p62, which binds to and inhibits an E3 ligase for histone H2A, RNF168. RNF168 then ubiquitinates H2A, which is necessary for recruiting DNA repair proteins such as breast cancer type 1 (BRCA1) and RAD51 to promote HR. Furthermore, autophagy targets Kruppel-associated box domain-containing zinc finger protein 1 (KAP1) for degradation (Tsuruma et al., 2008). KAP1 is a suppressor of signal transducer and activator of transcription 3 (STAT3), which upregulates BRCA1 during the DDR. Thus, autophagy activates the STAT3-BRCA1 axis to promote HR. Autophagy also positively regulates NER by enhancing DNA damage recognition through the Xeroderma pigmentosum complementation group C (XPC) and DNA damage binding protein 2 (DDB2) (Qiang et al., 2016). Autophagy deficiency stabilizes the transcription factor TWIST1, which activates the transcriptional repressor complex E2F4-RBL2. E2F4-RBL2 then suppresses XPC. TWIST1 also inhibits the histone acetyltransferase EP300, which impairs the recruitment of DDB2 to UV-induced DNA lesions (Qiang et al., 2014, 2016). In support of DNA repair functions of autophagy, autophagy-deficient cells display a high level of DNA damage under basal condition, likely because they tip the balance of DNA repair into an error-prone nonhomologous end joining (NHEJ) pathway (Gillespie & Ryan, 2016; Liu et al., 2015). Collectively, autophagy can suppress tumorigenesis by maintaining genome integrity.

3.1.4 The UPR and autophagy

Cancer cells generally have a high level of protein and lipid synthesis to match their proliferation rate, which leads to ER stress (Hsu et al., 2019). The UPR is triggered to relieve ER stress and sustain cell viability. The UPR consists of three branches of ER stress sensors: inositol requiring enzyme 1 (IRE1), protein kinase RNA-activated-like ER kinase (PERK), and activating transcription factor 6 (ATF6). These sensors suppress protein translation, increase chaperone proteins, and activate endoplasmic reticulum-associated protein degradation (ERAD) and autophagy in a coordinated manner.

Increasing evidence suggests that ER stress and UPR activation contribute to tumorigenesis in several types of cancer, including prostate, breast, and colon cancer (Dejeans, Barroso, Fernandez-Zapico, Samali, & Chevet, 2015). The UPR supports neoplastic transformation by increasing protein folding capacity and activating pro-survival pathways. The UPR allows tumor cells to overcome hypoxia and nutrient deprivation during tumor establishment by promoting angiogenesis and rewiring metabolic pathways. In addition, the UPR promotes metastasis by increasing epithelial-to-mesenchymal transition (EMT) and decreasing cell-cell contact. Lastly, the UPR protects cancer cells from chemotherapies by activating pro-survival pathways, increasing drug efflux, and maintaining cancer stem cells (Hsu et al., 2019).

The UPR is a potent stimulus of autophagy, and all three branches of ER stress sensors are involved in this process. PERK transcriptionally activates genes involved in autophagy, including LC3 and ATG5, via ATF4 and CHOP, respectively. PERK can also stimulate autophagy through the sestrin-mTOR pathway (Rashid, Yadav, Kim, & Chae, 2015). ATF6 can upregulate transcription of ATG3, ATG12, ATG5, and other autophagy genes. IRE1 activates the JNK pathway through the TRAF2-ASK1 axis, stimulating autophagy (Heath-Engel, Chang, & Shore, 2008; Wei, Pattingre, Sinha, Bassik, & Levine, 2008). Upon activation, autophagy relieves ER stress by degrading protein aggregates and frequently a portion of ER clusters that include unfolded proteins through ER-phagy (Loi, Fregno, Guerra, & Molinari, 2018). ER-phagy is mediated by a set of specialized autophagy receptors including FAM134B, RTN3L, ATL3, SEC62, CCPG1, and TEX264, each of which modulates a different part of the ER (Kirkin, 2020). Since autophagy can quickly relieve ER stress, it is plausible to assume that autophagy may prevent the pro-tumorigenic activities of the UPR. However, caution is needed to predict the role of autophagy during UPR-induced tumorigenesis, since autophagy can promote its pro-tumorigenic activities in a context-dependent manner. For example, autophagy induced by the UPR in a fully established cancer can support its metabolic demand and promote resistance to chemotherapy (see below).

3.1.5 Cell death and autophagy

Cell death plays an essential role in suppressing tumorigenesis by deleting pre-malignant cells. Apoptosis is the best characterized form of cell death, which depends on a caspase-dependent system to execute cell death. Accordingly, pre-malignant and malignant cells need to overcome apoptosis to become a fully developed cancer. Indeed, many apoptosis-related pathways are found to be mutated or dysregulated in most, if not all, types of human cancer, including the p53 pathway and anti-apoptotic BCL2 family of proteins (Pfeffer & Singh, 2018). When apoptosis is disabled, compensatory backup mechanisms ensure that pre-malignant or damaged cells are removed by non-apoptotic cell death, including autophagic cell death. The role of autophagy in cell death has been debated, yet it can be largely defined as follows based on recently published guidelines (Denton & Kumar, 2019): (1) autophagy-associated cell death, (2) autophagy-mediated cell death, and (3) autophagy-dependent cell death (ACD). In autophagy-associated cell death, autophagy is induced during cell death but plays a negligible role, as its inhibition does not affect the progress of cell death. Autophagy can trigger or support apoptosis in autophagy-mediated cell death, yet autophagy induces cell death independently of apoptosis or necrosis in ACD. Both types of autophagic cell death can act as a tumor-suppressive mechanism, but only ACD can cover apoptosis to remove premalignant cells when apoptosis is disabled.

Expression of oncogenic H-Ras^{V12} in ovarian epithelial cells induces caspase-independent cell death with features of autophagy, and autophagy inhibition rescues this ACD to promote the transforming potential of oncogenic forms of Ras (Elgendy, Sheridan, Brumatti, & Martin, 2011). In A549 lung cancer cells, treatment with resveratrol, a natural polyphenol that activates autophagy, induces cell death via autophagy induction in the absence of several apoptotic markers such as caspase activation (Dasari et al., 2017). Autophagy inhibition by knockdown of key autophagy components rescued cell death induced by resveratrol, indicating that ACD is executed upon resveratrol treatment. Treatment with *Z*-VAD-FMK, a pan-caspase inhibitor, induced cell death in human leukemia cells (Yu et al., 2004). Knockdown of ATG7 and Beclin 1 reversed this autophagic cell death. This set of data suggests that ACD can function as both a natural barrier to tumorigenesis and a treatment option for fully established cancer. How ACD induces cell death remains elusive except for autosis, which depends on the plasma membrane Na^+/K^+-ATPase but has not been associated with cancer (Liu et al., 2013). It is possible that excessive consumption of cytoplasmic components, including cellular organelles during ACD, leads to cell death. Alternatively, autophagy may selectively target specific survival factors for degradation to induce ACD (Denton & Kumar, 2019).

3.1.6 Immune regulation and autophagy

Autophagy suppresses tumorigenesis in a cell non-autonomous manner by modulating the immune surveillance of cancer cells. Autophagy has been shown to promote MHC class I-dependent cross-presentation of tumor antigens and activation of $CD8^+$ T cells (Crotzer & Blum, 2009).

Autophagy is induced in tumor cells undergoing tumorigenesis-related cellular stresses, which engulfs tumor antigens and is subsequently released into the extracellular environment. Dendritic cells (DCs) internalize secreted autophagosomes and present tumor peptides with MHC class I molecules to activate $CD8^+$ T cells. In addition, autophagy processes cytosolic and nuclear tumor antigens and subsequently delivers them to the lysosome for association with MHC class II molecules (Li et al., 2008; Munz, 2016). Further processing of cytosolic and nuclear tumor antigens may occur prior to binding to MHC class II molecules by lysosomal proteases. Displayed tumor antigens stimulate $CD4^+$ T cells and increase tumor immune surveillance. In addition, autophagy has been shown to increase the release of ATP from cancer cells during chemotherapy, which recruits DCs and T cells that exert a tumor-suppressive mechanism (Michaud et al., 2011).

3.2 Autophagy and tumor promotion

Genetic inhibition of autophagy in several tumor models generally causes less aggressive types of cancer, indicating that autophagy may be necessary for tumors to progress to malignancy. Indeed, increasing evidence suggests that autophagy turns on its heel once a tumor is established, maintaining the homeostatic state of cancer cells and promoting tumor progression.

3.2.1 Autophagy relieves metabolic stress in cancer cells

In most established tumors, autophagy is frequently hyperactivated to support cancer metabolism, which is essential for cancer cells that have a higher rate of proliferation and cell growth. This could be achieved by autophagic recycling of cellular components through general autophagy, ribophagy (autophagic degradation of the ribosomes), glycophagy (autophagic degradation of glycogen), or lipophagy (autophagic degradation of lipid droplets) to supply mitochondria with their substrates (e.g., amino acids and lipids), generating cellular energy under nutrient-poor conditions (Goldsmith, Levine, & Debnath, 2014). Consistent with this, inhibition of autophagy in most cancer cells is associated with a buildup of both structurally and functionally defective mitochondria. Mitochondrial defects could be due to either impaired metabolic substrate recycling or decreased mitophagy, or both (Galluzzi et al., 2015). Supplementation of pyruvate or glutamine rescues the survival of autophagy-deficient cells in several types of cancer, including pancreatic cancer, supporting this notion. Another cancer metabolism supported by autophagy is enhancing glycolysis. Cancer cells frequently shift their metabolism from oxidative phosphorylation to aerobic glycolysis for producing ATP from glucose, directing glycolytic intermediates toward biosynthetic pathways in a process known as the Warburg effect (Ward & Thompson, 2012). Inhibition of autophagy in HRas-transformed mouse embryonic fibroblasts (MEFs) or mouse breast cancer cells decreases glucose uptake and reduces de novo production of lactate, suggesting a role for autophagy in glycolytic flux (Lock et al., 2011). Thus, autophagy deficiency greatly sensitizes these transformed or cancer cells to reduced levels of glucose. Consistent with the metabolic support role of autophagy in cancer, activation of oncogenes, including expression of oncogenic Ras, upregulates basal levels of autophagy (Guo et al., 2011).

3.2.2 Autophagy maintains redox homeostasis in cancer cells

At the early stage of tumorigenesis when blood supply is limited, a large portion of the cancer region is deprived of nutrients and oxygen under hypoxic conditions. Under these conditions, the mitochondrial respiratory chain produces nitric oxide (NO), which further generates other ROS including reactive nitrogen species (RNS), reactive aldehyde-malondialdehyde (MDA), and 4-hydroxynonenal (4-HNE) (Cipak Gasparovic et al., 2019). Thus, cancer cells need to maintain redox homeostasis for their survival. In addition, genome instability frequently observed in cancer cells causes protein imbalance in the mitochondrial respiratory chain, contributing to mitochondrial dysfunction and oxidative stress (Ward & Thompson, 2012). Autophagy plays a crucial role in maintaining redox homeostasis under this condition, similar to its action in normal cells. Consistent with this, autophagy is observed to be hyperactivated in the hypoxic region of tumors, and its inhibition sensitizes tumor cells to cell death (Degenhardt et al., 2006). As discussed above, autophagy removes dysfunctional mitochondria, which are a main source of ROS via mitophagy, enhancing redox homeostasis (Weinberg et al., 2010). p62-mediated selective autophagy also plays a critical role in redox homeostasis in cancer, by degrading KEAP1 and subsequently activating the NRF2 anti-oxidant program. In activated KRas– and BRaf–driven lung cancers, autophagy contributes to ROS detoxification via the p62-KEAP1-NRF2 axis (DeNicola et al., 2011). Consistently, mutational signatures in multiple tumors indicate an activation signature for NRF2, which is opposite to an inactivation signature for KEAP1 (Hayes & McMahon, 2009).

3.2.3 Autophagy promotes therapeutic resistance to DNA-damaging chemotherapies

As most cancer cells are subject to severe genome instability, they need to disable certain DNA damage pathways for their survival. Genotoxic drugs used for chemotherapies target such vulnerability of cancer cells, increasing the burden of DNA damage to surpass their threshold to induce cell death (Swift & Golsteyn, 2014). Given its cytoprotective role during the DDR, it has been extensively studied that autophagy inhibition augments cytotoxicity in combination with several anticancer agents. Thus, readers interested in the role of autophagy during chemotherapies can refer to excellent recent reviews on this topic (Mulcahy Levy & Thorburn, 2020; Perez-Hernandez et al., 2019).

3.2.4 Autophagy maintains proteostasis in cancer cells for their survival

Tumor cells proliferate in an accelerated manner, requiring a high flux of protein and lipid synthesis to match their growth rate. This, in turn, causes a high level of proteotoxic stress. In addition, genome instability, such as aneuploidy, provides another important source of proteotoxic stress, as it can generate unbalanced protein complexes that can lead to protein aggregates. Autophagy can relieve such proteotoxic stress via aggrephagy (selective degradation of protein aggregates) or ER-phagy, aiding the survival of tumor cells. In aggrephagy, polyubiquitinated proteins coalesce to form aggresomal particles that are transported toward the microtubule organizing center (MTOC) in a histone deacetylase 6 (HDAC6)-dependent manner, where aggresomes are targeted for autophagic degradation (Losson, Schnekenburger, Dicato, & Diederich, 2020). Inhibition of HDAC6 using the specific inhibitor ACY-1215 might provide a therapeutic option for combinatorial treatments with inhibition of proteasome or other protein quality systems to induce cell death in cancer, as exemplified in multiple myeloma. The role of ER-phagy in tumorigenesis has not been revealed. However, the fact that FAM134B and SEC62, two autophagy receptors for ER-phagy, have been linked to several types of cancer indicates that ER homeostasis via autophagy plays a role in tumorigenesis (Wilkinson, 2020).

3.2.5 Autophagy modulates a cancer cell survival pathway

Autophagy can promote cancer cell survival under various stressful conditions as a stress-activated mechanism, as discussed above. In addition, autophagy can prevent cancer cell death by directly modulating factors related to apoptosis or necrosis. For example, autophagy promotes degradation of the active tyrosine kinase Src using an E3 ligase c-Cbl as an autophagy receptor, enhancing tumor cell survival in squamous carcinoma (Sandilands et al., 2011). Colon cancer cells deficient for the pro-apoptotic protein BAX are known to be resistant to TNF-related apoptosis-inducing ligand

(TRAIL)-induced cell death when autophagy is disabled (Sharma & Almasan, 2018). Mechanistically, TRAIL increases autophagy, which counterbalances TRAIL-mediated apoptosis in cancer cells by sequestrating caspase-8 into the autophagosome for subsequent lysosomal degradation. Autophagy can regulate tumor cell survival in a cell non-autonomous manner. Autophagy in normal endothelial cells in the tumor region mediates the secretion of high mobility group box 1 (HMGB1), which in turn interacts with the receptor for advanced glycosylation end products (RAGE) on pancreatic cancer cells to induce autophagy and promote survival and resistance to apoptosis (Kang et al., 2010).

3.2.6 Autophagy promotes immune evasion of cancer cells

Cancer cells need to overcome detection by the immune system for their survival. One such mechanism by which they avoid detection is the autophagy-dependent downregulation of tumor-specific antigen presentation. In pancreatic ductal adenocarcinoma (PDAC), MHC class I molecules that are required for antigen presentation are subject to NBR1-dependent selective autophagy, promoting immune evasion (Yamamoto et al., 2020). Inhibition of autophagy enhanced the surface display of MHC class I molecules, rescuing the susceptibility of PDAC to cytotoxic T lymphocytes (CTLs)-mediated cell death. Autophagy is also involved in modulating cytolytic granule-mediated tumor cell death (Thiery et al., 2011). Hypoxia activates autophagy in MCF-7 breast cancer cells, which in turn suppresses Natural killer (NK) cell-mediated cancer cell lysis. Mechanistically, autophagy targets granzyme B and perforin granules for lysosomal degradation in a p62-dependent manner (Baginska et al., 2013), thus promoting immune evasion of cancer cells. Autophagy can disrupt the immune synapse between NK cells and hypoxic melanoma cells by selectively degrading gap-junctional connexin 43, which in turn impairs NK-mediated tumor cell death (Tittarelli, Janji, Van Moer, Noman, & Chouaib, 2015). Autophagy indirectly contributes to tumor immune evasion by modulating proteasome-dependent degradation of STAT3, an important factor in the expression of several immunosuppressive genes (Noman et al., 2009). In lung carcinoma, hypoxia activates autophagy, degrading the autophagy substrate protein, p62. When autophagy is inhibited, p62 accumulates to promote STAT3 degradation in a proteasome-dependent manner, thus increasing the susceptibility of tumor cells to CTLs (Noman et al., 2011). Lastly, autophagy targets inflammasomes, reducing the secretion of cytokines/chemokines and thus preventing immune cell recruitment. Inhibition of autophagy causes a high level of IL-6 secretion, which is necessary for massive macrophage infiltration to suppress tumorigenesis (Degenhardt et al., 2006). Collectively, these findings support a role for autophagy in the immune evasion of cancer by modulating CTL and NK cells-mediated tumor cell death.

3.3 Cataloging properties of the stress support network associated with autophagy addiction for cancer treatment

Several studies have demonstrated that autophagy plays a dual role in tumorigenesis (Fig. 1). Thus, understanding the context-dependent roles of autophagy in tumorigenesis is essential for developing therapeutic strategies to target autophagy for cancer treatment. The timing of autophagy activation, what causes such autophagy activation, and its cellular targets should be considered for autophagy-based cancer therapies.

Cancer cells metabolically addicted to autophagy should be judged by the following criteria: (1) autophagy is activated at the early stage of tumorigenesis when blood supply is not reestablished around an emerging tumor region; (2) nutrient deprivation is accompanied by autophagy activation, in which case the rate of glucose transport might be a good surrogate marker, or the status of energy sensing pathways such as AMPK could be assessed upon autophagy inhibition; and (3) general autophagy, ribophagy, and lipophagy could occur in this situation. Although autophagy inhibition is favored for

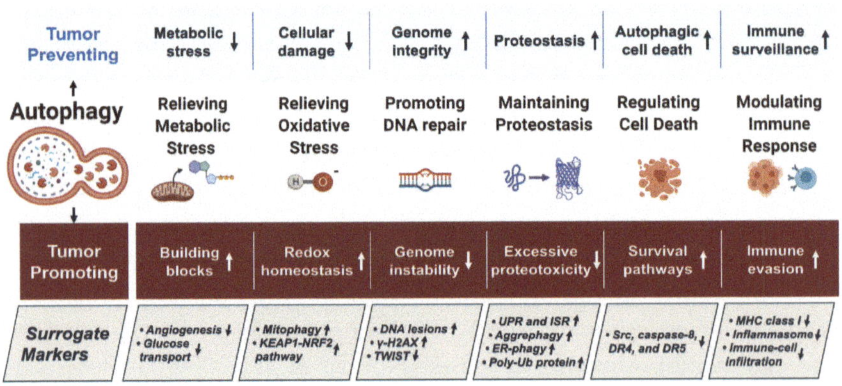

Fig. 1 Dual roles of autophagy in tumorigenesis. Autophagy prevents tumorigenesis by (1) relieving metabolic stress, (2) reducing cellular damage, (3) maintaining genome integrity, (4) promoting proteostasis, (5) serving as a backup cell death mechanism, and (6) enhancing immune surveillance. Autophagy promotes tumorigenesis by (1) providing cellular energy and building blocks, (2) maintaining redox homeostasis, (3) suppressing excessive DNA damage, (4) reducing proteotoxicity, (5) inhibiting cell death pathways, and (6) increasing immune evasion (created with BioRender.com).

cancer treatment when the above criteria are fulfilled, caution should be taken as excessive starvation caused by autophagy inhibition can promote metastasis.

Cancer cells addicted to autophagy for redox homeostasis should be judged by the following criteria: (1) signature of mitophagy such as autophagosome with engulfed mitochondria, or activation of the PINK-PARKIN pathway, which is known to modulate mitophagy (Narendra, Tanaka, Suen, & Youle, 2008), is enriched; and (2) the level of KEAP1 is reduced, and the activation signature of the NRF2 pathway is observed together with autophagy activation. Inhibition of general autophagy can increase ROS in normal cells, and thus targeting autophagy should be carefully conducted to target cancer cells addicted to autophagy for redox homeostasis. This could be achieved if the extent of autophagy inhibition is carefully controlled, or if tumor-specific selective autophagy of KEAP1 can be targeted.

Cancer cells addicted to autophagy for the DDR should be judged by the following criteria: (1) tumor mutational burden should be relatively high; (2) activation of molecules associated with the DDR, such as gamma-H2AX, is readily detected; and (3) selective autophagy targets that negatively modulate DNA damage repair, including TWIST, should be expressed at lower levels. In this case, autophagy inhibition combined with chemotherapies that cause additional DNA damage could be considered to enhance the effectiveness of treatment.

Cancer cells addicted to autophagy for proteostasis should be judged by the following criteria: (1) the UPR pathway and integrated stress response (ISR) are activated; (2) aggrephagy and ER-phagy are frequently observed, as assessed by aggresome formation and decreased levels of ER-phagy receptors, including FAM134B and SEC62, respectively; and (3) polyubiquitinated proteins are greatly accumulated. In this case, autophagy inhibition combined with proteasome inhibition could be explored for a better response.

Cancer cells addicted to autophagy for cell survival pathways should be judged by checking selective autophagy targets for modulating cell survival pathways, such as Src, caspase-8, DR4, and DR5, to determine whether they are downregulated at the protein level. If the apoptosis pathway is disabled, caution must be taken, as ACD can act as a back-up cell death pathway in such cancer cells. In this case, stimulating autophagy could be considered to enhance the effectiveness of treatment.

Cancer cells addicted to autophagy for immune surveillance should be judged by the following criteria: (1) expression of MHC class I molecules is downregulated; (2) inflammasome formation is greatly reduced, with a lower level of immune-stimulating inflammatory response; and (3) immune cell infiltration is lower. Since autophagy can also promote immune surveillance of cancer, caution should be taken when autophagy inhibition suppresses immune cell activation, including $CD4^+$ and $CD8^+$ T cells.

4. Senescence in tumors

4.1 Tumor-suppressive roles of senescence

Senescence was originally identified as a tumor suppressor mechanism that can restrict the expansion of damaged or premalignant cells by installing stable growth arrest, which is triggered after a finite number of replications. This type of senescence, called "replicative senescence," is mainly regulated by the DDR due to critically shortened telomeres (Harley, Futcher, & Greider, 1990). Later, other types of senescence related to tumorigenesis, including oncogene-induced senescence (OIS) and therapy-induced senescence (TIS), are also shown to be mainly regulated by the DDR (Wang, Kohli, & Demaria, 2020).

4.1.1 Oncogene-induced senescence

Tumors are primarily driven by driver mutations that confer growth advantage on the cells carrying them and have been positively selected during tumorigenesis. Oncogene activation and tumor suppressor inactivation are two representatives of such driver mutations (Braig et al., 2005; Chen et al., 2005). However, additional mutations besides driver mutations are generally necessary for malignant transformation, because aberrant activation of oncogenic signaling triggers oncogene-induced senescence (OIS). For example, overexpression of oncogenic Ras in primary human fibroblasts induces hyperactivation of proliferation followed by senescence, which depends on ROS generation and the DDR. Consistently, treatment with the antioxidant *N*-acetyl-cysteine (NAC) or inhibition of the DDR suppresses OIS (Di Micco et al., 2006; Serrano, Lin, McCurrach, Beach, & Lowe, 1997). Although OIS shares several characteristics with other types of senescence, including replicative senescence, OIS can also have unique features depending on how OIS is activated. For example, oncogenic Ras expression induces OIS with full activation of the SASP, yet a tumor suppressor PTEN deletion drives OIS without the SASP (Perez-Mancera, Young, & Narita, 2014). Since the SASP is closely associated with several aspects of tumorigenesis, such differences should be considered for senescence-based cancer treatments.

4.1.2 Therapy-induced senescence

Most, if not all, chemotherapeutic agents exert their tumor suppressive effects by inducing cancer cell death or senescence. Chemotherapeutics activate senescence mostly by inhibiting DNA repair pathways or by causing DNA lesions. Doxorubicin, etoposide, and camptothecin are inhibitors of topoisomerase, which induce DNA breaks and cause torsional stress during replication. These drugs are effective in several types of cancers, including lung cancers, lymphoma, and breast cancers (Thorn et al., 2011). Bleomycin, another widely used chemotherapeutic agent, directly oxidizes DNA and causes DNA double- and single-strand breaks. It is effective in several cancers, including Hodgkin's lymphoma, non-Hodgkin's lymphoma, and testicular cancers (Chen, Ghorai, Kenney, & Stubbe, 2008; Scheule, Perkins, Hamilton, & Holian, 1992). Radiotherapy induces DNA lesions by transferring a high amount of energy through electrically charged ions, causing senescence. These types of treatments lack specificity and thus can be generally used, regardless of cancer type (Baskar, Lee, Yeo, & Yeoh, 2012). However, whether the p53 pathway is intact in cancer cells is crucial for effective treatment, as such DNA damage agents act through the p53 pathway to induce senescence or apoptosis in most cases. For example, in glioblastoma carrying loss-of-function p53 proteins, radiotherapy is marginally effective in inducing senescence (Quick & Gewirtz, 2006).

Targeted therapy for activating senescence in cancer has recently been developed, which can reduce potential side effects frequently observed in traditional chemotherapy, and is similarly effective in reducing tumor growth. Palbociclib, ribociclib, and abemaciclib are inhibitors of CDK4 and CDK6, which are essential for cell cycle progression from G1 to S phase (O'Leary, Finn, & Turner, 2016). Inhibition of CDK4/6 mimics the state of cell cycle arrest induced by p16, including increased senescence-associated beta-galactosidase (SA-β-Gal) activity and stable growth arrest. However, it is possible that such CDK4/6 inhibitor-induced senescence lacks the activation of the SASP axis, analogous to the one induced by overexpression of p16 (Coppe et al., 2011). Because the SASP can either promote or suppress tumorigenesis and the efficacy of tumor immune surveillance in a context-dependent manner, caution should be taken when considering this promising senescence-based targeted therapy.

4.1.3 Paracrine senescence and immune surveillance

In contrast to quiescent cells that also freeze their cell cycle, senescent cells are metabolically active and display a massive secretory phenotype.

The SASP exerts a pleiotropic effect of senescence, which includes two non-cell autonomous, tumor suppressive mechanisms: paracrine senescence and immune surveillance (Perez-Mancera et al., 2014). The SASP induces paracrine senescence in normal cells, where TGF-β secreted from primary senescent cells plays a major role in regulating p15 and p21 in normal receiving cells (Zhang, Alexander, & Wang, 2017). This paracrine activity ensures senescence growth arrest of potentially premalignant, yet partially damaged cells that are closely located to the primarily damaged cells (e.g., oncogene-activated senescent cells by cellular stressors), impacting tumor suppression (Acosta et al., 2013).

Another non-autonomous tumor suppressive mechanism of senescence is promoting tumor immune surveillance. Most cells become senescent upon oncogenic activation, yet a few may escape OIS and become malignant. In this situation, senescent cells secrete the SASP, which includes many pro-inflammatory cytokines and chemokines that recruit immune cells to clear themselves and their neighboring premalignant/malignant cells (Childs, Baker, Kirkland, Campisi, & van Deursen, 2014; Lee & Schmitt, 2019). Innate immune response and CD4 T cell-mediated adaptive immune response are involved in such immune surveillance (Liu, Ding, & Meng, 2018). Thus, when the SASP or immune system is disabled, tumorigenesis is accelerated in the hepatocarcinoma mouse model (Kang et al., 2011). In addition, it was recently reported that the SASP promotes tumor angiogenesis, which in turn increases the delivery of immune checkpoint inhibitors to the tumor region. This reveals an unexpected therapeutic opportunity generated by the SASP (Ruscetti et al., 2020). In this study, the authors utilized a KRAS mutant PDAC model, which is characterized by extensive desmoplasia that promotes hypovascularity and resistance to chemotherapy and immunotherapy. Co-treatment with MEK and CDK4/6 inhibitors can induce RB-dependent senescence, suppressing the proliferation of PDACs. This senescence-inducing therapy also promotes tumor vascularization through the SASP, which in turn enhances drug delivery and efficacy of cytotoxic gemcitabine chemotherapy. In addition, the SASP stimulates the accumulation of CD8 T cells into tumor regions, greatly increasing the efficacy of the PD-1 checkpoint blockade for PDACs. It will be interesting to examine whether such therapy-induced senescence can sensitize other immunologically "cold" tumors to chemotherapies and immunotherapies through the SASP-dependent effects on the tumor vasculature and immune system.

4.2 Tumor-promoting roles of senescence

As senescence growth arrest is stable, tumor-promoting functions of senescence are generally emanated from its non-cell autonomous function, the SASP (Coppe et al., 2010).

4.2.1 Growth factors

The SASP includes a wide range of growth factors, such as insulin growth factor (IGF), epidermal growth factor (EGF), fibroblast growth factor (FGF), hepatocyte growth factor (HGF), amphiregulin (AREG), epiregulin (EREG), and neuregulin (NRG) (Coppe et al., 2010). These growth factors can stimulate the proliferation of neoplastic cells in various tissues (Witsch, Sela, & Yarden, 2010). Accordingly, coinjection of senescent fibroblasts, but not normal fibroblasts, stimulates proliferation of malignant epithelial cells in vivo. Consistently, targeting the SASP pathway abrogates tumor growth stimulated by senescent cells in vivo (Laberge et al., 2015).

4.2.2 Stress resistance

One characteristic of senescent cells is their stress resistance. Interestingly, this stress resistance can be relayed to their neighboring cells, including cancer cells, via the SASP, especially under therapy-induced stressful conditions. For example, IL-6 produced from senescent cells in the mouse thymus protects lymphoma cells from chemotherapy. Inhibition of IL-6 increases the efficacy of chemotherapy in a mouse lymphoma model (Gilbert & Hemann, 2010).

4.2.3 Tumor stemness

Cancer stem cells (CSCs) exhibit characteristics of both stem cells and cancer cells, including self-renewal and differentiation capacities. CSCs barely proliferate under normal conditions, yet increase their proliferation to repopulate cancer cells in a context-dependent manner, including cancer treatment (Nguyen, Vanner, Dirks, & Eaves, 2012). Recent studies suggest that senescence promotes the formation of CSCs and tumor relapse after targeted cancer therapy (Kurppa et al., 2020; Milanovic et al., 2018). Inhibition of the EGFR signaling by osimertinib and trametinib causes tumor stemness followed by relapse in EGFR mutant non-small cell lung cancers. CSCs display several characteristics of senescence, including high levels of SA-beta-Gal activity, a flattened morphology, and activation of the SASP. Abrogation of such a stemness state prevents tumor relapse after cancer treatment, suggesting a crucial role for senescence-like cancer stem cells in tumorigenesis (Kurppa et al., 2020). How CSCs escape from senescence and resume their proliferation is unclear. However, it is possible that senescence is not fully installed under this condition as senescence is not a static status, but instead evolves over time.

4.2.4 Angiogenesis and metastasis

The SASP promotes tumor angiogenesis by secreting vascular endothelial growth factor (VEGF), which supports tumor cells with oxygen and nutrients. In fact, senescent fibroblasts increase tumor vasculature when they are coinjected into mice with poorly malignant epithelial EpH4 cells (Coppe, Kauser, Campisi, & Beausejour, 2006). In addition, the SASP can also promote EMT in tumor cells, potentially increasing their invasion. Consistently, conditioned media from senescent fibroblasts induces EMT in breast cancer cells, which is partially mediated by IL-6 and IL-8 (Ortiz-Montero, Londono-Vallejo, & Vernot, 2017). Senescent cells also secrete various extracellular matrix (ECM) remodeling factors, including matrix metalloproteinase 1 (MMP1), MMP2, and MMP3, which directly influence metastasis (Coppe et al., 2010; Deryugina & Quigley, 2006). For example, MMP1 and MMP2 from senescent skin fibroblast cells can activate a serine/threonine protein kinase PAR-1 on tumorigenic keratinocytes and stimulate their invasion (Malaquin et al., 2013). In addition, MMPs can stimulate growth factor signaling in cancer cells by releasing growth factors entangled with ECMs. In fact, treatment with an MMP inhibitor GM6001 reverses the growth-promoting effects of senescent cells on xenograft tumors (Liu & Hornsby, 2007).

4.2.5 Immune suppressive microenvironment

The SASP can also create immune suppressive environments in a context-dependent manner that drives tumorigenesis. This is somewhat perplexing, as the SASP is shown to be critical for tumor immune surveillance (Childs et al., 2014; Lee & Schmitt, 2019). The types of senescent cells (e.g., senescent cancer cells, senescent stromal cells) and the stage of tumorigenesis seem to be critical for understanding such complexity (Perez-Mancera et al., 2014). For example, in the hepatocellular carcinoma model, senescent stromal cells secrete CCL2, which promotes the recruitment of immunosuppressive myeloid cells. This in turn restrains NK-driven tumor cell clearance (Eggert et al., 2016). Other SASP factors that exert immunosuppressive functions are IL-6, IL-10, and TGF-β in the

PTEN-loss prostate cancer model. These SASP factors inhibit NK cells, recruit regulatory T cells, and polarize macrophages to the M2 state, suppressing tumor immune surveillance (Toso et al., 2014).

4.3 Senolytics and senostatics as anti-cancer strategies

As in the case of autophagy, senescence also has dual roles in tumorigenesis (Fig. 2). Therefore, to improve the efficacy of senescence-based cancer therapies, it is necessary to consider how to decrease the deleterious effects of senescence in tumorigenesis. In other words, it is a balancing act between the beneficial effects of senescence (e.g., tumor cell senescence, paracrine senescence, and tumor immune surveillance) and its deleterious effects (e.g., stromal cell senescence, immune suppression, and resistance to therapy). Currently, two approaches are available for this purpose: senolytics and senostatics (Kang, 2019; Short, Fielder, Miwa, & von Zglinicki, 2019). Senolytics are agents that selectively kill senescent cells. Analogous to the ones used for selectively killing cancer cells, senolytics generally target survival pathways specific to senescent cells. For example, navitoclax (also known as ABT-263) targets an anti-apoptotic BCL2 family of proteins that increases during senescence, inducing apoptosis in senescent cells (Zhu et al., 2016). Dasatinib and quercetin induce apoptosis of senescent cells by targeting the upregulated dependence receptor and

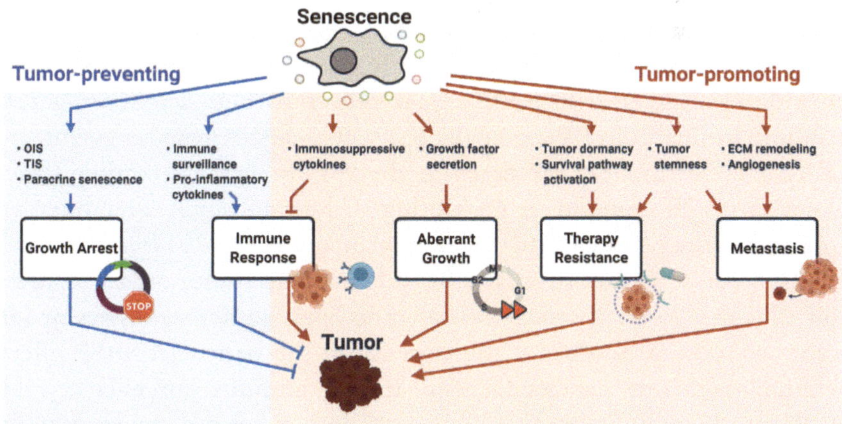

Fig. 2 Dual roles of senescence in tumorigenesis. Senescence prevents tumorigenesis by (1) inhibiting cell growth and (2) promoting immune surveillance. Senescence promotes tumorigenesis by (1) providing tumor cells with trophic support, (2) promoting resistance to cancer therapy, (3) creating tumor-suppressive microenvironment, and (4) enhancing invasion and metastasis (created with BioRender.com).

PI3K/Akt pathways (Zhu et al., 2015). Although these first generation of senolytics have their own limitations, such as off-target effects and inconsistent efficacy depending on the type of senescent cells, they display promising effects on several age-related disease models, including cancer metastasis, idiopathic pulmonary fibrosis, atherosclerosis, liver cirrhosis, glomerulosclerosis, diabetes, sarcopenia, and osteoarthritis (Kang, 2019; Short et al., 2019). Senostatics are agents that do not kill senescent cells but modulate the SASP, especially its proinflammatory secretome, to dampen the deleterious, non-cell autonomous effects of senescence. The development of senostatics is at a very early stage, and no specific senostatics are available. Currently, inhibitors of NF-κB, multiple flavonoids and polyphenols, and inhibitors of the mTOR pathway are proposed to act as senostatics (Short et al., 2019). However, caution should be taken when considering some of these inhibitors for cancer treatment, as they can also affect other critical biological processes. For example, inhibitors of the mTOR pathway activate general autophagy (Dunlop & Tee, 2014), which could enhance tumor survival in a context-dependent manner, as described above. As we have recently gained insights into how the SASP is regulated, more tailored therapeutic strategies for targeting the SASP are expected to be discovered in the near future.

When senolytics and senostatics are applied to improve cancer treatment, the following information should be carefully considered: (1) the types of senescent cells involved; (2) the stages of tumorigenesis; (3) the types of immune cells recruited by senescent cells; and (4) the age of patients. Senescence-inducing therapies (e.g., chemotherapy, radiotherapy) can induce senescence in both cancer cells and their neighboring cells, including stromal cells (Wang et al., 2020). The SASP from senescent stromal cells either induces resistance to therapy by either providing cancer cells with trophic support or promoting an immunosuppressive microenvironment (Coppe et al., 2010; Perez-Mancera et al., 2014). Thus, if the ratio between senescent stromal cells and senescent tumor cells is relatively high after treatment, it would be better to apply either senolytics or senostatics as soon as possible at the expense of the beneficial SASP effects, including paracrine senescence and tumor immune surveillance. The SASP can promote cancer cell invasion and metastasis, suggesting that senescence induction in fully established cancers may increase the possibility of devastating metastasis. Therefore, in this case, senolytics and senostatics should be applied immediately after cancer treatment. Alternatively, inhibition of metastasis should be considered together at the treatment of senescence-inducing therapy. However, at the early stage of cancer, it would be better to allow the beneficial effects of the SASP to be maximized first, following which senolytics and senostatics can be applied as a failsafe treatment. Another consideration is which immune modulatory axis of the SASP is ensued upon senescence-inducing therapy. The SASP includes both immune stimulating and suppressing factors, depending on senescent cell types, senescence-inducing stimuli, and the senescent stage. Thus, after senescence-inducing therapy, if immunosuppressive SASP factors such as IL-10 and TGF-β or tumor-promoting immune cells (e.g., M2 macrophages, regulatory T cells) are enriched in the tumor microenvironment, it would be better to apply senolytics and senostatics as soon as possible. Senolytics and senostatics can also be applied before cancer treatment to improve the tumor environment. For example, in aged patients, senescent stromal cells are already accumulated in the tumor region, decreasing the efficacy of senescence-inducing therapy (Baker et al., 2016). In this case, senolytics and senostatics can remove senescent stromal cells, suppressing the tumor-supporting microenvironment. Then, senescence-inducing therapy can be applied, and whether additional rounds of senolytics and senostatics are necessary can be determined after assessing the treatment response, as described above.

5. Interplay between autophagy and senescence

Many stressors that activate autophagy can also modulate senescence, and vice versa (Gewirtz, 2013). Indeed, autophagy has been shown to be a critical effector of senescence (Chan & Narita, 2019; Gorgoulis et al., 2019). This intertwined relationship should therefore be considered when autophagy and senescence are employed as therapeutic targets for cancer treatment.

5.1 Pro-senescence and anti-senescence roles of autophagy

The relationship between autophagy and senescence seems to be close to that between autophagy and cancer. Before senescence is induced (under normal conditions), autophagy acts to maintain cellular homeostasis, suppressing senescence in a passive manner. For example, autophagy targets dysfunctional mitochondria and protein aggregates for degradation to maintain redox homeostasis and proteostasis. In addition, autophagy maintains lysosomal integrity and resolves stress granules. Autophagy can also suppress senescence by directly targeting critical regulators of senescence.

For example, p62-dependent selective autophagy degrades a key regulator of the SASP, GATA4, suppressing the SASP and senescence (Kang & Elledge, 2016).

Once senescence is triggered, however, autophagy can contribute to senescence establishment. Autophagy functions as "non-senescence addiction" wherein senescence relies on autophagy, which is not a senescence-inducing process by itself. This dampens several senescence-associated stresses, thus maintaining the viability of senescent cells (Kwon, Kim, Jeoung, Kim, & Kang, 2017). In addition, autophagy provides amino acids to support the massive synthesis of SASP factors through the formation of TASCC (Narita et al., 2011). Lastly, autophagy can target negative regulators of senescence for degradation, thus actively causing senescence. Δ133p53α, an alternative splicing form of p53 (Horikawa et al., 2014), the nuclear lamina component Lamin B1 (Dou et al., 2015), and the histone deacetylase SIRT1 are such autophagy substrate targets (Xu et al., 2020). How such autophagy is modulated and why they need to be degraded during senescence are incompletely understood.

5.2 Context-specific regulation of autophagy and senescence for cancer treatment

As autophagy can either suppress or induce senescence in a context-dependent manner, targeting autophagy for cancer treatment should be carefully considered after examining whether it can modulate senescence under the same conditions. For example, autophagy inhibition could decrease the efficacy of chemotherapies to induce senescence. In addition, as autophagy is necessary for senescence maintenance, its inhibition after senescence-inducing treatment could escape severely damaged, potentially premalignant cells from senescence, rather accelerating, not suppressing, tumorigenesis.

Another consideration is the role of autophagy in controlling the SASP. Autophagy inhibition may potentiate activation of the SASP by accumulating GATA4 and sustaining cytoplasmic chromatin fragments that can be sensed by the cGAS-STING pathway (De Cecco et al., 2019; Dou et al., 2017; Gluck et al., 2017; Kang et al., 2015; Li & Chen, 2018; Yang, Wang, Ren, Chen, & Chen, 2017). Thus, it can boost paracrine senescence and immune surveillance after senescence-inducing treatment for cancer. Since GATA4 and the cGAS-STING pathways are mostly involved in the proinflammatory axis of the SASP, whether such modulation may increase immunosuppressive secretomes would be less of a concern; however, caution should be taken when applying these to treatment.

Autophagy modulation could also be considered for senolytics. Since senescent cells undergo very similar stress responses observed in cancer cells, including metabolic rewiring, proteotoxic stress, DNA damage, and oxidative stress, they might be vulnerable to autophagy inhibition (Soto-Gamez, Quax, & Demaria, 2019). In fact, autophagy inhibition sensitizes senescent cells to cell death by depleting cellular nutrients and energy (Dorr et al., 2013). This could provide additional options for enhancing senolytics during cancer treatment.

6. Concluding remarks and perspectives

The studies discussed above clearly demonstrate the complex functions of autophagy and senescence during tumorigenesis. Autophagy and senescence can either promote or suppress tumorigenesis in a context-dependent manner. Thus, we need to fully understand when, how, and why autophagy and senescence are activated during tumorigenesis and cancer treatment. For example, for over a decade, we have largely viewed senescence as stable growth arrest that suppresses tumorigenesis. Recent studies linking senescence to secretion have begun to highlight a completely novel non-cell autonomous role for senescence in tumorigenesis: paracrine senescence, immune surveillance and modulation, and trophic support. Further elucidation of autophagy and senescence programs, including their intertwined relationship, should provide insight into new therapeutic strategies for modulating autophagy and senescence in cancer.

Acknowledgments

C.K. is supported by the Suh Kyungbae Foundation (SUHF-17020068) and the National Research Foundation of Korea (NRF-2019R1C1C1006386). M.-S.K. is supported by the National Research Foundation of Korea (NRF-2017R1A6A3A11032039, NRF-2020R1I1A1A01072779).

References

Acosta, J. C., Banito, A., Wuestefeld, T., Georgilis, A., Janich, P., Morton, J. P., et al. (2013). A complex secretory program orchestrated by the inflammasome controls paracrine senescence. *Nature Cell Biology*, *15*(8), 978–990.

Acosta, J. C., O'Loghlen, A., Banito, A., Guijarro, M. V., Augert, A., Raguz, S., et al. (2008). Chemokine signaling via the CXCR2 receptor reinforces senescence. *Cell*, *133*(6), 1006–1018.

Aird, K. M., Iwasaki, O., Kossenkov, A. V., Tanizawa, H., Fatkhutdinov, N., Bitler, B. G., et al. (2016). HMGB2 orchestrates the chromatin landscape of senescence-associated secretory phenotype gene loci. *The Journal of Cell Biology*, *215*(3), 325–334.

Anding, A. L., & Baehrecke, E. H. (2017). Cleaning house: Selective autophagy of organelles. *Developmental Cell*, *41*(1), 10–22.

Baginska, J., Viry, E., Berchem, G., Poli, A., Noman, M. Z., van Moer, K., et al. (2013). Granzyme B degradation by autophagy decreases tumor cell susceptibility to natural killer-mediated lysis under hypoxia. *Proceedings of the National Academy of Sciences of the United States of America*, *110*(43), 17450–17455.

Baker, D. J., Childs, B. G., Durik, M., Wijers, M. E., Sieben, C. J., Zhong, J., et al. (2016). Naturally occurring p16(Ink4a)-positive cells shorten healthy lifespan. *Nature*, *530*(7589), 184–189.

Basisty, N., Kale, A., Jeon, O. H., Kuehnemann, C., Payne, T., Rao, C., et al. (2020). A proteomic atlas of senescence-associated secretomes for aging biomarker development. *PLoS Biology*, *18*(1), e3000599.

Baskar, R., Lee, K. A., Yeo, R., & Yeoh, K. W. (2012). Cancer and radiation therapy: Current advances and future directions. *International Journal of Medical Sciences*, *9*(3), 193–199.

Bataller, R., & Brenner, D. A. (2005). Liver fibrosis. *The Journal of Clinical Investigation*, *115*(2), 209–218.

Belaid, A., Ndiaye, P. D., Klionsky, D. J., Hofman, P., & Mograbi, B. (2013). Signalphagy: Scheduled signal termination by macroautophagy. *Autophagy*, *9*(10), 1629–1630.

Braig, M., Lee, S., Loddenkemper, C., Rudolph, C., Peters, A. H. F. M., Schlegelberger, B., et al. (2005). Oncogene-induced senescence as an initial barrier in lymphoma development. *Nature*, *436*(7051), 660–665.

Brighton, P. J., Maruyama, Y., Fishwick, K., Vrljicak, P., Tewary, S., Fujihara, R., et al. (2017). Clearance of senescent decidual cells by uterine natural killer cells in cycling human endometrium. *eLife*, *6*, e31274.

Budanov, A. V., & Karin, M. (2008). p53 target genes sestrin1 and sestrin2 connect genotoxic stress and mTOR signaling. *Cell*, *134*(3), 451–460.

Chan, A. S. L., & Narita, M. (2019). Short-term gain, long-term pain: The senescence life cycle and cancer. *Genes & Development*, *33*(3–4), 127–143.

Chen, J. Y., Ghorai, M. K., Kenney, G., & Stubbe, J. (2008). Mechanistic studies on bleomycin-mediated DNA damage: Multiple binding modes can result in double-stranded DNA cleavage. *Nucleic Acids Research*, *36*(11), 3781–3790.

Chen, Z. B., Trotman, L. C., Shaffer, D., Lin, H. K., Dotan, Z. A., Niki, M., et al. (2005). Crucial role of p53-dependent cellular senescence in suppression of Pten-deficient tumorigenesis. *Nature*, *436*(7051), 725–730.

Chien, Y., Scuoppo, C., Wang, X., Fang, X., Balgley, B., Bolden, J. E., et al. (2011). Control of the senescence-associated secretory phenotype by NF-kappaB promotes senescence and enhances chemosensitivity. *Genes & Development*, *25*(20), 2125–2136.

Childs, B. G., Baker, D. J., Kirkland, J. L., Campisi, J., & van Deursen, J. M. (2014). Senescence and apoptosis: Dueling or complementary cell fates? *EMBO Reports*, *15*(11), 1139–1153.

Chino, H., & Mizushima, N. (2020). ER-Phagy: Quality control and turnover of endoplasmic reticulum. *Trends in Cell Biology*, *30*(5), 384–398.

Cipak Gasparovic, A., Milkovic, L., Dandachi, N., Stanzer, S., Pezdirc, I., Vrancic, J., et al. (2019). Chronic oxidative stress promotes molecular changes associated with epithelial mesenchymal transition, NRF2, and breast Cancer stem cell phenotype. *Antioxidants (Basel)*, *8*(12), 633.

Coppe, J. P., Desprez, P. Y., Krtolica, A., & Campisi, J. (2010). The senescence-associated secretory phenotype: The dark side of tumor suppression. *Annual Review of Pathology*, *5*, 99–118.

Coppe, J. P., Kauser, K., Campisi, J., & Beausejour, C. M. (2006). Secretion of vascular endothelial growth factor by primary human fibroblasts at senescence. *Journal of Biological Chemistry*, *281*(40), 29568–29574.

Coppe, J. P., Patil, C. K., Rodier, F., Sun, Y., Munoz, D. P., Goldstein, J., et al. (2008). Senescence-associated secretory phenotypes reveal cell-nonautonomous functions of oncogenic RAS and the p53 tumor suppressor. *PLoS Biology*, *6*(12), 2853–2868.

Coppe, J. P., Rodier, F., Patil, C. K., Freund, A., Desprez, P. Y., & Campisi, J. (2011). Tumor suppressor and aging biomarker p16(INK4a) induces cellular senescence without the associated inflammatory secretory phenotype. *Journal of Biological Chemistry*, *286*(42), 36396–36403.

Crotzer, V. L., & Blum, J. S. (2009). Autophagy and its role in MHC-mediated antigen presentation. *Journal of Immunology*, *182*(6), 3335–3341.

Dasari, S. K., Bialik, S., Levin-Zaidman, S., Levin-Salomon, V., Merrill, A. H., Jr., Futerman, A. H., et al. (2017). Signalome-wide RNAi screen identifies GBA1 as a positive mediator of autophagic cell death. *Cell Death and Differentiation*, *24*(7), 1288–1302.

De Cecco, M., Ito, T., Petrashen, A. P., Elias, A. E., Skvir, N. J., Criscione, S. W., et al. (2019). L1 drives IFN in senescent cells and promotes age-associated inflammation. *Nature*, *566*(7742), 73.

Degenhardt, K., Mathew, R., Beaudoin, B., Bray, K., Anderson, D., Chen, G., et al. (2006). Autophagy promotes tumor cell survival and restricts necrosis, inflammation, and tumorigenesis. *Cancer Cell*, *10*(1), 51–64.

Dejeans, N., Barroso, K., Fernandez-Zapico, M. E., Samali, A., & Chevet, E. (2015). Novel roles of the unfolded protein response in the control of tumor development and aggressiveness. *Seminars in Cancer Biology*, *33*, 67–73.

Demaria, M., Ohtani, N., Youssef, S. A., Rodier, F., Toussaint, W., Mitchell, J. R., et al. (2014). An essential role for senescent cells in optimal wound healing through secretion of PDGF-AA. *Developmental Cell*, *31*(6), 722–733.

DeNicola, G. M., Karreth, F. A., Humpton, T. J., Gopinathan, A., Wei, C., Frese, K., et al. (2011). Oncogene-induced Nrf2 transcription promotes ROS detoxification and tumorigenesis. *Nature*, *475*(7354), 106–109.

Denton, D., & Kumar, S. (2019). Autophagy-dependent cell death. *Cell Death and Differentiation*, *26*(4), 605–616.

Deryugina, E. I., & Quigley, J. P. (2006). Matrix metalloproteinases and tumor metastasis. *Cancer and Metastasis Reviews*, *25*(1), 9–34.

DesMarais, T. L., & Costa, M. (2019). Mechanisms of chromium-induced toxicity. *Current Opinion in Toxicology*, *14*, 1–7.

Di Micco, R., Fumagalli, M., Cicalese, A., Piccinin, S., Gasparini, P., Luise, C., et al. (2006). Oncogene-induced senescence is a DNA damage response triggered by DNA hyper-replication. *Nature*, *444*(7119), 638–642.

Doria, A., Gatto, M., & Punzi, L. (2013). Autophagy in human health and disease. *New England Journal of Medicine*, *368*(19), 1845.

Dorr, J. R., Yu, Y., Milanovic, M., Beuster, G., Zasada, C., Dabritz, J. H., et al. (2013). Synthetic lethal metabolic targeting of cellular senescence in cancer therapy. *Nature*, *501*(7467), 421–425.

Dou, Z. X., Ghosh, K., Vizioli, M. G., Zhu, J. J., Sen, P., Wangensteen, K. J., et al. (2017). Cytoplasmic chromatin triggers inflammation in senescence and cancer. *Nature*, *550*(7676), 402–406.

Dou, Z., Xu, C., Donahue, G., Shimi, T., Pan, J. A., Zhu, J., et al. (2015). Autophagy mediates degradation of nuclear lamina. *Nature*, *527*(7576), 105–109.

Dunlop, E. A., & Tee, A. R. (2014). mTOR and autophagy: A dynamic relationship governed by nutrients and energy. *Seminars in Cell & Developmental Biology*, *36*, 121–129.

Egashira, M., Hirota, Y., Shimizu-Hirota, R., Saito-Fujita, T., Haraguchi, H., Matsumoto, L., et al. (2017). F4/80+ macrophages contribute to clearance of senescent cells in the mouse postpartum uterus. *Endocrinology*, *158*(7), 2344–2353.

Eggert, T., Wolter, K., Ji, J. L., Ma, C., Yevsa, T., Klotz, S., et al. (2016). Distinct functions of senescence-associated immune responses in liver tumor surveillance and tumor progression. *Cancer Cell*, *30*(4), 533–547.
Elgendy, M., Sheridan, C., Brumatti, G., & Martin, S. J. (2011). Oncogenic Ras-induced expression of Noxa and Beclin-1 promotes autophagic cell death and limits clonogenic survival. *Molecular Cell*, *42*(1), 23–35.
Eliopoulos, A. G., Havaki, S., & Gorgoulis, V. G. (2016). DNA damage response and autophagy: A meaningful partnership. *Frontiers in Genetics*, 7, 204.
Filomeni, G., De Zio, D., & Cecconi, F. (2015). Oxidative stress and autophagy: The clash between damage and metabolic needs. *Cell Death and Differentiation*, *22*(3), 377–388.
Galluzzi, L., Baehrecke, E. H., Ballabio, A., Boya, P., Bravo-San Pedro, J. M., Cecconi, F., et al. (2017). Molecular definitions of autophagy and related processes. *The EMBO Journal*, *36*(13), 1811–1836.
Galluzzi, L., Pietrocola, F., Bravo-San Pedro, J. M., Amaravadi, R. K., Baehrecke, E. H., Cecconi, F., et al. (2015). Autophagy in malignant transformation and cancer progression. *The EMBO Journal*, *34*(7), 856–880.
Gewirtz, D. A. (2013). Autophagy and senescence: A partnership in search of definition. *Autophagy*, *9*(5), 808–812.
Gilbert, L. A., & Hemann, M. T. (2010). DNA damage-mediated induction of a chemoresistant niche. *Cell*, *143*(3), 355–366.
Gillespie, D. A., & Ryan, K. M. (2016). Autophagy is critically required for DNA repair by homologous recombination. *Molecular & Cellular Oncology*, *3*(1), e1030538.
Gluck, S., Guey, B., Gulen, M. F., Wolter, K., Kang, T. W., Schmacke, N. A., et al. (2017). Innate immune sensing of cytosolic chromatin fragments through cGAS promotes senescence. *Nature Cell Biology*, *19*(9), 1061–1070.
Goldsmith, J., Levine, B., & Debnath, J. (2014). Autophagy and cancer metabolism. *Methods in Enzymology*, *542*, 25–57.
Gomes, L. R., Menck, C. F. M., & Leandro, G. S. (2017). Autophagy roles in the modulation of DNA repair pathways. *International Journal of Molecular Sciences*, *18*(11), 2351.
Gonzalez-Meljem, J. M., Apps, J. R., Fraser, H. C., & Martinez-Barbera, J. P. (2018). Paracrine roles of cellular senescence in promoting tumourigenesis. *British Journal of Cancer*, *118*(10), 1283–1288.
Gorgoulis, V., Adams, P. D., Alimonti, A., Bennett, D. C., Bischof, O., Bishop, C., et al. (2019). Cellular senescence: Defining a path forward. *Cell*, *179*(4), 813–827.
Guo, J. Y., Chen, H. Y., Mathew, R., Fan, J., Strohecker, A. M., Karsli-Uzunbas, G., et al. (2011). Activated Ras requires autophagy to maintain oxidative metabolism and tumorigenesis. *Genes & Development*, *25*(5), 460–470.
Hanahan, D., & Weinberg, R. A. (2000). The hallmarks of cancer. *Cell*, *100*(1), 57–70.
Hanahan, D., & Weinberg, R. A. (2011). Hallmarks of cancer: The next generation. *Cell*, *144*(5), 646–674.
Harley, C. B., Futcher, A. B., & Greider, C. W. (1990). Telomeres shorten during ageing of human fibroblasts. *Nature*, *345*(6274), 458–460.
Hayes, J. D., & McMahon, M. (2009). NRF2 and KEAP1 mutations: Permanent activation of an adaptive response in cancer. *Trends in Biochemical Sciences*, *34*(4), 176–188.
He, S., & Sharpless, N. E. (2017). Senescence in health and disease. *Cell*, *169*(6), 1000–1011.
Heath-Engel, H. M., Chang, N. C., & Shore, G. C. (2008). The endoplasmic reticulum in apoptosis and autophagy: Role of the BCL-2 protein family. *Oncogene*, 27(50), 6419–6433.
Hernandez-Segura, A., de Jong, T. V., Melov, S., Guryev, V., Campisi, J., & Demaria, M. (2017). Unmasking transcriptional heterogeneity in senescent cells. *Current Biology*, 27(17), 2652–2660 (e2654).

Hernandez-Segura, A., Nehme, J., & Demaria, M. (2018). Hallmarks of cellular senescence. *Trends in Cell Biology*, *28*(6), 436–453.

Herranz, N., Gallage, S., Mellone, M., Wuestefeld, T., Klotz, S., Hanley, C. J., et al. (2015). mTOR regulates MAPKAPK2 translation to control the senescence-associated secretory phenotype. *Nature Cell Biology*, *17*(9), 1205.

Herranz, N., & Gil, J. (2018). Mechanisms and functions of cellular senescence. *Journal of Clinical Investigation*, *128*(4), 1238–1246.

Hewitt, G., Carroll, B., Sarallah, R., Correia-Melo, C., Ogrodnik, M., Nelson, G., et al. (2016). SQSTM1/p62 mediates crosstalk between autophagy and the UPS in DNA repair. *Autophagy*, *12*(10), 1917–1930.

Higuchi-Sanabria, R., Frankino, P. A., Paul, J. W., 3rd, Tronnes, S. U., & Dillin, A. (2018). A futile Battle? Protein quality control and the stress of aging. *Developmental Cell*, *44*(2), 139–163.

Hoare, M., Ito, Y., Kang, T. W., Weekes, M. P., Matheson, N. J., Patten, D. A., et al. (2016). NOTCH1 mediates a switch between two distinct secretomes during senescence. *Nature Cell Biology*, *18*(9), 979–992.

Horikawa, I., Fujita, K., Jenkins, L. M., Hiyoshi, Y., Mondal, A. M., Vojtesek, B., et al. (2014). Autophagic degradation of the inhibitory p53 isoform Delta133p53alpha as a regulatory mechanism for p53-mediated senescence. *Nature Communications*, *5*, 4706.

Hsu, S. K., Chiu, C. C., Dahms, H. U., Chou, C. K., Cheng, C. M., Chang, W. T., et al. (2019). Unfolded protein response (UPR) in survival, dormancy, immunosuppression, metastasis, and treatments of Cancer cells. *International Journal of Molecular Sciences*, *20*(10), 2518.

Inami, Y., Waguri, S., Sakamoto, A., Kouno, T., Nakada, K., Hino, O., et al. (2011). Persistent activation of Nrf2 through p62 in hepatocellular carcinoma cells. *The Journal of Cell Biology*, *193*(2), 275–284.

Ji, C. H., & Kwon, Y. T. (2017). Crosstalk and interplay between the ubiquitin-proteasome system and autophagy. *Molecules and Cells*, *40*(7), 441–449.

Johansen, T., & Lamark, T. (2020). Selective autophagy: ATG8 family proteins, LIR motifs and cargo receptors. *Journal of Molecular Biology*, *432*(1), 80–103.

Kang, C. (2019). Senolytics and senostatics: A two-pronged approach to target cellular senescence for delaying aging and age-related diseases. *Molecules and Cells*, *42*(12), 821–827.

Kang, C., & Elledge, S. J. (2016). How autophagy both activates and inhibits cellular senescence. *Autophagy*, *12*(5), 898–899.

Kang, R., Tang, D., Schapiro, N. E., Livesey, K. M., Farkas, A., Loughran, P., et al. (2010). The receptor for advanced glycation end products (RAGE) sustains autophagy and limits apoptosis, promoting pancreatic tumor cell survival. *Cell Death and Differentiation*, *17*(4), 666–676.

Kang, C., Xu, Q. K., Martin, T. D., Li, M. Z., Demaria, M., Aron, L., et al. (2015). The DNA damage response induces inflammation and senescence by inhibiting autophagy of GATA4. *Science*, *349*(6255), aaa5612.

Kang, T. W., Yevsa, T., Woller, N., Hoenicke, L., Wuestefeld, T., Dauch, D., et al. (2011). Senescence surveillance of pre-malignant hepatocytes limits liver cancer development. *Nature*, *479*(7374), 547–551.

Kenzelmann Broz, D., Spano Mello, S., Bieging, K. T., Jiang, D., Dusek, R. L., Brady, C. A., et al. (2013). Global genomic profiling reveals an extensive p53-regulated autophagy program contributing to key p53 responses. *Genes & Development*, *27*(9), 1016–1031.

Kirkin, V. (2020). History of the selective autophagy research: How did it begin and where does it stand today? *Journal of Molecular Biology*, *432*(1), 3–27.

Kirkland, J. L., & Tchkonia, T. (2017). Cellular senescence: A translational perspective. *eBioMedicine*, *21*, 21–28.

Komatsu, M., Kurokawa, H., Waguri, S., Taguchi, K., Kobayashi, A., Ichimura, Y., et al. (2010). The selective autophagy substrate p62 activates the stress responsive transcription factor Nrf2 through inactivation of Keap1. *Nature Cell Biology*, *12*(3), 213–U217.

Krizhanovsky, V., Yon, M., Dickins, R. A., Hearn, S., Simon, J., Miething, C., et al. (2008). Senescence of activated stellate cells limits liver fibrosis. *Cell*, *134*(4), 657–667.

Kroemer, G., Marino, G., & Levine, B. (2010). Autophagy and the integrated stress response. *Molecular Cell*, *40*(2), 280–293.

Kuilman, T., Michaloglou, C., Mooi, W. J., & Peeper, D. S. (2010). The essence of senescence. *Genes & Development*, *24*(22), 2463–2479.

Kurppa, K. J., Liu, Y., To, C, Zhang, T. H., Fan, M. Y., Vajdi, A., et al. (2020). Treatment-induced tumor dormancy through YAP-mediated transcriptional reprogramming of the apoptotic pathway. *Cancer Cell*, *37*(1), 104.

Kwon, Y., Kim, J. W., Jeoung, J. A., Kim, M. S., & Kang, C. (2017). Autophagy is pro-senescence when seen in close-up, but anti-senescence in long-shot. *Molecules and Cells*, *40*(9), 607–612.

Laberge, R. M., Sun, Y., Orjalo, A. V., Patil, C. K., Freund, A., Zhou, L., et al. (2015). MTOR regulates the pro-tumorigenic senescence-associated secretory phenotype by promoting IL1A translation. *Nature Cell Biology*, *17*(8), 1049–1061.

Lee, S., & Schmitt, C. A. (2019). The dynamic nature of senescence in cancer. *Nature Cell Biology*, *21*(1), 94–101.

Leidal, A. M., Levine, B., & Debnath, J. (2018). Autophagy and the cell biology of age-related disease. *Nature Cell Biology*, *20*(12), 1338–1348.

Levine, B., & Klionsky, D. J. (2004). Development by self-digestion: Molecular mechanisms and biological functions of autophagy. *Developmental Cell*, *6*(4), 463–477.

Levine, B., & Kroemer, G. (2019). Biological functions of autophagy genes: A disease perspective. *Cell*, *176*(1–2), 11–42.

Li, T., & Chen, Z. J. (2018). The cGAS-cGAMP-STING pathway connects DNA damage to inflammation, senescence, and cancer. *The Journal of Experimental Medicine*, *215*(5), 1287–1299.

Li, Y., Wang, L. X., Yang, G., Hao, F., Urba, W. J., & Hu, H. M. (2008). Efficient cross-presentation depends on autophagy in tumor cells. *Cancer Research*, *68*(17), 6889–6895.

Liu, X. L., Ding, J., & Meng, L. H. (2018). Oncogene-induced senescence: A double edged sword in cancer. *Acta Pharmacologica Sinica*, *39*(10), 1553–1558.

Liu, D., & Hornsby, P. J. (2007). Senescent human fibroblasts increase the early growth of xenograft tumors via matrix metalloproteinase secretion. *Cancer Research*, *67*(7), 3117–3126.

Liu, Y., Shoji-Kawata, S., Sumpter, R. M., Jr., Wei, Y., Ginet, V., Zhang, L., et al. (2013). Autosis is a Na+,K+-ATPase-regulated form of cell death triggered by autophagy-inducing peptides, starvation, and hypoxia-ischemia. *Proceedings of the National Academy of Sciences of the United States of America*, *110*(51), 20364–20371.

Liu, E. Y., Xu, N., O'Prey, J., Lao, L. Y., Joshi, S., Long, J. S., et al. (2015). Loss of autophagy causes a synthetic lethal deficiency in DNA repair. *Proceedings of the National Academy of Sciences of the United States of America*, *112*(3), 773–778.

Lock, R., Roy, S., Kenific, C. M., Su, J. S., Salas, E., Ronen, S. M., et al. (2011). Autophagy facilitates glycolysis during Ras-mediated oncogenic transformation. *Molecular Biology of the Cell*, *22*(2), 165–178.

Loi, M., Fregno, I., Guerra, C., & Molinari, M. (2018). Eat it right: ER-phagy and recovER-phagy. *Biochemical Society Transactions*, *46*(3), 699–706.

Losson, H., Schnekenburger, M., Dicato, M., & Diederich, M. (2020). HDAC6-an emerging target against chronic myeloid leukemia? *Cancers (Basel)*, *12*(2), 318.

Lujambio, A., Akkari, L., Simon, J., Grace, D., Tschaharganeh, D. F., Bolden, J. E., et al. (2013). Non-cell-autonomous tumor suppression by p53. *Cell*, *153*(2), 449–460.

Malaquin, N., Vercamer, C., Bouali, F., Martien, S., Deruy, E., Wernert, N., et al. (2013). Senescent fibroblasts enhance early skin carcinogenic events via a paracrine MMP-PAR-1 Axis. *PLoS One*, *8*(5), e63607.

Mazzucco, A. E., Smogorzewska, A., Kang, C., Luo, J., Schlabach, M. R., Xu, Q. K., et al. (2017). Genetic interrogation of replicative senescence uncovers a dual role for USP28 in coordinating the p53 and GATA4 branches of the senescence program. *Genes & Development*, *31*(19), 1933–1938.

Mevorach, D., Trahtemberg, U., Krispin, A., Attalah, M., Zazoun, J., Tabib, A., et al. (2010). What do we mean when we write "senescence," "apoptosis," "necrosis," or "clearance of dying cells"? *Annals of the New York Academy of Sciences*, *1209*, 1–9.

Michaud, M., Martins, I., Sukkurwala, A. Q., Adjemian, S., Ma, Y., Pellegatti, P., et al. (2011). Autophagy-dependent anticancer immune responses induced by chemotherapeutic agents in mice. *Science*, *334*(6062), 1573–1577.

Milanovic, M., Fan, D. N. Y., Belenki, D., Dabritz, J. H. M., Zhao, Z., Yu, Y., et al. (2018). Senescence-associated reprogramming promotes cancer stemness. *Nature*, *553*(7686), 96–100.

Mitsuishi, Y., Taguchi, K., Kawatani, Y., Shibata, T., Nukiwa, T., Aburatani, H., et al. (2012). Nrf2 redirects glucose and glutamine into anabolic pathways in metabolic reprogramming. *Cancer Cell*, *22*(1), 66–79.

Mizushima, N., & Komatsu, M. (2011). Autophagy: Renovation of cells and tissues. *Cell*, *147*(4), 728–741.

Morin, A., Letouze, E., Gimenez-Roqueplo, A. P., & Favier, J. (2014). Oncometabolites-driven tumorigenesis: From genetics to targeted therapy. *International Journal of Cancer*, *135*(10), 2237–2248.

Mosteiro, L., Pantoja, C., Alcazar, N., Marion, R. M., Chondronasiou, D., Rovira, M., et al. (2016). Tissue damage and senescence provide critical signals for cellular reprogramming in vivo. *Science*, *354*(6315), aaf4445.

Mulcahy Levy, J. M., & Thorburn, A. (2020). Autophagy in cancer: Moving from understanding mechanism to improving therapy responses in patients. *Cell Death and Differentiation*, *27*(3), 843–857.

Munoz, D. P., Yannone, S. M., Daemen, A., Sun, Y., Vakar-Lopez, F., Kawahara, M., et al. (2019). Targetable mechanisms driving immunoevasion of persistent senescent cells link chemotherapy-resistant cancer to aging. *JCI Insight*, *5*, e124716.

Munoz-Espin, D., & Serrano, M. (2014). Cellular senescence: From physiology to pathology. *Nature Reviews. Molecular Cell Biology*, *15*(7), 482–496.

Munz, C. (2016). Autophagy beyond intracellular MHC class II antigen presentation. *Trends in Immunology*, *37*(11), 755–763.

Murrow, L., & Debnath, J. (2013). Autophagy as a stress-response and quality-control mechanism: Implications for cell injury and human disease. *Annual Review of Pathology: Mechanisms of Disease*, *8*(8), 105–137.

Narendra, D., Tanaka, A., Suen, D. F., & Youle, R. J. (2008). Parkin is recruited selectively to impaired mitochondria and promotes their autophagy. *Journal of Cell Biology*, *183*(5), 795–803.

Narita, M., Young, A. R., Arakawa, S., Samarajiwa, S. A., Nakashima, T., Yoshida, S., et al. (2011). Spatial coupling of mTOR and autophagy augments secretory phenotypes. *Science*, *332*(6032), 966–970.

Nguyen, L. V., Vanner, R., Dirks, P., & Eaves, C. J. (2012). Cancer stem cells: An evolving concept. *Nature Reviews Cancer*, *12*(2), 133–143.

Ni, H. M., Woolbright, B. L., Williams, J., Copple, B., Cui, W., Luyendyk, J. P., et al. (2014). Nrf2 promotes the development of fibrosis and tumorigenesis in mice with defective hepatic autophagy. *Journal of Hepatology*, *61*(3), 617–625.

Noman, M. Z., Buart, S., Van Pelt, J., Richon, C., Hasmim, M., Leleu, N., et al. (2009). The cooperative induction of hypoxia-inducible factor-1 alpha and STAT3 during hypoxia induced an impairment of tumor susceptibility to CTL-mediated cell lysis. *Journal of Immunology*, *182*(6), 3510–3521.

Noman, M. Z., Janji, B., Kaminska, B., Van Moer, K., Pierson, S., Przanowski, P., et al. (2011). Blocking hypoxia-induced autophagy in tumors restores cytotoxic T-cell activity and promotes regression. *Cancer Research*, *71*(18), 5976–5986.

O'Leary, B., Finn, R. S., & Turner, N. C. (2016). Treating cancer with selective CDK4/6 inhibitors. *Nature Reviews Clinical Oncology*, *13*(7), 417–430.

Ortiz-Montero, P., Londono-Vallejo, A., & Vernot, J. P. (2017). Senescence-associated IL-6 and IL-8 cytokines induce a self- and cross-reinforced senescence/inflammatory milieu strengthening tumorigenic capabilities in the MCF-7 breast cancer cell line. *Cell Communication and Signaling*, *15*, 1–18.

Pereira, B. I., Devine, O. P., Vukmanovic-Stejic, M., Chambers, E. S., Subramanian, P., Patel, N., et al. (2019). Senescent cells evade immune clearance via HLA-E-mediated NK and CD8(+) T cell inhibition. *Nature Communications*, *10*(1), 2387.

Perez-Hernandez, M., Arias, A., Martinez-Garcia, D., Perez-Tomas, R., Quesada, R., & Soto-Cerrato, V. (2019). Targeting autophagy for cancer treatment and tumor chemosensitization. *Cancers (Basel)*, *11*(10), 1599.

Perez-Mancera, P. A., Young, A. R. J., & Narita, M. (2014). Inside and out: The activities of senescence in cancer. *Nature Reviews Cancer*, *14*(8), 547–558.

Pfeffer, C. M., & Singh, A. T. K. (2018). Apoptosis: A target for anticancer therapy. *International Journal of Molecular Sciences*, *19*(2), 448.

Pohl, C., & Dikic, I. (2019). Cellular quality control by the ubiquitin-proteasome system and autophagy. *Science*, *366*(6467), 818.

Qiang, L., Zhao, B., Ming, M., Wang, N., He, T. C., Hwang, S., et al. (2014). Regulation of cell proliferation and migration by p62 through stabilization of Twist1. *Proceedings of the National Academy of Sciences of the United States of America*, *111*(25), 9241–9246.

Qiang, L., Zhao, B., Shah, P., Sample, A., Yang, S., & He, Y. Y. (2016). Autophagy positively regulates DNA damage recognition by nucleotide excision repair. *Autophagy*, *12*(2), 357–368.

Quick, Q. A., & Gewirtz, D. A. (2006). An accelerated senescence response to radiation in wild-type p53 glioblastoma multiforme cells. *Journal of Neurosurgery*, *105*(1), 111–118.

Rashid, H. O., Yadav, R. K., Kim, H. R., & Chae, H. J. (2015). ER stress: Autophagy induction, inhibition and selection. *Autophagy*, *11*(11), 1956–1977.

Ritschka, B., Storer, M., Mas, A., Heinzmann, F., Ortells, M. C., Morton, J. P., et al. (2017). The senescence-associated secretory phenotype induces cellular plasticity and tissue regeneration. *Genes & Development*, *31*(2), 172–183.

Ruscetti, M., Morris, J. P. t., Mezzadra, R., Russell, J., Leibold, J., Romesser, P. B., et al. (2020). Senescence-induced vascular remodeling creates therapeutic vulnerabilities in pancreas cancer. *Cell*, *181*(2), 424–441 (e421).

Sanchez-Martin, P., & Komatsu, M. (2020). Physiological stress response by selective autophagy. *Journal of Molecular Biology*, *432*(1), 53–62.

Sandilands, E., Serrels, B., McEwan, D. G., Morton, J. P., Macagno, J. P., McLeod, K., et al. (2011). Autophagic targeting of Src promotes cancer cell survival following reduced FAK signalling. *Nature Cell Biology*, *14*(1), 51–60.

Scheule, R. K., Perkins, R. C., Hamilton, R., & Holian, A. (1992). Bleomycin stimulation of cytokine secretion by the human alveolar macrophage. *American Journal of Physiology*, *262*(4), L386–L391.

Serrano, M., Lin, A. W., McCurrach, M. E., Beach, D., & Lowe, S. W. (1997). Oncogenic Ras provokes premature cell senescence associated with accumulation of p53 and p16INK4a. *Cell*, *88*(5), 593–602.

Sharma, A., & Almasan, A. (2018). Autophagy as a mechanism of Apo2L/TRAIL resistance. *Cancer Biology & Therapy*, *19*(9), 755–762.

Short, S., Fielder, E., Miwa, S., & von Zglinicki, T. (2019). Senolytics and senostatics as adjuvant tumour therapy. *eBioMedicine*, *41*, 683–692.

Soto-Gamez, A., Quax, W. J., & Demaria, M. (2019). Regulation of survival networks in senescent cells: From mechanisms to interventions. *Journal of Molecular Biology*, *431*(15), 2629–2643.

Swift, L. H., & Golsteyn, R. M. (2014). Genotoxic anti-cancer agents and their relationship to DNA damage, mitosis, and checkpoint adaptation in proliferating cancer cells. *International Journal of Molecular Sciences*, *15*(3), 3403–3431.

Takamura, A., Komatsu, M., Hara, T., Sakamoto, A., Kishi, C., Waguri, S., et al. (2011). Autophagy-deficient mice develop multiple liver tumors. *Genes & Development*, *25*(8), 795–800.

Takasugi, M., Okada, R., Takahashi, A., Virya Chen, D., Watanabe, S., & Hara, E. (2017). Small extracellular vesicles secreted from senescent cells promote cancer cell proliferation through EphA2. *Nature Communications*, *8*, 15729.

Tasdemir, N., Banito, A., Roe, J. S., Alonso-Curbelo, D., Camiolo, M., Tschaharganeh, D. F., et al. (2016). BRD4 connects enhancer remodeling to senescence immune surveillance. *Cancer Discovery*, *6*(6), 612–629.

Thiery, J., Keefe, D., Boulant, S., Boucrot, E., Walch, M., Martinvalet, D., et al. (2011). Perforin pores in the endosomal membrane trigger the release of endocytosed granzyme B into the cytosol of target cells. *Nature Immunology*, *12*(8), 770–777.

Thorn, C. F., Oshiro, C., Marsh, S., Hernandez-Boussard, T., McLeod, H., Klein, T. E., et al. (2011). Doxorubicin pathways: Pharmacodynamics and adverse effects. *Pharmacogenetics and Genomics*, *21*(7), 440–446.

Tittarelli, A., Janji, B., Van Moer, K., Noman, M. Z., & Chouaib, S. (2015). The selective degradation of synaptic Connexin 43 protein by hypoxia-induced autophagy impairs natural killer cell-mediated tumor cell killing. *The Journal of Biological Chemistry*, *290*(39), 23670–23679.

Toso, A., Revandkar, A., Di Mitri, D., Guccini, I., Proietti, M., Sarti, M., et al. (2014). Enhancing chemotherapy efficacy in Pten-deficient prostate tumors by activating the senescence-associated antitumor immunity. *Cell Reports*, *9*(1), 75–89.

Tsuruma, R., Ohbayashi, N., Kamitani, S., Ikeda, O., Sato, N., Muromoto, R., et al. (2008). Physical and functional interactions between STAT3 and KAP1. *Oncogene*, *27*(21), 3054–3059.

van Deursen, J. M. (2019). Senolytic therapies for healthy longevity. *Science*, *364*(6441), 636–637.

Wang, B. S., Kohli, J., & Demaria, M. (2020). Senescent cells in Cancer therapy: Friends or foes? *Trends in Cancer*, *6*(10), 838–857.

Wang, Y., Zhang, N., Zhang, L., Li, R., Fu, W., Ma, K., et al. (2016). Autophagy regulates chromatin ubiquitination in DNA damage response through elimination of SQSTM1/p62. *Molecular Cell*, *63*(1), 34–48.

Ward, P. S., & Thompson, C. B. (2012). Metabolic reprogramming: A cancer hallmark even Warburg did not anticipate. *Cancer Cell*, *21*(3), 297–308.

Wei, Y., Pattingre, S., Sinha, S., Bassik, M., & Levine, B. (2008). JNK1-mediated phosphorylation of Bcl-2 regulates starvation-induced autophagy. *Molecular Cell*, *30*(6), 678–688.

Weinberg, F., Hamanaka, R., Wheaton, W. W., Weinberg, S., Joseph, J., Lopez, M., et al. (2010). Mitochondrial metabolism and ROS generation are essential for Kras-mediated tumorigenicity. *Proceedings of the National Academy of Sciences of the United States of America*, *107*(19), 8788–8793.

White, E. (2015). The role for autophagy in cancer. *The Journal of Clinical Investigation*, *125*(1), 42–46.

Wilkinson, S. (2020). Emerging principles of selective ER autophagy. *Journal of Molecular Biology*, *432*(1), 185–205.
Witsch, E., Sela, M., & Yarden, Y. (2010). Roles for growth factors in Cancer progression. *Physiology*, *25*(2), 85–101.
Wolfson, R. L., Chantranupong, L., Saxton, R. A., Shen, K., Scaria, S. M., Cantor, J. R., et al. (2016). Sestrin2 is a leucine sensor for the mTORC1 pathway. *Science*, *351*(6268), 43–48.
Xu, C., Wang, L., Fozouni, P., Evjen, G., Chandra, V., Jiang, J., et al. (2020). SIRT1 is downregulated by autophagy in senescence and ageing. *Nature Cell Biology*, *22*(10), 1170–1179.
Yamamoto, K., Venida, A., Yano, J., Biancur, D. E., Kakiuchi, M., Gupta, S., et al. (2020). Autophagy promotes immune evasion of pancreatic cancer by degrading MHC-I. *Nature*, *581*(7806), 100–105.
Yang, H., Wang, H., Ren, J., Chen, Q., & Chen, Z. J. (2017). cGAS is essential for cellular senescence. *Proceedings of the National Academy of Sciences of the United States of America*, *114*(23), E4612–E4620.
Youle, R. J., & Narendra, D. P. (2011). Mechanisms of mitophagy. *Nature Reviews Molecular Cell Biology*, *12*(1), 9–14.
Young, A. R., Narita, M., & Narita, M. (2011). Spatio-temporal association between mTOR and autophagy during cellular senescence. *Autophagy*, 7(11), 1387–1388.
Yu, L., Alva, A., Su, H., Dutt, P., Freundt, E., Welsh, S., et al. (2004). Regulation of an ATG7-beclin 1 program of autophagic cell death by caspase-8. *Science*, *304*(5676), 1500–1502.
Zhang, Y., Alexander, P. B., & Wang, X. F. (2017). TGF-beta family signaling in the control of cell proliferation and survival. *Cold Spring Harbor Perspectives in Biology*, *9*(4), a022145.
Zhu, Y., Tchkonia, T., Fuhrmann-Stroissnigg, H., Dai, H. M., Ling, Y. Y., Stout, M. B., et al. (2016). Identification of a novel senolytic agent, navitoclax, targeting the Bcl-2 family of anti-apoptotic factors. *Aging Cell*, *15*(3), 428–435.
Zhu, Y., Tchkonia, T., Pirtskhalava, T., Gower, A. C., Ding, H. S., Giorgadze, N., et al. (2015). The Achilles' heel of senescent cells: From transcriptome to senolytic drugs. *Aging Cell*, *14*(4), 644–658.

CHAPTER THREE

Autophagy and senescence, converging roles in pathophysiology as seen through mouse models

Andrew R.J. Young*, Liam D. Cassidy, and Masashi Narita*
Cancer Research UK Cambridge Research Institute, University of Cambridge, Li Ka Shing Centre, Cambridge, United Kingdom
*Corresponding authors: e-mail address: andrew.young@cruk.cam.ac.uk; masashi.narita@cruk.cam.ac.uk

Contents

Abstract

Both senescence and autophagy have been strongly linked to aging and also cancer development. Numerous molecular, cellular, and physiological changes are known to correlate with an increasing age, yet our understanding of what underlies these changes or how they combine to give rise to the various pathologies associated with aging is still unclear. Levels of autophagy activity are known to decrease with advancing age, in a variety of organisms including mammals. Whereas senescent cells are known to accumulate in our bodies with age. Herein we review evidence from some elegant genetic mouse models linking senescence and also autophagy to aging and cancer. It is especially interesting to note the convergence in the pathological phenotypes of these two processes, senescence and autophagy, in these mouse models.

Advances in Cancer Research, Volume 150
ISSN 0065-230X
https://doi.org/10.1016/bs.acr.2021.02.001

1. Introduction

Since the subject of this entire volume is autophagy and senescence this introduction will be brief, covering topics only in sufficient depth for this chapter to standalone. Please see the citations mentioned herein for more in depth information.

1.1 Senescence

1.1.1 Replicative senescence and oncogene-induced senescence (OIS)

Cellular senescence was initially described in normal (diploid) human fibroblasts, which undergo a "permanent" proliferative arrest after long-term cell culture in vitro (Hayflick, 1965; Hayflick & Moorhead, 1961). This phenomenon, known as replicative senescence, is mainly attributed to the progressive shortening of telomeres occurring with each subsequent cell division, provoking a persistent DNA damage response and the activation of tumor-suppressive signals, including activation of the p53 and the $p16^{INK4A}$-Rb pathways (Shay & Wright, 2005). Prolonged activation of these signals leads to stable proliferative arrest: cellular senescence.

In addition to telomere shortening, a number of stimuli are known to induce cellular senescence, including DNA damage, oxidative stress and aberrant oncogene activation (Campisi, 2013). Oncogene-induced senescence (OIS) was also initially described in cultured diploid human fibroblasts, where the ectopic expression of oncogenic HRAS-G12V or its downstream effectors paradoxically led to senescence rather than proliferation (Lin et al., 1998; Serrano, Lin, McCurrach, Beach, & Lowe, 1997; Zhu, Woods, McMahon, & Bishop, 1998). However, this phenomenon is now known to go way beyond fibroblasts in culture as in 2005 a series of studies (Braig et al., 2005; Chen et al., 2005; Collado et al., 2005; Denchi, Attwooll, Pasini, & Helin, 2005; Michaloglou et al., 2005) identified senescent cells within premalignant tumors arising from several different organs. Since then, around 50 oncogenes have been shown to trigger senescence (Gorgoulis & Halazonetis, 2010) and senescence, oncogene induced or otherwise, has been shown to occur in a variety of tissues and different pathophysiological contexts (Muñoz-Espín & Serrano, 2014). Premature senescence is not only limited to the activation of oncogenes either as loss of tumor suppressor genes has also been associated with senescence in premalignant tumors, for instance PTEN (Chen et al., 2005), RB (Shamma et al., 2009), neurofibromin 1

(NF1) (Courtois-Cox et al., 2006) and inositol polyphosphate-4-phosphatase, type II (INPP4B) (Gewinner et al., 2009), reviewed in Gorgoulis and Halazonetis (2010).

1.1.2 The senescence-associated secretory phenotype (SASP)

The culture medium of senescent cells had long been known to be thick with secreted proteins (Krtolica & Campisi, 2002; Shelton, Chang, Whittier, Choi, & Funk, 1999). Functional consequences of these secreted proteins were first proposed by Judith Campisi's group (Krtolica & Campisi, 2002). They showed that secreted factors from senescent fibroblasts promoted the transformation of premalignant, but not normal, mammary epithelial cells (Krtolica, Parrinello, Lockett, Desprez, & Campisi, 2001; Parrinello, Coppe, Krtolica, & Campisi, 2005). A series of papers in the latter half of the first decade of this millennium then showed functional effects of the SASP on the senescence phenotype itself (Acosta et al., 2008; Augert et al., 2009; Coppé et al., 2008; Kuilman et al., 2008; Wajapeyee, Serra, Zhu, Mahalingam, & Green, 2008; Young & Narita, 2009). This series of papers systematically characterized the constituents of the SASP, by for instance microarray analysis, and went on to show that there were autonomous feedback loops involved in the SASP, whereby SASP factors could reinforce senescence in the cells from which they had originated: cell-autonomous effects. There were also paracrine effects whereby those cells surrounding the initiating cells were induced to become senescent by SASP factors (Acosta et al., 2008; Young & Narita, 2009).

Thus, the implications for SASP in tumorigenesis essentially go two ways, forming a double edged sword if you will. As described in the next section, the SASP can have anticancer functions through its ability to attract cells of the immune system to clear both damaged (e.g. DNA damaged) and tumor cells. However, if the senescent cells are not cleared, then chronic inflammation can promote a pro-tumorigenic microenvironment (Young & Narita, 2009).

1.1.3 Beneficial effects of senescence

Senescence has been shown to play a role in fibrosis, as first shown by Krizhanovsky et al. (2008) in the context of liver damage. Persistent liver damage caused by alcohol or even through the laying down of fat reserves, non-alcoholic fatty liver disease (NAFLD), may initiate—presumably aberrant—fibrosis to varying degrees in the liver. Krizhanovsky et al. showed that acute liver damage induced hepatic stellate cells to proliferate and produce extracellular matrix (ECM) but then undergo senescence. In mice lacking senescence effectors, these hepatic stellate cells continued to proliferate and produced ECM leading to tissue fibrosis. Liver injury thus seemed to induce a senescence response, which limited liver fibrosis. Lujambio et al. (2013) went on to show that in chronic liver damage, mice lacking senescence effectors in hepatic stellate cells developed cirrhotic livers, a disease which is often characterized by sweeping bands of fibrosis. They showed that the senescence compromised hepatic stellate cells enhanced transformation of adjacent epithelial cells into hepatocellular carcinoma (Lujambio et al., 2013).

In the context of the skin where fibrosis facilitates wound closure, senescence was shown to accelerate wound closure by inducing myofibroblast differentiation through the secretion of platelet-derived growth factor AA (PDGF-AA) (Demaria et al., 2014). Selective elimination of $p16^{INK4a}$-positive senescent cells delayed wound closure, and topical treatment of these senescence-free wounds with recombinant PDGF-AA rescued the delayed wound closure and induced myofibroblast differentiation in the absence of senescence.

Senescent cells can also recruit immune cells to facilitate their own elimination from the tissue/organ. Various immune cells are able to facilitate this process. For instance, SASP factors recruit natural killer cells of the innate immune system to eliminate senescent cells (Iannello, Thompson, Ardolino, Lowe, & Raulet, 2013; Krizhanovsky et al., 2008). Adaptive immunity has also been shown to be involved, whereby premalignant but senescent murine hepatocytes secreting chemokines and cytokines promoted senescence surveillance, requiring an intact CD4+ T-cell-mediated adaptive immune response. Impaired immune surveillance of these premalignant senescent hepatocytes led to the development of hepatocellular carcinomas (Kang et al., 2011). A key factor in this process was shown to be CCL2 which recruited immature myeloid cells (iMC), which differentiated into macrophages and cleared the premalignant senescent cells. However, if senescent cells were present within established tumors then the iMC recruitment instead promoted the growth of these hepatocellular carcinomas through NK cell inhibition (Eggert et al., 2016). Thus, depending on the context, senescence can suppress tumorigenesis in vivo through immune surveillance and/or promote tumorigenesis through modulating the tumor immune response.

1.1.4 Negative effects of senescence

As eluded to above, senescence can have detrimental effects through, for instance, a persistent of SASP factors and associated chronic inflammation. Osteoarthritis (OA) occurs when the cartilage that lines the surfaces of articulating joints, providing a low-friction surface, deteriorates. Although the mechanism of cartilage deterioration is chiefly thought to involve mechanical factors, altered synthesis of the cartilage matrix by chondrocytes also plays a role. The dysregulation of chondrocytes in OA is related to chronic inflammation, such as increased receptor levels for certain cytokines, including TNFα (Maksymowych et al., 2012). Dysregulation of chondrocytes changes the quantity and composition of the ECM secreted by the cells resulting in a reduced proteoglycan content and a decreased compressive modulus of the cartilage, which in turn causes the joint to experience greater strains when exposed to mechanical stress (Maldonado & Nam, 2013). Several markers of senescence, including $p16^{INK4A}$ and p21 expression, are found in multiple tissues in arthritic joints, including cartilage, subchondral bone, and the synovium (Jeon, David, Campisi, & Elisseeff, 2018). Deletion of these p16 positive cells alleviated OA in a mouse model of the disease (Jeon et al., 2017).

Idiopathic pulmonary fibrosis (IPF) is characterized by interstitial parenchymal modifications resulting in lung volume abnormalities and impaired diffusion capacity (Chilosi, Carloni, Rossi, & Poletti, 2013). IPF is associated with age and long-term tobacco smoking. IPF is also associated with an increased risk of cancer development (Ballester, Milara, & Cortijo, 2019). IPF can either present as sporadic disease or as heritable familial forms, such as familial interstitial pneumonia (FIP) (García-Sancho et al., 2011). In 15%–20% of cases of FIP, genetic lesions affect the telomerase complex (Armanios, 2011). Hence, it was suggested that a mechanism akin to cellular senescence might be responsible for the early damage to the alveolar parenchyma (Armanios, 2011). Expression of markers of senescence such as p16, p21 and senescence-associated beta-galactosidase (SA-βgal) have since been demonstrated in IPF. These markers are expressed strongly at sites of alveolar damage and hyperplasia, as well as in fibroblast foci localized in the discrete clusters of bronchiolar basal cells (Chilosi et al., 2013). In mice, treatment with the DNA damaging agent bleomycin recapitulates many features of human IPF, including the expression of markers of senescence (Aoshiba, Tsuji, & Nagai, 2003). IPF shows evidence of a chronic inflammatory reaction and lymphoid follicle accumulation. Chronic inflammatory infiltrates are common around alveolar structures exhibiting senescent features (Chilosi et al., 2013). In IPF then, it would seem that despite, or perhaps facilitated by, an active SASP parenchymal senescence without resolution can lead to chronic obstruction of the lungs.

Markers of senescence, such as SA-βgal, SASP components and upregulation of p53 and p21, can be found in the adipose tissue of obese subjects, both in mouse models and in humans. For instance, a mouse model of accelerated aging, caused by a hypomorphic allele of the Bub1b (budding uninhibited by benzimidazoles 1 homolog beta) gene has high levels of SA-βgal activity in its adipose tissue (Baker et al., 2008).

Senescence has also been shown to contribute to type 2 diabetes. Insulin resistance is known to be associated with obesity and aging. Insulin resistance is initially compensated by an expansion of the β-cell compartment but this can eventually lead to proliferative exhaustion and the loss of β-cell mass. GWAS studies have found that polymorphisms that are in close proximity to the p16(CDKN2A)–p15(CDKN2B) locus, two cell cycle inhibitors with key roles in senescence, are linked to type 2 diabetes (Diabetes Genetics Initiative of Broad Institute of Harvard and MIT, Lund University, and Novartis Institutes of BioMedical Research, et al., 2007). Further evidence for a connection between obesity, diabetes and senescence came from studies in mice, which when fed on a high-fat diet for 12 months are strongly positive for SA-βgal activity in their β-cell islets (Sone & Kagawa, 2005). Senescent β-cells are also a good example of the potential effects of the hyperfunction of senescent cells. Senescent β-cells overproduce insulin (Helman et al., 2016) and may thereby activate adipocytes and other cells, causing their hyperfunction too, which in turn can lead to metabolic syndromes (Extermann, 2014).

Senescence also aggravates sarcopenia: loss of muscle function is one of the most prevalent aging-associated pathologies (Muñoz-Espín & Serrano, 2014). As we shall see Section 2 the accumulation of senescent cells is associated with aging.

1.2 Autophagy

1.2.1 Brief introduction

When people refer to autophagy, they usually mean macroautophagy, which is a term coined to refer to the *en masse* sequestration of cytosol, and even organelles, into a membrane-bound compartment, an autophagosome, for its subsequent delivery to the endosomal/lysosomal system and, therein, its degradation and recycling (de Duve, 2005). Other types of autophagy include microautophagy which occurs via invagination of the lysosomal membrane and chaperone mediated autophagy, where proteins are specifically targeted for degradation by chaperones and hence trafficked to, and inserted, into the lysosome. Macroautophagy was first described by Clark (1957) using electron microscopy (EM). It was generally assumed to be a non-specific process at the time since the autophagosomes could be seen to contain a mixture of cytosol and organelles in a fairly random manner (Holtzman, 1989). However, while *en masse* degradation certainly does occur there is also increasing evidence that has since been generated that autophagy can also be much more specific for certain cargos through selective autophagy receptors (SARs), for instance through adapter proteins such as p62/SQSTM1 for ubiquitinated proteins and PINK1/Parkin which target mitochondrial for degradation: mitophagy (Jin & Youle, 2012). The selective elimination of mitochondria from cells is of the utmost interest due to the central role that mitochondria play in programmed cell death. SARs have also been found to exist on the endoplasmic reticulum (ER) and the nucleus, leading to ER-phagy and even nucleophagy. In total, around 30 SARs are now known, enabling a wide range of specific forms of autophagy: xenophagy (the removal of foreign entities, for instance, bacteria), lipophagy (lipids), ribophagy (ribosomes), ferritinophagy (ferritin/iron), glycophagy (glycogen) and pexophagy (peroxisomes) (Kirkin & Rogov, 2019).

Autophagy can occur both at a basal, constitutive (non-induced), level but it can also be upregulated by for instance, as classically studied, nutrient starvation. It is also apparent that the activation of oncogenes such as RAS can upregulate autophagy (Maiuri et al., 2009).

1.2.2 ATG proteins

Autophagy research took the leap into the molecular biology age in the mid-1990s, when screens in yeast identified proteins required for autophagosome formation and maturation (Harding, Morano, Scott, & Klionsky, 1995; Thumm et al., 1994; Tsukada & Ohsumi, 1993), the classification of which was later unified so that they were called the *Atg* genes (Klionsky et al., 2003). Some of the ATG proteins were identified to take part in one (or both) of two ubiquitiation-like conjugation reactions. In one pathway ATG12 is coupled to ATG5, a process which requires the E1-like enzyme ATG7 and the E2-like enzyme ATG10 (Mizushima, Sugita, Yoshimori, & Ohsumi, 1998). ATG5–12 then binds to ATG16 to form a complex and this complex can be seen, by electron microscopy for instance, to be required reasonably early in autophagosome formation

(Mizushima et al., 2001; Suzuki et al., 2001). In a second ubiquitin-like conjugation pathway Atg8 becomes membrane associated through its conjugation to the lipid phosphatidylethanolamine (PE) (Ichimura et al., 2000; Kirisako et al., 1999, 2000). Lipidation of Atg8 requires the E1-like enzyme Atg7 and the E2-like enzyme Atg3 (Tanida et al., 1999). Also, before lipidation can occur Atg8 is cleaved by cysteine protease Atg4 (Tanida et al., 2004). Lipidation of Atg8 causes it to associate to autophagosomal membranes and Atg8 is thus a marker of autophagosomes. There are several Atg8 homologs in mammals, including GATE-16, GABARAP and most famously LC3. The lipidated form of LC3 is designated LC3-II and the cytosolic form as LC3-I.

A further autophagy complex brings together the ATG proteins and a phosphatidylinositol-3-kinase (PI3K) central to membrane trafficking Vps34. Atg6 (Beclin1) functions as a subunit of the Vps34 PI3K complex (Liang et al., 1999).

1.2.3 Initial autophagy studies in mice reveal neurological defects

Mice systemically expressing GFP-LC3 gave great insight into the levels of autophagy naturally occurring at a basal level in vivo (Mizushima, Yamamoto, Matsui, Yoshimori, & Ohsumi, 2004). Liver, muscle (including the heart), pancreas, thymus and kidney all showed a constitutive basal level of autophagy. The first autophagy gene for which a knockout was generated was *Atg5*. *Atg5* homozygous knockout mice displayed minimal abnormalities at birth but died within 1 day of delivery (Kuma et al., 2004). Although they displayed a lack of milk in their stomachs, even hand fed using a gavage they did not survive more than 2 days after birth. By crossing to the GFP-LC3 mice it was shown that autophagy was upregulated in a variety of tissues within 3 h after birth, thus indicating a requirement for autophagy after detachment from the placenta, which of course was not able to be called upon in the *Atg5* knockout animals.

Atg7 homozygous knockout mice had significantly lower body weight after birth and also died within 1 day (Komatsu et al., 2005). In conditional knockout adult mice, autophagosomes were seen to be smaller with no observable mitochondria inside. An accumulation of sick or damaged mitochondria could be seen throughout the cytoplasm. Concentric ring like structures were prevalent as were ubiquitin aggregates. Indeed Komatsu et al. (2006) went on to show that mice lacking *Atg7* specifically in the central nervous system displayed a notable accumulation of polyubiquitinated proteins in autophagy-deficient neurons as inclusion bodies, which increased in size and number with age. There was, however, no obvious alteration in proteasome function. This polyubiquitinated aggregate accumulation was thought to cause the neurodegeneration seen in these mice. A consecutive paper using mice deficient for *Atg*5 specifically in neural cell also found polyubiquitinated aggregates associated with neurodegeneration (Hara et al., 2006).

There are approximately 20 core autophagy-related genes involved in autophagosome formation in mammals. To date, around 14 of them have been knocked out in mice (Kuma, Komatsu, & Mizushima, 2017). In general *Atg* knockout mice die either in utero or are born alive but die within 1 day of birth, making long-term longitudinal studies impossible.

Conditional *Atg*-deficient mice have also been generated, e.g. Ubc-CreERT2, a tamoxifen-inducible Cre allowing for deletion of floxed genes throughout the body. *Atg*7flox/flox;CreERT2 mice given tamoxifen at 8 weeks old showed many histological abnormalities 6–12 weeks post deletion, with more than 90% of these mice dying by 12 weeks post deletion. The vast majority of these mice again died because of neurodegeneration or bacterial infection (Karsli-Uzunbas et al., 2014).

Expression of exogenous *Atg*5 only in the brain under control of the Eno2 promoter was sufficient to rescue *Atg*5 knockout mice from post-natal death (Yoshii et al., 2016). The majority of these rescued mice survive for between 8 weeks and 8 months. Although the cause of lethality in these mice was not established, their increased lifespan would seem to indicate that these mice did not all succumb to neurological deficiencies, which was the main cause of death of *Atg*7 knockout mice. In histological analyses, *Eno2-Atg*5 mice show similar abnormalities to *Atg*7 knockout mice. Additional previously uncharacterized abnormalities observed in *Eno2-Atg*5 mice were morphological changes related to inflammation in many tissues (liver, small intestine, kidney, lung, adipose tissues, spleen, lymph nodes), atrophy in organs regulated by sex hormones (testis, seminal vesicle, ovary, uterus, pituitary gland, salivary gland) and iron-deficiency anemia. These finding indicate that there may be causes, other than accumulation of aggregated ubiquitinated proteins, for the abnormalities observed in other organs. Utilization of these and other mouse models whereby autophagy deficiency can be monitored long term may well therefore provide new insights into the physiological roles of autophagy.

2. Senescence and autophagy in aging and cancer

2.1 Introduction

Aging is a highly complex process that is associated with the gradual decline in tissue and organismal function, which ultimately results in death. Numerous molecular, cellular, and physiological changes are known to correlate with an increasing age, yet our understanding of what underlies these changes or how they combine to give rise to the various pathologies associated with aging is still unclear. These pathologies include, but are not limited to: sarcopenia (manifesting as a loss of muscle strength) (Larsson et al., 2019), atherosclerosis (Wang & Bennett, 2012) and heart failure (Lakatta & Levy, 2003), osteoporosis (loss of bone density) (Pietschmann, Rauner, Sipos, & Kerschan-Schindl, 2009), macular degeneration (Lim, Mitchell, Seddon, Holz, & Wong, 2012), pulmonary insufficiency (Kovacs, Lowery, Kuhlmann, & Brubaker, 2013), renal failure (O'Sullivan, Hughes, & Ferenbach, 2017), neurodegeneration (including prominent neurodegenerative diseases such as Alzheimer's and Parkinson's diseases) (Kukull et al., 2002; Reeve, Simcox, & Turnbull, 2014) and, of course, cancer development (de Magalhães, 2013).

With regards to this chapter, both senescence and autophagy have been strongly linked to aging as well as cancer development. Herein we review evidence from some elegant genetic mouse models linking senescence and also autophagy to aging and cancer. It is especially interesting to note the convergence in the pathological phenotypes of these two processes, senescence and autophagy, in these mouse models.

2.2 Senescence, aging and cancer

Studies looking at the expression of markers of senescence, such as $p16^{Ink4a}$, have provided evidence that senescent cells accumulate within our tissues with age (Burd et al., 2013; Herbig, Ferreira, Condel, Carey, & Sedivy, 2006; Jeyapalan, Ferreira, Sedivy, & Herbig, 2007; Krishnamurthy et al., 2006). However, these data should be interpreted with caution, since as yet there is not one definitive marker of senescence. Due to this lack of a specific marker, currently senescence detection in vivo must be supported by the presence of multiple markers in combination. These include, but are not limited to, markers such as double stranded DNA breaks (DSBs), components of the SASP, chromosomal alterations (such as the SAHF

(Narita et al., 2003) and macroH2A (Zhang et al., 2005)), telomere associated DNA damage, and the presence of cell cycle inhibitors such as p21 and $p16^{INK4a}$, usually in combination with SA-βgal activity (Dimri et al., 1995).

All that said, some of the most intriguing data on senescence and aging comes from studies in which senescent cells are depleted using genetic systems driven by the $p16^{Ink4a}$ promoter. Multiple iterations of this have now been made (Baker et al., 2011; Demaria et al., 2014; Grosse et al., 2020; Omori et al., 2020). Two of the most extensively published are those from the labs of Jan van Deursen and Judith Campisi. In the van Deursen INK-ATTAC model, administration of a synthetic drug induces dimerization of a membrane-bound myristoylated FK506-binding protein–caspase 8 (FKBP–Casp8) fusion protein leading to apoptosis of p16 expressing cells (Baker et al., 2011). The Campisi lab's mouse model takes advantage of the 3MR (trimodality reporter) fusion protein, which contains luciferase monomeric red fluorescent protein (mRFP), and truncated herpes simplex virus 1 (HSV-1) thymidine kinase (HSV-TK) engineered into a bacterial artificial chromosome (BAC) containing approximately 50 kb of the murine $p16^{Ink4a}$ locus, such that the $p16^{Ink4a}$ promoter drives 3MR expression (Demaria et al., 2014). Upon addition of ganciclovir (GCV), a nucleoside analog that has a high affinity for HSV-TK but a low affinity for the cellular TK, HSV-TK converts GCV into a toxic DNA chain terminator. Thus, in nondividing senescent cells, GCV fragments mitochondrial DNA, causing death by apoptosis.

In a series of elegant experiments, van Deursen's lab went on to determine the effect of clearance of naturally occurring p16 positive cells on health and lifespan (Baker et al., 2011). As has been seen for p16 itself (Burd et al., 2013), expression of the INK-ATTAC transcripts under the $p16^{Ink4a}$ promoter markedly increased with age in all the tissues examined. Importantly, senescent cells could be removed from most tissues by INK-ATTAC system activation; however the colon and liver appeared to be exceptions. On average, p16 positive cell clearance led to the mice living around 25% longer than normal, although maximum possible lifespan was unaffected. Depending on the genetic background of the mice used in their experiments, they saw a 18–25% increase in tumor free survival. The INK-ATTAC system had no effect on the incidence or spectrum of macroscopically detectable tumors at autopsy; however tumor latency was significantly increased, with treated mice living for nearly 200 days longer on average than untreated mice (which lived for 750 days—median survival).

While it is tempting to speculate that the p16-positive cells depleted by the INK-ATTAC system were putative tumor cells, thus leading to a reduced and delayed tumor burden, evidence from Judith Campisi's lab also hints at another possibility. While no data exists on the effects of depleting p16-postive cells in the p16-3MR mice during natural aging and spontaneous tumorigenesis, this model has been used to study the role of senescence in tumorigenesis directly (Alimirah et al., 2020), using a chemically induced skin carcinoma model. Painting mouse skin with a single subcarcinogenic dose of the carcinogen 7,12-dimethylbenz[α]anthracene (DMBA) leads to tumor initiation, with subsequent twice weekly follow up with a tumor promoting chemical (12-O-tetradecanoyl-phorbol-13-acetate) TPA leading to the formation of papillomata and, eventually, squamous cell carcinomas (SCCs) wherein the vast majority of these papillomata express oncogenic H-ras (Balmain & Pragnell, 1983; Balmain, Ramsden, Bowden, & Smith, 1984). Senescent cells had been reported previously in DMBA/TPA induced papillomas (Ritschka et al., 2017; Young et al., 2009). However, making use of the luciferase cassette of the p16-3MR model, Alimirah et al. (2020) found that the initiator, DMBA, itself did not lead to senescent cells within the skin. Instead senescent cells were only present upon the addition of the tumor promoter TPA. Providing ganciclovir to activate the 3MR system and kill off the senescent cells resulted in the tumors not becoming as large as the controls. Interested to see what would happen if the mice had senescent cells with the tumors to begin with, these investigators chose to treat the mice with a high dose of doxorubicin (doxo; a topoisomerase II inhibitor and DNA damaging agent). Mice pre-treated with doxo before DMBA/TPA had tumors that were much larger than those that were not pre-treated. Pre-treated tumors were on average 183 mm^3 by 21 weeks post DMBA treatment, compared to 29 mm^3 for the non-pre-treated group. Senescent cells could be detected in the surrounding stroma of the tumors as well as within the tumors themselves. These data may suggest that senescent cells promoted tumor growth, and that their removal could be detrimental to tumor development. However, further work is needed to determine the exact mechanism by which doxorubicin mediates its effect in this system.

2.3 Autophagy, aging and cancer

The first genetic evidence for a connection between autophagy and aging was shown in the worm *Caenorhabditis elegans*. The first mutants that could extend lifespan had been discovered in *C. elegans* (Kenyon, 2011), and one of the first mutants to be discovered was daf-2 (Kenyon, Chang, Gensch, Rudner, & Tabtiang, 1993) (the gene encoding insulin-like growth factor 1 (IGF-1) (Kimura, Tissenbaum, Liu, & Ruvkun, 1997)). The knockdown of beclin 1 in daf-2 mutants drastically reduced their lifespan (Meléndez et al., 2003). This is especially interesting as, since those early studies in worms, mutations in many insulin and IGF-1-pathway genes have also been shown to extend the lifespan of mammals and several have been linked to human longevity. In *Drosophila melanogaster*, inhibition of mTOR signaling, which thereby promotes autophagy, extends lifespan, as it does in yeast and *C. elegans* (Bjedov et al., 2010; Kapahi et al., 2004). In addition, pharmacological approaches using other well-known activators of autophagy, resveratrol (Lagouge et al., 2006), spermidine (Eisenberg et al., 2009) and rapamycin (the mTOR inhibitor and immunosuppressant (Harrison et al., 2009)) provided an indication that the activation of autophagy, as well as other processes, may be associated with longevity in eukaryotes and mice (Eisenberg et al., 2009; Harrison et al., 2009; Lagouge et al., 2006). However, the effect of these treatments can be pleiotropic. Ideally, genetic evidence in mammalian systems was needed to really nail the relationship between autophagy and aging. However, evidence for a role of autophagy in longevity was lacking since, as we saw in the introduction, autophagy gene knockouts in mice resulted in either embryonic or neonatal lethality and even systemic conditional knockouts suffered from neurological defects. Thus, critical evidence that autophagy contributes to longevity in mammals was therefore until relatively recently unavailable.

Overexpression of *Atg5* is considered to enable the promotion of autophagy, suggesting that ATG5 protein levels are rate limiting in some way for autophagosome formation. Pyo et al. (2013) took this one stage further and made a transgenic mouse overexpressing ATG5 from a CAGGs promoter (a combination of a cytomegalovirus enhancer and a chicken β-actin promoter; overall then, quite a strong promoter). Enhanced levels of autophagy were seen in all the tissues examined from these transgenic mice: heart, lung, brain, liver and muscle. Remarkably, electron microscopy studies revealed that the numbers of autophagosomes seen within the livers of *Atg5* transgenic mice were threefold higher than their wt counterparts. *Atg5* transgenic mice lived for a staggering 119 days longer than their wt counterparts, or 17% longer. Aged *Atg5* transgenic mice were also lean, although weighing similar amounts at a young age; after 12 months of age, the *Atg5* transgenic weighed on average 12% less than their wt counterparts. Although the propensity of these mice to develop tumors either more or less readily than their wt counterparts was not addressed in this study, the authors did observe that the mice exhibited a resistance to oxidative stress. The ratio of reduced glutathione to oxidized glutathione (GSH/GSSG), which represents the levels of intracellular antioxidants responsible for oxidative damage, was decreased during aging of both wt and *Atg*5 transgenic mice in an age-dependent manner; its decrease was slow in the liver tissues of 12- and 24-month-old *Atg*5 transgenic mice compared with wt mice. In addition, the mitochondrial oxygen consumption rate was increased in MEFs derived from *Atg*5 transgenic mice. This resistance to oxidative stress could well be suggestive that these mice could be less prone to DNA damage from any oxidative stress, but was not specifically addressed in this study.

A further seminal study to examine the effect of increased autophagy on mammalian health came from Beth Levine's lab (Fernández et al., 2018). This group developed a mouse with a Phe-to-Ala knock-in substitution mutation in the BH3 domain of beclin 1 (F121A) that decreases the binding of two negative regulators of autophagy (BCL2 and BCL-XL; famous for their involvement in apoptosis) to beclin 1. They observed increased levels of GFP-LC3 puncta in all tissues examined, including skeletal muscle, heart, kidney and liver. These beclin 1 F121A knock-in mice showed a significant lifespan extension compared to wild-type littermate controls; median survival was 26 months for wild-type mice but 29 months in the knock-in mice. Interestingly, unlike with the ablation of senescent cells via the INK-ATTAC system, here both median survival and maximum lifespan seemed to be improved. However, the increase in median lifespan was only two thirds of that seen with the *Atg*5 transgenic mice (Pyo et al., 2013). It could be interesting therefore to combine the two alleles, *Atg*5 transgenic and the beclin 1 F121A and see if there might be an additive effect upon lifespan extension.

The beclin 1 F121A knock-in mutation delayed aging-associated phenotypes in the kidney and heart, including fibrosis. Most interestingly, beclin 1 F121A knock-in mice with increased basal autophagy had a decreased incidence of age-related spontaneous cancers, which they classified as lymphoproliferative disease (LPD), lymphomas, and non-lymphoid malignancies including histiocytic sarcomas, a type of liver cancer. They saw a significant decrease in age-related spontaneous tumorigenesis in the beclin 1 F121A mice, either when comparing all malignancies, which dropped from around 30% to just over 10%, but also observed a drop in non-lymphoid malignancies alone. Fernández et al. also crossed the beclin 1 F121A knock-in mutation mice to a model of premature aging, the Klotho mice. Klotho is a single-pass membrane-bound protein that can be cleaved and released into the circulation; it is highly expressed in the kidney, and it regulates aging. Klotho expression decreases with aging in mice and humans (Kuro-o, 2009). Klotho hypomorphic or knockout mice have a syndrome resembling aging, including premature lethality—death occurs around 12 weeks after birth. Remarkably beclin 1 F121A rescued the premature lethality and the Klotho/beclin 1 F121A mice lived until the end of their observation period, 45 weeks. Other symptoms of the Klotho mice are infertility and growth retardation and the beclin 1 F121A rescued these effects as well.

In our recent publication, we addressed the autophagy and aging question from the other way around, using our recently developed *Atg*5 RNAi mouse (*Atg*5i) (Cassidy et al., 2018) to ask whether a systemic reduction in basal autophagic flux is enough to promote aging and age-related phenotypes in mammals. To circumvent the limitations imposed by knockout mice, we made use of a doxycycline-inducible short hairpin RNA (shRNA) to achieve body-wide temporal control of autophagy. The original tetracycline gene expression control systems were "Dox-off," i.e. in the absence of tetracycline, or an analog such as doxycycline (dox), a tet-transactivator protein (tTA) forms a dimer that that binds to a tet-responsive element (TRE) and causes expression of the gene following the TRE. When dox is given in this system, it binds to tTA and causes it to dissociate from the TRE. However, in mouse models where temporal control is desired, a "Dox-on" reverse tTA system is preferred, since treating a mouse with dox to induced upregulation of gene expression is more rapid than waiting for the depletion of an inhibitor. rtTA simply has four amino acid substitutions in the Tc repressor domain of tTA that reverse the affect of the dox, at the cost of a decrease in the sensitivity of the system to dox (Saunders, 2010). For us, however, the use of the "Dox-on" system was fortuitous; dox is poorly permeable across the blood-brain-barrier (BBB) (Dow et al., 2012) and, as such, our system showed no activation in the brain, hence avoiding the neurotoxic phenotypes that led to the premature death of adult whole-body knockout models of autophagy (e.g. tamoxifen-inducible systemic *Atg*7 knockout in adult mice leads to death between 2 and 3 months (Karsli-Uzunbas et al., 2014)).

A common caveat of many mouse models is that genetic manipulations are often present during embryogenesis. Therefore, any phenotypes that the mice develop may be a combination of both developmental and tissue homoeostasis effects. In all our experiments we aged our *Atg*5i mice until they reached adulthood, at eight-weeks old, to avoid developmental effects, before placing them on a dox diet to inhibit autophagy. Due to dox's poor penetrance of the BBB, our mice survived any premature death due to the neuronal toxicity previously observed in autophagy knockout mice, enabling us to perform longitudinal studies that were previously unachievable with autophagy gene knockout mice. Our autophagy-inhibited mice displayed dramatically reduced lifespans compared to normal mice. The median survival for *Atg*5i mice was around 6 months old (185 days for males and 207 days for females) (Cassidy et al., 2020). Furthermore, our *Atg*5i mice displayed evidence of aging like phenotypes, such as graying, kyphosis, loss of muscle mass and evidence of autoimmunity. All the *Atg*5i mice had hepatomegaly and splenomegaly, consistent with phenotypes previously published for tissue specific knockout mice. *Atg*5i mice also exhibited sclerotic and enlarged glomeruli, indicative of a degenerative kidney disorder. In addition, *Atg*5i mice had cardiomyopathy, with enlarged, degenerate and vacuolated cardiomyocytes and also cardiac fibrosis (Cassidy et al., 2020).

*Atg*5i mice had increased levels of key senescence markers: p16, p21, and p53. Nuclear accumulation of p21 was particularly evident in those hepatocytes with enlarged morphology. In addition, whole mount livers from *Atg*5i mice stained positive for SA-ß-gal (Cassidy et al., 2020). Telomere-associated γ-H2AX foci (TAF) represent persistent damage in telomeric regions, which are resistant to repair machinery and have been shown to correlate with senescence, increasing age and mitochondrial dysfunction (Correia-Melo et al., 2016; Hewitt et al., 2012; Jurk et al., 2014). *Atg*5i mice had a significant increase in both the abundance and frequency of TAF in liver, lung and heart tissue (Cassidy et al., 2020). Taking our data together, our study provides evidence that loss of autophagy systemically induces molecular, cellular and organismal phenotypes that are strongly associated with increasing age. Aging is a complex and multifaceted process that is associated with some degree of stochasticity; this process at least appeared to be recapitulated in our *Atg*5i mice. The main advantage of our *Atg*5i mouse being based on a doxycycline-inducible shRNA was that we could remove dox and thereby address whether the effects induced by autophagy knockdown were reversible. We waited until 4 months after dox treatment began, whereupon the *Atg*5i mice universally displayed aging phenotypes, including frailty. At this timepoint, we removed dox from the diet and restored autophagy levels. Strikingly, the mice showed a gross recovery in frailty and a dramatic recovery in their health- and lifespan. Lifespan was significantly extended; median survival was 493 days for mice taken off dox versus 185 days since treatment began for mice always on dox (Cassidy et al., 2020). The livers of mice taken off dox displayed a complete reversal of hepatomegaly and the kidneys of these mice also appeared to recover from autophagy inhibition, with sclerotic and enlarged glomeruli becoming undetectable. However, the kyphosis and graying phenotypes were largely, if not completely, irreversible, while the molecular damage that had accumulated (such as telomere associated DNA damage) was also irredeemable. Analysis of skeletal muscle from mice taken off dox showed little sign of recovery. In addition, the cardiac fibrosis observed in the *Atg*5i mice appeared to still be present 4 months post dox removal (Cassidy et al., 2020). Together these data suggest that tissue and pathology specific limitations may apply to the capacity of our mice to recover upon autophagy restoration, which may have resulted from the tissue and cellular damage induced upon autophagy inhibition. Crucially, while some tissues, such as the liver, appear to recover, they still exhibit age-associated pathologies at the molecular level. Thus, although encouraging, it is important to note that not all damage induced by autophagy loss was reversible in our model.

In marked contrast to *Atg*5i mice that were maintained on dox throughout their adult life, the cause of death when autophagy was restored was predominantly associated with the development of tumors, occurring with an increased frequency and at earlier timepoints than in wild-type mice. These tumors display no evidence of continued autophagy inhibition via immunohistochemical analysis (Cassidy et al., 2020). Thus, our data suggest that a temporary period of autophagy inhibition is enough to irreversibly accelerate spontaneous cancer development.

However, it is unknown how autophagy inhibition, and subsequent restoration drives cancer in this model. One possibility is that the increase in DNA damage, both telomere and non-telomere associated, is a source of mutational burden in the *Atg*5i mice. Sequencing both the cancers that arise, or "normal" and premalignant tissue, for evidence of this mutational burden in these cohorts may support this theory. The second possibility is that *Atg*5i mice have an altered physiological environment that is more reminiscent of a naturally aged individual, for instance, with an increased senescence load and the associated SASP. As described in the introduction, senescent cells in the microenvironment can promote tumor development through the SASP. Here it would be of great interest to see if senolytics can shift the incidence of tumors. Additionally, further studies incorporating oncogene loss and tumor suppressor inactivation in tissue specific contexts with temporal autophagy loss would greatly assist in uncovering the role of autophagy in cancer evolution.

Of note, similar aging phenotypes were observed in a second shRNA mouse *Atg*5i-2 (median survival 240 days) (Cassidy et al., 2020). While *Atg*5i mice displayed a remarkably strong knockdown efficiency, producing phenotypes reminiscent of those seen in genetic knockout models (hepatomegaly and splenomegaly), the *Atg*5i-2 mice appeared hypomorphic in nature (Cassidy et al., 2020). In particular, while they displayed a block in autophagy and an increase in markers such as p62/Sqstm1, although to a lower degree than the *Atg*5i mice, they lacked the development of hepatomegaly and splenomegaly (Cassidy et al., 2020). As in humans, aging is associated with a decrease in autophagic flux, not a complete blockade; this makes the *Atg*5i-2 mice an exceptionally interesting tool for future studies, enabling researchers to avoid modeling the extreme and non-physiological phenotypes that manifest upon complete autophagy loss. Here, however, it would be extremely important to recapitulate the tumor results seen in the original *Atg*5i mice.

Overall, our study provided missing data further linking autophagy and mammalian aging, while at the same time exploring the timing and efficacy of autophagy restoration therapies on aging pathologies. Importantly, our work also provided supportive evidence for a biphasic role of autophagy in cancer development. In normal cells, a reduction in autophagy is associated with the accumulation of pro-tumorigenic damage and possibly the establishment of premalignant disease without further progression, at least in the liver as previously reported (Takamura et al., 2011). However, once irreversible damage has accumulated, a tumorous cell that has re-established autophagic flux would undoubtedly have a fitness advantage, as autophagy is important for basic cellular homeostasis. However, it should be noted that our work does not conclusively prove this hypothesis and further studies would be required. Our data also raises the specter of long-term cancer risk in the setting of early life autophagy inhibition, such as childhood obesity, or during the use of autophagy inhibition therapies (Levy & Thorburn, 2020). Further studies, both in mice and epidemiologically in humans, would add welcome real-world data to support or refute these possibilities.

2.4 The complex relationship of autophagy and senescence

To understand the roles of autophagy, it may be useful to subclassify autophagy into "basal autophagy" and "induced autophagy" (Mizushima, 2007). Basal, low level autophagy is important for the quality control of cellular components and energy homeostasis. However, autophagy levels can also be upregulated not only upon metabolic stress (e.g. starvation) but also upon cytotoxic stress (e.g. DNA damage and oncogenic stress): "induced autophagy." In the basal context, autophagy is an important cytoprotective program helping to maintaining cellular health and quality control. With increasing age, it is well established that autophagy rates decline which, in theory, should result in more cellular damage and the induction of cellular senescence. In line with this concept, failure of autophagy in physiologically aged satellite cells has been shown to cause entry into senescence by loss of proteostasis (García-Prat et al., 2016).

The regenerative capacity of skeletal muscle relies on long-lived Pax7-expressing muscle stem cells, called satellite cells, which are normally in a quiescent state (a G0, reversible, cell cycle arrest). In response to tissue damage, these cells re-activate, re-entering the cell cycle to both replenish the stem cell population and also to differentiate into new myofiber cells. However, the number of muscle stem cells is known to decline with age, satellite cells becoming senescent in sarcopenic muscle at geriatric age (over 28 months in mice) (Sousa-Victor et al., 2014). García-Prat et al. (2016) showed an accumulation of non-degraded autophagic vesicles with p62/SQSTM1 and ubiquitin-positive inclusions in aged satellite cells. Autophagic flux was found to be compromised in the aged muscle stem cells. A 2-week treatment with rapamycin or spermidine, known autophagy inducers, was able to reduce the number of ubiquitin-positive inclusions and to reduce the expression of senescence markers p16 and γ-H2Ax seen in these cells in aged mice. As is the case for *Atg5*, overexpression of *Atg7* can also enhance autophagy levels. Introducing ectopic expression of *Atg7* in geriatric satellite cells reduced markers of senescence and rescued the associated proliferative defects seen in these stem cells. Overexpression of ATG7 alone in geriatric satellite cells (28months+) was even able to rescue their intrinsic regenerative capacity, allowing for the formation of new muscle fibers. Conversely Pax7-cre driven knockout of *Atg7* caused the satellite muscle stem cell pool to be severely reduced with those remaining showing signs of senescence including p15, p16, p21, γ-H2AX and SA-β-gal staining. García-Prat et al. attributed the reason for the autophagy defect causing stem cell senescence to be reduced mitophagy and therefore an accumulation of reactive oxygen species (ROS) (García-Prat et al., 2016). Treatment with a Vitamin-E analog was able to alleviate p62/SQSTM1 aggregate formation and reduce the senescence of aged satellite cells. Presumably the p62/SQSTM1 positive aggregates of damaged proteins were choking the autophagic pathway.

Very similar results, in terms of the accumulation of ROS in autophagy defective stem cells, have also been seen for hematopoietic stem cells (HSCs). Ho et al. (2017) saw that Mx1-cre driven *Atg*12 knockout HSCs showed increased levels of ROS. However, Ho et al. did not look at levels of senescence markers in their *Atg*12 knockout HSCs. They maintained that autophagy was required simply to maintain the health of the aged HSCs. Interestingly, *Atg*12 knockout HSCs had higher levels of oxidative phosphorylation than quiescent HSCs, and exhibited features of an activated HSC state, with increased cell size, NADH, ATP, and glucose uptake. *Atg*12 knockout HSCs also had elevated protein synthesis rates and increased cell cycle activity. Both *Atg*12 knockout HSCs and old HSCs lost quiescence and instead underwent a pro-myeloid differentiation. We too saw such a pro-myeloid skewing of the immune system in our *Atg*5i mouse (Cassidy et al., 2020). *Atg*12 knockout HSCs showed a significantly altered DNA methylation profile, consistent with the changes in fate decisions observed at steady state in autophagy-deficient HSCs, results that suggested to the authors that epigenetic remodeling through DNA demethylation was an early consequence of increased oxidative metabolism, and might play a direct role in HSC loss of stemness and accelerated myeloid differentiation. They suggested that autophagy played an essential role in clearing metabolically activated mitochondria and allowing HSCs to maintain a mostly glycolytic, quiescent state. The role of autophagy in HSCs was different in different subsets, however; while most of the blood cells including most of the HSC compartment showed no activation or a decline in autophagy with age, a fraction of old HSCs had increased autophagy levels. This seemed to preserve a subset of old HSCs from replication stress (γ-H2AX) and associated cellular aging, perhaps by protecting them from senescence.

The papers above highlight the relationship of basal autophagy with senescence, specifically the role of autophagy in the quality control of cellular components, removing damaged proteins and damaged organelles (e.g. mitochondria) and thereby acting as an anti-senescence mechanism by helping to avoid the intracellular damage that could lead to senescence. This kind of damage usually takes a while to manifest. In the context of aging it can take a lifetime. However, if autophagy is blocked or reduced in situations where, e.g. cytotoxic stress has led to a transient upregulation of autophagy levels, then this may have consequences through the inhibition of induced autophagy before the build-up of problems from basal autophagy have time to manifest.

Autophagy is activated during the dynamic phase of OIS in human diploid fibroblasts (HDFs), which is accompanied by a concerted upregulation of genes involved in both autophagy and lysosomal biogenesis (Young et al., 2009). Inhibition of autophagy, through the knockdown of key autophagy genes (*Atg*5/7), suppressed the production of the key SASP factors IL6 and IL8 and delayed the establishment of senescence. It was later identified that the subcellular localization of autophagy and mTOR plays a key role in the simultaneous activation of both anabolic and catabolic processes during senescence (Narita et al., 2011). mTOR is a nutrient-sensitive kinase that promotes global protein translation and ribosome biogenesis. In OIS in HDFs autolysosomes accumulate, forming a compartment where the sequestration of mTOR correlates with enhanced autophagosome formation as well as cytokine production (Narita et al., 2011). The spatial proximity of autolysosomes and mTOR into a "Tor autophagy spatial coupling compartment" (TASCC) may provide a mechanism to connect a cell's catabolic processes (i.e. autophagy) to its anabolic processes (i.e. protein synthesis) for the simultaneous activation of protein degradation and local protein synthesis (protein degradation providing amino acids, activating mTOR and in turn stimulating protein synthesis, in certain contexts). Thus, the TASCC efficiently coordinates the *en masse* synthesis of the SASP proteins which occurs during oncogene-induced senescence. As outlined in the introduction, through the SASP senescent cells actively communicate with their surroundings and immune system components, having a profound impact on the tissue microenvironment. The SASP is often the double edged sword of senescence; timely control of the SASP and senescence resolution is critical for tissue homeostasis as the SASP is capable of leading to chronic inflammation and even of inducing pre-disposed cells into tumorigenesis if the senescent cells are not removed from the tissue first. Proper execution of the senescence lifecycle, facilitated by induced autophagy and SASP production, is therefore critical to senescence, whereas senescence-associated pathologies may be a consequence of a deviation from an intact senescence lifecycle (Chan & Narita, 2019).

In addition to conceptualizing autophagy as basal and induced, thinking of autophagy in terms of selective and general protein degradation activities may also be helpful to explaining its context depended roles in senescence. During a search for specific regulators of the SASP, Kang et al. (2015) came across GATA4, a transcription factor that activates the NF-κB pathway, thereby regulating the SASP. GATA4 protein, but not mRNA, increased during cellular senescence, mainly due to the increased protein stability.

Surprisingly, this regulation was mediated by selective autophagy, not the ubiquitin proteasome system. Under normal conditions, GATA4 is degraded through selective autophagy by binding to p62/SQSTM1. Upon senescence, however, selective autophagy of GATA4 is disabled and GATA4 becomes stabilized to activate the SASP program and cellular senescence (Kang et al., 2015). GATA4 required a "general" autophagy activity to fully induce cellular senescence (Kang et al., 2015). Thus a specialized type of selective autophagy actively suppresses cellular senescence through the degradation of the SASP regulator, GATA4, whereas general autophagy supports the senescence transition, possibly through the TASCC. Thus, GATA4 may go some way toward resolving the apparently contradictory roles of autophagy in the regulation of cellular senescence.

3. Beyond senescence; autophagy and cell death

If cells manage to bypass replicative senescence, through loss of p16 or Rb, they enter into a phase known as crisis (Shay & Wright, 2005). Cells continue to proliferate with damaged, critically short, telomeres and eventually the telomeres can fuse, which leads to all kinds of trouble, including mis-segregation of chromosomes during mitosis and thus aneuploidy. Most of the cells that enter into this catastrophic event of crisis do not emerge from the other side, instead dying off. However, a few do survive and essentially emerge from the other side as the seeds of tumors. Such post-crisis cells exhibit characteristics of malignant transformation, such as an unstable genome and loss of checkpoint control (Shay & Wright, 2005).

In order for cells to bypass senescence and make it to crisis, cells in vitro need to have their Rb and p53 pathways disabled using, e.g. the SV40 large T antigen. Nassour et al. (2019) found that these cells were full of vacuoles, which by electron microscopy could be seen to have features of autophagosomes; the cells contained intact cytosol and also single-membrane autolysosomes, containing digested cellular components. Although suppression of autophagy using shRNAs against *ATG3*, *ATG5* and *ATG7* did not affect the proliferation rate, sh*ATG* cells did continue to proliferate and to bypass crisis whereas control cells failed to emerge from crisis and simply died off. The sh*ATG* cells that had bypassed crisis accumulated high levels of chromosomal aberrations, such as non-reciprocal translocations, deletions and aneuploidy.

What triggered the autophagy-associated cell death seen in crisis? The authors determined that whereas intrachromosomal breaks activated apoptosis, it was telomere damage that led to the activation of autophagy. The authors also observed an increase in cytoplasmic chromatin fragments in cells that had bypassed crisis. Cytoplasmic DNA can trigger the DNA-sensing cGAS–STING pathway and the authors found that cGAS and STING were essential for crisis associated cell death as cells lacking cGAS or STING proliferated beyond crisis (Nassour et al., 2019).

Autophagy has also been described to play a role in cell death in other contexts. Wakita et al. (2020), looking at the cell death associated with senolysis, showed that although levels of the cleaved form of caspase 3, a sign of apoptosis, correlated well with those of senolysis, autophagic cell death would also seem to play a role since treatment of senescent HDFs with autophagy inhibitors, such as bafilomycin A1 or chloroquine, substantially reduced their senolytic (ARV825)-induced senolysis. At the same time, treatment with bafilomycin A1 or chloroquine was accompanied by a substantial reduction of the activated caspase 3 levels. These results suggested to the authors that what was happening was autophagy induced apoptotic cell death. Also, ARV825 reduced obesity-associated hepatocellular carcinoma development in mice. The authors did not investigate how it was that autophagy was able to trigger apoptosis. One possibility may be a kind of partial degradation of mitochondria that exposes the caspase 3 and triggers the apoptosis.

Dörr et al. (2014), studying therapy induced senescence (TIS) in lymphomas, observed that TIS cells were more sensitive to a cocktail of lysosomal protease inhibitors, blocking lysosomal activity and thus autophagic flux, which lead to the selective elimination of the senescent cells through caspase-12- and caspase-3-mediated endoplasmic-reticulum-related apoptosis. Their model was transplanted Eμ-myc lymphomas, wherein tumor-bearing mice were given the DNA damaging anticancer drug cyclophosphamide (CTX). They found that these TIS cells had a very high glycolytic turnover akin to the Warburg effect. They were even able to specifically, non-invasively, detect TIS cells in mice via a combination of FLT-PET and FDG-PET scans, TIS cells being negative and positive respectively, for these markers. The TIS cells displayed a greatly expanded lysosomal compartment accompanied by structural and biochemical features of autophagy, which were not present in senescence incapable (Suv39h1 knockout; suv39h1 is histone-methyltransferase, required for senescence in these cells) lymphoma cells. Treatment of the senescent lymphoma cells with bafilomycin A1, concanamycin A (specific inhibitors of lysosomal V-ATPases), a cocktail of lysosomal protease inhibitors or knockdown of p62/SQSTM1, caused TIS-cell death at a significantly higher rate. Thus lysosomal function would seem to be essential for the survival of TIS cells. The authors postulated that the high demands placed upon the cell by having to synthesize secretory proteins *en masse* to facilitate the SASP made these cells vulnerable to a build-up of misfolded, proteins and hence endoplasmic reticulum stress, as indicated by the induction of the unfolded protein response (UPR) marker ATF4. Inactivation of NF-κB, a key regulator of the SASP, in their Eμ-myc TIS lymphomas caused the cells to be much less susceptible to bafilomycin-A1 induced cell death. They demonstrated that bafilomycin A1 treatment of TIS lymphomas was a potential therapeutic. In their mouse models, exposure of CTX-senescent Eμ-myc lymphomas to bafilomycin A1 resulted in rapid apoptotic tumor regression, and a significantly extended overall survival, which was not seen with non-senescent (Suv39h1 knockout) lymphomas.

The race is on to develop senolytics—especially in order to allow treatments that would allow us to rejuvenate ourselves in old age. To clear out senescent cells and allow repopulation from non-senescent stem cell pools. Also to clear out those senescent cells that have not been cleared by the immune system, either because the immune system declines or the "eat me" signals from the senescence cells themselves decline. Therefore, since autophagy may play a role in the triggering the cell death of cells treated with senolytics it could be very interesting indeed, if this turned out to be a more wide spread phenomenon.

4. Summary

As described in this chapter, senescent cells accumulate within our tissues with age (Burd et al., 2013; Herbig et al., 2006; Jeyapalan et al., 2007; Krishnamurthy et al., 2006). Elegant mouse models have shown that if these senescent cells are removed from most tissues this leads to the mice living around 25% longer on average than normal, although maximum possible lifespan was unaffected (Baker et al., 2016). Crucially also these mice saw up to a 25% increase in tumor free survival (Baker et al., 2016). Conversely levels of autophagy activity are known to decrease with advancing age, in a variety of organisms including mammals. Restoring that decline in autophagy levels results in a significant extension of lifespan, either through *Atg5* transgenic mice, which lived for 17% longer than their wt counterparts, or *Beclin 1* (*Atg6*) mutant mice, which lived for 11% longer on average than their wt counterparts. *Beclin 1* mutant mice also saw a drop in spontaneous tumorigenesis from 30% to 10%. Declining autophagy rates with age may well result in more cellular damage and could lead to the induction of cellular senescence. In line with this concept, failure of autophagy in physiologically aged satellite cells has been shown to cause entry into senescence by loss of proteostasis (García-Prat et al., 2016). In our recent study we also saw autophagy reduction led to senescence, and we were able to characterize that best in the livers of our mice (Cassidy et al., 2020). However, whether the connection between declining autophagy rates and senescence is generally applicable across all tissues of the body will require further investigation, as autophagy has been shown to have context dependent roles in senescence. We believe that conceptualizing autophagy as basal and induced, as well as thinking of autophagy in terms of its selective and general protein degradation activities may be helpful to explaining its context depended roles in senescence. We watch and wait for new in vivo studies connecting autophagy and senescence with great anticipation.

References

Acosta, J. C., O'Loghlen, A., Banito, A., Guijarro, M. V., Augert, A., Raguz, S., et al. (2008). Chemokine signaling via the CXCR2 receptor reinforces senescence. *Cell, 133*(6), 1006–1018. https://doi.org/10.1016/j.cell.2008.03.038.

Alimirah, F., Pulido, T., Valdovinos, A., Alptekin, S., Chang, E., Jones, E., et al. (2020). Cellular senescence promotes skin carcinogenesis through p38MAPK and p44/p42MAPK signaling. *Cancer Research, 80*(17), 3606–3619. canres.0108.2020 https://doi.org/10.1158/0008-5472.can-20-0108.

Aoshiba, K., Tsuji, T., & Nagai, A. (2003). Bleomycin induces cellular senescence in alveolar epithelial cells. *European Respiratory Journal, 22*(3), 436–443. https://doi.org/10.1183/09031936.03.00011903.

Armanios, M. (2011). Telomerase and idiopathic pulmonary fibrosis. *Mutation Research, 730*(1–2), 52–58. https://doi.org/10.1016/j.mrfmmm.2011.10.013.

Augert, A., Payré, C., de Launoit, Y., Gil, J., Lambeau, G., & Bernard, D. (2009). The M-type receptor PLA2R regulates senescence through the p53 pathway. *EMBO Reports, 10*(3), 271–277. https://doi.org/10.1038/embor.2008.255.

Baker, D. J., Childs, B. G., Durik, M., Wijers, M. E., Sieben, C. J., Zhong, J., et al. (2016). Naturally occurring p16Ink4a-positive cells shorten healthy lifespan. *Nature, 530*(7589), 184–189. https://doi.org/10.1038/nature16932.

Baker, D. J., Perez-Terzic, C., Jin, F., Pitel, K. S., Pitel, K., Niederländer, N. J., et al. (2008). Opposing roles for p16Ink4a and p19Arf in senescence and ageing caused by BubR1 insufficiency. *Nature Cell Biology, 10*(7), 825–836. https://doi.org/10.1038/ncb1744.

Baker, D. J., Wijshake, T., Tchkonia, T., LeBrasseur, N. K., Childs, B. G., van de Sluis, B., et al. (2011). Clearance of p16Ink4a-positive senescent cells delays ageing-associated disorders. *Nature, 479*(7372), 232–236. https://doi.org/10.1038/nature10600.

Ballester, B., Milara, J., & Cortijo, J. (2019). Idiopathic pulmonary fibrosis and lung cancer: Mechanisms and molecular targets. *International Journal of Molecular Sciences, 20*(3), 593. https://doi.org/10.3390/ijms20030593.

Balmain, A., & Pragnell, I. B. (1983). Mouse skin carcinomas induced in vivo by chemical carcinogens have a transforming Harvey-ras oncogene. *Nature*, *303*(5912), 72–74. https://doi.org/10.1038/303072a0.

Balmain, A., Ramsden, M., Bowden, G. T., & Smith, J. (1984). Activation of the mouse cellular Harvey-ras gene in chemically induced benign skin papillomas. *Nature*, *307*(5952), 658–660. https://doi.org/10.1038/307658a0.

Bjedov, I., Toivonen, J. M., Kerr, F., Slack, C., Jacobson, J., Foley, A., et al. (2010). Mechanisms of life span extension by rapamycin in the fruit fly Drosophila melanogaster. *Cell Metabolism*, *11*(1), 35–46. https://doi.org/10.1016/j.cmet.2009.11.010.

Braig, M., Lee, S., Loddenkemper, C., Rudolph, C., Peters, A. H. F. M., Schlegelberger, B., et al. (2005). Oncogene-induced senescence as an initial barrier in lymphoma development. *Nature*, *436*(7051), 660–665. https://doi.org/10.1038/nature03841.

Burd, C. E., Sorrentino, J. A., Clark, K. S., Darr, D. B., Krishnamurthy, J., Deal, A. M., et al. (2013). Monitoring tumorigenesis and senescence in vivo with a p16INK4a-luciferase model. *Cell*, *152*(1–2), 340–351. https://doi.org/10.1016/j.cell.2012.12.010.

Campisi, J. (2013). Aging, cellular senescence, and cancer. *Annual Review of Physiology*, *75*(1), 685–705. https://doi.org/10.1146/annurev-physiol-030212-183653.

Cassidy, L. D., Young, A. R., Pérez-Mancera, P. A., Nimmervoll, B., Jaulim, A., Chen, H.-C., et al. (2018). A novel Atg5-shRNA mouse model enables temporal control of autophagy in vivo. *Autophagy*, *14*(7), 1256–1266. https://doi.org/10.1080/15548627.2018.1458172.

Cassidy, L. D., Young, A. R. J., Young, C. N. J., Soilleux, E. J., Fielder, E., Weigand, B. M., et al. (2020). Temporal inhibition of autophagy reveals segmental reversal of ageing with increased cancer risk. *Nature Communications*, *11*(1), 307. https://doi.org/10.1038/s41467-019-14187-x.

Chan, A. S. L., & Narita, M. (2019). Short-term gain, long-term pain: The senescence life cycle and cancer. *Genes & Development*, *33*(3–4), 127–143. https://doi.org/10.1101/gad.320937.118.

Chen, Z., Trotman, L. C., Shaffer, D., Lin, H.-K., Dotan, Z. A., Niki, M., et al. (2005). Crucial role of p53-dependent cellular senescence in suppression of Pten-deficient tumorigenesis. *Nature*, *436*(7051), 725–730. https://doi.org/10.1038/nature03918.

Chilosi, M., Carloni, A., Rossi, A., & Poletti, V. (2013). Premature lung aging and cellular senescence in the pathogenesis of idiopathic pulmonary fibrosis and COPD/emphysema. *Translational Research*, *162*(3), 156–173. https://doi.org/10.1016/j.trsl.2013.06.004.

Clark, S. L. (1957). Cellular differentiation in the kidneys of newborn mice studied with the electron microscope. *The Journal of Biophysical and Biochemical Cytology*, *3*(3), 349–362. https://doi.org/10.1083/jcb.3.3.349.

Collado, M., Gil, J., Efeyan, A., Guerra, C., Schuhmacher, A. J., Barradas, M., et al. (2005). Senescence in premalignant tumours. *Nature*, *436*(7051), 642. https://doi.org/10.1038/436642a.

Coppé, J.-P., Patil, C. K., Rodier, F., Sun, Y., Muñoz, D. P., Goldstein, J., et al. (2008). Senescence-associated secretory phenotypes reveal cell-nonautonomous functions of oncogenic RAS and the p53 tumor suppressor. *PLoS Biology*, *6*(12), e301. https://doi.org/10.1371/journal.pbio.0060301.

Correia-Melo, C., Marques, F. D., Anderson, R., Hewitt, G., Hewitt, R., Cole, J., et al. (2016). Mitochondria are required for pro-ageing features of the senescent phenotype. *The EMBO Journal*, *35*(7), 724–742. https://doi.org/10.15252/embj.201592862.

Courtois-Cox, S., Williams, S. M. G., Reczek, E. E., Johnson, B. W., McGillicuddy, L. T., Johannessen, C. M., et al. (2006). A negative feedback signaling network underlies oncogene-induced senescence. *Cancer Cell*, *10*(6), 459–472. https://doi.org/10.1016/j.ccr.2006.10.003.

de Duve, C. (2005). The lysosome turns fifty. *Nature Cell Biology*, 7(9), 847–879. https://doi.org/10.1038/ncb0905-847.
de Magalhães, J. P. (2013). How ageing processes influence cancer. *Nature Reviews Cancer*, *13*(5), 357–365. https://doi.org/10.1038/nrc3497.
Demaria, M., Ohtani, N., Youssef, S. A., Rodier, F., Toussaint, W., Mitchell, J. R., et al. (2014). An essential role for senescent cells in optimal wound healing through secretion of PDGF-AA. *Developmental Cell*, *31*(6), 722–733. https://doi.org/10.1016/j.devcel.2014.11.012.
Denchi, E. L., Attwooll, C., Pasini, D., & Helin, K. (2005). Deregulated E2F activity induces hyperplasia and senescence-like features in the mouse pituitary gland†. *Molecular and Cellular Biology*, *25*(7), 2660–2672. https://doi.org/10.1128/mcb.25.7.2660-2672.2005.
Diabetes Genetics Initiative of Broad Institute of Harvard and MIT, Lund University, and Novartis Institutes of BioMedical Research, Saxena, R., Voight, B. F., Lyssenko, V., Burtt, N. P., de Bakker, P. I. W., et al. (2007). Genome-wide association analysis identifies loci for type 2 diabetes and triglyceride levels. *Science*, *316*(5829), 1331–1336. https://doi.org/10.1126/science.1142358.
Dimri, G. P., Lee, X., Basile, G., Acosta, M., Scott, G., Roskelley, C., et al. (1995). A biomarker that identifies senescent human cells in culture and in aging skin in vivo. *Proceedings of the National Academy of Sciences*, *92*(20), 9363–9367. https://doi.org/10.1073/pnas.92.20.9363.
Dörr, J. R., Yu, Y., Milanovic, M., Beuster, G., Zasada, C., Däbritz, J. H. M., et al. (2014). Synthetic lethal metabolic targeting of cellular senescence in cancer therapy. *Nature*, *501*(7467), 421–425. https://doi.org/10.1038/nature12437.
Dow, L. E., Premsrirut, P. K., Zuber, J., Fellmann, C., McJunkin, K., Miething, C., et al. (2012). A pipeline for the generation of shRNA transgenic mice. *Nature Protocols*, 7(2), 374–393. https://doi.org/10.1038/nprot.2011.446.
Eggert, T., Wolter, K., Ji, J., Ma, C., Yevsa, T., Klotz, S., et al. (2016). Distinct functions of senescence-associated immune responses in liver tumor surveillance and tumor progression. *Cancer Cell*, *30*(4), 533–547. https://doi.org/10.1016/j.ccell.2016.09.003.
Eisenberg, T., Knauer, H., Schauer, A., Büttner, S., Ruckenstuhl, C., Carmona-Gutierrez, D., et al. (2009). Induction of autophagy by spermidine promotes longevity. *Nature Cell Biology*, *11*(11), 1305–1314. https://doi.org/10.1038/ncb1975.
Extermann, M. (2014). Metabolic syndrome, aging, and cancer. *Critical Reviews in Oncogenesis*, *18*(6), 515–529. https://doi.org/10.1615/critrevoncog.2014010612.
Fernández, Á. F., Sebti, S., Wei, Y., Zou, Z., Shi, M., McMillan, K. L., et al. (2018). Disruption of the beclin 1–BCL2 autophagy regulatory complex promotes longevity in mice. *Nature*, *558*(7708), 136–140. https://doi.org/10.1038/s41586-018-0162-7.
García-Prat, L., Martínez-Vicente, M., Perdiguero, E., Ortet, L., Rodríguez-Ubreva, J., Rebollo, E., et al. (2016). Autophagy maintains stemness by preventing senescence. *Nature*, *529*(7584), 37–42. https://doi.org/10.1038/nature16187.
García-Sancho, C., Buendía-Roldán, I., Fernández-Plata, M. R., Navarro, C., Pérez-Padilla, R., Vargas, M. H., et al. (2011). Familial pulmonary fibrosis is the strongest risk factor for idiopathic pulmonary fibrosis. *Respiratory Medicine*, *105*(12), 1902–1907. https://doi.org/10.1016/j.rmed.2011.08.022.
Gewinner, C., Wang, Z. C., Richardson, A., Teruya-Feldstein, J., Etemadmoghadam, D., Bowtell, D., et al. (2009). Evidence that inositol polyphosphate 4-phosphatase type ii is a tumor suppressor that inhibits PI3K signaling. *Cancer Cell*, *16*(2), 115–125. https://doi.org/10.1016/j.ccr.2009.06.006.
Gorgoulis, V. G., & Halazonetis, T. D. (2010). Oncogene-induced senescence: The bright and dark side of the response. *Current Opinion in Cell Biology*, *22*(6), 816–827. https://doi.org/10.1016/j.ceb.2010.07.013.

Grosse, L., Wagner, N., Emelyanov, A., Molina, C., Lacas-Gervais, S., Wagner, K.-D., et al. (2020). Defined p16 high senescent cell types are indispensable for mouse healthspan. *Cell Metabolism*, *32*(1), 87–99.e6. https://doi.org/10.1016/j.cmet.2020.05.002.
Hara, T., Nakamura, K., Matsui, M., Yamamoto, A., Nakahara, Y., Suzuki-Migishima, R., et al. (2006). Suppression of basal autophagy in neural cells causes neurodegenerative disease in mice. *Nature*, *441*(7095), 885–889. https://doi.org/10.1038/nature04724.
Harding, T. M., Morano, K. A., Scott, S. V., & Klionsky, D. J. (1995). Isolation and characterization of yeast mutants in the cytoplasm to vacuole protein targeting pathway. *The Journal of Cell Biology*, *131*(3), 591–602. https://doi.org/10.1083/jcb.131.3.591.
Harrison, D. E., Strong, R., Sharp, Z. D., Nelson, J. F., Astle, C. M., Flurkey, K., et al. (2009). Rapamycin fed late in life extends lifespan in genetically heterogeneous mice. *Nature*, *460*(7253), 392–395. https://doi.org/10.1038/nature08221.
Hayflick, L. (1965). The limited in vitro lifetime of human diploid cell strains. *Experimental Cell Research*, *37*(3), 614–636. https://doi.org/10.1016/0014-4827(65)90211-9.
Hayflick, L., & Moorhead, P. S. (1961). The serial cultivation of human diploid cell strains. *Experimental Cell Research*, *25*(3), 585–621. https://doi.org/10.1016/0014-4827(61)90192-6.
Helman, A., Klochendler, A., Azazmeh, N., Gabai, Y., Horwitz, E., Anzi, S., et al. (2016). p16Ink4a-induced senescence of pancreatic beta cells enhances insulin secretion. *Nature Medicine*, *22*(4), 412–420. https://doi.org/10.1038/nm.4054.
Herbig, U., Ferreira, M., Condel, L., Carey, D., & Sedivy, J. M. (2006). Cellular senescence in aging primates. *Science*, *311*(5765), 1257. https://doi.org/10.1126/science.1122446.
Hewitt, G., Jurk, D., Marques, F. D. M., Correia-Melo, C., Hardy, T., Gackowska, A., et al. (2012). Telomeres are favoured targets of a persistent DNA damage response in ageing and stress-induced senescence. *Nature Communications*, *3*(1), 708. https://doi.org/10.1038/ncomms1708.
Ho, T. T., Warr, M. R., Adelman, E. R., Lansinger, O. M., Flach, J., Verovskaya, E. V., et al. (2017). Autophagy maintains the metabolism and function of young and old stem cells. *Nature*, *543*(7644), 205–210. https://doi.org/10.1038/nature21388.
Holtzman, E. (1989). *Lysosomes*. Springer. https://doi.org/10.1007/978-1-4899-2540-4.
Iannello, A., Thompson, T. W., Ardolino, M., Lowe, S. W., & Raulet, D. H. (2013). p53-dependent chemokine production by senescent tumor cells supports NKG2D-dependent tumor elimination by natural killer cells. *The Journal of Experimental Medicine*, *210*(10), 2057–2069. https://doi.org/10.1084/jem.20130783.
Ichimura, Y., Kirisako, T., Takao, T., Satomi, Y., Shimonishi, Y., Ishihara, N., et al. (2000). A ubiquitin-like system mediates protein lipidation. *Nature*, *408*(6811), 488–492. https://doi.org/10.1038/35044114.
Jeon, O. H., David, N., Campisi, J., & Elisseeff, J. H. (2018). Senescent cells and osteoarthritis: A painful connection. *Journal of Clinical Investigation*, *128*(4), 1229–1237. https://doi.org/10.1172/jci95147.
Jeon, O. H., Kim, C., Laberge, R.-M., Demaria, M., Rathod, S., Vasserot, A. P., et al. (2017). Local clearance of senescent cells attenuates the development of post-traumatic osteoarthritis and creates a pro-regenerative environment. *Nature Medicine*, *23*(6), 775–781. https://doi.org/10.1038/nm.4324.
Jeyapalan, J. C., Ferreira, M., Sedivy, J. M., & Herbig, U. (2007). Accumulation of senescent cells in mitotic tissue of aging primates. *Mechanisms of Ageing and Development*, *128*(1), 36–44. https://doi.org/10.1016/j.mad.2006.11.008.
Jin, S. M., & Youle, R. J. (2012). PINK1- and Parkin-mediated mitophagy at a glance. *Journal of Cell Science*, *125*(Pt. 4), 795–799. https://doi.org/10.1242/jcs.093849.
Jurk, D., Wilson, C., Passos, J. F., Oakley, F., Correia-Melo, C., Greaves, L., et al. (2014). Chronic inflammation induces telomere dysfunction and accelerates ageing in mice. *Nature Communications*, *5*(1), 4172. https://doi.org/10.1038/ncomms5172.

Kang, C., Xu, Q., Martin, T. D., Li, M. Z., Demaria, M., Aron, L., et al. (2015). The DNA damage response induces inflammation and senescence by inhibiting autophagy of GATA4. *Science, 349*(6255), aaa5612. https://doi.org/10.1126/science.aaa5612.

Kang, T.-W., Yevsa, T., Woller, N., Hoenicke, L., Wuestefeld, T., Dauch, D., et al. (2011). Senescence surveillance of pre-malignant hepatocytes limits liver cancer development. *Nature, 479*(7374), 547–551. https://doi.org/10.1038/nature10599.

Kapahi, P., Zid, B. M., Harper, T., Koslover, D., Sapin, V., & Benzer, S. (2004). Regulation of lifespan in Drosophila by modulation of genes in the TOR signaling pathway. *Current Biology, 14*(10), 885–890. https://doi.org/10.1016/j.cub.2004.03.059.

Karsli-Uzunbas, G., Guo, J. Y., Price, S., Teng, X., Laddha, S. V., Khor, S., et al. (2014). Autophagy is required for glucose homeostasis and lung tumor maintenance. *Cancer Discovery, 4*(8), 914–927. https://doi.org/10.1158/2159-8290.cd-14-0363.

Kenyon, C. (2011). The first long-lived mutants: Discovery of the insulin/IGF-1 pathway for ageing. *Philosophical Transactions of the Royal Society B: Biological Sciences, 366*(1561), 9–16. https://doi.org/10.1098/rstb.2010.0276.

Kenyon, C., Chang, J., Gensch, E., Rudner, A., & Tabtiang, R. (1993). A C. elegans mutant that lives twice as long as wild type. *Nature, 366*(6454), 461–464. https://doi.org/10.1038/366461a0.

Kimura, K. D., Tissenbaum, H. A., Liu, Y., & Ruvkun, G. (1997). *daf-2*, an insulin receptor-like gene that regulates longevity and diapause in *Caenorhabditis elegans*. *Science, 277*(5328), 942–946. https://doi.org/10.1126/science.277.5328.942.

Kirisako, T., Baba, M., Ishihara, N., Miyazawa, K., Ohsumi, M., Yoshimori, T., et al. (1999). Formation process of autophagosome is traced with Apg8/Aut7p in yeast. *Journal of Cell Biology, 147*(2), 435–446. https://doi.org/10.1083/jcb.147.2.435.

Kirisako, T., Ichimura, Y., Okada, H., Kabeya, Y., Mizushima, N., Yoshimori, T., et al. (2000). The reversible modification regulates the membrane-binding state of Apg8/Aut7 essential for autophagy and the cytoplasm to vacuole targeting pathway. *Journal of Cell Biology, 151*(2), 263–276. https://doi.org/10.1083/jcb.151.2.263.

Kirkin, V., & Rogov, V. V. (2019). A diversity of selective autophagy receptors determines the specificity of the autophagy pathway. *Molecular Cell, 76*(2), 268–285. https://doi.org/10.1016/j.molcel.2019.09.005.

Klionsky, D. J., Cregg, J. M., Dunn, W. A., Emr, S. D., Sakai, Y., Sandoval, I. V., et al. (2003). A unified nomenclature for yeast autophagy-related genes. *Developmental Cell, 5*(4), 539–545. https://doi.org/10.1016/s1534-5807(03)00296-x.

Komatsu, M., Waguri, S., Chiba, T., Murata, S., Iwata, J., Tanida, I., et al. (2006). Loss of autophagy in the central nervous system causes neurodegeneration in mice. *Nature, 441*(7095), 880–884. https://doi.org/10.1038/nature04723.

Komatsu, M., Waguri, S., Ueno, T., Iwata, J., Murata, S., Tanida, I., et al. (2005). Impairment of starvation-induced and constitutive autophagy in Atg7-deficient mice. *Journal of Cell Biology, 169*(3), 425–434. https://doi.org/10.1083/jcb.200412022.

Kovacs, E., Lowery, E., Kuhlmann, E., & Brubaker, A. (2013). The aging lung. *Clinical Interventions in Aging, 8*, 1489–1496. https://doi.org/10.2147/cia.s51152.

Krishnamurthy, J., Ramsey, M. R., Ligon, K. L., Torrice, C., Koh, A., Bonner-Weir, S., et al. (2006). p16INK4a induces an age-dependent decline in islet regenerative potential. *Nature, 443*(7110), 453–457. https://doi.org/10.1038/nature05092.

Krizhanovsky, V., Yon, M., Dickins, R. A., Hearn, S., Simon, J., Miething, C., et al. (2008). Senescence of activated stellate cells limits liver fibrosis. *Cell, 134*(4), 657–667. https://doi.org/10.1016/j.cell.2008.06.049.

Krtolica, A., & Campisi, J. (2002). Cancer and aging: A model for the cancer promoting effects of the aging stroma. *The International Journal of Biochemistry & Cell Biology, 34*(11), 1401–1414. https://doi.org/10.1016/s1357-2725(02)00053-5.

Krtolica, A., Parrinello, S., Lockett, S., Desprez, P.-Y., & Campisi, J. (2001). Senescent fibroblasts promote epithelial cell growth and tumorigenesis: A link between cancer and aging. *Proceedings of the National Academy of Sciences*, *98*(21), 12072–12077. https://doi.org/10.1073/pnas.211053698.

Kuilman, T., Michaloglou, C., Vredeveld, L. C. W., Douma, S., van Doorn, R., Desmet, C. J., et al. (2008). Oncogene-induced senescence relayed by an interleukin-dependent inflammatory network. *Cell*, *133*(6), 1019–1031. https://doi.org/10.1016/j.cell.2008.03.039.

Kukull, W. A., Higdon, R., Bowen, J. D., McCormick, W. C., Teri, L., Schellenberg, G. D., et al. (2002). Dementia and Alzheimer disease incidence: A prospective cohort study. *Archives of Neurology*, *59*(11), 1737. https://doi.org/10.1001/archneur.59.11.1737.

Kuma, A., Hatano, M., Matsui, M., Yamamoto, A., Nakaya, H., Yoshimori, T., et al. (2004). The role of autophagy during the early neonatal starvation period. *Nature*, *432*(7020), 1032–1036. https://doi.org/10.1038/nature03029.

Kuma, A., Komatsu, M., & Mizushima, N. (2017). Autophagy-monitoring and autophagy-deficient mice. *Autophagy*, *13*(10), 1619–1628. https://doi.org/10.1080/15548627.2017.1343770.

Kuro-o, M. (2009). Klotho and aging. *Biochimica et Biophysica Acta (BBA)—General Subjects*, *1790*(10), 1049–1058. https://doi.org/10.1016/j.bbagen.2009.02.005.

Lagouge, M., Argmann, C., Gerhart-Hines, Z., Meziane, H., Lerin, C., Daussin, F., et al. (2006). Resveratrol improves mitochondrial function and protects against metabolic disease by activating SIRT1 and PGC-1α. *Cell*, *127*(6), 1109–1122. https://doi.org/10.1016/j.cell.2006.11.013.

Lakatta, E. G., & Levy, D. (2003). Arterial and cardiac aging: Major shareholders in cardiovascular disease enterprises. *Circulation*, *107*(2), 346–354. https://doi.org/10.1161/01.cir.0000048893.62841.f7.

Larsson, L., Degens, H., Li, M., Salviati, L., Lee, Y.i., Thompson, W., et al. (2019). Sarcopenia: Aging-related loss of muscle mass and function. *Physiological Reviews*, *99*(1), 427–511. https://doi.org/10.1152/physrev.00061.2017.

Levy, J. M. M., & Thorburn, A. (2020). Autophagy in cancer: Moving from understanding mechanism to improving therapy responses in patients. *Cell Death and Differentiation*, *27*(3), 843–857. https://doi.org/10.1038/s41418-019-0474-7.

Liang, X. H., Jackson, S., Seaman, M., Brown, K., Kempkes, B., Hibshoosh, H., et al. (1999). Induction of autophagy and inhibition of tumorigenesis by beclin 1. *Nature*, *402*(6762), 672–676. https://doi.org/10.1038/45257.

Lim, L. S., Mitchell, P., Seddon, J. M., Holz, F. G., & Wong, T. Y. (2012). Age-related macular degeneration. *The Lancet*, *379*(9827), 1728–1738. https://doi.org/10.1016/s0140-6736(12)60282-7.

Lin, A. W., Barradas, M., Stone, J. C., van Aelst, L., Serrano, M., & Lowe, S. W. (1998). Premature senescence involving p53 and p16 is activated in response to constitutive MEK/MAPK mitogenic signaling. *Genes & Development*, *12*(19), 3008–3019. https://doi.org/10.1101/gad.12.19.3008.

Lujambio, A., Akkari, L., Simon, J., Grace, D., Tschaharganeh, D. F., Bolden, J. E., et al. (2013). Non-cell-autonomous tumor suppression by p53. *Cell*, *153*(2), 449–460. https://doi.org/10.1016/j.cell.2013.03.020.

Maiuri, M. C., Tasdemir, E., Criollo, A., Morselli, E., Vicencio, J. M., Carnuccio, R., et al. (2009). Control of autophagy by oncogenes and tumor suppressor genes. *Cell Death and Differentiation*, *16*(1), 87–93. https://doi.org/10.1038/cdd.2008.131.

Maksymowych, W. P., Russell, A. S., Chiu, P., Yan, A., Jones, N., Clare, T., et al. (2012). Targeting tumour necrosis factor alleviates signs and symptoms of inflammatory osteoarthritis of the knee. *Arthritis Research & Therapy*, *14*(5), R206. https://doi.org/10.1186/ar4044.

Maldonado, M., & Nam, J. (2013). The role of changes in extracellular matrix of cartilage in the presence of inflammation on the pathology of osteoarthritis. *BioMed Research International, 2013*, 1–10. https://doi.org/10.1155/2013/284873.

Meléndez, A., Tallóczy, Z., Seaman, M., Eskelinen, E.-L., Hall, D. H., & Levine, B. (2003). Autophagy genes are essential for Dauer development and life-span extension in C. elegans. *Science, 301*(5638), 1387–1391. https://doi.org/10.1126/science.1087782.

Michaloglou, C., Vredeveld, L. C. W., Soengas, M. S., Denoyelle, C., Kuilman, T., van der Horst, C. M. A. M., et al. (2005). BRAFE600-associated senescence-like cell cycle arrest of human naevi. *Nature, 436*(7051), 720–724. https://doi.org/10.1038/nature03890.

Mizushima, N. (2007). Autophagy: Process and function. *Genes & Development, 21*(22), 2861–2873. https://doi.org/10.1101/gad.1599207.

Mizushima, N., Sugita, H., Yoshimori, T., & Ohsumi, Y. (1998). A new protein conjugation system in human: The counterpart of the yeast Apg12p conjugation system essential for autophagy. *Journal of Biological Chemistry, 273*(51), 33889–33892. https://doi.org/10.1074/jbc.273.51.33889.

Mizushima, N., Yamamoto, A., Hatano, M., Kobayashi, Y., Kabeya, Y., Suzuki, K., et al. (2001). Dissection of autophagosome formation using Apg5-deficient mouse embryonic stem cells. *Journal of Cell Biology, 152*(4), 657–668. https://doi.org/10.1083/jcb.152.4.657.

Mizushima, N., Yamamoto, A., Matsui, M., Yoshimori, T., & Ohsumi, Y. (2004). In vivo analysis of autophagy in response to nutrient starvation using transgenic mice expressing a fluorescent autophagosome marker. *Molecular Biology of the Cell, 15*(3), 1101–1111. https://doi.org/10.1091/mbc.e03-09-0704.

Muñoz-Espín, D., & Serrano, M. (2014). Cellular senescence: From physiology to pathology. *Nature Reviews Molecular Cell Biology, 15*(7), 482–496. https://doi.org/10.1038/nrm3823.

Narita, M., Nuñez, S., Heard, E., Narita, M., Lin, A. W., Hearn, S. A., et al. (2003). Rb-mediated heterochromatin formation and silencing of E2F target genes during cellular senescence. *Cell, 113*(6), 703–716. https://doi.org/10.1016/s0092-8674(03)00401-x.

Narita, M., Young, A. R. J., Arakawa, S., Samarajiwa, S. A., Nakashima, T., Yoshida, S., et al. (2011). Spatial coupling of mTOR and autophagy augments secretory phenotypes. *Science, 332*(6032), 966–970. https://doi.org/10.1126/science.1205407.

Nassour, J., Radford, R., Correia, A., Fusté, J. M., Schoell, B., Jauch, A., et al. (2019). Autophagic cell death restricts chromosomal instability during replicative crisis. *Nature, 565*(7741), 659–663. https://doi.org/10.1038/s41586-019-0885-0.

O'Sullivan, E. D., Hughes, J., & Ferenbach, D. A. (2017). Renal aging: Causes and consequences. *Journal of the American Society of Nephrology, 28*(2), 407–420. https://doi.org/10.1681/asn.2015121308.

Omori, S., Wang, T.-W., Johmura, Y., Kanai, T., Nakano, Y., Kido, T., et al. (2020). Generation of a p16 reporter mouse and its use to characterize and target p16high cells in vivo. *Cell Metabolism, 32*(5), 814–828.e6. https://doi.org/10.1016/j.cmet.2020.09.006.

Parrinello, S., Coppe, J.-P., Krtolica, A., & Campisi, J. (2005). Stromal-epithelial interactions in aging and cancer: Senescent fibroblasts alter epithelial cell differentiation. *Journal of Cell Science, 118*(3), 485–496. https://doi.org/10.1242/jcs.01635.

Pietschmann, P., Rauner, M., Sipos, W., & Kerschan-Schindl, K. (2009). Osteoporosis: An age-related and gender-specific disease—A mini-review. *Gerontology, 55*(1), 3–12. https://doi.org/10.1159/000166209.

Pyo, J.-O., Yoo, S.-M., Ahn, H.-H., Nah, J., Hong, S.-H., Kam, T.-I., et al. (2013). Overexpression of Atg5 in mice activates autophagy and extends lifespan. *Nature Communications, 4*(1), 2300. https://doi.org/10.1038/ncomms3300.

Reeve, A., Simcox, E., & Turnbull, D. (2014). Ageing and Parkinson's disease: Why is advancing age the biggest risk factor? *Ageing Research Reviews, 14*, 19–30. https://doi.org/10.1016/j.arr.2014.01.004.

Ritschka, B., Storer, M., Mas, A., Heinzmann, F., Ortells, M. C., Morton, J. P., et al. (2017). The senescence-associated secretory phenotype induces cellular plasticity and tissue regeneration. *Genes & Development*, *31*(2), 172–183. https://doi.org/10.1101/gad.290635.116.

Saunders, T. L. (2010). Transgenic mouse methods and protocols. *Methods in Molecular Biology (Clifton, N.J.)*, *693*, 103–115. https://doi.org/10.1007/978-1-60761-974-1_7.

Serrano, M., Lin, A. W., McCurrach, M. E., Beach, D., & Lowe, S. W. (1997). Oncogenic ras provokes premature cell senescence associated with accumulation of p53 and p16INK4a. *Cell*, *88*(5), 593–602. https://doi.org/10.1016/s0092-8674(00)81902-9.

Shamma, A., Takegami, Y., Miki, T., Kitajima, S., Noda, M., Obara, T., et al. (2009). Rb regulates DNA damage response and cellular senescence through E2F-dependent suppression of N-Ras isoprenylation. *Cancer Cell*, *15*(4), 255–269. https://doi.org/10.1016/j.ccr.2009.03.001.

Shay, J. W., & Wright, W. E. (2005). Senescence and immortalization: Role of telomeres and telomerase. *Carcinogenesis*, *26*(5), 867–874. https://doi.org/10.1093/carcin/bgh296.

Shelton, D. N., Chang, E., Whittier, P. S., Choi, D., & Funk, W. D. (1999). Microarray analysis of replicative senescence. *Current Biology*, *9*(17), 939–945. https://doi.org/10.1016/s0960-9822(99)80420-5.

Sone, H., & Kagawa, Y. (2005). Pancreatic beta cell senescence contributes to the pathogenesis of type 2 diabetes in high-fat diet-induced diabetic mice. *Diabetologia*, *48*(1), 58–67. https://doi.org/10.1007/s00125-004-1605-2.

Sousa-Victor, P., Gutarra, S., García-Prat, L., Rodriguez-Ubreva, J., Ortet, L., Ruiz-Bonilla, V., et al. (2014). Geriatric muscle stem cells switch reversible quiescence into senescence. *Nature*, *506*(7488), 316–321. https://doi.org/10.1038/nature13013.

Suzuki, K., Kirisako, T., Kamada, Y., Mizushima, N., Noda, T., & Ohsumi, Y. (2001). The pre-autophagosomal structure organized by concerted functions of APG genes is essential for autophagosome formation. *The EMBO Journal*, *20*(21), 5971–5981. https://doi.org/10.1093/emboj/20.21.5971.

Takamura, A., Komatsu, M., Hara, T., Sakamoto, A., Kishi, C., Waguri, S., et al. (2011). Autophagy-deficient mice develop multiple liver tumors. *Genes & Development*, *25*(8), 795–800. https://doi.org/10.1101/gad.2016211.

Tanida, I., Mizushima, N., Kiyooka, M., Ohsumi, M., Ueno, T., Ohsumi, Y., et al. (1999). Apg7p/Cvt2p: A novel protein-activating enzyme essential for autophagy. *Molecular Biology of the Cell*, *10*(5), 1367–1379. https://doi.org/10.1091/mbc.10.5.1367.

Tanida, I., Sou, Y., Ezaki, J., Minematsu-Ikeguchi, N., Ueno, T., & Kominami, E. (2004). HsAtg4B/HsApg4B/Autophagin-1 cleaves the carboxyl termini of three human Atg8 homologues and delipidates microtubule-associated protein light chain 3- and GABA A receptor-associated protein-phospholipid conjugates. *Journal of Biological Chemistry*, *279*(35), 36268–36276. https://doi.org/10.1074/jbc.m401461200.

Thumm, M., Egner, R., Koch, B., Schlumpberger, M., Straub, M., Veenhuis, M., et al. (1994). Isolation of autophagocytosis mutants of Saccharomyces cerevisiae. *FEBS Letters*, *349*(2), 275–280. https://doi.org/10.1016/0014-5793(94)00672-5.

Tsukada, M., & Ohsumi, Y. (1993). Isolation and characterization of autophagy-defective mutants of Saccharomyces cerevisiae. *FEBS Letters*, *333*(1–2), 169–174. https://doi.org/10.1016/0014-5793(93)80398-e.

Wajapeyee, N., Serra, R. W., Zhu, X., Mahalingam, M., & Green, M. R. (2008). Oncogenic BRAF induces senescence and apoptosis through pathways mediated by the secreted protein IGFBP7. *Cell*, *132*(3), 363–374. https://doi.org/10.1016/j.cell.2007.12.032.

Wakita, M., Takahashi, A., Sano, O., Loo, T. M., Imai, Y., Narukawa, M., et al. (2020). A BET family protein degrader provokes senolysis by targeting NHEJ and autophagy in senescent cells. *Nature Communications*, *11*(1), 1935. https://doi.org/10.1038/s41467-020-15719-6.

Wang, J. C., & Bennett, M. (2012). Aging and atherosclerosis. *Circulation Research, 111*(2), 245–259. https://doi.org/10.1161/circresaha.111.261388.

Yoshii, S. R., Kuma, A., Akashi, T., Hara, T., Yamamoto, A., Kurikawa, Y., et al. (2016). Systemic analysis of Atg5-null mice rescued from neonatal lethality by transgenic ATG5 expression in neurons. *Developmental Cell, 39*(1), 116–130. https://doi.org/10.1016/j.devcel.2016.09.001.

Young, A. R. J., & Narita, M. (2009). SASP reflects senescence. *EMBO Reports, 10*(3), 228–230. https://doi.org/10.1038/embor.2009.22.

Young, A. R. J., Narita, M., Ferreira, M., Kirschner, K., Sadaie, M., Darot, J. F. J., et al. (2009). Autophagy mediates the mitotic senescence transition. *Genes & Development, 23*(7), 798–803. https://doi.org/10.1101/gad.519709.

Zhang, R., Poustovoitov, M. V., Ye, X., Santos, H. A., Chen, W., Daganzo, S. M., et al. (2005). Formation of MacroH2A-containing senescence-associated heterochromatin foci and senescence driven by ASF1a and HIRA. *Developmental Cell, 8*(1), 19–30. https://doi.org/10.1016/j.devcel.2004.10.019.

Zhu, J., Woods, D., McMahon, M., & Bishop, J. M. (1998). Senescence of human fibroblasts induced by oncogenic Raf. *Genes & Development, 12*(19), 2997–3007. https://doi.org/10.1101/gad.12.19.2997.

CHAPTER FOUR

Autophagy and senescence: Insights from normal and cancer stem cells

Sarmistha Talukdar[a,b], Swadesh K. Das[a,b,c], Luni Emdad[a,b,c], and Paul B. Fisher[a,b,c,*]

[a]Department of Human and Molecular Genetics, Virginia Commonwealth University, School of Medicine, Richmond, VA, United States
[b]VCU Institute of Molecular Medicine, Virginia Commonwealth University, School of Medicine, Richmond, VA, United States
[c]VCU Massey Cancer Center, Virginia Commonwealth University, School of Medicine, Richmond, VA, United States
*Corresponding author: e-mail address: paul.fisher@vcuhealth.org

Contents

Abstract

Autophagy is a fundamental cellular process, which allows cells to adapt to metabolic stress through the degradation and recycling of intracellular components to generate macromolecular precursors and produce energy. Autophagy is also critical in maintaining cellular/tissue homeostasis, as well preserving immunity and preventing human disease. Deregulation of autophagic processes is associated with cancer,

Advances in Cancer Research, Volume 150
ISSN 0065-230X
https://doi.org/10.1016/bs.acr.2021.01.005

neurodegeneration, muscle and heart disease, infectious diseases and aging. Research on a variety of stem cell types establish that autophagy plays critical roles in normal and cancer stem cell quiescence, activation, differentiation, and self-renewal. Considering its critical function in regulating the metabolic state of stem cells, autophagy plays a dual role in the regulation of normal and cancer stem cell senescence, and cellular responses to various therapeutic strategies. The relationships between autophagy, senescence, dormancy and apoptosis frequently focus on responses to various forms of stress. These are interrelated processes that profoundly affect normal and abnormal human physiology that require further elucidation in cancer stem cells. This review provides a current perspective on autophagy and senescence in both normal and cancer stem cells.

1. Introduction

Autophagy is a highly conserved cellular process, responsible for the degradation and recycling of damaged and/or not required proteins, protein aggregates, dysfunctional/long-lived organelles, intracellular pathogens, and storage nutrients (glycogen and lipid droplets) (Chun & Kim, 2018). This is the major cellular pathway, continuing throughout the formation of cytosolic vesicles, called autophagosomes, for delivery to lysosomes. Autophagy is a key regulator of cellular homeostasis and tissue re-modeling during development because of its ability to recycle cellular components (Rodolfo, Di Bartolomeo, & Cecconi, 2016). Impairment of this process is linked to the pathogenesis of various diseases, such as cancer and neurodegeneration, to the response to bacterial and viral infections, and to aging (Levine & Kroemer, 2008).

The process of biological aging, is a fundamental and complex feature of living organisms, which is controlled by genetic and environmental factors occurring over ones lifetime (Dodig, Čepelak, & Pavić, 2019). Among the many changes that arise in the body during aging, three characteristics that are most important include inflammation, immune aging and senescence (cellular aging, biological aging) (Dodig et al., 2019). Senescence is a potentially irreversible form of long-term cell-cycle arrest, caused by excessive intracellular or extracellular stress or damage (Dodig et al., 2019). This cell-cycle arrest prohibits the proliferation of damaged cells, removes accumulated toxic factors and delimits potential growth of malignant cells (Dodig et al., 2019). Aging refers to degenerative changes that occur in all organisms without directly inducing death, while senescence refers to the developmental stage at which close to death' symptoms become apparent

(Schmeer, Kretz, Wengerodt, Stojiljkovic, & Witte, 2019). Autophagy plays a crucial role in senescence in both normal and cancer stem cells. In this review we discuss some of the ways autophagy and senescence are tightly regulated with respect to diverse functions of stemness.

2. The overlap between stem cells and cancer stem cells

Stem cells (SCs) are an infrequent population of cells, present in every tissue, which possess the capacity for self-renewal and the ability to differentiate into a variety of specialized cell types. The concept of stemness is not only pertinent to embryonic stem cells (ESCs), but also too adult stem cells, cancer stem cells (CSCs) and induced pluripotent stem cells (iPSCs) (Yu, Pestell, Lisanti, & Pestell, 2012). Through self-renewal, SCs are created from an original mother SC, and these cells continue to maintain an undifferentiated state. However, in the process of differentiation, SCs give rise to a mature cell type. During embryonic development, ESCs possess the ability to differentiate into any and all tissue types. Adult SCs are more restricted in their differentiation abilities, and these play an important role in the replenishing and repair of adult tissues (Lawson, Blatch, & Edkins, 2009; Yu et al., 2012). Somatic cells can be reprogramed into stem-like cells, called iPSCs, and are similar to ESCs in their characteristics (Takahashi & Yamanaka, 2006; Yu et al., 2012). It has been demonstrated that a subpopulation of stem-like cells exists within tumors. These are called CSCs, which display the features of both SCs and cancer cells. They are capable of self-renewal and differentiation, but additionally they also have the ability to give rise to heterogeneous lineages of cancer cells that comprise the bulk of a tumor with the capacity to form tumors when transplanted into an animal host. CSCs can be recognized from other cancer cells by symmetry of their cell division and variations in their cell surface markers and gene expression profiles (Rosen & Jordan, 2009; Yu et al., 2012).

CSCs can hypothetically originate from stem, progenitor, or differentiated cells. Cancer is also referred to as the disorder of CSCs by some investigators (Ratajczak, Bujko, Mack, Kucia, & Ratajczak, 2018) due to the fact that tumors can be initiated in normal SCs in various tissues. CSCs and normal SCs share many similar properties (Asaduzzaman et al., 2016; Bapat, 2010; Cetin & Topcul, 2012; Valk-Lingbeek, Bruggeman, & van Lohuizen, 2004). The most important feature of CSCs is self-renewal through asymmetric cell divisions. This self-renewal process also occurs by similar molecular and epigenetic pathways in both normal SCs and

CSCs. A number of extensively studied self-renewal mechanisms are regulated by Wnt, Nanog, Notch, Hedgehog, and BMI-1, and these same regulatory pathways are utilized by both normal SCs and CSCs (Asaduzzaman et al., 2016). Both types of SCs also express similar cell surface markers that correspond with stemness such as CD-133, CD-44, and c-kit. There is also similarity in the epigenetic regulation of CSCs and SCs (Asaduzzaman et al., 2016).

There is often a limited, overlap between the pluripotent stem cell miRNA signature and CSC miRNA profiles. The miRNA circuitry of ESCs includes pluripotency-promoting miRNAs (miR-290, miR-302, miR-371 families), induced by pluripotency transcription factors (Oct-4, Sox2, Nanog, Tcf3, Kfl4), which also induce Lin28 and c-Myc. In contrast, anti-stemness miRNAs (miR-134, miR-296, miR-200c, miR-203, miR-183 and Let7 miRNAs) oppose the action of the pluripotent ESC miRNAs (Martinez & Gregory, 2010; Yu et al., 2012). The characterization of normal SCs and CSCs also fit within the same model. These different cells are isolated by a similar mechanism that utilizes fluorescence-activated cell sorting (Mayol, Loeuillet, Hérodin, & Wion, 2009), which is based on cell surface markers. Normal SCs and CSCs also share the ability to give rise to a hierarchical of heterogeneous cell populations, comprised of initiating cells, progenitor cells and differentiated cells. Because of their unique characteristics, they often display similar patterns of independence from external growth factors due to their intrinsic ability to secrete many of these growth factors. Both normal SCs and CSCs are reported to stimulate angiogenesis, for normal SCs this plays a central role in wound healing and for CSCs this contributes to cancer progression. It has been suggested that tumorigenesis and tissue regeneration are to some extent related processes involving analogous mechanisms, including stem cell migration, recruitment and secretion of chemotactic factors promoting cell motility (Balkwill & Mantovani, 2001; Dvorak, 1986; Ratajczak et al., 2018). This concept is further supported by the fact that, cancer often originates in response to tissue/organ injury or chronic tissue inflammation, and experiments implicate misappropriation of homeostatic mechanisms that regulate normal tissue repair processes and SC renewal (Coussens & Werb, 2002; Ratajczak et al., 2018).

In spite of these similarities there are some distinct differences between normal SCs and CSCs, e.g., CSCs are tumorigenic and display aberrant division combined with deregulated self-renewal and differentiation. CSCs also have abnormal karyotypes and enhanced mitotic activity in comparison with normal SCs. Apart from these traits both normal SCs and CSCs are dependent on protective autophagy for their survival and maintenance. In addition, despite similarities normal SCs and CSCs exhibit some interesting differences relative to their dependence on autophagy.

3. Autophagy and stem cells

Christian de Duve in 1963 invented the term autophagy, which literally translates to "self-eating" to describe the presence of double-membrane vesicles containing different cytoplasmic constituents within the cell (Chang, 2020; Klionsky, 2008; Yang & Klionsky, 2010). These vesicles are known as autophagosomes and are phenotypic hallmarks of macroautophagy, the most prevalent and extensively characterized form of autophagy (Parzych & Klionsky, 2014). Autophagy is a sequential multistep process consisting of induction of a phagophore, nucleation of a phagophore structure, maturation of the autophagosome, autophagosome fusion with the lysosome, and lastly the degradation and recycling of nutrients (Chang, 2020; Mizushima, 2007). Under homeostatic conditions, cells maintain a constitutive basal level of autophagy as a method of turning over cellular components including organelles (Komatsu et al., 2007). Autophagy also occurs as a response to cellular stress such as nutrient deprivation, oxidative stress, DNA damage, endoplasmic reticulum stress, hypoxia, and infection (Chang, 2020).

Over the past 20 years, our comprehension of the characteristics, roles and regulation of autophagy in somatic cells has expanded significantly (Rodolfo et al., 2016); however, further clarifications of these autophagic properties in different types of SCs is necessary. Autophagy plays a prominent role in the quality control and maintenance of cellular homeostasis, both defining factors for SCs during an organism's lifetime (Rodolfo et al., 2016). All properties of SCs including their ability of self-renewal, pluripotency, quiescence, long life span, along with their requirement for remodeling in order to differentiate, suggest that they may be more dependent on autophagy than other cell types (Phadwal, Watson, & Simon, 2012).

3.1 Role of autophagy in maintenance of stem cells

In various SC systems, autophagy plays a critical role in SC maintenance (Fig. 1A) (Chen, He, & Lu, 2018), describe a detailed list of molecules involved in autophagy regulation of SCs (Fig. 1B). Autophagy contributes to homeostatic maintenance through regulation of the levels of reactive oxygen species (ROS). Autophagy prevents accumulation of damaged mitochondria and the production of ROS in SCs. Autophagy also regulates mitochondrial content of SCs to adapt to their ongoing metabolic requirements. In addition, autophagy can regulate the cellular metabolic state of SCs thereby contributing to the critical responsibilities of SCs. Thus, autophagy protects cells against metabolic stress and prevents DNA damage, genome instability, cell death and senescence (Chang, 2020). Because of its ability to modulate epigenetic and metabolic programs, autophagy is also capable of mediating a cell's fate along with regulation of quiescence, activation, differentiation, and self-renewal in SCs (Chang, 2020).

3.2 Role of autophagy in development, differentiation, and aging of stem cells

SCs are widely distributed in postnatal organs and tissues. Somatic or normal adult SCs are indispensable in mammalian development, tissue renewal, and certain disease processes (Chen, He, et al., 2018). Despite the large database derived from studies of somatic cells, cancer cells, and various disease models; the role of autophagy in the regulation of stem cell biology is still

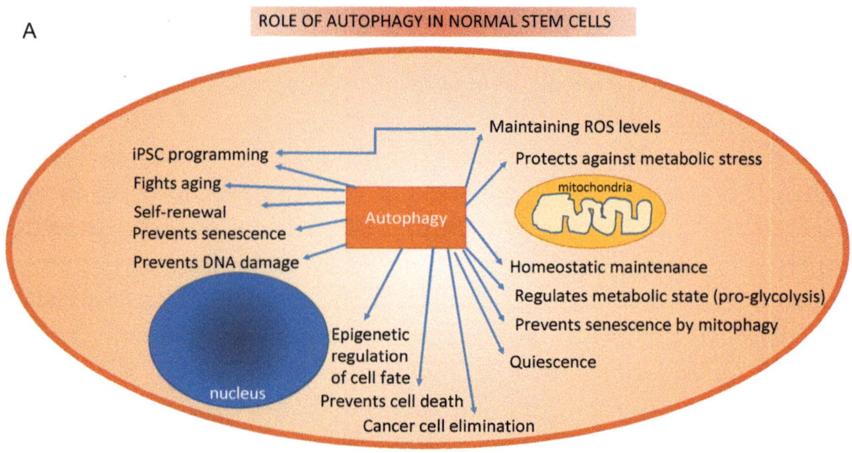

Fig. 1 Autophagy involvement in self-renewal and differentiation of stem cells (SCs). (A) Schematic representation of how autophagy regulates the survival and maintenance of normal SCs. (B) Autophagy involvement in various types of self-renewal and differentiation in SCs. HSCs, hematopoietic stem cells; BMSCs, bone marrow-derived mesenchymal stem cells; SCs, satellite cells; HPCs, hepatic progenitor cells; CSCs, cardiac stem cells; NSCs, neural stem cells; ASCs, adipose-derived stem cells; ISCs, intestinal stem cells; iPSCs, induced pluripotent stem cells. *From Chen, X., He, Y., & Lu, F. (2018). Autophagy in stem cell biology: A perspective on stem cell self-renewal and differentiation.* Stem Cells International, *2018, 1–12. doi: https://doi.org/10.1155/2018/9131397.*

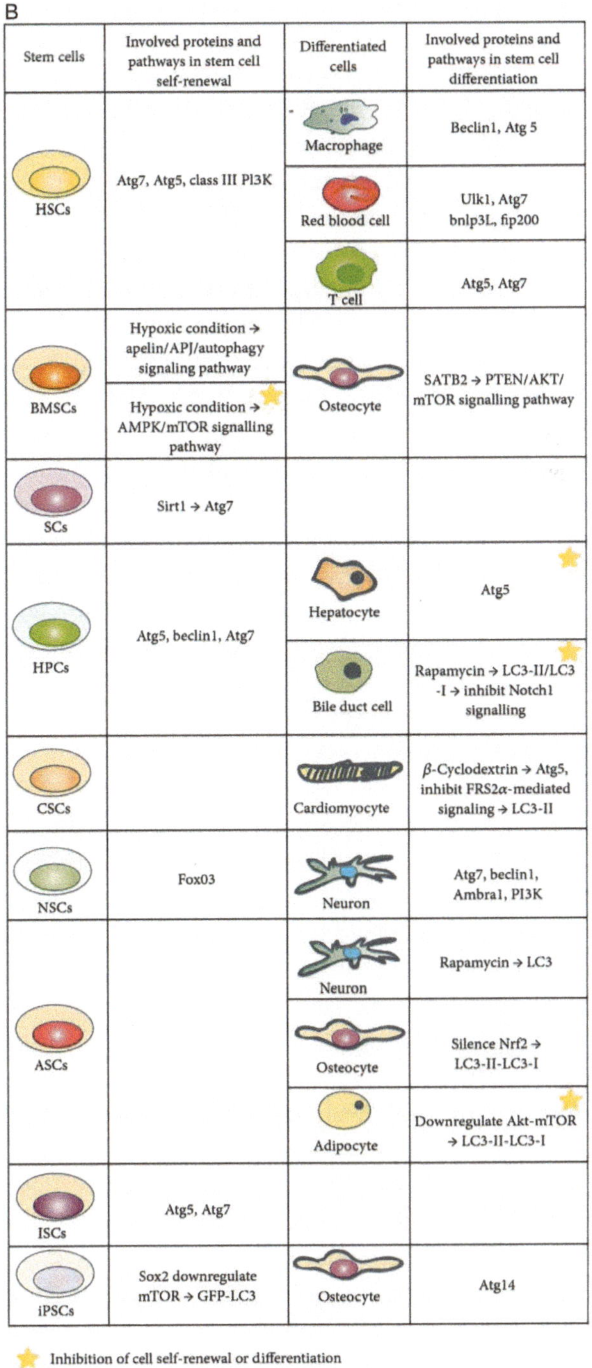
B
Stem cells
Involved proteins and pathways in stem cell self-renewal
Differentiated cells
Involved proteins and pathways in stem cell differentiation
HSCs
Atg7, Atg5, class III PI3K
Macrophage
Beclin1, Atg 5
Red blood cell
Ulk1, Atg7
bnlp3L, fip200
T cell
Atg5, Atg7
BMSCs
Hypoxic condition → apelin/APJ/autophagy signaling pathway
Hypoxic condition → AMPK/mTOR signalling pathway
Osteocyte
SATB2 → PTEN/AKT/ mTOR signalling pathway
SCs
Sirt1 → Atg7
HPCs
Atg5, beclin1, Atg7
Hepatocyte
Atg5
Bile duct cell
Rapamycin → LC3-II/LC3 -I → inhibit Notch1 signalling
CSCs
Cardiomyocyte
β-Cyclodextrin → Atg5, inhibit FRS2α-mediated signaling → LC3-II
NSCs
Fox03
Neuron
Atg7, beclin1, Ambra1, PI3K
ASCs
Neuron
Rapamycin → LC3
Osteocyte
Silence Nrf2 → LC3-II-LC3-I
Adipocyte
Downregulate Akt-mTOR → LC3-II-LC3-I
ISCs
Atg5, Atg7
iPSCs
Sox2 downregulate mTOR → GFP-LC3
Osteocyte
Atg14
Inhibition of cell self-renewal or differentiation

Fig.1—Cont'd

poorly understood. Established aspects of autophagy are that self-renewal and differentiation of SCs are functionally linked with protein turnover and lysosome-mediated degradation of organelles (Boya, Codogno, & Rodriguez-Muela, 2018; Chen, He, et al., 2018; Phadwal et al., 2012; Zeng & Zhou, 2008). The autophagic process contributes directly to cellular morphology and function, through control of protein turnover (Boya et al., 2018; Chen, He, et al., 2018; Lum, DeBerardinis, & Thompson, 2005; Phadwal et al., 2012). Recent studies demonstrate that SC self-renewal and differentiation depend on the activation of autophagy (Boya et al., 2018; Chen, He, et al., 2018; Phadwal et al., 2012; Vessoni, Muotri, & Okamoto, 2011). After receiving appropriate environmental cues, autophagy efficiently transports sets of transcription factors, adhesion molecules, or secreted factors, which are central for SC self-renewal and differentiation (Chen, He, et al., 2018). The importance of autophagy in different types of normal SCs is shown in Tables 1 and 2 (Lei et al., 2017; Phadwal et al., 2012). Most of these responses are protective and homeostatic in nature and they contribute to overall tissue health.

Several recent studies demonstrate that autophagy contributes to the control of metabolism (Boya et al., 2018; Esteban-Martínez, Sierra-Filardi, & Boya, 2017; Kaur & Debnath, 2015) and therefore can regulate the identity and function of SCs (Boya et al., 2018). Autophagy contributes to these activities in three major ways. First, the processes of self-renewal and differentiation of SCs both require a strict temporal control of cellular remodeling that involves protein turnover and lysosomal degradation of organelles, in which autophagy plays a seminal role. Second, the elimination and turnover of damaged macromolecules and organelles are essential to preserve pluripotency of long-lived SCs during their lengthy periods of quiescence, and this turnover is also mediated by autophagy. Third, a basal level of autophagy in SCs contributes to cellular homeostasis by its ability to quality control, eliminate altered and damaged intracellular proteins and organelles, and regulate cellular remodeling through degradation of structural components (Guan et al., 2014).

Apart from maintaining the health of the SCs, autophagic-specific factors can also influence the "fate-direction" of these multipotent SCs toward specific cell lineages (Phadwal et al., 2012) as can be observed in the process of mesenchymal SC differentiation or in the process of reprogramming somatic cells to become induced pluripotent SCs (iPSCs). It is established that pluripotency potential and genomic stability are two critical factors that are required for the successful reprogramming of somatic cells (Kim, Choi, Choi, & Do, 2011). ROS is one of the fundamental triggers in reprogramming barriers like senescence (Salama, Sadaie, Hoare, & Narita, 2014),

Table 1 Role of autophagy in development, differentiation, and aging of stem cells.

Type of stem cells	Role	References
Embryonic stem cells	Clearance of dead cells during embryonic morphogenesis, efficient removal of apoptotic cells, removal of defective mitochondria and ubiquitinated proteins, maintain proliferation, degrades midbodies in ESC differentiation	Mizushima et al. (2001), Todd, Gomathinayagam, and Sankar (2010), Fimia et al. (2007), Kuo et al. (2011), Qu et al. (2007), and Yue, Jin, Yang, Levine, and Heintz (2003)
Adult stem cells	Cellular remodeling during differentiation and development, maintain quiescence	Singh et al. (2009) and Takano-Ohmuro, Mukaida, Kominami, and Morioka (2000)
Hematopoietic stem cells	Homeostatic proliferation, effective mitochondrial removal, preventing DNA damage, self-renewal, lineage fate	Gurumurthy, et al. (2010), Ito et al. (2004), Liu et al. (2010), Meenhuis et al. (2011), Mortensen et al. (2011a, 2011b), and Pua, Dzhagalov, Chuck, Mizushima, and He (2007)
Mesenchymal stem cells	Cytoprotective response, differentiation	Sanchez et al. (2011) and Ugland, Boquest, Naderi, Collas, and Blomhoff (2008)
Neuronal stem cells	Differentiation, survival under nutrient stress	Zeng and Zhou (2008)
Stem cell aging	HSC proliferation and function, stem cell senescence, immunosenescence, myelogenesis, removal of defective mitochondria, proteolysis, reducing oncogenesis	Chen, Liu, Liu, and Zheng (2009), Downes et al. (2004), Ertl, Chen, Astle, Duffy, and Harrison (2008), Jia and Levine (2007), and Murakami et al. (2004)

From Phadwal, K., Watson, A. S., & Simon, A. K. (2012). Tightrope act: Autophagy in stem cell renewal, differentiation, proliferation, and aging. *Cellular and Molecular Life Sciences, 70,* 89–103. doi:10.1007/s00018-012-1032-3.

which is intrinsically linked with autophagy (Scherz-Shouval & Elazar, 2011). Confirmation of this association derives from experiments demonstrating that rapamycin (an inducer of autophagy) augments the reprogramming of somatic cells to iPSCs. Reprogramming can induce

Table 2 Summary of the different roles of autophagy in different normal stem cells.

Cells	Role of autophagy	References
Mesenchymal stem cells	Decrease in ROS generation and maintenance of stemness	Hou, Han, et al. (2013)
Satellite cells	Maintenance of stemness by preventing senescence	García-Prat et al. (2016)
Cardiac stem cells Abbreviations: NSCs	Promotion of myocardial differentiation	Zhang et al. (2012)
Hematopoietic stem cells	Sustenance of hematopoiesis by direct targeting Notch	Cao, Li, Hao, Vadgama, and Wu (2019)
NSCs	Contributions to the maintenance and functions of NSCs through regulation of oxidative state	Wang, Liang, Bian, Zhu, and Guan (2013a)
Limbal epithelial stem cells	Stem cell maintenance and differentiation	Dhamodaran et al. (2015)
Human embryonic stem cells	Pluripotency-associated proteins	Cho et al. (2014)

From Lei, Y., Zhang, D., Yu, J., Dong, H., Zhang, J., & Yang, S. (2017). Targeting autophagy in cancer stem cells as an anticancer therapy. *Cancer Letters*, 393, 33–39. doi: https://doi.org/10.1016/j.canlet.2017.02.012.

senescence and cell cycle arrest, but these processes maintain iPSCs in a quiescent state rather than pushing them to senescence by allowing these cells to retain the capacity to resume proliferation (Chen et al., 2005; Phadwal et al., 2012).

As discussed earlier in this review, autophagy contributes to differentiation of various SCs and their progenies (Guan et al., 2014). Accumulating evidence supports an active role of autophagy in the regulation of ESCs, several tissue SCs (in particular hematopoietic SCs (HSCs)), as well as specific CSCs. Autophagy promoting molecules such as rapamycin, are also reported to assist osteoblastic differentiation of human ESCs. cAMP, another regulator of autophagy, promotes differentiation of mesenchymal SCs into neuron-like cells (Guan et al., 2014; Warr et al., 2013) and the inhibition of ERK kinase MEK (via U0126 and siRNA) significantly reduces cAMP-mediated autophagy. Autophagy deficient hematopoietic stem and progenitor cells (HSPCs) from Atg7 −/− mice are unable to form secondary colonies during colony forming cell (CFC) assays, suggesting that they have a defect in self-renewal. Addition of rapamycin can impair cell differentiation in some cases, e.g., neuronal differentiation, indicating that this process requires a precise balance of autophagic activities in these cells, as neither complete inhibition nor induction (inhibition of mTOR) of autophagy restores proper differentiation of NSCs (Boya et al., 2018; Phadwal et al., 2012). Thus, it is apparent that the exceptional characteristics of SCs (self-renewal, pluripotency, differentiation and quiescence) are reliant on the activation of the autophagic process (Guan et al., 2014).

3.3 Role of autophagy in stem cell therapeutics

Autophagy through its central role in SC biology strongly impacts on developing SC-based therapies. Modulating autophagy in mesenchymal SCs effectively protects them against hypoxia- or ischemia-induced injury (Hu, Zhao, Wu, & Li, 2019). Inducing autophagy may improve aged SC function, promoting reprogramming of somatic cells into iPSCs, facilitating direct conversion of somatic cells into tissue-specific SCs, and improving SC transplantation therapies (Chang, 2020). However, it is worth mentioning that autophagy modulation can also have undesirable effects. Deregulated autophagy is implicated in a plethora of pathologies, including neurodegenerative, metabolic and immune diseases (Boya et al., 2018; Boya, Esteban-Martinez, Serrano-Puebla, Gomez-Sintes, & Villarejo-Zori, 2016; Deretic et al., 2015; Galluzzi et al., 2015; Menzies et al., 2017; Stienstra et al., 2014). Focused research in the future should pay particular attention to the balance between autophagy and the possibility of therapeutic genetic modulation of this pathway (utilizing master switch regulators like GATA-1, TFEB and FOXO3) or the use of pharmacological inhibitors to achieve similar outcomes (Phadwal et al., 2012).

4. Autophagy and cancer

Autophagy is a dual-functioning cellular process that can be cytoprotective or capable of inducing programmed cell death similar to apoptosis and necrosis (Bhutia et al., 2013). Under metabolic stress, autophagy is protective and maintains cell survival under external stresses, such as starvation, energy deficiency, hypoxia, and chemotherapy/radiotherapy. Defects in autophagy signaling coincide with increased tumorigenesis (Mathew, Karantza-Wadsworth, & White, 2007). The functional relevance of autophagy in tumor formation and progression is controversial due to its dual role at various stages of cancer. Autophagy can promote tumor suppression during cancer initiation and protect tumors during progression (Bhutia et al., 2013). Autophagy-associated cell death can also act as a tumor suppressor, which is confirmed by the fact that several autophagy-related genes are deleted in cancers (Bhutia et al., 2013; Yun & Lee, 2018). Loss of autophagy causes genomic instability and necrosis that is associated with inflammation in mouse tumor models (Degenhardt et al., 2006). On the other hand, autophagy increases survival of tumor cells under metabolic stress and increases metastatic cancer progression by enhancing tumor cell survival under environmental stress (Xu, Xia, & Pan, 2013). Unraveling this intricate balance of molecular regulation combined with the multiple diverse roles of autophagy will be critical in designing potentially clinically effective cancer therapies (Bhutia et al., 2013). This is extremely important for cancer therapeutic development as the relationship of autophagy with cancer is threefold; functioning at the level of cancer initiation, later stages of cancer progression and cancer treatment (Levine, 2007; Li, He, & Ma, 2020).

Cancer initiation can begin with an oncogenic mutation that enhances cellular proliferation (Li et al., 2020). These mutant abnormal cells and their daughter cells grow and divide at an uncontrolled and rapid rate during the hyperplasia stage. At a later dysplasia stage, these rapidly growing cells undergo changes in architecture and behavior, exhibiting increased growth potential, with the tumor consisting of more immature than mature cells (Li et al., 2020). These cells grow rapidly, lose their tissue identity, and proliferate without constraint. The evolving tumor cells start invading through the basal membrane into neighboring areas, and enter the blood circulation system from their primary tumor site. These migrating cancer cells can reach distant sites through the lymphatic system and blood circulation forming metastases (Li et al., 2020). Inactivation of autophagy-specific genes, such as BECLIN 1, results in increased tumorigenesis in mice, and enforced expression of autophagy-specific genes (BECLIN 1, ATG5) inhibit the formation of human breast tumors in mouse models (Wechman et al., 2018). These results highlight the tumor-suppressing properties of autophagy. In support of this possibility, tumor-suppressor genes that are frequently mutated in human cancer (p53, PTEN) activate autophagy, and genes that are frequently activated in cancer, such as those encoding class I PI3K and AKT, inactivate autophagy (Levine, 2007). Autophagy has also been firmly linked to apoptosis and programmed cell death, and understanding these interactions is relevant in improving cancer therapy and patient outcomes (Levine, 2007; Li et al., 2020). On the other hand, pro-survival autophagy in later tumor stages of cancer enhances certain forms of therapy-resistance, anoikis-resistance and stemness. The contradictory pro-survival and pro-death functions of autophagy make any application of this process to cancer treatment more complex (Fig. 2) (Levine, 2007; Li et al., 2020).

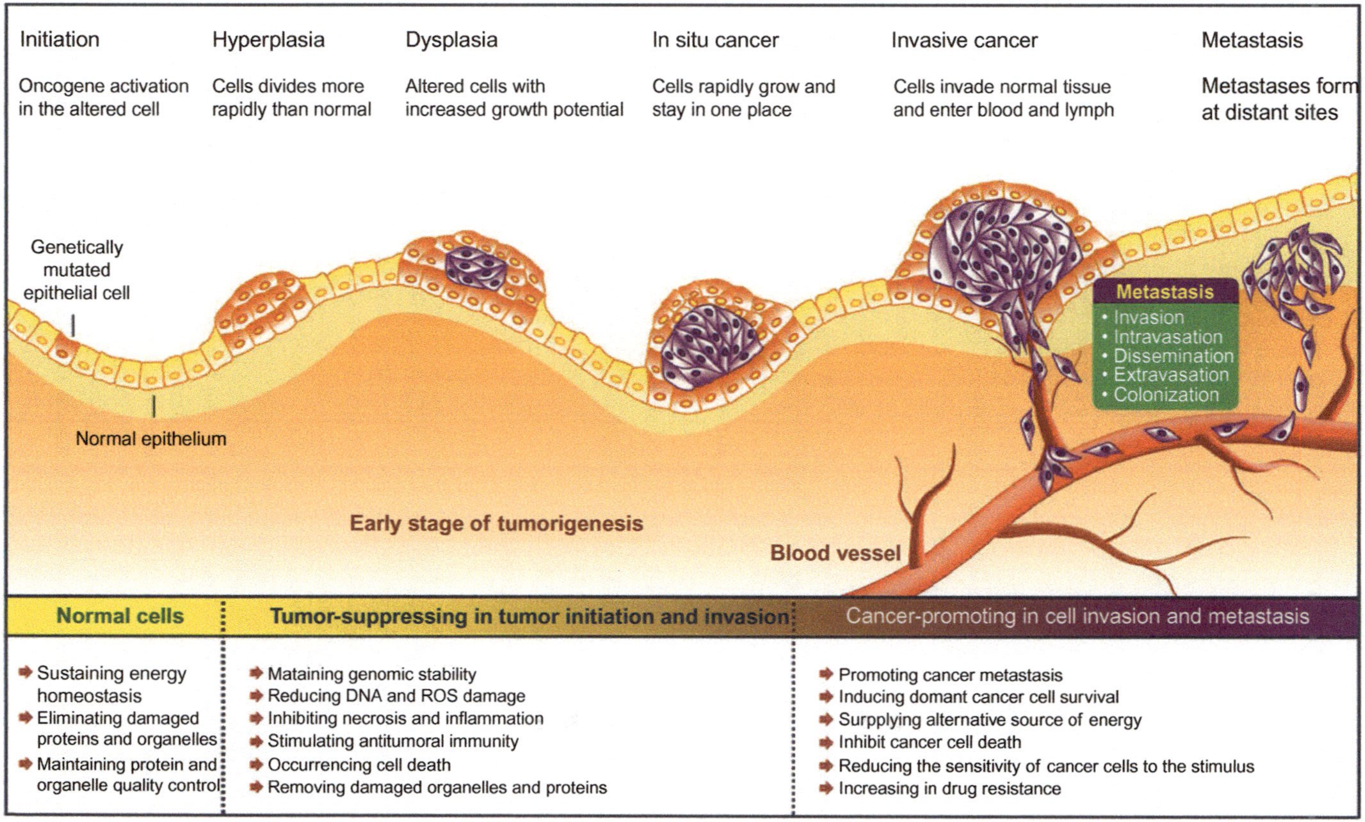

Fig. 2 See figure legend on next page.

5. Autophagy and cancer stem cells (CSCs)

Like tissue SCs, CSCs also show autophagy-dependence (Fig. 3), e.g., CSCs from primary human ductal carcinoma in situ of the breast are greatly dependent on autophagy for mammosphere formation, invasive properties and survival both in vitro and in vivo (Espina et al., 2010; Smith & Macleod, 2019). Beclin1 and autophagic flux are highly expressed in CSC-enriched mammospheres compared to cancer cells in the bulk population or when cancer cells are grown in 2D-culture conditions (Gong et al., 2013; Smith & Macleod, 2019). Beclin1 and autophagy are also reported to be critical for CSC maintenance and tumorigenesis in vivo (Gong et al., 2013; Smith & Macleod, 2019). Similarly, CD44+CD24−/low breast CSCs are reported to be highly dependent on autophagic flux for survival and stem-like properties, including reduced expression of CD24, increased CD44 expression, vimentin expression and a mesenchymal phenotype induced by TGF-β (Cufí et al., 2011; Smith & Macleod, 2019). Two different shRNA screens identified a key role for autophagy in maintaining breast CSCs. Beclin-1/ATG6 were identified from an shRNA screen for genes that modulate breast CSC plasticity (Gupta et al., 2011; Smith & Macleod, 2019) and ATG4A was identified from a screen for genes required for spheroid formation (Smith & Macleod, 2019; Wolf et al., 2013), which additionally promotes increases in the number of CSCs and in vivo tumorigenicity (Smith & Macleod, 2019; Wolf et al., 2013). There are also additional genes that were identified from this screen that are involved in

Fig. 2 Dual role of autophagy in tumorigenesis. Tumorigenesis can begin with an oncogene mutation in the epithelial cell that promotes cell division. The genetically altered or abnormal cells and their descendants grow and divide at an uncontrolled and rapid rate during the Hyperplasia stage. At the Dysplasia stage, the overgrowing cells change their original form and behavior, have increased growth potential, and consist of more immature than mature cells. During in situ cancer, the cells grow rapidly, but do not undergo the process of maturation, lose their tissue identity, and grow without regulation. In the malignant tumor (invasive cancer), the overgrowing cells invade neighboring areas and blood circulation systems from the primary tumor site by rupturing the basal membrane. Metastases occur when cancer cells reach to distant parts of the body through the lymphatic system and the blood circulation. Autophagy plays dual roles during tumorigenesis including a tumor-suppressing role during the early stages and a cancer-promoting role during the late stages. *From Li, X., He, S., & Ma, B. (2020). Autophagy and autophagy-related proteins in cancer.* Molecular Cancer, *19, 12. doi: https://doi.org/10.1186/s12943-020-1138-4.*

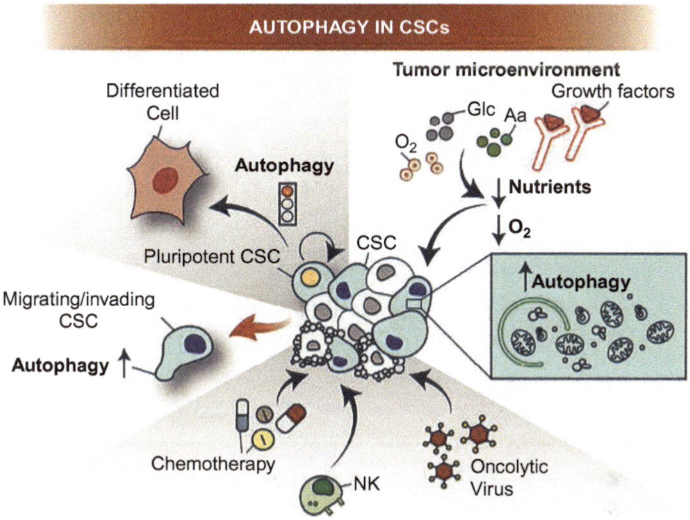

Fig. 3 The roles of autophagy in cancer stem cells (CSCs). Tumor cells are heterogeneous and include CSC populations. CSCs are often characterized by high levels of autophagy that (1) maintain pluripotency; (2) cope with low nutrients and oxygen levels (hypoxia) in the tumor microenvironment; (3) regulate CSC migration and invasion; (4) promote resistance to chemotherapy, (5) help to escape immunosurveillance; (6) support oncovirus capabilities to infect, replicate and kill CSCs. In this scenario, autophagy manipulation is found to be crucial for the effective targeting of cancer cells. Aa, Amino acids; Glc, Glucose; NK, Natural Killer; O_2, Oxygen. *From Nazio, F., Bordi, M., Cianfanelli, V., Locatelli, F., & Cecconi, F. (2019). Autophagy and cancer stem cells: molecular mechanisms and therapeutic applications.* Cell Death & Differentiation, *26, 690–702. doi: https://doi.org/10.1038/s41418-019-0292-y.*

stemness including the JAK–STAT signaling pathway (Smith & Macleod, 2019; Wolf et al., 2013), which is consistent with the fact that STAT3 phosphorylation/activation has also been identified as a molecular readout of autophagy-dependency in cancer (Maycotte et al., 2014; Smith & Macleod, 2019). In addition, CSCs secrete IL-6 (which signals through gp130 to JAK–STAT), which is also autophagy-dependent and essential for CSC maintenance (Maycotte et al., 2014; Smith & Macleod, 2019). Stresses such as hypoxia and TGF-β, which induce EMT and stemness, also induce autophagy (Caramel et al., 2013; Kiyono et al., 2009; Li et al., 2013; Smith & Macleod, 2019). Other transcription factors, including the core stemness factors SOX2 and NANOG, have also been linked to autophagy induction (Sharif et al., 2017; Smith & Macleod, 2019). Stem regulatory genes such as NANOG, are known to bind to the BNIP3L promoter, leading to induction of autophagy as a response to hypoxia and immune surveillance (Hasmim et al., 2017; Smith & Macleod, 2019). Nazio, Bordi, Cianfanelli, Locatelli, and Cecconi (2019) describe the autophagy signatures in CSCs (Table 3). Autophagy has also been implicated in maintaining CSCs in cancer types other than breast, such as pancreatic cancer (Rausch et al., 2012; Smith & Macleod, 2019; Viale et al., 2014), bladder cancer (Ojha, Singh, & Bhattacharyya, 2016; Smith & Macleod, 2019), colorectal cancer (Kantara et al., 2014; Smith & Macleod, 2019), prostate cancer (Rahman et al., 2020), chronic myeloid leukemia (Bellodi et al., 2009; Smith & Macleod, 2019) and glioblastoma (Galavotti et al., 2013; Smith & Macleod, 2019; Talukdar et al., 2018). It is important to critically unravel the extent of underlying pathways that induce autophagy in CSCs, and probing the mechanisms by which autophagy promotes stemness, which may be conserved from one cancer type to another. Mitophagy is also reported to promote preferential segregation of younger, more functional mitochondria to daughter CSCs and older mitochondria to daughter non-stem cells (Katajisto et al., 2015; Smith & Macleod, 2019). On a similar note, suppression of mitophagy (or increased mitochondrial biogenesis) augments mitochondrial respiration to promote differentiation and loss of stemness (Smith & Macleod, 2019). Protective autophagy in glioma CSCs promotes resistance to anoikis, and programmed death resulting from anchorage-independent growth. Melanoma differentiation associated gene-9 (MDA-9)/Syntenin is critical in maintaining protective autophagy in GSCs, thereby contributing to anoikis-resistance and this occurs through the regulation of BCL2 and EGFR, which control autophagy (Talukdar et al., 2018). Activation of PARP1 can downregulate the mTOR signaling pathway (Cao et al., 2019; Chen et al., 2015), thereby promoting autophagy following ionizing radiation or starvation.

NAD + consumption by PARP1 produces a Ca^{2+} mobilizing messenger that increases intracellular Ca^{2+} signaling through transient receptor potential melastatin 2 channels, which are also reported to enhance autophagy (Nazio et al., 2019). However, further studies need to investigate whether NAD + metabolism induced by PARP1 contributes to autophagy initiation in CSCs. CSCs are highly dependent on glycolytic and/or oxidative metabolism (Sancho, Barneda, & Heeschen, 2016). In nutrient-poor states, autophagy is often utilized as an alternative energy source. The aberrant metabolic autophagy-dependent adaptation of CSCs to its microenvironment may provide an explanation for the metabolic differences observed in CSCs. However, the exact role of autophagy in plasticity and metabolic reprogramming is still poorly defined (Nazio et al., 2019). Recent studies on

Table 3 Autophagy signature in cancer stem cells (CSCs).

Cancer stem cells	Upregulated autophagy gene expression	Targeted genes and treatments	Function of targeting/ treatment
Breast CSCs	ATG5-12 LC3B (Chaterjee & van Golen, 2011) BECLIN 1 (Gong et al., 2013)	mTOR inhibitor FIP200, ATG7, LC3 ATG4C and ATG12 KD CQ Salinomycin	Induction of metastatic resistance Decreased pluripotency, CSCs maintenance, migration Inhibition of tumor formation Reduction of breast CSCs number
Ovarian CSCs	ATG5 (Song et al., 2013)	BafA1 or ATG5 KD	Reduction of self-renewal
Glioblastoma CSCs	BECLIN 1, LC3 (Zhang et al., 2016)	ATG4B KD in combination with radiotherapy	Slowed tumor growth
Pancreatic CSCs	HIF-1a BECLIN 1 LC3B (Sharif et al., 2017)	HIF-1-a KD 3-MA	Promotion of the dynamic equilibrium between CSCs and non-CSCs
Chronic myeloid leukemia (CML) CSCs	(Altman et al., 2011; Karvela et al., 2016; Rothe et al., 2014)	ATG5 and ATG7 KD CQ and Lys05	Increased TKI-induced cell death
Acute myeloid leukemia (AML) CSCs	FIS1 (Pickles, Vigié, & Youle, 2018)	ATG7 and LC3B KD CQ or BafA1 FIS1 KD	Overcome hypoxia-induced resistance Attenuated mitophagy and impairment of self-renewal potential
Gastric CSCs	LC3B (Cufí et al., 2011)	CQ in combination with 5-FU	Cell death
Hepatic CSCs	ATG5, ATG7, BECLIN 1 (Gong et al., 2013)	CQ ATG5 KD BafA1, 3-MA CCCP	Increased apoptosis and decreased clonogenic capacity of CD133+ Reduction of hepatic CSC populations

From Nazio, F., Bordi, M., Cianfanelli, V., Locatelli, F., & Cecconi, F. (2019). Autophagy and cancer stem cells: molecular mechanisms and therapeutic applications. *Cell Death & Differentiation*, 26, 690–702. doi: https://doi.org/10.1038/s41418-019-0292-y.

autophagy-dependent CSC metabolism in terms of redox state, lipid metabolism, and use of alternative fuels, such as amino acids or ketone bodies, may provide a means of identifying important vulnerabilities thereby uncovering new therapeutic opportunities (Hou, Zhang, et al., 2013; Nazio et al., 2019). In addition to metabolic control of CSCs, autophagy also affects CSC migration, invasion, and chemo-, radio- and immune-resistance to therapy (Huang et al., 2014; Nazio et al., 2019). The modulation of autophagy/mitophagy and its impact on CSC generation, differentiation, plasticity, migration/invasion and pharmacological, viral and immune-resistance represent important areas worthy of further exploration (Nazio et al., 2019).

6. Role of autophagy in senescence and dormancy

Senescence is biologically described as a cellular condition in which cells have lost their proliferative capacity, yet they maintain metabolic activity. This is a genetically-based program as a response to stress signals that prevents further proliferation in damaged cells, serving as a potent tumor suppressive mechanism (Childs, Durik, Baker, & van Deursen, 2015; Kuilman, Michaloglou, Mooi, & Peeper, 2010; López-Otín, Blasco, Partridge, Serrano, & Kroemer, 2013; Muñoz-Espín & Serrano, 2014; Rastaldo, Vitale, & Giachino, 2020). Senescent cells accumulate in various tissues and organs due to aging, and in this process alter their neighboring tissue due to an active secretome. Elimination of senescent cells is reported to extend healthy lifespan and ameliorate several chronic diseases (Moreno-Blas et al., 2019). Some researchers maintain that accelerated senescence of CSCs signifies an adaptive response to withstand cell death, thereby reinforcing enhanced expression of stemness and DNA repair (Erenpreisa, Salmina, & Cragg, 2017).

Autophagy and senescence share several overlapping characteristics (Fig. 4), which indicate that both processes may collaterally protect a cell from the toxicity of external stresses such as radiation and chemotherapy and internal stresses such as telomere shortening and oncogene activation (Gewirtz, 2013). An increased presence of autophagic vacuoles and senescence-associated β-galactosidase activity has been observed in aging fibroblasts (Gerland et al., 2003; Gewirtz, 2013). Molecular markers of autophagy and senescence coexist in bile duct cells of patients with primary biliary cirrhosis, and biliary epithelial cells isolated from mice treated with either hydrogen peroxide or etoposide (Gewirtz, 2013; Sasaki, Miyakoshi,

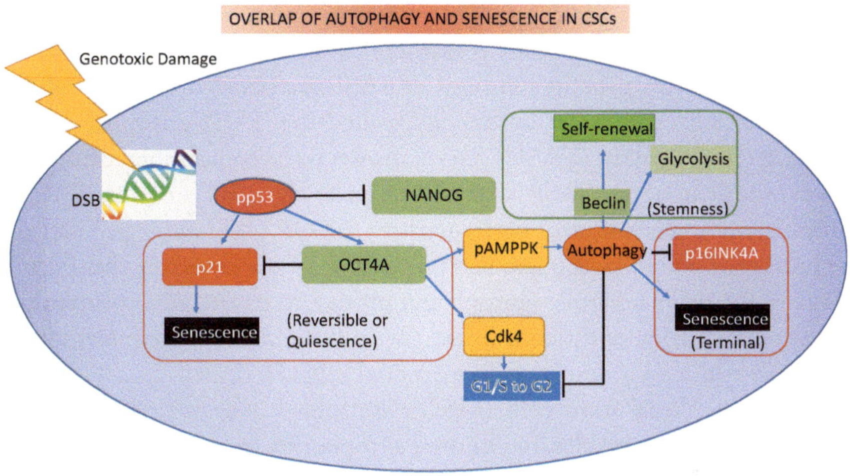

Fig. 4 Schematic representation showing the overlapping pathways shared by both autophagy and senescence.

Sato, & Nakanuma, 2010; Sasaki, Miyakoshi, Sato, & Nakanuma, 2012). Autophagic vesicles are also observed in senescent keratinocytes (Deruy et al., 2010; Gewirtz, 2013; Gosselin et al., 2009) senescent endothelial cells (Gewirtz, 2013; Patschan et al., 2008) as well as in senescent human dental pulp cells (Gewirtz, 2013; Li, Zhu, Jiang, & Peng, 2012). Several studies in both normal and malignant cells demonstrate that autophagy and senescence are closely related but not necessarily interdependent responses; and additionally, interference with autophagy delays but does not eliminate senescence (Gewirtz, 2013).

Autophagy was initially believed to be a suppressor of cellular senescence. Autophagy eliminated damaged macromolecules or organelles, which have the capacity to inhibit senescence. However, recent studies also indicate that autophagy promotes cellular senescence by facilitating the synthesis of senescence-associated secretory proteins (Kwon, Kim, Jeoung, Kim, & Kang, 2017). These seemingly opposite roles of autophagy emphasize the complexity of autophagic regulation and its impact on cellular senescence, including different types of autophagy or a unique spatiotemporal activation of autophagy (Kwon et al., 2017). Autophagy has also been shown to be activated during oncogene-induced senescence, and knockdown of several key autophagy regulators delays oncogene-induced senescence, especially the senescence-associated secretory phenotype (SASP) (Levine & Kroemer, 2008; Rubinsztein, Gestwicki, Murphy, & Klionsky, 2007;

Young et al., 2009). Since autophagy mainly modulates the SASP at the post-transcriptional level, many researchers believe autophagy may provide building blocks for protein synthesis of SASP factors (Young et al., 2009). A very specialized form of autophagy, called the TOR-autophagy spatial coupling compartment (TASCC), was shown to be responsible for the protein synthesis of some SASP factors (Narita et al., 2011). At the TASCC, autophagy delivers amino acids to activate mTOR, which in turn activates the local production of IL6 and IL8 proteins. In these contexts, the TASCC connects the cell's catabolic (autophagy) process to its anabolic (protein synthesis) process to adeptly coordinate the massive SASP machinery during oncogene-induced senescence (Kwon et al., 2017).

In mesenchymal stem cells (MSC), autophagy was at first reported to suppress senescence by clearing injured cytoplasmic organelles and impaired macromolecules, however, recent studies show that autophagy can promote MSC senescence by inducing the production of SASP. These inconsistent actions of autophagy possibly indicate a complicated autophagic regulation of MSC senescence (Rastaldo et al., 2020). Pathogenesis of age-related macular degeneration exhibits patterns of autophagy dysfunction, cellular senescence, and abnormal immune-inflammatory responses (Wang et al., 2019). Functional autophagy appears to inhibit neuronal senescence and autophagy impairment induces cortical cell senescence, while stimulation of autophagy inhibits cortical cell senescence (Moreno-Blas et al., 2019).

As discussed above, autophagy also has dual roles relative to the regulation of senescence. Under normal healthy conditions, autophagy eases stressors that can cause cellular senescence such as dysfunctional mitochondria, ROS, and ER stress. Autophagy also preserves the lysosomal integrity in cells, which also passively prevents senescence (Kwon et al., 2017). p62-dependent autophagy, which explicitly degrades GATA4 a main regulator of the SASP, leads to the suppression of cellular senescence (Kwon et al., 2017) (Fig. 5). Conversely, autophagy function can help ameliorate senescence-associated stresses and thus maintain the viability of senescent cells. Autophagy can also target Δ133p53α or lamin B leading to senescence in cells (Figs. 5 and 6) (Kwon et al., 2017).

Oncogene inactivation can cause a conversion of proliferating tumor cells to a dormant state that is associated with cellular differentiation and/or loss of a cell's ability to self-replicate (Bellovin, Das, & Felsher, 2013). However, this dormant state is reversible, and tumor cells can regain the ability to self-renew upon oncogene reactivation (Bellovin et al., 2013). This oncogene addiction that mediates dormancy might involve regulation

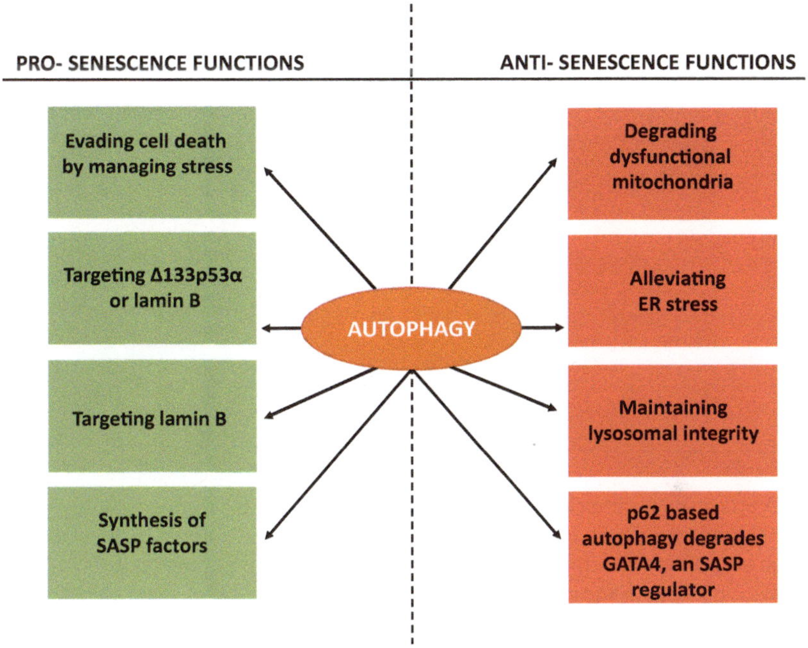

Fig. 5 Dual roles of autophagy in the regulation of cellular senescence. (Left) Under normal conditions, autophagy alleviates stressors that can cause cellular senescence such as dysfunctional mitochondria, ROS, and ER stress. Autophagy also maintains the lysosomal integrity, thus acting as an anti-senescence mechanism in a passive manner. p62-dependent autophagy, meanwhile, specifically degrades GATA4, a main regulator of the SASP, thereby actively suppressing cellular senescence. (Right) Autophagy functions as a 'non-senescence addiction' that can manage several senescence-associated stresses and thus maintain the viability of senescent cells. In addition, autophagy targets Δ133p53α or lamin B to cause cellular senescence. Lastly, autophagy may provide amino acids to support the massive synthesis of some SASP factors through the TASCC. *From Kwon, Y., Kim, J. W., Jeoung J.A., Kim M.-S., & Kang C. (2017). Autophagy is pro-senescence when seen in close-up, but anti-senescence in long-shot.* Molecules and Cells, *doi: 10.14348/molcells.2017.0151.*

of a cell's decision between self-renewal and cellular senescence (Bellovin et al., 2013). Oncogene addiction can be conceptualized as a response that exemplifies the balance between programs of self-renewal and cellular senescence (Fig. 7). Cellular senescence is distinct from dormancy in that senescence is predominantly irreversible (Bellovin et al., 2013). Oncogene inactivation through cellular senescence, can prevent self-renewal of cancer cells, and suppress tumor progression leading to tumor dormancy. Thus this supports the theoretical concept that cellular senescence actually determines tumor dormancy (Bellovin et al., 2013). Accordingly, the decision whether

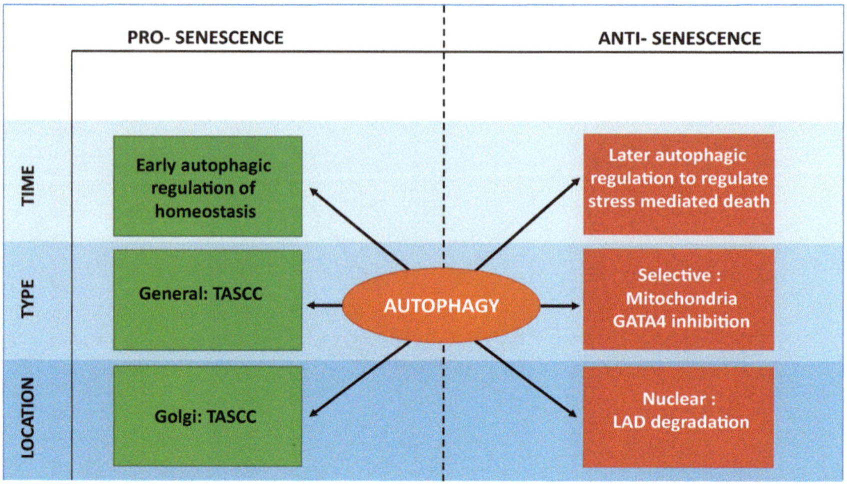

Fig. 6 A prototype toolbox for unraveling a complex relationship between autophagy and cellular senescence. Three tools are provided: (1) the type of autophagy involved (type), (2) when autophagy acts (timing), and (3) where autophagy acts (location). *From Kwon, Y., Kim, J. W., Kim, M.-S., & Kang, C. (2017). Autophagy is pro-senescence when seen in close-up, but anti-senescence in long-shot.* Molecules and Cells, *doi: 10.14348/molcells.2017.0151.*

oncogene inactivation will cause tumor elimination or tumor dormancy also appears to hinge on the balance between programs of self-renewal and cellular senescence (Bellovin et al., 2013). Oncogene addiction is capable of causing cellular senescence through at least four different mechanisms: (1) Induction of expression of cell cycle arrest proteins including p15(INK4b), p16(INK4a) and p21(mda-6/Waf-1/CIP1) (Bellovin et al., 2013; Wu et al., 2007); (2) Restoration of autocrine programs that induce cellular senescence, such as TGF-beta (TGF-β) signaling (Reimann et al., 2010; van Riggelen et al., 2010); (3) Unopposed MAPK signaling (Bellovin et al., 2013; Lin et al., 1998; Zhuang et al., 2008); and (4) Immune mechanisms that appear to be mediated through secreted cytokines such as TSP-1 (Bellovin et al., 2013; Rakhra et al., 2010; Xue et al., 2007). Another interesting example of the elaborate relationship between dormancy and autophagy is that cancer cells with low metastatic potential can be induced into a reversible state of dormancy by farnesyl transferase inhibitors (FTIs) (Chaterjee & van Golen, 2011). These dormant cells also express certain known SC markers such as aldehyde dehydrogenase I (ALDHI) and cluster of differentiation 44 (CD44) (Chaterjee & van Golen, 2011). Autophagy markers Atg5, Atg12, and LC3-B are expressed in these dormant SC-like breast cancer cells (Chaterjee & van Golen, 2011). c-jun

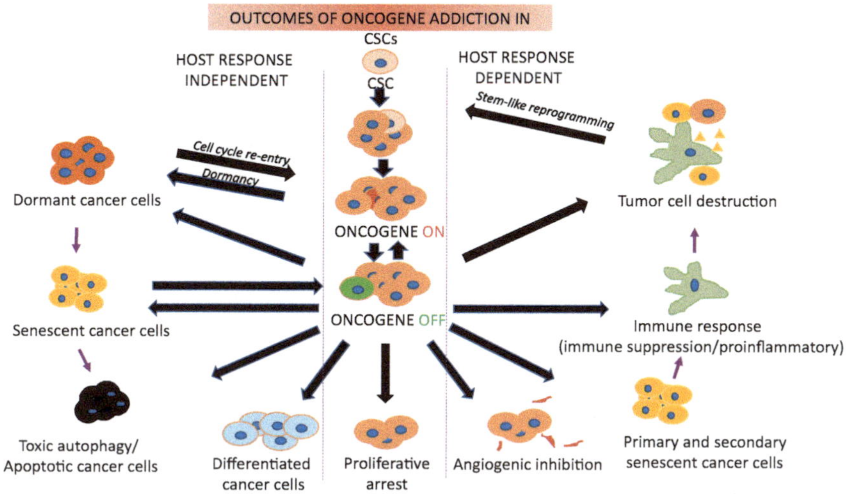

Fig. 7 Schematic diagram showing the various outcomes of oncogene addiction. CSCs can initiate tumors, which can mutate to turn on oncogenes, making them highly proliferative and resistant to stress. Oncogene inactivation can restore safety switches and lead to several outcomes: no effect, complete, or partial tumor reversion, proliferative arrest, senescence, tumor destruction, anti-angiogenesis signaling, dormancy, differentiation, or relapse. These responses can be either tumor intrinsic or host dependent. Oncogene-sensitized pre-tumor cells enter primary senescence, which may further spread to adjacent tumor cells as paracrine or secondary senescence via SASP production. In early states the senescent cells can cause immune suppression. In later stages, innate immune cells, such as macrophages and natural killer cells, attracted by SASP factors migrate to the area and clear some senescent cells. The surviving senescent cells with latent stem-like reprogramming may convert into tumor-reinitiating cells following spontaneous cell-cycle re-entry. *Concepts on oncogene addiction and senescence in the figure are adapted from Felsher, D. W. (2008). Tumor dormancy and oncogene addiction.* APMIS, 116*(7–8), 629–637. doi:10.1111/j.1600-0463.2008.01037.x and Lee, S., & Schmitt, C. A. (2019). The dynamic nature of senescence in cancer.* Nature Cell Biology, 21, *94–101. doi:10.1038/s41556-018-0249-2.*

NH2 terminal kinase (JNK/SAPK) is upregulated in these dormant SC-like breast cancer cells and is also reported to be responsible for increasing autophagy (Chaterjee & van Golen, 2011). Inhibition of autophagy by 3-methyladenine (3-MA) also reverses this dormant phenotype.

7. Role of senescence in cancer stem cells

Aging is one of the most important regulators of malignant transformation (Blasco, 2005; Franco et al., 2015; Liu & Rando, 2011; Mimeault & Batra, 2009; Rando, 2006; Welch et al., 2012). Aging causes all kinds of cells, including SCs, to accrue genetic and/or epigenetic mutations, thereby affecting the cellular, molecular and physiological functionality of tissues (Franco et al., 2015; Gravina & Vijg, 2010; Liu & Rando, 2011; Mimeault & Batra, 2009; Welch et al., 2012). Aging also results in a higher risk of tumorigenesis, especially when associated with deregulated cell signaling and fluctuations in the microenvironment of SCs. Aging can increase resistance to cellular senescence and programmed cell death, which could potentially lead to the transformation of SCs into CSCs. Over time adult SCs can accrue several age-related modifications, such as telomeric DNA reduction, DNA repair deficiency, chromosome rearrangements, genotoxic effects, and accumulation of genetic/epigenetic alterations (Franco et al., 2015; Mimeault & Batra, 2009). Aging causes the depletion of SC pools either due to senescence or apoptotic death, and/or their reduced functionality, which causes reduced tissue regeneration potential and diminished homeostatic ability (Franco et al., 2015; Rando, 2006; Raveh-Amit, Berzsenyi, Vas, Ye, & Dinnyes, 2013; Waterstrat & Van Zant, 2009). These aging-mediated deleterious effects lead to the development of age-related human diseases, such as neurodegenerative disorders, heart failure, arthritis, diabetes, cancer, and others (Franco et al., 2015; Liu & Rando, 2011; López-Otín et al., 2013; Rossi, Jamieson, & Weissman, 2008). Aging can cause transformation of stem/progenitor cells into CSCs over time, leading to tumor initiation and progression (Dick, 2008; Franco et al., 2015; Mimeault & Batra, 2009; Nguyen, Vanner, Dirks, & Eaves, 2012; Reya, Morrison, Clarke, & Weissman, 2001; Welch et al., 2012). Senescence and apoptosis are important protective mechanisms that can inhibit tumor formation however, aging-associated effects can also lead to elimination of the protective nature of these mechanisms. These aging-associated effects can be broadly divided into age-related genetic and epigenetic events that lead to the generation of CSCs. Since age-related effects on stem/progenitor cells are difficult to detect, further research is needed to comprehend this relationship (Franco et al., 2015). Cancer treatment-induced senescence is also reported to be associated with SC reprogramming of non-stem cancer cells. This is a p53/H3K9me3-mediated senescence process that allows the expression of stem-regulatory genes (Milanovic et al., 2018).

Accelerated senescence of CSCs is an adaptive response that permits cells to cope with cell death. The key regulator of this process is the p53 tumor suppressor (TP53), which causes a protracted dual state, with both senescence and self-renewal as impending outcomes. This is attained by the simultaneous activation of both OCT4A (POU5F1) and $p21^{mda\text{-}6/Waf\text{-}1/Cip1}$ (Fig. 4) (Erenpreisa et al., 2017). Autophagy also plays an important role in this process. OCT4A is generated through autophagy and it reduces the unequal activity of $p21^{mda-6/Waf-1/Cip1}$, thereby inhibiting apoptosis or terminal senescence. OCT4A also sequestrates the cell cycle inhibitor/senescence inducer $p16^{INK4A}$ in the cytoplasm. Since autophagic capacity maintains the balance in these TP53-dependent senescence reversal processes, it leads to apoptosis/terminal senescence if DNA damage is not ultimately repaired. In TP53 mutants the CSC-like state is augmented when the stressed cells can overcome the tetraploidy barrier (Erenpreisa et al., 2017). These TP53 mutant cells can acquire additional DNA repair abilities through mitotic slippage and entrance into a sequence of ploidy cycles (Erenpreisa et al., 2017). Withaferin A (WA) a potent inhibitor of CSC stemness leads to cellular senescence mainly through $p21^{mda-6/Waf-1/Cip1}$ expression (Nishi et al., 2014).

The degree of senescence/inflammation-associated deviation from the homeostatic state is also capable of generating a balance that allows the cells to distinguish beneficial effects from deleterious effects of in vivo reprogramming (Menendez & Alarcón, 2017). These mechanisms can be broadly divided into three types: (1) Transient activation of NF-κB-related innate immunity and senescence-associated inflammatory molecules (e.g., IL-6) allow the cells to repair cellular reprogramming in response to acute inflammatory events; (2) Para-inflammation switches are capable of promoting a long, yet reversible, refractoriness to repairing cellular reprogramming; and (3) Chronic senescence-associated inflammatory signaling can confine these cells in an extremely plastic epigenetic state that disables an ability to repair differentiation (Menendez & Alarcón, 2017). In the stochastic cancer model, all cells are considered to be tumorigenic, and senescence is seen as the result of random mutations (La Porta, Zapperi, & Sethna, 2012). In the hierarchical CSC model, cells are considered to divide indefinitely while other cells eventually become senescent. Utilizing these models, researchers have predicted cancer growth and evolution of senescence (La Porta et al., 2012). However, the concept of a cellular reprogramming-centered view of epigenetic plasticity as an essential aspect of a tissue's ability to undergo successful repair, aging degeneration or malignant transformation has potential to provide some interesting insights. These insights might challenge stochastic ideas of the current deterministic genetic model for most chronic diseases (Menendez & Alarcón, 2017). Enhancing senescence is believed by some researchers to not be a beneficial therapeutic strategy to fight cancer, unless the CSCs are specifically targeted (La Porta et al., 2012). But reprogramming centered strategies could potentially help in the development of novel therapeutic strategies for physiological aging and cancer.

8. Role of autophagy and senescence in resistance to cancer stem cell therapy

8.1 Chemotherapy

Autophagy performs an important role in anticancer therapy and therapy-resistance. Autophagy induced by chemotherapeutic agents can reduce apoptosis caused by these drugs, facilitating cancer cell survival and contributing to tumor cell chemoresistance. Autophagy promotes cancer cell survival and resistance to chemotherapeutic drugs, and can induce stem-like properties in cancer. Moreover, suppressing the expression of autophagy can promote apoptosis and sensitize cancer to chemotherapy (Yun & Lee, 2018).

Autophagy and senescence are both extensively studied responses to cancer therapy that often occur in parallel, culminating in growth arrest in tumor cells (Gewirtz, 2014a, 2014b). However, it remains to be determined whether this growth arrest is reversible (Gewirtz, 2014a, 2014b). Pharmacologic inhibition studies of autophagy with chloroquine and 3-methyl-adenine and genetic ablation of the autophagy-related genes ATG5 and ATG7 have provided some insights, however, inhibition of autophagy by pharmacological and genetic approaches could not entirely extinguish the senescence response, which was only reduced and/or delayed. Some reports suggest that autophagy accelerates the development of the senescent phenotype (Gewirtz, 2014a, 2014b). These processes are coordinated as senescence can occur when autophagy is abrogated (Goehe et al., 2012). Tumor cells are simultaneously induced into senescence and autophagy by etoposide, doxorubicin, or radiation therapy; and neither pharmacological nor genetic inhibition of early autophagy is capable of altering the ability of cells to recover from senescence. Moreover, some of these studies indicate that re-emergence from senescence is not necessarily dependent on autophagy. In summary, these results show that the current efforts to inhibit autophagy for therapeutic benefit could have unexpected negative consequences (Saleh et al., 2020).

Wang et al. describe chemotherapeutic agents that induce senescence (Wang, Kohli, & Demaria, 2020) (Table 4). This is complicated by the fact that senescence can have pro-tumorigenic effects in cancer patients (Fig. 8). However, senescence can also be used to target cancer. Some of the proinflammatory SASP factors show the potential to cause immune clearance of cancer cells. Recovery of TP53 functions in lymphomas, sarcomas, and liver carcinomas can cause cellular senescence and tumor regression

Table 4 Summary of senescence-inducing cancer therapy.

Category	Drug	Cell type	Marker	Treatment	References
Chemotherapy	Doxorubicin	Fibrosarcoma (HT1080)	(i) Elevated SA-β-galactosidase activities (ii) G1 cell cycle arrest (iii) Increased micronuclei formation	40 nM for 4 days	Chang et al. (1999)
		Breast cancer (MCF7)	(i) Elevated SA-β-galactosidase activities (ii) Increased micronuclei formation	50 nM for 2 days	
		Breast cancer (MDA-MB-231)	(i) Increased micronuclei formation	50 nM for 2 days	
		Colon cancer (HCT116)	(i) Elevated SA-β-galactosidase activities (ii) G1 cell cycle arrest (iii) Increased micronuclei formation	50 nM for 4 days	
		Prostate cancer (PC3)	(i) G1 cell cycle arrest (ii) Increased micronuclei formation	100 nM for 4 days	
		Prostate cancer (LNCaP)	(i) Elevated SA-β-galactosidase activities (ii) G1 cell cycle arrest (iii) Increased micronuclei formation	100–500 nM for 4 days	
		Ovarian cancer (A2780)	(i) Elevated SA-β-galactosidase activities (ii) G1 cell cycle arrest (iii) Increased micronuclei formation	20–50 nM for 4 days	
		Glioma (U251)	(i) Elevated SA-β-galactosidase activities (ii) Increased micronuclei formation	200 nM for 4 days	

Continued

Table 4 Summary of senescence-inducing cancer therapy.—cont'd

Category	Drug	Cell type	Marker	Treatment	References
		Leukemia (K562)	(i) Elevated SA-β-galactosidase activities	50 nM for 4 days	Yang et al. (2012)
		Melanoma (SK-MEL-103)	(i) Elevated SA-β-galactosidase activities	1 μM for 7 days	Muñoz-Espín et al. (2018)
		Vascular smooth muscle cells	(i) Elevated SA-β-galactosidase activities (ii) Morphology (enlarged) (iii) G1 cell cycle arrest (iv) p53/p21 upregulation (v) p16 upregulation (vi) SASP induction (vii) Increased DDR	100 nM for 24 h	Bielak-Zmijewska et al. (2014)
		Prostate stromal cells (PSC27)	(i) Elevated SA-β-galactosidase activities (ii) Reduced BrdU incorporation	10 μM for 24 h	Chen, Long, et al. (2018)
		Rat alveolar epithelial cells (L2)	(i) Elevated SA-β-galactosidase activities (ii) p53/p21 upregulation	50 nM for 24 h	Jiang et al. (2017)
		Human fibroblasts (IMR90)	(i) Elevated SA-β-galactosidase activities (ii) p16 upregulation (iii) SASP induction	100 nM for two times 24 h 24 h drug holiday in between	Baar et al. (2017)
		Human fibroblasts (WI38)	(i) Elevated SA-β-galactosidase activities (ii) p53 upregulation	100 or 500 ng/ml for 4 days	

		Mouse dermal fibroblast	(i) Elevated SA-β-galactosidase activities (ii) Reduced EdU incorporation (iii) p21 upregulation (iv) p16 upregulation (v) *LMNB1* reduction (vi) SASP induction (vii) DDR	250 nM for 24 h	Demaria et al. (2017)
		Reporter p16-3MR mice	(i) p16 upregulation in tissues (ii) Increased bioluminescence (iii) DDR markers (iv) SASP in serum (v) p21 staining in the heart and skin	10 mg/kg once	
	Etoposide	Fibrosarcoma (HT1080)	(i) Elevated SA-β-galactosidase activities (ii) G1 cell cycle arrest (iii) Increased micronuclei formation	900 nM once	Chang et al. (1999)
		Fibroblasts (IMR90)	(i) Elevated SA-β-galactosidase activities (ii) p53 upregulation	100 μM once	Chien et al. (2011)
	Camptothecin	Colon cancer (HCT116)	(i) Elevated SA-β-galactosidase activities (ii) Reduced colony formation capacity (iii) p53/p21 upregulation	20 nM for 96 h	Han et al. (2002)
		Lung cancer (H1299)	(i) Elevated SA-β-galactosidase activities	30 nM for 24 h or 48 h	Roberson, Kussick, Vallieres, Chen, and Wu (2005)

Continued

Table 4 Summary of senescence-inducing cancer therapy.—cont'd

Category	Drug	Cell type	Marker	Treatment	References
	Cisplatin	Nasopharyngeal carcinoma (CNE1)	(i) Elevated SA-β-galactosidase activities	0.5 or 2 μg/ml for 24 or 48 h	Wang et al. (1998)
		Fibrosarcoma (HT1080)	(i) Elevated SA-β-galactosidase activities (ii) G1 cell cycle arrest (iii) Increased micronuclei formation	2.2 μM	Chang et al. (1999)
		Immortalized skin fibroblasts (hTERT-BJ)	(i) Elevated SA-β-galactosidase activities (ii) p53 upregulation (iii) IL-6 upregulation	100 μM for 24–72 h	Peiris-Pagès, Sotgia, and Lisanti (2015)
		Reporter p16-3MR mice	(i) Increased bioluminescence (ii) p16 upregulation in skin and lung	2.3 mg/kg for 3 consecutive days	Demaria et al. (2017)
	Paclitaxel	Breast cancer (MCF7)	(i) Reduced proliferation (ii) Increased DDR (iii) Elevated SA-β-galactosidase activities	0.03, 0.3, 1 μM for 5 days	Hu et al. (2016)
		Endothelial cell (HUVEC)	(i) Elevated SA-β-galactosidase activities	2.5 or 5 nM for 24 h	Ota et al. (2009)
		Immortalized skin fibroblasts (hTERT-BJ)	(i) Elevated SA-β-galactosidase activities (ii) p53 upregulation (iii) IL-6 upregulation	100 nM for 24 to 72 h	Peiris-Pagès et al. (2015)

		Human bone marrow mononuclear cells (hBMNCs)	(i) Reduced proliferation (ii) Increased DDR (iii) Elevated SA-β-galactosidase activities	0.03, 0.3, 1 μM for 5 days	Hu et al. (2016)
		Mouse dermal fibroblast	(i) Elevated SA-β-galactosidase activities (ii) Reduced population doubling (iii) p16 upregulation (iv) *LMNB1* reduction (v) SASP induction (vi) Increased DDR	50 nM for 24 h or 48 h	Demaria et al. (2017)
		Reporter p16-3MR mice	(i) Increased bioluminescence (ii) p16 upregulation in skin and lung	10 mg/kg for 3 consecutive days	
	Bleomycin	Lung cancer (A549)	(i) Elevated SA-β-galactosidase activities (ii) Reduced BrdU incorporation (iii) Increased cell size (iv) p53/p21 upregulation	50 μg/ml for 120 h 50 mU/ml for 48 h 0.05–50 μg/ml for 72 h	Linge et al. (2007), Qiu et al. (2019), and Tian et al. (2019)
		Mouse lung tissue	(i) Elevated SA-β-galactosidase activities (ii) Upregulated p21 staining (iii) Reduced p-RB staining (iv) Increased DDR (v) Increased SASP	5 U/kg 50 mg/kg 2.5 mg/kg	Aoshiba et al. (2013), Qiu et al. (2019), and Shivshankar et al. (2012)
		Rat alveolar epithelial cell/lung tissue	(i) Elevated SA-β-galactosidase activities (ii) Reduced BrdU incorporation	5 μg/ml for 24 h	Jiang et al. (2017)

Continued

Table 4 Summary of senescence-inducing cancer therapy.—cont'd

Category	Drug	Cell type	Marker	Treatment	References
		Human alveolar epithelial cell	(i) Elevated SA-β-galactosidase activities (ii) p21 upregulation (iii) p16 upregulation	0.1–10 μg/ml for 5 days	Qiu et al. (2019) and Tian et al. (2019)
		Skin fibroblasts (HCA2)	(i) SASP induction	10 μg/ml for 24 h	Liu and Hornsby (2007)
		Prostate stromal cells (PSC27)	(i) Elevated SA-β-galactosidase activities (ii) Reduced BrdU incorporation (iii) SASP induction	50 μg/ml for 24 h	Chen, Long, et al. (2018)
	Temozolomide	Glioma	(i) Elevated SA-β-galactosidase activities (ii) Reduced colony formation capacity (iii) p53/p21 upregulation	10 μg/ml for 24 h 100 μM for 24 h	Günther, Pawlak, Damasceno, Arnold, and Terzis (2003), Hirose, Katayama, Mirzoeva, Berger, and Pieper (2005), Ohba, Hirose, Kawase, and Sano (2009), and Knizhnik et al. (2013)
		Melanoma	(i) Reduced colony formation capacity (ii) p53/p21 upregulation	25 to 250 μM for 12 days	Mhaidat et al. (2007)
		Colorectal cancer	(i) Elevated SA-β-galactosidase activities	100–1000 μM for 48 h	Jaiswal et al. (2015)
		Reporter p16-3MR mice	(i) Increased bioluminescence (ii) p16 upregulation in skin and lung	10 mg/kg for 3 consecutive days	Demaria et al. (2017)

Radiotherapy	Fibrosarcoma (HT1080)	(i) Elevated SA-β-galactosidase activities (ii) G1 cell cycle arrest (iii) Increased micronuclei formation	1300 rad	Chang et al. (1999)
	Breast cancer (MCF7)	(i) Elevated SA-β-galactosidase activities	Five times 2 Gy	DeMasters et al. (2004)
	Breast cancer (MCF7 and MDA-231)	(i) Elevated SA-β-galactosidase activities (ii) p53 and p21 upregulation	10 Gy	Jones et al. (2005)
	Glioblastoma (U87)	(i) Elevated SA-β-galactosidase activities	2 Gy	Quick and Gewirtz (2006)
	Endothelial cells (HUVEC and BAEC)	(i) Elevated SA-β-galactosidase activities (ii) Reduced BrdU incorporation	8 Gy	Igarashi, Sakimoto, Kataoka, Ohta, and Miura (2007)
	Keloid fibroblasts	(i) Enlarged cell size (ii) Elevated SA-β-galactosidase activities (iii) G1 cell cycle arrest (iv) p16, p21 upregulation	2, 4, 8 Gy	Ji et al. (2015)
	Fibroblasts (HCA2)	(i) Elevated SA-β-galactosidase activities (ii) SASP induction (iii) Colony formation assay (irreversible)	10 Gy	Laberge et al. (2015)
	Fibroblasts (IMR90 and WI38)	(i) Elevated SA-β-galactosidase activities (ii) Reduced BrdU incorporation	2.5, 5, 7.5, and 10 Gy	Chang et al. (2016)

Continued

Table 4 Summary of senescence-inducing cancer therapy.—cont'd

Category	Drug	Cell type	Marker	Treatment	References
		Mouse inguinal fat	(i) p16 upregulation in organs (ii) Elevated SA-β-galactosidase activities	10 Gy local irradiation on one leg	Zhu et al. (2015)
		Fibroblasts from naked mole rats	(i) Elevated SA-β-galactosidase activities (ii) p21 upregulation (iii) Reduced BrdU incorporation	10 Gy or 20 Gy	Zhao et al. (2018)
		Mouse tibialis anterior muscle	(i) Elevated SA-β-galactosidase activities	10 Gy local irradiation	Chiche et al. (2017)
		Reporter p16-3MR mice	(i) Increased bioluminescence (ii) p16 upregulation in organs (iii) SASP induction in organs	7 Gy total-body irradiation	Demaria et al. (2017)
		Mouse bone	(i) p16 upregulation in organs (ii) SASP induction in organs	20 Gy local irradiation	Yao et al. (2020)
CDK4/6 inhibitors	Palbociclib	Breast cancer (MCF7)	(i) Elevated SA-β-galactosidase activities (ii) Reduced Proliferation	0.1, 0.3, 1 μM for 5 days	Hu et al. (2016)
		Breast cancer (MCF7 and T47D)	(i) Elevated SA-β-galactosidase activities (ii) Reduced proliferation (reversible)	1 or 5 μM for 6 days 5 μM shows off-target effects	Vijayaraghavan et al. (2017)
		Melanoma (MEL10)	(i) Elevated SA-β-galactosidase activities (ii) Upregulated cyclin-D1	0.5 μM for 7 days	Leontieva and Blagosklonny (2013)

	Melanoma (1205Lu)	(i) G1 growth arrest (FACS) (ii) Reduced p-Rb (iii) Elevated SA-β-galactosidase activities (iv) SASP induction (v) Did not check reversibility	1 μM for 8 days	Yoshida, Lee, and Diehl (2016)
	Melanoma (SK-MEL-103)	(i) Elevated SA-β-galactosidase activities	1 μM for 14 days	Muñoz-Espín et al. (2018)
	Melanoma (SK-MEL-103)	(i) Elevated SA-β-galactosidase activities (ii) Colony formation assay	1 μM for 7 days	Yang et al. (2012)
	Liposarcoma (LS8817, LS141, LS0082)	(i) Elevated SA-β-galactosidase activities (ii) Reduced BrdU incorporation	1 μM for 2 days	Kovatcheva et al. (2015)
	Liposarcoma (U2OS)	(i) Elevated SA-β-galactosidase activities	1 μM for 7 days	Kovatcheva et al. (2017)
	Liposarcoma (LS8107)	(i) Elevated SA-β-galactosidase activities	1 μM for 7 days	Klein et al. (2018)
	Gastric cancer (AGS and MKN-45)	(i) Elevated SA-β-galactosidase activities (ii) Reduced proliferation	0.5 μM for 2 or 4 days	Valenzuela et al. (2017)
	Hepatocellular carcinoma (Huh7 and skHep1)	(i) Elevated SA-β-galactosidase activities (ii) Reduced proliferation (drug on) (iii) Colony formation assay (drug off) (irreversible) (iv) Cyclin D1 upregulation	0.1, 0.5, 1 μM for 10 or 14 days	Bollard et al. (2017)

Continued

Table 4 Summary of senescence-inducing cancer therapy.—cont'd

Category	Drug	Cell type	Marker	Treatment	References
		hBMNCs	(i) Reduced proliferation (reversible) (ii) No DDR (iii) No SA-β-galactosidase activities	0.1, 0.3, 1 μM for 5 days	Hu et al. (2016)
		Mouse embryonic fibroblast	(i) Elevated SA-β-galactosidase activities (ii) G1 cell cycle arrest (drug on) (iii) No DDR (iv) p16 upregulation (v) p21 upregulation (vi) SASP induction	4 μM for two times 4 days	Guan et al. (2017)
		Human fibroblasts (IMR90)	(i) No SASP induction detected	10 μM for 7 days	Hari et al. (2019)
	Ribociclib	Neuroblastoma	(i) Elevated SA-β-galactosidase activities	500 nM for 6 days	Rader et al. (2013)
		Ewing sarcoma (SKNEP1)	(i) Elevated SA-β-galactosidase activities (ii) Cyclin D3 upregulation	1 μM for 5 days	Guenther et al. (2019)
	Abemaciclib	Breast cancer (MDA-MB-453)	(i) Elevated SA-β-galactosidase activities	25 nM for 48 h	Goel et al. (2016)
		Breast cancer xenograft (MDA-MB-453)	(i) Elevated SA-β-galactosidase activities (ii) Reduced $KI67^+$ cells	90 mg/kg daily for 5 days	Goel et al. (2017)
		Breast cancer (MDA-MB-453 and BT474)	(i) Elevated SA-β-galactosidase activities (ii) Reduced proliferation (reversible)	250 or 500 nM for 7 days	Goel et al. (2017)

		Breast cancer transgenic model (MMTV-rtTA/tetO-HER2)	(i) Elevated SA-β-galactosidase activities (ii) NO SASP induction detected	75–90 mg/kg daily for 12 days	Goel et al. (2017)
		Breast cancer (MDA-MB-453)	(i) Elevated SA-β-galactosidase activities	500 nM for 72 h	Gong et al. (2017)
		Breast cancer (MCF7)	(i) Elevated SA-β-galactosidase activities	500 nM for 4 doubling time	Torres-Guzmán et al. (2017)
		Breast cancer (MCF7)	(i) G1 cell cycle arrest (ii) No SASP induction detected	500 nM for 6 h	Schaer et al. (2018)
Epigenetic modulators	5-Aza-2′ deoxycytidine	Osteosarcoma (U2OS)	(i) Elevated SA-β-galactosidase activities (ii) p16 upregulation	5 μM or 10 μM	Widodo et al. (2007)
		Lung mesothelioma (H28)	(i) Elevated SA-β-galactosidase activities (ii) p21/p27 upregulation (iii) Reduced proliferation (iv) DDR	1 μM or 10 μM for 2 days	Amatori et al. (2011)
		Liver cancer (HepG2, Hep3B, and Huh-7)	(i) Elevated SA-β-galactosidase activities (ii) SASP induction (iii) Reduced proliferation	20 μM for 24 h 12 μM for 72 h	Borghesan et al. (2016) and Venturelli et al. (2013)

Continued

Table 4 Summary of senescence-inducing cancer therapy.—cont'd

Category	Drug	Cell type	Marker	Treatment	References
	SAHA	Colon cancer (HCT116)	(i) Elevated SA-β-galactosidase activities (ii) Reduced proliferation (iii) Colony formation assay	2 μM for 24 h	Xu et al. (2005)
		Leukemia (MOLM-7, HL-60, and JURL-MK1)	(i) Reduced proliferation (ii) Elevated SA-β-galactosidase activities	0.5–10 μM for 72 h	Elknerova et al. (2011)
		Urothelial carcinoma	(i) Re-plated colony formation assay (ii) G1 cell cycle arrest	2.5 μM for 48 h	Kaletsch et al. (2018)
		Glioblastoma stem cell	(i) Elevated SA-β-galactosidase activities	2.5 μM for 7 days	Chiao, Cheng, Yang, Shen, and Ko (2013)
		Adenoid cystic carcinoma primary cell	(i) Elevated SA-β-galactosidase activities (ii) p16 upregulation (iii) DDR	1 or 1.4 μM for 120 h	Almeida et al. (2017)
Immunotherapy	Rituximab (CD20-targeting antibody)	Human lymphoma (EHEB, RC-K8, SD1)	(i) Elevated SA-β-galactosidase activities (ii) DDR (iii) SASP induction (iv) p53 and p21 induction	20 μg/ml	Däbritz et al. (2016)

From Wang, B., Kohli, J., & Demaria, M. (2020). Senescent cells in cancer therapy: friends or foes? *Trends in Cancer*, 6, 838–857. doi: https://doi.org/10.1016/j.trecan.2020.05.004.

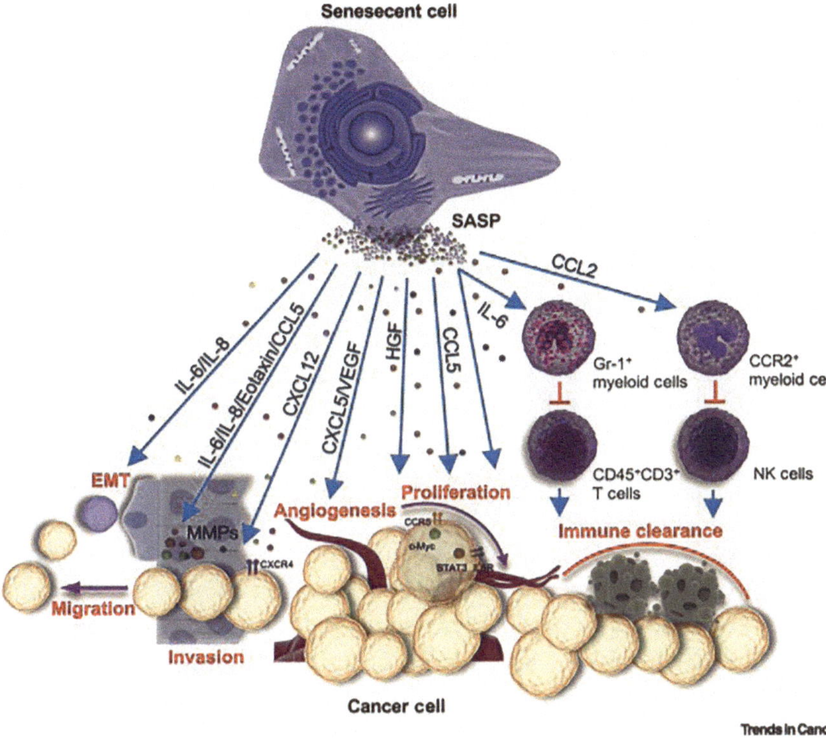

Fig. 8 Pro-tumorigenic effects of senescent cells. Senescence-associated secretory phenotype (SASP) factors are engaged in senescent cells through the activation of the NF-κB, C/EBPβ, and p38MAPK pathways. SASP factors contribute to various aspects of cancer progression: IL-6 and IL-8 promote EMT; IL-6–IL-6R–STAT3/CCL5–CCR5–c-Myc/HGF contribute to the proliferation of cancer cells; CXCL5 and VEGF promote angiogenesis; CXCL12–CXCR4/IL-6/IL-8/eotaxin/CCL5/MMPs promote the invasion and migration of cancer cells; and IL-6 suppresses the CD45+CD3+T cell-mediated immune clearance of cancer cells by activating Gr-1+myeloid cells. CCL2 suppresses the natural killer (NK) cell-mediated immune clearance of cancer cells by activating CCR2+myeloid cells. The SASP is a collection of several factors that can act collectively. CCL, (C-C motif) chemokine ligand; CCR, (C-C motif) chemokine receptor; CXCL, (C-X-C motif) chemokine ligand; CXCR, (C-X-C motif) chemokine receptor; EMT, epithelial–mesenchymal transition; HGF, hepatocyte growth factor; IL, interleukin; MMP, matrix metalloproteinase; VEGF, vascular endothelial growth factor. *From Wang, B., Kohli, J., & Demaria, M. (2020). Senescent cells in cancer therapy: friends or foes?* Trends in Cancer, *6, 838–857. doi: https://doi.org/10.1016/j.trecan.2020.05.004.*

(Ventura et al., 2007; Wang et al., 2020 ; Xue et al., 2007) which is primarily mediated by the activation of chronic inflammation and the innate immune system by senescent stromal cells (Wang et al., 2020; Xue et al., 2007). CD4+ T cells are capable of detecting premalignant senescent hepatocytes preventing the development of liver carcinomas (Kang et al., 2011; Wang et al., 2020), which indicate that senescence and its secretory phenotypes can possibly inhibit malignancy in the early stages of cancer progression (Wang et al., 2020).

To carefully study the relationship between chemotherapy, autophagy and senescence, studies should be performed in cells lacking the ability to undergo apoptosis. This is necessary since apoptosis is frequently induced when autophagy is inhibited, which can cause a misinterpretation of data in defining the relationship between autophagy potentially leading to senescence (Gewirtz, 2014a, 2014b). Knizhnik et al. report that the antitumor drug, temozolomide (TMZ), promotes both autophagy and senescence in glioma cells (Knizhnik et al., 2013). Inhibition of autophagy by 3-MA suppresses senescence in these cells for as long as 144h. However, autophagy declined after the 96h time point following TMZ treatment, whereas senescence steadily increased over the 144h time period of the study, suggesting a temporal absence of dependence of senescence on autophagy. Lee et al. reported that knockdown of the monocarboxylate transporter 2 resulted in both autophagy and senescence in human colorectal cancer cells through ROS generation (Lee, Lee, Kang, & Park, 2012). Suppression of autophagy using three different inhibitors (3-MA, Bafilomycin and chloroquine) did not alter senescence induction, showing that there are some stages where these two pathways diverge from each other (Gewirtz, 2014a, 2014b).

8.2 Radiotherapy

Radiation is a standard-of-care anti-cancer therapy that exerts its anti-tumor DNA-damaging effects through various direct and indirect mechanisms (Patel, Sohal, Manjili, Harrell, & Gewirtz, 2020). Radiation has served as an effective mode of treatment for several cancer types, and it has been utilized for both curative and palliative treatment strategies. However, resistance to radiation represents a fundamental barrier to treatment (Patel et al., 2020). Radiation is delivered to cancer cells to cause their death, however, in reality radiation induces multiple responses. One of which is autophagy, which plays a very important role in both sensitization and resistance to radiation. Tam et al. demonstrate (Table 5) how autophagy interacts with radiation in different types of cancer (Tam, Wu, & Law, 2017. Other responses to radiation include mitotic catastrophe, apoptotic cell death, and senescence, and it is important to remember that autophagy and senescence may also in specific contexts promote cell survival (Maier, Hartmann, Wenz, & Herskind, 2016; Patel et al., 2020). Frequently, autophagy

Table 5 Effects of autophagy modifications on improving radiosensitivity or radiotherapy efficacy.

Cancer type	Cell line	Autophagy agent (induction (+)/ inhibition (−))	Autophagy pathway affected	Animal study (Yes (+)/No (−))	References
Glioblastoma	T98G + U373MG	Rapamycin (+)	PI3K-Akt-mTOR (mTOR inhibitor)	−	Palumbo et al. (2012)
	SU2	NVP-BEZ235 (+)	PI3K-Akt-mTOR (PI3K/mTOR inhibitor)	−	Wang et al. (2013b)
	U373MG	Chloroquine (−)	UPR (PERK)	+	Rouschop et al. (2010)
Oral cancer	OC3 + SAS	Rapamycin (+)	PI3K-Akt-mTOR (mTOR inhibitor)	−	Wu et al. (2014)
Lung cancer	H460	RAD001 (+)	PI3K-Akt-mTOR (mTOR inhibitor)	+	Kim et al. (2008)
	H460	Rapamycin (+)	PI3K-Akt-mTOR (mTOR inhibitor)	+	Kim et al. (2009)
	CDDP-Resistant H460	NVP-BEZ235 (+)	PI3K-Akt-mTOR (PI3K/mTOR inhibitor)	−	Kim et al. (2014)
Breast cancer	MDA-MB-23 + MCF-7	RAD001 (+)	PI3K-Akt-mTOR (mTOR inhibitor)	−	Albert, Kim, Cao, and Lu (2006)
	MCF-7	Rapamycin (+)	PI3K-Akt-mTOR (mTOR inhibitor)	−	Paglin et al. (2005)
Oesophageal cancer	EC109	Tunicamycin (+)	PI3K-Akt-mTOR (ER stressor)	+	Pang, He, Liu, Wang, and Zhang (2013)
Pancreatic cancer	MIA PaCa-2 + PANC-1	MG132 (+)	MAPK (JNK) (Proteasome inhibitor/ ER stressor)	+	Chiu et al. (2015)
Colorectal cancer	HCT-116	BCG/CWS (+)	MAPK (JNK/ERK)	+	Yuk et al. (2010)
	HCT-116 + HT-29	Chloroquine (−)	UPR (PERK)	+	Rouschop et al. (2010)
Prostate cancer	Biopsy specimens	MG132 (+)	MAPK (JNK) (Proteasome inhibitor/ ER stressor)	−	Koukourakis et al. (2015)
	DU145 + PC3	RAD001 (+)	PI3K-Akt-mTOR (mTOR inhibitor)	−	Cao et al. (2006)

From Tam, S. Y., Wu, V. W., & Law, H. K. (2017). Influence of autophagy on the efficacy of radiotherapy. *Radiation Oncology (London, England)*, 12(1), 57. doi:10.1186/s13014-017-0795-y.

produces a cytoprotective effect as a first response to deal with the damage acquired from chemotherapy and radiation treatment. This protective autophagy model has led to the development of several current clinical trials, which are specifically designed to radiosensitize cancer cells through the use of autophagy inhibitors. While some trials show promise, these trials have not provided consistent results (Patel et al., 2020). Autophagy has also been shown to produce a cytotoxic or nonprotective response, which may be partially responsible for the inconsistency of some of the clinical outcomes. Furthermore, autophagy can also produce important immune modulatory functions (Patel et al., 2020; Schlie, Spowart, Hughson, Townsend, & Lum, 2011). Radiation-induced senescence can be followed by a phase of active proliferation in a subset of cells, which can contribute to resistance to therapy and possibly also disease recurrence (Patel et al., 2020). Senescence is sometimes accompanied by a unique secretory phenotype which is capable of either promoting or inhibiting immune functions, depending on the cytokines and chemokine profiles of these cells (Huang et al., 2014; Patel et al., 2020). Senescence-associated secretory phenotypes (SASPs) are reported to support tumor progression (Huang et al., 2014). Inhibition of autophagy using chloroquine or by downregulating autophagy-related protein 7 (ATG7) upregulates apoptosis and protects colorectal cancer from Photosan-II (PS-II)-mediated photodynamic therapy (PS-PDT) (Xiong et al., 2017). Huang et al. showed that radiation induces senescence in PTTG1-depleted cancer cells (Huang et al., 2014). They found that autophagy was required for the radiation-induced senescence in PTTG1-depleted breast cancer cells. Inhibition of autophagy caused the cells to switch from radiation-induced senescence to apoptosis. Post irradiation cancer cells in many subtypes show induction of autophagy, development of radioresistance, and enrichment of CSC subpopulations (Lee et al., 2017).

8.3 Using autophagy and senescence to target cancer stem cells

As stated above, it is now well established that autophagy plays an important role in anticancer therapy (Eritja et al., 2017; Galluzzi, Bravo-San Pedro, Demaria, Formenti, & Kroemer, 2017; Lei et al., 2017) and therapy-resistance (Belounis et al., 2016; Lei et al., 2017; Zhang & McCarty, 2017). Autophagy induced by chemotherapy and radiotherapeutic agents are capable of causing apoptosis, or facilitating cancer cell survival and contributing to tumor cell resistance (Cojoc, Mäbert, Muders, & Dubrovska, 2015; Zhang & McCarty, 2017). In specific cellular contexts inhibiting cytoprotective autophagy that arises following radiation therapy increases radiosensitivity and confers longer relapse-free survival by eliminating tumor-initiating CSCs (Liang, El-Zein, & Dave, 2015), while in other cases it has the opposite effect. CSCs inherently possess the ability to resist anti-cancer therapies (Talukdar et al., 2019), and as we described earlier in this review it is postulated that autophagy and its intricate and controversial relationship with senescence could play very important roles in this phenomenon. Lei et al. described how autophagy can be targeted in CSCs as a form of anticancer therapy (Lei et al., 2017). Currently, there are some anticancer therapies targeting CSCs that are undergoing clinical trials (Lei et al., 2017). Ojha et al. details these clinical trials that utilize combinations of autophagy inhibitors (Table 6) (Ojha, Bhattacharyya, & Singh, 2015). By modulating autophagic pathways, it might be possible to design more effective anticancer strategies. However, further studies are needed to clarify how to precisely manipulate autophagic and senescent pathways uniquely in cancer cells before such new therapies can be developed and exploited as an anti-cancer strategy (Ojha et al., 2015).

Table 6 Examples of clinical trials combining the autophagy inhibitor hydroxychloroquine as an adjunct to anticancer therapies.

Tumor	Interventions	Clinical trial number
Multiple myeloma	HCQ + bortezomib	NCT00568880
Brain, central nervous system tumors	HCQ + temozolomide/radiation therapy	NCT00486603
Prostate cancer	HCQ + docetaxel	NCT00786682
Prostate cancer	HCQ (after local therapy)	NCT00726596
Breast cancer	HCQ + ixabepilone	NCT00765765
Breast cancer	HCQ + ixabepilone	NCT00765765 B
Breast cancer	HCQ	NCT01292408
Ductal carcinoma in situ	CQ + tamoxifen	NCT01023477
Lung cancer	HCQ + bevacizumab/carboplatin paclitaxel	NCT00728845
Pancreas cancer	HCQ + gemcitabine	NCT01128296
Pancreatic cancer	HCQ + gemcitabine/abraxane	NCT01506973
Pancreatic cancer	HCQ + capecitabine + photon radiation	NCT01494155
Renal cancer	HCQ (patients with resectable renal cell carcinoma)	NCT01144169

Continued

Table 6 Examples of clinical trials combining the autophagy inhibitor hydroxychloroquine as an adjunct to anticancer therapies.—cont'd

Tumor	Interventions	Clinical trial number
Renal cancer	HCQ (patients with resectable renal cell carcinoma)	NCT01144169R
Renal cell carcinoma	HCQ + interleukin-2	NCT0155036
Renal cell carcinoma	HCQ and RAD001	NCT01510119
Advanced solid tumors or prostate or renal cancer	HCQ + Akt inhibitor/MK2206 (MK-2206)	NCT01480154
Adult solid tumors	HCQ + temozolomide	NCT00714181
Adult solid tumors	HCQ + temsirolimus	NCT00909831
Adult solid tumors	HCQ + vorinostat	NCT01023737
Advanced cancer	HCQ + sunitinib	NCT00813423
Metastatic solid tumors	HCQ + temsirolimus	NCT00909831
Advanced solid tumors	HCQ + vorinostat	NCT01023737
Colorectal cancer	HCQ + XELOX + bevacizumab	NCT01006369
Colorectal cancer	HCQ + FOLFOX/bevacizumab	NCT01206530
Metastatic colorectal cancer	HCQ + capecitabine, oxaliplatin, and bevacizumab	NCT01006369
Melanoma	HCQ (after surgery)	NCT00962845
Solid tumors undergoing radiation therapy for bone metastases	HCQ	NCT01417403
NSCLC	CQ + cisplatin/etoposide	NCT00969306
NSCLC	HCQ + gefitinib	NCT00809237
NSCLC	HCQ + paclitaxel and carboplatin	NCT01649947
Advanced or recurrent NSCLC	HCQ + carboplatin, paclitaxel, bevacizumab	NCT00933803
Advanced NSCLC and (EGFR) mutations	HCQ + erlotinib	NCT00977470
Chronic myeloid leukemia	HCQ + imatinib	NCT01227135

From Ojha, R., Bhattacharyya, S., & Singh, S. K. (2015). Autophagy in cancer stem cells: A potential link between chemoresistance, recurrence, and metastasis. *Biores Open Access,* 4, 97–108. doi: https://doi.org/10.1089/biores.2014.0035.

9. Conclusions

SCs and CSCs share many similar characteristics including a complex interrelationship with autophagy and senescence. In addition to maintaining the health of SCs, autophagic-specific factors can also influence the differentiation of these multipotent SCs toward specific cell lineages, as well as impact aging-related properties, including senescence. CSCs can self-renew, differentiate and form tumors. The presence of CSCs in tumors significantly contributes to chemoresistance and/or radioresistance, tumor recurrence and metastasis. Autophagy promotes these negative characteristics of CSCs that are often unregulated in CSCs and display dual functions relative to inducing senescence and dormancy. It is currently assumed that autophagy helps to maintain the stemness of CSCs and results in the failure of anticancer therapies. Therapy-induced senescence can cause cancer metastasis and relapse and several adverse responses to cancer treatments. Pharmacological interference with detrimental senescence might improve the efficacy of cancer treatments and enhance the quality of life in treated patients (Wang et al., 2020). New insights continue to be garnered relative to senescence-mediated plasticity of CSCs (Milanovic et al., 2018). Accordingly, autophagy induction may provide a desirable outcome for CSCs thereby decreasing their resistance to anticancer therapies. However, there are still many caveats that must be examined before actually applying this concept for practical clinical applications. Through an intricate and dual role in regulating senescence and dormancy in CSCs further investigations are required to exploit these properties as anticancer therapies. Insights into the overlapping yet distinct features of CSCs from normal SCs can prove extremely valuable in developing appropriate therapeutic agents that uniquely impact CSCs, thereby avoiding therapy-resistance.

10. Future perspectives

A further comprehension of autophagy and senescence in CSCs could usher in a paradigm shift in cancer therapy. Novel therapeutics selectively cytotoxic to senescent cells (senolytics) may prove to prolong remission by delaying disease recurrence in patients. Accurate assessment of primary responses to radiation may provide potential new targets that can be manipulated for therapeutic benefit to sensitize cancer cells to radiotherapy, while sparing normal tissue (Huang et al., 2014). A combination of autophagy-based treatments with senolytics and existing chemotherapies and radiation therapy may prove beneficial in the targeting CSCs. In these contexts, targeting senescent cells for elimination could provide unique therapeutic benefits in patients with cancer, including preventing cancer dormancy and tumor recurrence.

References

Albert, J. M., Kim, K. W., Cao, C., & Lu, B. (2006). Targeting the Akt/mammalian target of rapamycin pathway for radiosensitization of breast cancer. *Molecular Cancer Therapeutics*, *5*(5), 1183–1189. https://doi.org/10.1158/1535-7163.

Almeida, L. O., Guimarães, D. M., Martins, M. D., Martins, M. A. T., Warner, K. A., Nör, J. E., et al. (2017). Unlocking the chromatin of adenoid cystic carcinomas using HDAC inhibitors sensitize cancer stem cells to cisplatin and induces tumor senescence. *Stem Cell Research*, *21*, 94–105. https://doi.org/10.1016/j.scr.2017.04.003.

Altman, B. J., Jacobs, S. R., Mason, E. F., Michalek, R. D., MacIntyre, A. N., Coloff, J. L., et al. (2011). Autophagy is essential to suppress cell stress and to allow BCR-Abl-mediated leukemogenesis. *Oncogene*, *30*(16), 1855–1867. https://doi.org/10.1038/onc.2010.561.

Amatori, S., Bagaloni, I., Viti, D., & Fanelli, M. (2011). Premature senescence induced by DNA demethylating agent (decitabine) as therapeutic option for malignant pleural mesothelioma. *Lung Cancer*, *71*(1), 113–115. https://doi.org/10.1016/j.lungcan.2010.10.016.

Aoshiba, K., Tsuji, T., Kameyama, S., Itoh, M., Semba, S., Yamaguchi, K., et al. (2013). Senescence-associated secretory phenotype in a mouse model of bleomycin-induced lung injury. *Experimental and Toxicologic Pathology*, *65*(7–8), 1053–1062. https://doi.org/10.1016/j.etp.2013.04.001.

Asaduzzaman, S. M., Islam, R., Rahman, K. M. T., Akhtar, N., Jamil, H. M., & Rahman, M. (2016). STEM CELL AND CANCER STEM CELL: A tale of two cells. *Progress in Stem Cell*, *3*, 97. https://doi.org/10.15419/psc.v3i02.124.

Baar, M. P., Brandt, R. M. C., Putavet, D. A., Klein, J. D. D., Derks, K. W. J., Bourgeois, B. R. M., et al. (2017). Targeted apoptosis of senescent cells restores tissue homeostasis in response to chemotoxicity and aging. *Cell*, *169*(1), 132–147.e16. https://doi.org/10.1016/j.cell.2017.02.031.

Balkwill, F., & Mantovani, A. (2001). Inflammation and cancer: Back to Virchow? *Lancet*, *357*, 539–545. https://doi.org/10.1016/S0140-6736(00)04046-0.

Bapat, S. A. (2010). Human ovarian cancer stem cells. *Reproduction*, *140*, 33–41. https://doi.org/10.1530/REP-09-0389.

Bellodi, C., Lidonnici, M. R., Hamilton, A., Helgason, G. V., Soliera, A. R., Ronchetti, M., et al. (2009). Targeting autophagy potentiates tyrosine kinase inhibitor-induced cell death in Philadelphia chromosome-positive cells, including primary CML stem cells. *The Journal of Clinical Investigation*, *119*, 1109–1123. https://doi.org/10.1172/JCI35660.

Bellovin, D. I., Das, B., & Felsher, D. W. (2013). Tumor dormancy, oncogene addiction, cellular senescence, and self-renewal programs. *Advances in Experimental Medicine and Biology*, *734*, 91–107. https://doi.org/10.1007/978-1-4614-1445-2_6.

Belounis, A., Nyalendo, C., Le Gall, R., Imbriglio, T. V., Mahma, M., Teira, P., et al. (2016). Autophagy is associated with chemoresistance in neuroblastoma. *BMC Cancer*, *16*, 891. https://doi.org/10.1186/s12885-016-2906-9.

Bhutia, S. K., Mukhopadhyay, S., Sinha, N., Das, D. N., Panda, P. K., Patra, S. K., et al. (2013). Autophagy: Cancer's friend or foe? *Advances in Cancer Research*, *118*, 61–95. https://doi.org/10.1016/B978-0-12-407173-5.00003-0.

Bielak-Zmijewska, A., Wnuk, M., Przybylska, D., Grabowska, W., Lewinska, A., Alster, O., et al. (2014). A comparison of replicative senescence and doxorubicin-induced premature senescence of vascular smooth muscle cells isolated from human aorta. *Biogerontology, 15*(1), 47–64. https://doi.org/10.1007/s10522-013-9477-9.

Blasco, M. A. (2005). Telomeres and human disease: Ageing, cancer and beyond. *Nature Reviews Genetics, 6*, 611–622. https://doi.org/10.1038/nrg1656.

Bollard, J., Miguela, V., Ruiz de Galarreta, M., Venkatesh, A., Bian, C. B., Roberto, M. P., et al. (2017). Palbociclib (PD-0332991), a selective CDK4/6 inhibitor, restricts tumour growth in preclinical models of hepatocellular carcinoma. *Gut, 66*(7), 1286–1296. https://doi.org/10.1136/gutjnl-2016-312268.

Borghesan, M., Fusilli, C., Rappa, F., Panebianco, C., Rizzo, G., Oben, J. A., et al. (2016). DNA hypomethylation and histone variant macroH2A1 synergistically attenuate chemotherapy-induced senescence to promote hepatocellular carcinoma progression. *Cancer Research, 76*(3), 594–606. https://doi.org/10.1158/0008-5472.

Boya, P., Codogno, P., & Rodriguez-Muela, N. (2018). Autophagy in stem cells: Repair, remodelling and metabolic reprogramming. *Development, 145*. https://doi.org/10.1242/dev.146506, dev146506.

Boya, P., Esteban-Martinez, L., Serrano-Puebla, A., Gomez-Sintes, R., & Villarejo-Zori, B. (2016). Autophagy in the eye: Development, degeneration, and aging. *Progress in Retinal and Eye Research, 55*, 206–245. https://doi.org/10.1016/j.preteyeres.2016.08.001.

Cao, C., Subhawong, T., Albert, J. M., Kim, K. W., Geng, L., Sekhar, K. R., et al. (2006). Inhibition of mammalian target of rapamycin or apoptotic pathway induces autophagy and radiosensitizes pten null prostate cancer cells. *Cancer Research, 66*(20), 10040–10047. https://doi.org/10.1158/0008-5472.

Cao, W., Li, J., Hao, Q., Vadgama, J. V., & Wu, Y. (2019). AMP-activated protein kinase: A potential therapeutic target for triple-negative breast cancer. *Breast Cancer Research, 21*(1), 29. https://doi.org/10.1186/s13058-019-1107-2.

Caramel, J., Papadogeorgakis, E., Hill, L., Browne, G. J., Richard, G., Wierinckx, A., et al. (2013). A switch in the expression of embryonic EMT-inducers drives the development of malignant melanoma. *Cancer Cell, 24*, 466–480. https://doi.org/10.1016/j.ccr.2013.08.018.

Cetin, I., & Topcul, M. (2012). Cancer stem cells in oncology. *Journal of BUON, 17*, 644–648. doi: PMID: 23335519.

Chang, B. D., Broude, E. V., Dokmanovic, M., Zhu, H., Ruth, A., Xuan, Y., et al. (1999). A senescence-like phenotype distinguishes tumor cells that undergo terminal proliferation arrest after exposure to anticancer agents. *Cancer Research, 59*(15), 3761–3767.

Chang, J., Wang, Y., Shao, L., Laberge, R. M., Demaria, M., Campisi, J., et al. (2016). Clearance of senescent cells by ABT263 rejuvenates aged hematopoietic stem cells in mice. *Nature Medicine, 22*(1), 78–83. https://doi.org/10.1038/nm.4010.

Chang, N. C. (2020). Autophagy and stem cells: Self-eating for self-renewal. *Frontiers in Cell and Development Biology, 8*, 138. https://doi.org/10.3389/fcell.2020.00138.

Chaterjee, M., & van Golen, K. L. (2011). Breast cancer stem cells survive periods of farnesyl-transferase inhibitor-induced dormancy by undergoing autophagy. *Bone Marrow Research, 2011*, 362938. https://doi.org/10.1155/2011/362938.

Chen, C., Liu, Y., Liu, Y., & Zheng, P. (2009). mTOR regulation and therapeutic rejuvenation of aging hematopoietic stem cells. *Science Signaling, 2*(98), ra75. https://doi.org/10.1126/scisignal.2000559.

Chen, F., Long, Q., Fu, D., Zhu, D., Ji, Y., Han, L., et al. (2018). Targeting SPINK1 in the damaged tumour microenvironment alleviates therapeutic resistance. *Nature Communications, 9*(1), 4315. https://doi.org/10.1038/s41467-018-06860-4.

Chen, X., He, Y., & Lu, F. (2018). Autophagy in stem cell biology: A perspective on stem cell self-renewal and differentiation. *Stem Cells International, 2018*, 1–12. https://doi.org/10.1155/2018/9131397.

Chen, Z., Trotman, L. C., Shaffer, D., Lin, H. K., Dotan, Z. A., Niki, M., et al. (2005). Crucial role of p53-dependent cellular senescence in suppression of Pten-deficient tumorigenesis. *Nature, 436*, 725–730. https://doi.org/10.1038/nature03918.

Chen, Z.-T., Zhao, W. E. I., Qu, S., Li, L., Lu, X.-D., Su, F., et al. (2015). PARP-1 promotes autophagy via the AMPK/mTOR pathway in CNE-2 human nasopharyngeal carcinoma cells following ionizing radiation, while inhibition of autophagy contributes to the radiation sensitization of CNE-2 cells. *Molecular Medicine Reports, 12*, 1868–1876. https://doi.org/10.3892/mmr.2015.3604.

Chiao, M. T., Cheng, W. Y., Yang, Y. C., Shen, C. C., & Ko, J. L. (2013). Suberoylanilide hydroxamic acid (SAHA) causes tumor growth slowdown and triggers autophagy in glioblastoma stem cells. *Autophagy, 9*(10), 1509–1526. https://doi.org/10.4161/auto.25664.

Chiche, A., Le Roux, I., von Joest, M., Sakai, H., Aguín, S. B., Cazin, C., et al. (2017). Injury-induced senescence enables in vivo reprogramming in skeletal muscle. *Cell Stem Cell, 20*(3), 407–414. https://doi.org/10.1016/j.stem.2016.11.020.

Chien, Y., Scuoppo, C., Wang, X., Fang, X., Balgley, B., Bolden, J. E., et al. (2011). Control of the senescence-associated secretory phenotype by NF-κB promotes senescence and enhances chemosensitivity. *Genes & Development, 25*(20), 2125–2136. https://doi.org/10.1101/gad.17276711.

Childs, B. G., Durik, M., Baker, D. J., & van Deursen, J. M. (2015). Cellular senescence in aging and age-related disease: From mechanisms to therapy. *Nature Medicine, 21*, 1424–1435. https://doi.org/10.1038/nm.4000.

Chiu, H. W., Lin, S. W., Lin, L. C., Hsu, Y. H., Lin, Y. F., Ho, S. Y., et al. (2015). Synergistic antitumor effects of radiation and proteasome inhibitor treatment in pancreatic cancer through the induction of autophagy and the downregulation of TRAF6. *Cancer Letters, 365*(2), 229–239. https://doi.org/10.1016/j.canlet.2015.05.025.

Cho, Y. H., Han, K. M., Kim, D., Lee, J., Lee, S. H., Choi, K. W., et al. (2014). Autophagy regulates homeostasis of pluripotency-associated proteins in hESCs. *Stem Cells, 32*(2), 424–435. https://doi.org/10.1002/stem.1589.

Chun, Y., & Kim, J. (2018). Autophagy: An essential degradation program for cellular homeostasis and life. *Cell, 7*, 278. https://doi.org/10.3390/cells7120278.

Cojoc, M., Mäbert, K., Muders, M. H., & Dubrovska, A. (2015). A role for cancer stem cells in therapy resistance: Cellular and molecular mechanisms. *Seminars in Cancer Biology, 31*, 16–27. https://doi.org/10.1016/j.semcancer.2014.06.004.

Coussens, L. M., & Werb, Z. (2002). Inflammation and cancer. *Nature, 420*, 860–867. https://doi.org/10.1038/nature01322.

Cufí, S., Vazquez-Martin, A., Oliveras-Ferraros, C., Martin-Castillo, B., Vellon, L., & Menendez, J. A. (2011). Autophagy positively regulates the CD44(+) CD24(−/low) breast cancer stem-like phenotype. *Cell Cycle, 10*(22), 3871–3885. https://doi.org/10.4161/cc.10.22.17976.

Däbritz, J. H., Yu, Y., Milanovic, M., Schönlein, M., Rosenfeldt, M. T., Dörr, J. R., et al. (2016). CD20-targeting immunotherapy promotes cellular senescence in B-cell lymphoma. *Molecular Cancer Therapeutics, 15*(5), 1074–1081. https://doi.org/10.1158/1535-7163.

Degenhardt, K., Mathew, R., Beaudoin, B., Bray, K., Anderson, D., Chen, G., et al. (2006). Autophagy promotes tumor cell survival and restricts necrosis, inflammation, and tumorigenesis. *Cancer Cell, 10*, 51–64. https://doi.org/10.1016/j.ccr.2006.06.001.

Demaria, M., O'Leary, M. N., Chang, J., Shao, L., Liu, S., Alimirah, F., et al. (2017). Cellular senescence promotes adverse effects of chemotherapy and cancer relapse. *Cancer Discovery, 7*(2), 165–176. https://doi.org/10.1158/2159-8290.

DeMasters, G. A., Gupta, M. S., Jones, K. R., Cabot, M., Wang, H., Gennings, C., et al. (2004). Potentiation of cell killing by fractionated radiation and suppression of proliferative recovery in MCF-7 breast tumor cells by the vitamin D3 analog EB 1089. *Journal of Steroid Biochemistry and Molecular Biology*, *92*(5), 365–374. https://doi.org/10.1016/j.jsbmb.2004.07.011.

Deretic, V., Kimura, T., Timmins, G., Moseley, P., Chauhan, S., & Mandell, M. (2015). Immunologic manifestations of autophagy. *The Journal of Clinical Investigation*, *125*, 75–84. https://doi.org/10.1172/jci73945.

Deruy, E., Gosselin, K., Vercamer, C., Martien, S., Bouali, F., Slomianny, C., et al. (2010). MnSOD upregulation induces autophagic programmed cell death in senescent keratinocytes. *PLoS One*, *5*. https://doi.org/10.1371/journal.pone.0012712, 0012712.

Dhamodaran, K., Subramani, M., Jeyabalan, N., Ponnalagu, M., Chevour, P., Shetty, R., et al. (2015). Characterization of ex vivo cultured limbal, conjunctival, and oral mucosal cells: A comparative study with implications in transplantation medicine. *Molecular Vision*, *21*, 828–845. https://pubmed.ncbi.nlm.nih.gov/26283864.

Dick, J. E. (2008). Stem cell concepts renew cancer research. *Blood*, *112*, 4793–4807. https://doi.org/10.1182/blood-2008-08-077941.

Dodig, S., Čepelak, I., & Pavić, I. (2019). Hallmarks of senescence and aging. *Biochemia Medica*, *29*, 483–497. https://doi.org/10.11613/bm.2019.030501.

Downes, C. P., Walker, S., McConnachie, G., Lindsay, Y., Batty, I. H., & Leslie, N. R. (2004). Acute regulation of the tumour suppressor phosphatase, PTEN, by anionic lipids and reactive oxygen species. *Biochemical Society Transactions*, *32*(Part 2), 338–342. https://doi.org/10.1042/bst0320338.

Dvorak, H. F. (1986). Tumors: Wounds that do not heal. Similarities between tumor stroma generation and wound healing. *The New England Journal of Medicine*, *315*, 1650–1659. https://doi.org/10.1056/NEJM198612253152606.

Elknerova, K., Myslivcova, D., Lacinova, Z., Marinov, I., Uherkova, L., & Stöckbauer, P. (2011). Epigenetic modulation of gene expression of human leukemia cell lines – induction of cell death and senescence. *Neoplasma*, *58*(1), 35–44. https://doi.org/10.4149/neo_2011_01_35.

Erenpreisa, J., Salmina, K., & Cragg, M. S. (2017). Accelerated senescence of cancer stem cells: A failure to thrive or a route to survival? In J. Dorszewska, & W. Kozubski (Eds.), *Senescence: Physiology or pathology* (pp. 387–398). INTECH. https://doi.org/10.5772/intechopen.68582.

Eritja, N., Chen, B.-J., Rodríguez-Barrueco, R., Santacana, M., Gatius, S., Vidal, A., et al. (2017). Autophagy orchestrates adaptive responses to targeted therapy in endometrial cancer. *Autophagy*, *13*, 608–624. https://doi.org/10.1080/15548627.2016.1271512.

Ertl, R. P., Chen, J., Astle, C. M., Duffy, T. M., & Harrison, D. E. (2008). Effects of dietary restriction on hematopoietic stem-cell aging are genetically regulated. *Blood*, *111*(3), 1709–1716. https://doi.org/10.1182/blood-2007-01-069807.

Espina, V., Mariani, B. D., Gallagher, R. I., Tran, K., Banks, S., Wiedemann, J., et al. (2010). Malignant precursor cells pre-exist in human breast DCIS and require autophagy for survival. *PLoS One*, *5*, 0010240. https://doi.org/10.1371/journal.pone.0010240.

Esteban-Martínez, L., Sierra-Filardi, E., & Boya, P. (2017). Mitophagy, metabolism, and cell fate. *Molecular & Cellular Oncology*, *4*, e1353854. https://doi.org/10.1080/23723556.2017.1353854.

Fimia, G. M., Stoykova, A., Romagnoli, A., Giunta, L., Di Bartolomeo, S., Nardacci, R., et al. (2007). Ambra1 regulates autophagy and development of the nervous system. *Nature*, *447*(7148), 1121–1125. https://doi.org/10.1038/nature05925.

Franco, S. S., Raveh-Amit, H., Kobolák, J., Alqahtani, M. H., Mobasheri, A., & Dinnyes, A. (2015). The crossroads between cancer stem cells and aging. *BMC Cancer*, *15*(Suppl. 1), S1. https://doi.org/10.1186/1471-2407-15-s1-s1.

Galavotti, S., Bartesaghi, S., Faccenda, D., Shaked-Rabi, M., Sanzone, S., McEvoy, A., et al. (2013). The autophagy-associated factors DRAM1 and p62 regulate cell migration and invasion in glioblastoma stem cells. *Oncogene*, *32*, 699–712. https://doi.org/10.1038/onc.2012.111.

Galluzzi, L., Bravo-San Pedro, J. M., Demaria, S., Formenti, S. C., & Kroemer, G. (2017). Activating autophagy to potentiate immunogenic chemotherapy and radiation therapy. *Nature Reviews. Clinical Oncology*, *14*, 247–258. https://doi.org/10.1038/nrclinonc.2016.183.

Galluzzi, L., Pietrocola, F., Bravo-San Pedro, J. M., Amaravadi, R. K., Baehrecke, E. H., Cecconi, F., et al. (2015). Autophagy in malignant transformation and cancer progression. *The EMBO Journal*, *34*, 856–880. https://doi.org/10.15252/embj.201490784.

García-Prat, L., Martínez-Vicente, M., Perdiguero, E., Ortet, L., Rodríguez-Ubreva, J., Rebollo, E., et al. (2016). Autophagy maintains stemness by preventing senescence. *Nature*, *529*(7584), 37–42. https://doi.org/10.1038/nature16187.

Gerland, L. M., Peyrol, S., Lallemand, C., Branche, R., Magaud, J. P., & Ffrench, M. (2003). Association of increased autophagic inclusions labeled for beta-galactosidase with fibroblastic aging. *Experimental Gerontology*, *38*, 887–895. https://doi.org/10.1016/s0531-5565(03)00132-3.

Gewirtz, D. A. (2013). Autophagy and senescence: A partnership in search of definition. *Autophagy*, *9*, 808–812. https://doi.org/10.4161/auto.23922.

Gewirtz, D. A. (2014a). Autophagy and senescence. *Autophagy*, *9*, 808–812. https://doi.org/10.4161/auto.23922.

Gewirtz, D. A. (2014b). Autophagy and senescence in cancer therapy. *Journal of Cellular Physiology*, *229*, 6–9. https://doi.org/10.1002/jcp.24420.

Goehe, R. W., Di, X., Sharma, K., Bristol, M. L., Henderson, S. C., Valerie, K., et al. (2012). The autophagy-senescence connection in chemotherapy: Must tumor cells (self) eat before they sleep? *Journal of Pharmacology and Experimental Therapeutics*, *343*, 763–778. https://doi.org/10.1124/jpet.112.197590.

Goel, S., DeCristo, M. J., Watt, A. C., BrinJones, H., Sceneay, J., Li, B. B., et al. (2017). CDK4/6 inhibition triggers anti-tumour immunity. *Nature*, *548*(7668), 471–475. https://doi.org/10.1038/nature23465.

Goel, S., Wang, Q., Watt, A. C., Tolaney, S. M., Dillon, D. A., Li, W., et al. (2016). Overcoming therapeutic resistance in HER2-positive breast cancers with CDK4/6 inhibitors. *Cancer Cell*, *29*(3), 255–269. https://doi.org/10.1016/j.ccell.2016.02.006.

Gong, C., Bauvy, C., Tonelli, G., Yue, W., Deloménie, C., Nicolas, V., et al. (2013). Beclin 1 and autophagy are required for the tumorigenicity of breast cancer stem-like/progenitor cells. *Oncogene*, *32*, 2261–2272. https://doi.org/10.1038/onc.2012.252.

Gong, X., Litchfield, L. M., Webster, Y., Chio, L. C., Wong, S. S., Stewart, T. R., et al. (2017). Genomic aberrations that activate D-type cyclins are associated with enhanced sensitivity to the CDK4 and CDK6 inhibitor abemaciclib. *Cancer Cell*, *32*(6), 761–776. https://doi.org/10.1016/j.ccell.2017.11.006.

Gosselin, K., Deruy, E., Martien, S., Vercamer, C., Bouali, F., Dujardin, T., et al. (2009). Senescent keratinocytes die by autophagic programmed cell death. *The American Journal of Pathology*, *174*, 423–435. https://doi.org/10.2353/ajpath.2009.080332.

Gravina, S., & Vijg, J. (2010). Epigenetic factors in aging and longevity. *Pflügers Archiv*, *459*, 247–258. https://doi.org/10.1007/s00424-009-0730-7.

Guan, J.-L., Simon, A. K., Prescott, M., Menendez, J. A., Liu, F., Wang, F., et al. (2014). Autophagy in stem cells. *Autophagy*, *9*, 830–849. https://doi.org/10.4161/auto.24132.

Guan, X., LaPak, K. M., Hennessey, R. C., Yu, C. Y., Shakya, R., Zhang, J., et al. (2017). Stromal senescence by prolonged CDK4/6 inhibition potentiates tumor growth. *Molecular Cancer Research*, *15*(3), 237–249. https://doi.org/10.1158/1541-7786.

Guenther, L. M., Dharia, N. V., Ross, L., Conway, A., Robichaud, A. L., Catlett, J. L., 2nd, et al. (2019). A combination CDK4/6 and IGF1R inhibitor strategy for ewing sarcoma. *Clinical Cancer Research*, *25*(4), 1343–1357. https://doi.org/10.1158/1078-0432.CCR-18-0372.

Günther, W., Pawlak, E., Damasceno, R., Arnold, H., & Terzis, A. J. (2003). Temozolomide induces apoptosis and senescence in glioma cells cultured as multicellular spheroids. *British Journal of Cancer*, *88*(3), 463–469. https://doi.org/10.1038/sj.bjc.6600711.

Gupta, P. B., Fillmore, C. M., Jiang, G., Shapira, S. D., Tao, K., Kuperwasser, C., et al. (2011). Stochastic state transitions give rise to phenotypic equilibrium in populations of cancer cells. *Cell, 146*, 633–644. https://doi.org/10.1016/j.cell.2011.07.026.

Gurumurthy, S., Xie, S. Z., Alagesan, B., Kim, J., Yusuf, R. Z., Saez, B., et al. (2010). The Lkb1 metabolic sensor maintains haematopoietic stem cell survival. *Nature, 468*(7324), 659–663. https://doi.org/10.1038/nature09572.

Han, Z., Wei, W., Dunaway, S., Darnowski, J. W., Calabresi, P., Sedivy, J., et al. (2002). Role of p21 in apoptosis and senescence of human colon cancer cells treated with camptothecin. *Journal of Biological Chemistry, 277*(19), 17154–17160. https://doi.org/10.1074/jbc.M112401200.

Hari, P., Millar, F. R., Tarrats, N., Birch, J., Quintanilla, A., Rink, C. J., et al. (2019). The innate immune sensor toll-like receptor 2 controls the senescence-associated secretory phenotype. *Science Advances, 5*(6). https://doi.org/10.1126/sciadv.aaw0254.

Hasmim, M., Janji, B., Khaled, M., Noman, M. Z., Louache, F., Bordereaux, D., et al. (2017). Cutting edge: NANOG activates autophagy under hypoxic stress by binding to BNIP3L promoter. *The Journal of Immunology, 198*, 1423. https://doi.org/10.4049/jimmunol.1600981.

Hirose, Y., Katayama, M., Mirzoeva, O. K., Berger, M. S., & Pieper, R. O. (2005). Akt activation suppresses Chk2-mediated, methylating agent-induced G2 arrest and protects from temozolomide-induced mitotic catastrophe and cellular senescence. *Cancer Research, 65*(11), 4861–4869. https://doi.org/10.1158/0008-5472.

Hou, J., Han, Z. P., Jing, Y. Y., Yang, X., Zhang, S. S., Sun, K., et al. (2013). Autophagy prevents irradiation injury and maintains stemness through decreasing ros generation in mesenchymal stem cells. *Cell Death & Disease, 4*(10), e844. https://doi.org/10.1038/cddis.2013.338.

Hou, W., Zhang, Q., Yan, Z., Chen, R., Zeh Iii, H. J., Kang, R., et al. (2013). Strange attractors: DAMPs and autophagy link tumor cell death and immunity. *Cell Death & Disease, 4*, 493. https://doi.org/10.1038/cddis.2013.493.

Hu, C., Zhao, L., Wu, D., & Li, L. (2019). Modulating autophagy in mesenchymal stem cells effectively protects against hypoxia- or ischemia-induced injury. *Stem Cell Research & Therapy, 10*, 120. https://doi.org/10.1186/s13287-019-1225-x.

Hu, W., Sung, T., Jessen, B. A., Thibault, S., Finkelstein, M. B., Khan, N. K., et al. (2016). Mechanistic investigation of bone marrow suppression associated with palbociclib and its differentiation from cytotoxic chemotherapies. *Clinical Cancer Research, 22*(8), 2000–2008. https://doi.org/10.1158/1078-0432.CCR-15-1421.

Huang, Y.-H., Yang, P.-M., Chuah, Q.-Y., Lee, Y.-J., Hsieh, Y.-F., Peng, C.-W., et al. (2014). Autophagy promotes radiation-induced senescence but inhibits bystander effects in human breast cancer cells. *Autophagy, 10*, 1212–1228. https://doi.org/10.4161/auto.28772.

Igarashi, K., Sakimoto, I., Kataoka, K., Ohta, K., & Miura, M. (2007). Radiation-induced senescence-like phenotype in proliferating and plateau-phase vascular endothelial cells. *Experimental Cell Research, 313*(15), 3326–3336. https://doi.org/10.1016/j.yexcr.2007.06.001.

Ito, K., Hirao, A., Arai, F., Matsuoka, S., Takubo, K., Hamaguchi, I., et al. (2004). Regulation of oxidative stress by ATM is required for self-renewal of haematopoietic stem cells. *Nature, 431*(7011), 997–1002. https://doi.org/10.1038/nature02989.

Jaiswal, A. S., Panda, H., Law, B. K., Sharma, J., Jani, J., Hromas, R., et al. (2015). NSC666715 and its analogs inhibit strand-displacement activity of DNA polymerase β and potentiate temozolomide-induced DNA damage, senescence and apoptosis in colorectal cancer cells. *PLoS One, 10*(5). https://doi.org/10.1371/journal.pone.0123808.

Ji, J., Tian, Y., Zhu, Y. Q., Zhang, L. Y., Ji, S. J., Huan, J., et al. (2015). Ionizing irradiation inhibits keloid fibroblast cell proliferation and induces premature cellular senescence. *Journal of Dermatology, 42*(1), 56–63. https://doi.org/10.1111/1346-8138.12702.

Jia, K., & Levine, B. (2007). Autophagy is required for dietary restriction-mediated life span extension in *C. elegans*. *Autophagy*, *3*(6), 597–599. https://doi.org/10.4161/auto.4989.

Jiang, C., Liu, G., Luckhardt, T., Antony, V., Zhou, Y., Carter, A. B., et al. (2017). Serpine 1 induces alveolar type II cell senescence through activating p53-p21-Rb pathway in fibrotic lung disease. *Aging Cell*, *16*(5), 1114–1124. https://doi.org/10.1111/acel.12643.

Jones, K. R., Elmore, L. W., Jackson-Cook, C., Demasters, G., Povirk, L. F., Holt, S. E., et al. (2005). P53-dependent accelerated senescence induced by ionizing radiation in breast tumour cells. *International Journal of Radiation Biology*, *81*(6), 445–458. https://doi.org/10.1080/09553000500168549.

Kaletsch, A., Pinkerneil, M., Hoffmann, M. J., Jaguva Vasudevan, A. A., Wang, C., Hansen, F. K., et al. (2018). Effects of novel HDAC inhibitors on urothelial carcinoma cells. *Clinical Epigenetics*, *10*(1). https://doi.org/10.1186/s13148-018-0531-y.

Kang, T.-W., Yevsa, T., Woller, N., Hoenicke, L., Wuestefeld, T., Dauch, D., et al. (2011). Senescence surveillance of pre-malignant hepatocytes limits liver cancer development. *Nature*, *479*, 547–551. https://doi.org/10.1038/nature10599.

Kantara, C., O'Connell, M., Sarkar, S., Moya, S., Ullrich, R., & Singh, P. (2014). Curcumin promotes autophagic survival of a subset of colon cancer stem cells, which are ablated by DCLK1-siRNA. *Cancer Research*, *74*, 2487–2498. https://doi.org/10.1158/0008-5472.CAN-13-3536.

Karvela, M., Baquero, P., Kuntz, E. M., Mukhopadhyay, A., Mitchell, R., Allan, E. K., et al. (2016). ATG7 regulates energy metabolism, differentiation and survival of philadelphia-chromosome-positive cells. *Autophagy*, *12*(6), 936–948. https://doi.org/10.1080/15548627.2016.1162359.

Katajisto, P., Döhla, J., Chaffer, C. L., Pentinmikko, N., Marjanovic, N., Iqbal, S., et al. (2015). Stem cells. Asymmetric apportioning of aged mitochondria between daughter cells is required for stemness. *Science*, *348*, 340–343. https://doi.org/10.1126/science.1260384.

Kaur, J., & Debnath, J. (2015). Autophagy at the crossroads of catabolism and anabolism. *Nature Reviews Molecular Cell Biology*, *16*, 461–472. https://doi.org/10.1038/nrm4024.

Kim, J. S., Choi, H. W., Choi, S., & Do, J. T. (2011). Reprogrammed pluripotent stem cells from somatic cells. *International Journal of Stem Cells*, *4*, 1–8. https://doi.org/10.15283/ijsc.2011.4.1.1.

Kim, K. W., Hwang, M., Moretti, L., Jaboin, J. J., Cha, Y. I., & Lu, B. (2008). Autophagy upregulation by inhibitors of caspase-3 and mTOR enhances radiotherapy in a mouse model of lung cancer. *Autophagy*, *4*(5), 659–668. https://doi.org/10.4161/auto.6058.

Kim, K. W., Moretti, L., Mitchell, L. R., Jung, D. K., & Lu, B. (2009). Combined Bcl-2/mammalian target of rapamycin inhibition leads to enhanced radiosensitization via induction of apoptosis and autophagy in non-small cell lung tumor xenograft model. *Clinical Cancer Research*, *15*(19), 6096–6105. https://doi.org/10.1158/1078-0432.CCR-09-0589.

Kim, K. W., Myers, C. J., Jung, D. K., & Lu, B. (2014). NVP-BEZ-235 enhances radiosensitization via blockade of the PI3K/mTOR pathway in cisplatin-resistant non-small cell lung carcinoma. *Genes Cancer*, *5*(7–8), 293–302. https://doi.org/10.18632/genesandcancer.27.

Kiyono, K., Suzuki, H. I., Matsuyama, H., Morishita, Y., Komuro, A., Kano, M. R., et al. (2009). Autophagy is activated by TGF-β and potentiates TGF-β–mediated growth inhibition in human hepatocellular carcinoma cells. *Cancer Research*, *69*, 8844. https://doi.org/10.1158/0008-5472.can-08-4401.

Klein, M. E., Dickson, M. A., Antonescu, C., Qin, L. X., Dooley, S. J., Barlas, A., et al. (2018). PDLIM7 and CDH18 regulate the turnover of MDM2 during CDK4/6 inhibitor therapy-induced senescence. *Oncogene*, *37*(37), 5066–5078. https://doi.org/10.1038/s41388-018-0332-y.

Klionsky, D. J. (2008). Autophagy revisited: A conversation with Christian de Duve. *Autophagy*, *4*, 740–743. https://doi.org/10.4161/auto.6398.

Knizhnik, A. V., Roos, W. P., Nikolova, T., Quiros, S., Tomaszowski, K.-H., Christmann, M., et al. (2013). Survival and death strategies in glioma cells: Autophagy, senescence and apoptosis triggered by a single type of temozolomide-induced DNA damage. *PLoS One*, *8*(1), e55665. https://doi.org/10.1371/journal.pone.0055665.

Komatsu, M., Ueno, T., Waguri, S., Uchiyama, Y., Kominami, E., & Tanaka, K. (2007). Constitutive autophagy: Vital role in clearance of unfavorable proteins in neurons. *Cell Death and Differentiation*, *14*, 887–894. https://doi.org/10.1038/sj.cdd.4402120.

Koukourakis, M. I., Kalamida, D., Mitrakas, A., Pouliliou, S., Kalamida, S., Sivridis, E., et al. (2015). Intensified autophagy compromises the efficacy of radiotherapy against prostate cancer. *Biochemical and Biophysical Research Communications*, *461*(2), 268–274. https://doi.org/10.1016/j.bbrc.2015.04.014.

Kovatcheva, M., Liao, W., Klein, M. E., Robine, N., Geiger, H., Crago, A. M., et al. (2017). ATRX is a regulator of therapy induced senescence in human cells. *Nature Communications*, *8*(1). https://doi.org/10.1038/s41467-017-00540-5.

Kovatcheva, M., Liu, D. D., Dickson, M. A., Klein, M. E., O'Connor, R., Wilder, F. O., et al. (2015). MDM2 turnover and expression of ATRX determine the choice between quiescence and senescence in response to CDK4 inhibition. *Oncotarget*, *6*(10), 8226–8243. https://doi.org/10.18632/oncotarget.3364.

Kuilman, T., Michaloglou, C., Mooi, W. J., & Peeper, D. S. (2010). The essence of senescence. *Genes & Development*, *24*, 2463–2479. https://doi.org/10.1101/gad.1971610.

Kuo, T.-C., Chen, C.-T., Baron, D., Onder, T. T., Loewer, S., Almeida, S., et al. (2011). Midbody accumulation through evasion of autophagy contributes to cellular reprogramming and tumorigenicity. *Nature Cell Biology*, *13*(10), 1214–1223. https://doi.org/10.1038/ncb2332.

Kwon, Y., Kim, J. W., Jeoung, J. A., Kim, M.-S., & Kang, C. (2017). Autophagy is pro-senescence when seen in close-up, but anti-senescence in long-shot. *Molecules and Cells*, *40*, 607–612. https://doi.org/10.14348/molcells.2017.0151.

Laberge, R. M., Sun, Y., Orjalo, A. V., Patil, C. K., Freund, A., Zhou, L., et al. (2015). mTOR regulates the pro-tumorigenic senescence-associated secretory phenotype by promoting IL1A translation. *Nature Cell Biology*, *17*(8), 1049–1061. https://doi.org/10.1038/ncb3195.

La Porta, C. A., Zapperi, S., & Sethna, J. P. (2012). Senescent cells in growing tumors: Population dynamics and cancer stem cells. *PLoS Computational Biology*, *8*, e1002316. https://doi.org/10.1371/journal.pcbi.1002316.

Lawson, J. C., Blatch, G. L., & Edkins, A. L. (2009). Cancer stem cells in breast cancer and metastasis. *Breast Cancer Research and Treatment*, *118*, 241–254. https://doi.org/10.1007/s10549-009-0524-9.

Lee, I., Lee, S.-J., Kang, W. K., & Park, C. (2012). Inhibition of monocarboxylate transporter 2 induces senescence-associated mitochondrial dysfunction and suppresses progression of colorectal malignancies in vivo. *Molecular Cancer Therapeutics*, *11*, 2342. https://doi.org/10.1158/1535-7163.mct-12-0488.

Lee, S. Y., Jeong, E. K., Ju, M. K., Jeon, H. M., Kim, M. Y., Kim, C. H., et al. (2017). Induction of metastasis, cancer stem cell phenotype, and oncogenic metabolism in cancer cells by ionizing radiation. *Molecular Cancer*, *16*, 10. https://doi.org/10.1186/s12943-016-0577-4.

Lei, Y., Zhang, D., Yu, J., Dong, H., Zhang, J., & Yang, S. (2017). Targeting autophagy in cancer stem cells as an anticancer therapy. *Cancer Letters*, *393*, 33–39. https://doi.org/10.1016/j.canlet.2017.02.012.

Leontieva, O. V., & Blagosklonny, M. V. (2013). CDK4/6-inhibiting drug substitutes for p21 and p16 in senescence: Duration of cell cycle arrest and mtor activity determine geroconversion. *Cell Cycle*, *12*(18), 3063–3069. https://doi.org/10.4161/cc.26130.
Levine, B. (2007). Autophagy and cancer. *Nature*, *446*, 745–747. https://doi.org/10.1038/446745a.
Levine, B., & Kroemer, G. (2008). Autophagy in the pathogenesis of disease. *Cell*, *132*, 27–42. https://doi.org/10.1016/j.cell.2007.12.018.
Li, J., Yang, B., Zhou, Q., Wu, Y., Shang, D., Guo, Y., et al. (2013). Autophagy promotes hepatocellular carcinoma cell invasion through activation of epithelial-mesenchymal transition. *Carcinogenesis*, *34*, 1343–1351. https://doi.org/10.1093/carcin/bgt063.
Li, L., Zhu, Y. Q., Jiang, L., & Peng, W. (2012). Increased autophagic activity in senescent human dental pulp cells. *International Endodontic Journal*, *45*, 1074–1079. https://doi.org/10.1111/j.1365-2591.2012.02064.
Li, X., He, S., & Ma, B. (2020). Autophagy and autophagy-related proteins in cancer. *Molecular Cancer*, *19*, 12. https://doi.org/10.1186/s12943-020-1138-4.
Liang, D. H., El-Zein, R., & Dave, B. (2015). Autophagy inhibition to increase radiosensitization in breast cancer. *Journal of Nuclear Medicine & Radiation Therapy*, *6*, 254. https://doi.org/10.4172/2155-9619.1000254.
Lin, A. W., Barradas, M., Stone, J. C., van Aelst, L., Serrano, M., & Lowe, S. W. (1998). Premature senescence involving p53 and p16 is activated in response to constitutive MEK/MAPK mitogenic signaling. *Genes & Development*, *12*, 3008–3019. https://doi.org/10.1101/gad.12.19.3008.
Linge, A., Weinhold, K., Bläsche, R., Kasper, M., & Barth, K. (2007). Downregulation of caveolin-1 affects bleomycin-induced growth arrest and cellular senescence in A549 cells. *International Journal of Biochemistry & Cell Biology*, *39*(10), 1964–1974. https://doi.org/10.1016/j.biocel.2007.05.018.
Liu, D., & Hornsby, P. J. (2007). Senescent human fibroblasts increase the early growth of xenograft tumors via matrix metalloproteinase secretion. *Cancer Research*, *67*(7), 3117–3126. https://doi.org/10.1158/0008-5472.
Liu, F., Lee, J. Y., Wei, H., Tanabe, O., Engel, J. D., Morrison, S. J., et al. (2010). FIP200 is required for the cell-autonomous maintenance of fetal hematopoietic stem cells. *Blood*, *116*(23), 4806–4814. https://doi.org/10.1182/blood-2010-06-288589.
Liu, L., & Rando, T. A. (2011). Manifestations and mechanisms of stem cell aging. *The Journal of Cell Biology*, *193*, 257–266. https://doi.org/10.1101/gad.12.19.3008.
López-Otín, C., Blasco, M. A., Partridge, L., Serrano, M., & Kroemer, G. (2013). The hallmarks of aging. *Cell*, *153*, 1194–1217. https://doi.org/10.1016/j.cell.2013.05.039.
Lum, J. J., DeBerardinis, R. J., & Thompson, C. B. (2005). Autophagy in metazoans: Cell survival in the land of plenty. *Nature Reviews. Molecular Cell Biology*, *6*, 439–448. https://doi.org/10.1038/nrm1660.
Maier, P., Hartmann, L., Wenz, F., & Herskind, C. (2016). Cellular pathways in response to ionizing radiation and their targetability for tumor radiosensitization. *International Journal of Molecular Sciences*, *17*, 102. https://doi.org/10.3390/ijms17010102.
Martinez, N. J., & Gregory, R. I. (2010). MicroRNA gene regulatory pathways in the establishment and maintenance of ESC identity. *Cell Stem Cell*, *7*, 31–35. https://doi.org/10.1016/j.stem.2010.06.011.
Mathew, R., Karantza-Wadsworth, V., & White, E. (2007). Role of autophagy in cancer. *Nature Reviews Cancer*, *7*, 961–967. https://doi.org/10.1038/nrc2254.
Maycotte, P., Gearheart, C. M., Barnard, R., Aryal, S., Mulcahy Levy, J. M., Fosmire, S. P., et al. (2014). STAT3-mediated autophagy dependence identifies subtypes of breast cancer where autophagy inhibition can be efficacious. *Cancer Research*, *74*, 2579–2590. https://doi.org/10.1158/0008-5472.CAN-13-3470.
Mayol, J. F., Loeuillet, C., Hérodin, F., & Wion, D. (2009). Characterisation of normal and cancer stem cells: One experimental paradigm for two kinds of stem cells. *BioEssays*, *31*, 993–1001. https://doi.org/10.1002/bies.200900041.

Meenhuis, A., van Veelen, P. A., de Looper, H., van Boxtel, N., van den Berge, I. J., Sun, S. M., et al. (2011). MiR-17/20/93/106 promote hematopoietic cell expansion by targeting sequestosome 1-regulated pathways in mice. *Blood, 118*(4), 916–925. https://doi.org/10.1182/blood-2011-02-336487.

Menendez, J. A., & Alarcón, T. (2017). Senescence-inflammatory regulation of reparative cellular reprogramming in aging and cancer. *Frontiers in Cell and Development Biology, 5*, 49. https://doi.org/10.3389/fcell.2017.00049.

Menzies, F. M., Fleming, A., Caricasole, A., Bento, C. F., Andrews, S. P., Ashkenazi, A., et al. (2017). Autophagy and neurodegeneration: Pathogenic mechanisms and therapeutic opportunities. *Neuron, 93*, 1015–1034. https://doi.org/10.1016/j.neuron.2017.01.022.

Mhaidat, N. M., Zhang, X. D., Allen, J., Avery-Kiejda, K. A., Scott, R. J., & Hersey, P. (2007). Temozolomide induces senescence but not apoptosis in human melanoma cells. *British Journal of Cancer, 97*(9), 1225–1233. https://doi.org/10.1038/sj.bjc.6604017.

Milanovic, M., Fan, D. N. Y., Belenki, D., Dabritz, J. H. M., Zhao, Z., Yu, Y., et al. (2018). Senescence-associated reprogramming promotes cancer stemness. *Nature, 553*, 96–100. https://doi.org/10.1038/nature25167.

Mimeault, M., & Batra, S. K. (2009). Recent insights into the molecular mechanisms involved in aging and the malignant transformation of adult stem/progenitor cells and their therapeutic implications. *Ageing Research Reviews, 8*, 94–112. https://doi.org/10.1016/j.arr.2008.12.001.

Mizushima, N. (2007). Autophagy: Process and function. *Genes & Development, 21*, 2861–2873. https://doi.org/10.1101/gad.1599207.

Mizushima, N., Yamamoto, A., Hatano, M., Kobayashi, Y., Kabeya, Y., Suzuki, K., et al. (2001). Dissection of autophagosome formation using Apg5-deficient mouse embryonic stem cells. *Journal of Cell Biology, 152*(4), 657–668. https://doi.org/10.1083/jcb.152.4.657.

Moreno-Blas, D., Gorostieta-Salas, E., Pommer-Alba, A., Muciño-Hernández, G., Gerónimo-Olvera, C., Maciel-Barón, L. A., et al. (2019). Cortical neurons develop a senescence-like phenotype promoted by dysfunctional autophagy. *Aging (Albany NY), 11*, 6176–6198. https://doi.org/10.18632/aging.102181.

Mortensen, M., Soilleux, E. J., Djordjevic, G., Tripp, R., Lutteropp, M., Sadighi-Akha, E., et al. (2011a). The autophagy protein Atg7 is essential for hematopoietic stem cell maintenance. *The Journal of Experimental Medicine, 208*(3), 455–467. https://doi.org/10.1084/jem.20101145.

Mortensen, M., Watson, A. S., & Simon, A. K. (2011b). Lack of autophagy in the hematopoietic system leads to loss of hematopoietic stem cell function and dysregulated myeloid proliferation. *Autophagy*, 7(9), 1069–1070. https://doi.org/10.4161/auto.7.9.15886.

Muñoz-Espín, D., & Serrano, M. (2014). Cellular senescence: From physiology to pathology. *Nature Reviews Molecular Cell Biology, 15*, 482–496. https://doi.org/10.1038/nrm3823.

Muñoz-Espín, D., Rovira, M., Galiana, I., Giménez, C., Lozano-Torres, B., Paez-Ribes, M., et al. (2018). A versatile drug delivery system targeting senescent cells. *EMBO Molecular Medicine, 10*(9), 201809355. https://doi.org/10.15252/emmm.201809355.

Murakami, M., Ichisaka, T., Maeda, M., Oshiro, N., Hara, K., Edenhofer, F., et al. (2004). mTOR is essential for growth and proliferation in early mouse embryos and embryonic stem cells. *Molecular and Cellular Biology, 24*(15), 6710–6718. https://doi.org/10.1128/mcb.24.15.6710-6718.2004.

Narita, M., Young, A. R., Arakawa, S., Samarajiwa, S. A., Nakashima, T., Yoshida, S., et al. (2011). Spatial coupling of mTOR and autophagy augments secretory phenotypes. *Science, 332*, 966–970. https://doi.org/10.1126/science.1205407.

Nazio, F., Bordi, M., Cianfanelli, V., Locatelli, F., & Cecconi, F. (2019). Autophagy and cancer stem cells: Molecular mechanisms and therapeutic applications. *Cell Death and Differentiation, 26*, 690–702. https://doi.org/10.1038/s41418-019-0292-y.

Nguyen, L. V., Vanner, R., Dirks, P., & Eaves, C. J. (2012). Cancer stem cells: An evolving concept. *Nature Reviews. Cancer*, *12*(2), 133–143. https://doi.org/10.1038/nrc3184.

Nishi, M., Akutsu, H., Kudoh, A., Kimura, H., Yamamoto, N., Umezawa, A., et al. (2014). Induced cancer stem-like cells as a model for biological screening and discovery of agents targeting phenotypic traits of cancer stem cell. *Oncotarget*, *5*, 8665–8680. https://doi.org/10.18632/oncotarget.2356.

Ohba, S., Hirose, Y., Kawase, T., & Sano, H. (2009). Inhibition of c-jun N-terminal kinase enhances temozolomide-induced cytotoxicity in human glioma cells. *Journal of Neuro-Oncology*, *95*(3), 307–316. https://doi.org/10.1007/s11060-009-9929-x.

Ojha, R., Bhattacharyya, S., & Singh, S. K. (2015). Autophagy in cancer stem cells: A potential Link between chemoresistance, recurrence, and metastasis. *Biores Open Access*, *4*, 97–108. https://doi.org/10.1089/biores.2014.0035.

Ojha, R., Singh, S. K., & Bhattacharyya, S. (2016). JAK-mediated autophagy regulates stemness and cell survival in cisplatin resistant bladder cancer cells. *Biochimica et Biophysica Acta*, *11*, 27. https://doi.org/10.1016/j.bbagen.2016.07.021.

Ota, H., Eto, M., Ako, J., Ogawa, S., Iijima, K., Akishita, M., et al. (2009). Sirolimus and everolimus induce endothelial cellular senescence via sirtuin 1 down-regulation: Therapeutic implication of cilostazol after drug-eluting stent implantation. *Journal of the American College of Cardiology*, *53*(24), 2298–2305. https://doi.org/10.1016/j.jacc.2009.01.072.

Paglin, S., Lee, N. Y., Nakar, C., Fitzgerald, M., Plotkin, J., Deuel, B., et al. (2005). Rapamycin-sensitive pathway regulates mitochondrial membrane potential, autophagy, and survival in irradiated MCF-7 cells. *Cancer Research*, *65*(23), 11061–11070. https://doi.org/10.1158/0008-5472.

Palumbo, S., Pirtoli, L., Tini, P., Cevenini, G., Calderaro, F., Toscano, M., et al. (2012). Different involvement of autophagy in human malignant glioma cell lines undergoing irradiation and temozolomide combined treatments. *Journal of Cellular Biochemistry*, *113*(7), 2308–2318. https://doi.org/10.1002/jcb.24102.

Pang, X. L., He, G., Liu, Y. B., Wang, Y., & Zhang, B. (2013). Endoplasmic reticulum stress sensitizes human esophageal cancer cell to radiation. *World Journal of Gastroenterology*, *19*(11), 1736–1748. https://doi.org/10.3748/wjg.v19.i11.1736.

Parzych, K. R., & Klionsky, D. J. (2014). An overview of autophagy: Morphology, mechanism, and regulation. *Antioxidants & Redox Signaling*, *20*, 460–473. https://doi.org/10.1089/ars.2013.5371.

Patel, N. H., Sohal, S. S., Manjili, M. H., Harrell, J. C., & Gewirtz, D. A. (2020). The roles of autophagy and senescence in the tumor cell response to radiation. *Radiation Research*, *194*, 103. https://doi.org/10.1667/rade-20-00009.

Patschan, S., Chen, J., Polotskaia, A., Mendelev, N., Cheng, J., Patschan, D., et al. (2008). Lipid mediators of autophagy in stress-induced premature senescence of endothelial cells. *American Journal of Physiology. Heart and Circulatory Physiology*, *294*, 18. https://doi.org/10.1152/ajpheart.00713.2007.

Peiris-Pagès, M., Sotgia, F., & Lisanti, M. P. (2015). Chemotherapy induces the cancer-associated fibroblast phenotype, activating paracrine Hedgehog-GLI signalling in breast cancer cells. *Oncotarget*, *6*(13), 10728–10745. https://doi.org/10.18632/oncotarget.3828.

Phadwal, K., Watson, A. S., & Simon, A. K. (2012). Tightrope act: Autophagy in stem cell renewal, differentiation, proliferation, and aging. *Cellular and Molecular Life Sciences*, *70*, 89–103. https://doi.org/10.1007/s00018-012-1032-3.

Pickles, S., Vigié, P., & Youle, R. J. (2018). Mitophagy and quality control mechanisms in mitochondrial maintenance. *Current Biology*, *28*(4), R170–R185. https://doi.org/10.1016/j.cub.2018.01.004.

Pua, H. H., Dzhagalov, I., Chuck, M., Mizushima, N., & He, Y.-W. (2007). A critical role for the autophagy gene Atg5 in T cell survival and proliferation. *The Journal of Experimental Medicine*, *204*(1), 25–31. https://doi.org/10.1084/jem.20061303.

Qiu, T., Tian, Y., Gao, Y., Ma, M., Li, H., Liu, X., et al. (2019). PTEN loss regulates alveolar epithelial cell senescence in pulmonary fibrosis depending on Akt activation. *Aging*, *11*(18), 7492–7509. https://doi.org/10.18632/aging.102262.

Qu, X., Zou, Z., Sun, Q., Luby-Phelps, K., Cheng, P., Hogan, R. N., et al. (2007). Autophagy gene-dependent clearance of apoptotic cells during embryonic development. *Cell*, *128*(5), 931–946. https://doi.org/10.1016/j.cell.2006.12.044.

Quick, Q. A., & Gewirtz, D. A. (2006). An accelerated senescence response to radiation in wild-type p53 glioblastoma multiforme cells. *Journal of Neurosurgery*, *105*(1), 111–118. https://doi.org/10.3171/jns.2006.105.1.111.

Rader, J., Russell, M. R., Hart, L. S., Nakazawa, M. S., Belcastro, L. T., Martinez, D., et al. (2013). Dual CDK4/CDK6 inhibition induces cell-cycle arrest and senescence in neuroblastoma. *Clinical Cancer Research*, *19*(22), 6173. https://doi.org/10.1158/1078-0432.ccr-13-1675.

Rahman, M. A., Saha, S. K., Rahman, M. S., Uddin, M. J., Uddin, M. S., Pang, M. G., et al. (2020). Molecular insights into therapeutic potential of autophagy modulation by natural products for cancer stem cells. *Frontiers in Cell and Development Biology*, *8*, 283. https://doi.org/10.3389/fcell.2020.00283. PMID: 32391363; PMCID: PMC7193248.

Rakhra, K., Bachireddy, P., Zabuawala, T., Zeiser, R., Xu, L., Kopelman, A., et al. (2010). CD4(+) T cells contribute to the remodeling of the microenvironment required for sustained tumor regression upon oncogene inactivation. *Cancer Cell*, *18*, 485–498. https://doi.org/10.1016/j.ccr.2010.10.002.

Rando, T. A. (2006). Stem cells, ageing and the quest for immortality. *Nature*, *441*, 1080–1086. https://doi.org/10.1038/nature04958.

Rastaldo, R., Vitale, E., & Giachino, C. (2020). Dual role of autophagy in regulation of mesenchymal stem cell senescence. *Frontiers in Cell and Development Biology*, *8*, 276. https://doi.org/10.3389/fcell.2020.00276.

Ratajczak, M. Z., Bujko, K., Mack, A., Kucia, M., & Ratajczak, J. (2018). Cancer from the perspective of stem cells and misappropriated tissue regeneration mechanisms. *Leukemia*, *32*, 2519–2526. https://doi.org/10.1038/s41375-018-0294-7.

Rausch, V., Liu, L., Apel, A., Rettig, T., Gladkich, J., Labsch, S., et al. (2012). Autophagy mediates survival of pancreatic tumour-initiating cells in a hypoxic microenvironment. *The Journal of Pathology*, *227*, 325–335. https://doi.org/10.1002/path.3994.

Raveh-Amit, H., Berzsenyi, S., Vas, V., Ye, D., & Dinnyes, A. (2013). Tissue resident stem cells: till death do us part. *Biogerontology*, *14*, 573–590. https://doi.org/10.1007/s10522-013-9469-9.

Reimann, M., Lee, S., Loddenkemper, C., Dörr, J. R., Tabor, V., Aichele, P., et al. (2010). Tumor stroma-derived TGF-beta limits myc-driven lymphomagenesis via Suv39h1-dependent senescence. *Cancer Cell*, *17*, 262–272. https://doi.org/10.1016/j.ccr.2009.12.043.

Reya, T., Morrison, S. J., Clarke, M. F., & Weissman, I. L. (2001). Stem cells, cancer, and cancer stem cells. *Nature*, *414*, 105–111. https://doi.org/10.1038/35102167.

Roberson, R. S., Kussick, S. J., Vallieres, E., Chen, S. Y., & Wu, D. Y. (2005). Escape from therapy-induced accelerated cellular senescence in p53-null lung cancer cells and in human lung cancers. *Cancer Research*, *65*(7), 2795–2803. https://doi.org/10.1158/0008-5472.

Rodolfo, C., Di Bartolomeo, S., & Cecconi, F. (2016). Autophagy in stem and progenitor cells. *Cellular and Molecular Life Sciences*, *73*, 475–496. https://doi.org/10.1007/s00018-015-2071-3.

Rosen, J. M., & Jordan, C. T. (2009). The increasing complexity of the cancer stem cell paradigm. *Science*, *324*, 1670–1673. https://doi.org/10.1126/science.1171837.
Rossi, D. J., Jamieson, C. H., & Weissman, I. L. (2008). Stems cells and the pathways to aging and cancer. *Cell*, *132*, 681–696. https://doi.org/10.1016/j.cell.2008.01.036.
Rothe, K., Lin, H., Lin, K. B., Leung, A., Wang, H. M., Malekesmaeili, M., et al. (2014). The core autophagy protein ATG4B is a potential biomarker and therapeutic target in CML stem/progenitor cells. *Blood*, *123*(23), 3622–3634. https://doi.org/10.1182/blood-2013-07-516807.
Rouschop, K. M., van den Beucken, T., Dubois, L., Niessen, H., Bussink, J., Savelkouls, K., et al. (2010). The unfolded protein response protects human tumor cells during hypoxia through regulation of the autophagy genes MAP1LC3B and ATG5. *Journal of Clinical Investigation*, *120*(1), 127–141. https://doi.org/10.1172/JCI40027.
Rubinsztein, D. C., Gestwicki, J. E., Murphy, L. O., & Klionsky, D. J. (2007). Potential therapeutic applications of autophagy. *Nature Reviews. Drug Discovery*, *6*, 304–312. https://doi.org/10.1038/nrd2272.
Salama, R., Sadaie, M., Hoare, M., & Narita, M. (2014). Cellular senescence and its effector programs. *Genes & Development*, *28*, 99–114. https://doi.org/10.1101/gad.235184.113.
Saleh, T., Bloukh, S., Carpenter, V. J., Alwohoush, E., Bakeer, J., Darwish, S., et al. (2020). Therapy-induced senescence: An "old" friend becomes the enemy. *Cancers*, *12*, 822. https://doi.org/10.3390/cancers12040822.
Sanchez, C. G., Penfornis, P., Oskowitz, A. Z., Boonjindasup, A. G., Cai, D. Z., Dhule, S. S., et al. (2011). Activation of autophagy in mesenchymal stem cells provides tumor stromal support. *Carcinogenesis*, *32*(7), 964–972. https://doi.org/10.1093/carcin/bgr029.
Sancho, P., Barneda, D., & Heeschen, C. (2016). Hallmarks of cancer stem cell metabolism. *British Journal of Cancer*, *114*, 1305–1312. https://doi.org/10.1038/bjc.2016.152.
Sasaki, M., Miyakoshi, M., Sato, Y., & Nakanuma, Y. (2010). Autophagy mediates the process of cellular senescence characterizing bile duct damages in primary biliary cirrhosis. *Laboratory Investigation*, *90*, 835–843. https://doi.org/10.1038/labinvest.2010.56.
Sasaki, M., Miyakoshi, M., Sato, Y., & Nakanuma, Y. (2012). Autophagy may precede cellular senescence of bile ductular cells in ductular reaction in primary biliary cirrhosis. *Digestive Diseases and Sciences*, *57*, 660–666. https://doi.org/10.1007/s10620-011-1929-y.
Schaer, D. A., Beckmann, R. P., Dempsey, J. A., Huber, L., Forest, A., Amaladas, N., et al. (2018). The CDK4/6 inhibitor abemaciclib induces a T cell inflamed tumor microenvironment and enhances the efficacy of PD-L1 checkpoint blockade. *Cell Reports*, *22*(11), 2978–2994. https://doi.org/10.1016/j.celrep.2018.02.053.
Scherz-Shouval, R., & Elazar, Z. (2011). Regulation of autophagy by ROS: Physiology and pathology. *Trends in Biochemical Sciences*, *36*, 30–38. https://doi.org/10.1016/j.tibs.2010.07.007.
Schlie, K., Spowart, J. E., Hughson, L. R. K., Townsend, K. N., & Lum, J. J. (2011). When cells suffocate: Autophagy in cancer and immune cells under low oxygen. *International Journal of Cell Biology*, *2011*, 1–13. https://doi.org/10.1155/2011/470597.
Schmeer, C., Kretz, A., Wengerodt, D., Stojiljkovic, M., & Witte, O. W. (2019). Dissecting aging and senescence-current concepts and open lessons. *Cell*, *8*(11), 1446. https://doi.org/10.3390/cells8111446. PMID: 31731770; PMCID: PMC6912776.
Sharif, T., Martell, E., Dai, C., Kennedy, B. E., Murphy, P., Clements, D. R., et al. (2017). Autophagic homeostasis is required for the pluripotency of cancer stem cells. *Autophagy*, *13*(2), 264–284. https://doi.org/10.1080/15548627.2016.1260808.
Shivshankar, P., Brampton, C., Miyasato, S., Kasper, M., Thannickal, V. J., & Le Saux, C. J. (2012). Caveolin-1 deficiency protects from pulmonary fibrosis by modulating epithelial cell senescence in mice. *American Journal of Respiratory Cell and Molecular Biology*, *47*(1), 28–36. https://doi.org/10.1165/rcmb.2011-0349OC.

Singh, R., Xiang, Y., Wang, Y., Baikati, K., Cuervo, A. M., Luu, Y. K., et al. (2009). Autophagy regulates adipose mass and differentiation in mice. *The Journal of Clinical Investigation*, *119*(11), 3329–3339. https://doi.org/10.1172/jci39228.

Smith, A. G., & Macleod, K. F. (2019). Autophagy, cancer stem cells and drug resistance. *The Journal of Pathology*, *247*, 708–718. https://doi.org/10.1002/path.5222.

Song, Y. J., Zhang, S. S., Guo, X. L., Sun, K., Han, Z. P., Li, R., et al. (2013). Autophagy contributes to the survival of CD133+ liver cancer stem cells in the hypoxic and nutrient-deprived tumor microenvironment. *Cancer Letters*, *339*(1), 70–81. https://doi.org/10.1016/j.canlet.2013.07.021.

Stienstra, R., Haim, Y., Riahi, Y., Netea, M., Rudich, A., & Leibowitz, G. (2014). Autophagy in adipose tissue and the beta cell: Implications for obesity and diabetes. *Diabetologia*, *57*, 1505–1516. https://doi.org/10.1007/s00125-014-3255-3.

Takahashi, K., & Yamanaka, S. (2006). Induction of pluripotent stem cells from mouse embryonic and adult fibroblast cultures by defined factors. *Cell*, *126*, 663–676. https://doi.org/10.1016/j.cell.2006.07.024.

Takano-Ohmuro, H., Mukaida, M., Kominami, E., & Morioka, K. (2000). Autophagy in embryonic erythroid cells: Its role in maturation. *European Journal of Cell Biology*, *79*(10), 759–764. https://doi.org/10.1078/0171-9335-00096.

Talukdar, S., Das, S. K., Pradhan, A. K., Emdad, L., Windle, J. J., Sarkar, D., et al. (2019). MDA-9/syntenin (SDCBP) is a critical regulator of chemoresistance, survival and stemness in prostate Cancer stem cells. *Cancers*, *12*, 53. https://doi.org/10.3390/cancers12010053.

Talukdar, S., Pradhan, A. K., Bhoopathi, P., Shen, X.-N., August, L. A., Windle, J. J., et al. (2018). MDA-9/Syntenin regulates protective autophagy in anoikis-resistant glioma stem cells. *PNAS*, *115*, 5768–5773. https://doi.org/10.1073/pnas.1721650115.

Tam, S. Y., Wu, V. W., & Law, H. K. (2017). Influence of autophagy on the efficacy of radiotherapy. *Radiation Oncology*, *12*, 017–0795. https://doi.org/10.1186/s13014-017-0795-y.

Tian, Y., Li, H., Qiu, T., Dai, J., Zhang, Y., Chen, J., et al. (2019). Loss of pten induces lung fibrosis via alveolar epithelial cell senescence depending on NF-κB activation. *Aging Cell*, *18*(1), 12. https://doi.org/10.1111/acel.12858.

Todd, L. R., Gomathinayagam, R., & Sankar, U. (2010). A novel Gfer-Drp1 link in preserving mitochondrial dynamics and function in pluripotent stem cells. *Autophagy*, *6*(6), 821–822. https://doi.org/10.1091/mbc.E09-11-0937.

Torres-Guzmán, R., Calsina, B., Hermoso, A., Baquero, C., Alvarez, B., Amat, J., et al. (2017). Preclinical characterization of abemaciclib in hormone receptor positive breast cancer. *Oncotarget*, *8*(41), 69493–69507. https://doi.org/10.18632/oncotarget.17778.

Ugland, H., Boquest, A. C., Naderi, S., Collas, P., & Blomhoff, H. K. (2008). cAMP-mediated induction of cyclin E sensitizes growth-arrested adipose stem cells to DNA damage-induced apoptosis. *Molecular Biology of the Cell*, *19*(12), 5082–5092. https://doi.org/10.1091/mbc.e08-01-0094.

Valenzuela, C. A., Vargas, L., Martinez, V., Bravo, S., & Brown, N. E. (2017). Palbociclib-induced autophagy and senescence in gastric cancer cells. *Experimental Cell Research*, *360*(2), 390–396. https://doi.org/10.1016/j.yexcr.2017.09.031.

Valk-Lingbeek, M. E., Bruggeman, S. W., & van Lohuizen, M. (2004). Stem cells and cancer; the polycomb connection. *Cell*, *118*, 409–418. https://doi.org/10.1016/j.cell.2004.08.005.

van Riggelen, J., Müller, J., Otto, T., Beuger, V., Yetil, A., Choi, P. S., et al. (2010). The interaction between Myc and Miz1 is required to antagonize TGFbeta-dependent autocrine signaling during lymphoma formation and maintenance. *Genes & Development*, *24*, 1281–1294. https://doi.org/10.1101/gad.585710.

Ventura, A., Kirsch, D. G., McLaughlin, M. E., Tuveson, D. A., Grimm, J., Lintault, L., et al. (2007). Restoration of p53 function leads to tumour regression in vivo. *Nature*, *445*, 661–665. https://doi.org/10.1038/nature05541.

Venturelli, S., Berger, A., Weiland, T., Essmann, F., Waibel, M., Nuebling, T., et al. (2013). Differential induction of apoptosis and senescence by the DNA methyltransferase inhibitors 5-azacytidine and 5-aza-2′-deoxycytidine in solid tumor cells. *Molecular Cancer Therapeutics*, *12*(10), 2226–2236. https://doi.org/10.1158/1535-7163.

Vessoni, A. T., Muotri, A. R., & Okamoto, O. K. (2011). Autophagy in stem cell maintenance and differentiation. *Stem Cells and Development*, *21*, 513–520. https://doi.org/10.1089/scd.2011.0526.

Viale, A., Pettazzoni, P., Lyssiotis, C. A., Ying, H., Sánchez, N., Marchesini, M., et al. (2014). Oncogene ablation-resistant pancreatic cancer cells depend on mitochondrial function. *Nature*, *514*, 628–632. https://doi.org/10.1038/nature13611.

Vijayaraghavan, S., Karakas, C., Doostan, I., Chen, X., Bui, T., Yi, M., et al. (2017). CDK4/6 and autophagy inhibitors synergistically induce senescence in Rb positive cytoplasmic cyclin e negative cancers. *Nature Communications*, *8*, 15916. https://doi.org/10.1038/ncomms15916.

Wang, B., Kohli, J., & Demaria, M. (2020). Senescent cells in cancer therapy: Friends or foes? *Trends in Cancer*, *6*, 838–857. https://doi.org/10.1016/j.trecan.2020.05.004.

Wang, C., Liang, C.-C., Bian, Z. C., Zhu, Y., & Guan, J.-L. (2013a). FIP200 is required for maintenance and differentiation of postnatal neural stem cells. *Nature Neuroscience*, *16*(5), 532–542. https://doi.org/10.1038/nn.3365.

Wang, S., Wang, X., Cheng, Y., Ouyang, W., Sang, X., Liu, J., et al. (2019). Autophagy dysfunction, cellular senescence, and abnormal immune-inflammatory responses in AMD: From mechanisms to therapeutic potential. *Oxidative Medicine and Cellular Longevity*, *2019*, 3632169. https://doi.org/10.1155/2019/3632169.

Wang, W. J., Long, L. M., Yang, N., Zhang, Q. Q., Ji, W. J., Zhao, J. H., et al. (2013b). NVP-BEZ235, a novel dual PI3K/mTOR inhibitor, enhances the radiosensitivity of human glioma stem cells in vitro. *Acta Pharmacologica Sinica*, *34*(5), 681–690. https://doi.org/10.1038/aps.2013.22.

Wang, X., Wong, S. C., Pan, J., Tsao, S. W., Fung, K. H., Kwong, D. L., et al. (1998). Evidence of cisplatin-induced senescent-like growth arrest in nasopharyngeal carcinoma cells. *Cancer Research*, *58*(22), 5019–5022.

Warr, M. R., Binnewies, M., Flach, J., Reynaud, D., Garg, T., Malhotra, R., et al. (2013). FOXO3A directs a protective autophagy program in haematopoietic stem cells. *Nature*, *494*, 323–327. https://doi.org/10.1038/nature11895.

Waterstrat, A., & Van Zant, G. (2009). Effects of aging on hematopoietic stem and progenitor cells. *Current Opinion in Immunology*, *21*, 408–413. https://doi.org/10.1016/j.coi.2009.05.002.

Wechman, S. L., Pradhan, A. K., DeSalle, R., Das, S. K., Emdad, L., Sarkar, D., et al. (2018). New insights into beclin-1: Evolution and pan-malignancy inhibitor activity. *Advances in Cancer Research*, *137*, 77–114. https://doi.org/10.1016/bs.acr.2017.11.002.

Welch, J. S., Ley, T. J., Link, D. C., Miller, C. A., Larson, D. E., Koboldt, D. C., et al. (2012). The origin and evolution of mutations in acute myeloid leukemia. *Cell*, *150*, 264–278. https://doi.org/10.1016/j.cell.2012.06.023.

Widodo, N., Deocaris, C. C., Kaur, K., Hasan, K., Yaguchi, T., Yamasaki, K., et al. (2007). Stress chaperones, mortalin, and pex19p mediate 5-aza-2′ deoxycytidine-induced senescence of cancer cells by DNA methylation-independent pathway. *The Journals of Gerontology. Series A, Biological Sciences and Medical Sciences*, *62*(3), 246–255. https://doi.org/10.1093/gerona/62.3.246.

Wolf, J., Dewi, D. L., Fredebohm, J., Müller-Decker, K., Flechtenmacher, C., Hoheisel, J. D., et al. (2013). A mammosphere formation RNAi screen reveals that ATG4A promotes a breast cancer stem-like phenotype. *Breast Cancer Research*, *15*, R109. https://doi.org/10.1186/bcr3576.

Wu, C.-H., van Riggelen, J., Yetil, A., Fan, A. C., Bachireddy, P., & Felsher, D. W. (2007). Cellular senescence is an important mechanism of tumor regression upon c-Myc inactivation. *Proceedings of the National Academy of Sciences of the United States of America, 104*, 13028–13033. https://doi.org/10.1073/pnas.0701953104.

Wu, S. Y., Liu, Y. W., Wang, Y. K., Lin, T. H., Li, Y. Z., Chen, S. H., et al. (2014). Ionizing radiation induces autophagy in human oral squamous cell carcinoma. *Journal of Buon, 19*(1), 137–144.

Xiong, L., Liu, Z., Ouyang, G., Lin, L., Huang, H., Kang, H., et al. (2017). Autophagy inhibition enhances photocytotoxicity of Photosan-II in human colorectal cancer cells. *Oncotarget, 8*, 6419–6432. https://doi.org/10.18632/oncotarget.14117.

Xu, W. S., Perez, G., Ngo, L., Gui, C. Y., & Marks, P. A. (2005). Induction of polyploidy by histone deacetylase inhibitor: A pathway for antitumor effects. *Cancer Research, 65*(17), 7832–7839. https://doi.org/10.1158/0008-5472.

Xu, Y., Xia, X., & Pan, H. (2013). Active autophagy in the tumor microenvironment: A novel mechanism for cancer metastasis. *Oncology Letters, 5*, 411–416. https://doi.org/10.3892/ol.2012.1015.

Xue, W., Zender, L., Miething, C., Dickins, R. A., Hernando, E., Krizhanovsky, V., et al. (2007). Senescence and tumour clearance is triggered by p53 restoration in murine liver carcinomas. *Nature, 445*, 656–660. https://doi.org/10.1038/nature05529.

Yang, M. Y., Lin, P. M., Liu, Y. C., Hsiao, H. H., Yang, W. C., Hsu, J. F., et al. (2012). Induction of cellular senescence by doxorubicin is associated with upregulated miR-375 and induction of autophagy in K562 cells. *PLoS One*, 7(5), 11. https://doi.org/10.1371/journal.pone.0037205.

Yang, Z., & Klionsky, D. J. (2010). Eaten alive: A history of macroautophagy. *Nature Cell Biology, 12*, 814–822. https://doi.org/10.1038/ncb0910-814.

Yao, Z., Murali, B., Ren, Q., Luo, X., Faget, D. V., Cole, T., et al. (2020). Therapy-induced senescence drives bone loss. *Cancer Research, 80*(5), 1171–1182. https://doi.org/10.1158/0008-5472.CAN-19-2348.

Yoshida, A., Lee, E. K., & Diehl, J. A. (2016). Induction of therapeutic senescence in vemurafenib-resistant melanoma by extended inhibition of CDK4/6. *Cancer Research, 76*(10), 2990–3002. https://doi.org/10.1158/0008-5472.

Young, A. R., Narita, M., Ferreira, M., Kirschner, K., Sadaie, M., Darot, J. F., et al. (2009). Autophagy mediates the mitotic senescence transition. *Genes & Development, 23*, 798–803. https://doi.org/10.1101/gad.519709.

Yu, Z., Pestell, T. G., Lisanti, M. P., & Pestell, R. G. (2012). Cancer stem cells. *The International Journal of Biochemistry & Cell Biology, 44*, 2144–2151. https://doi.org/10.1016/j.biocel.2012.08.022.

Yue, Z., Jin, S., Yang, C., Levine, A. J., & Heintz, N. (2003). Beclin 1, an autophagy gene essential for early embryonic development, is a haploinsufficient tumor suppressor. *Proceedings of the National Academy of Sciences of the United States of America, 100*(25), 15077–15082. https://doi.org/10.1073/pnas.2436255100.

Yuk, J. M., Shin, D. M., Song, K. S., Lim, K., Kim, K. H., Lee, S. H., et al. (2010). Bacillus calmette-guerin cell wall cytoskeleton enhances colon cancer radiosensitivity through autophagy. *Autophagy, 6*(1), 46–60. https://doi.org/10.4161/auto.6.1.10325.

Yun, C. W., & Lee, S. H. (2018). The roles of autophagy in Cancer. *International Journal of Molecular Sciences, 19*, 3466. https://doi.org/10.3390/ijms19113466.

Zeng, M., & Zhou, J. N. (2008). Roles of autophagy and mTOR signaling in neuronal differentiation of mouse neuroblastoma cells. *Cellular Signalling, 20*(4), 659–665. https://doi.org/10.1016/j.cellsig.2007.11.015.

Zhang, D., Zhao, Q., Sun, H., Yin, L., Wu, J., Xu, J., et al. (2016). Defective autophagy leads to the suppression of stem-like features of CD271+ osteosarcoma cells. *Journal of Biomedical Science, 23*(1), 82. https://doi.org/10.1186/s12929-016-0297-5.

Zhang, H., & McCarty, N. (2017). Tampering with cancer chemoresistance by targeting the TGM2-IL6-autophagy regulatory network. *Autophagy, 13*, 627–628. https://doi.org/10.1080/15548627.2016.1271516.

Zhang, J., Liu, J., Huang, Y., Chang, J. Y., Liu, L., McKeehan, W. L., et al. (2012). FRS2α-mediated FGF signals suppress premature differentiation of cardiac stem cells through regulating autophagy activity. *Circulation Research, 110*(4), 29. https://doi.org/10.1161/CIRCRESAHA.111.255950.

Zhao, Y., Tyshkovskiy, A., Muñoz-Espín, D., Tian, X., Serrano, M., de Magalhaes, J. P., et al. (2018). Naked mole rats can undergo developmental, oncogene-induced and DNA damage-induced cellular senescence. *Proceedings of the National Academy of Sciences, 115*(8), 1801. https://doi.org/10.1073/pnas.1721160115.

Zhu, Y., Tchkonia, T., Pirtskhalava, T., Gower, A. C., Ding, H., Giorgadze, N., et al. (2015). The achilles' heel of senescent cells: From transcriptome to senolytic drugs. *Aging Cell, 14*(4), 644–658. https://doi.org/10.1111/acel.12344.

Zhuang, D., Mannava, S., Grachtchouk, V., Tang, W. H., Patil, S., Wawrzyniak, J. A., et al. (2008). C-MYC overexpression is required for continuous suppression of oncogene-induced senescence in melanoma cells. *Oncogene, 27*, 6623–6634. https://doi.org/10.1038/onc.2008.258.

CHAPTER FIVE

The role of autophagy in escaping therapy-induced polyploidy/senescence

Magdalena Dudkowska†, **Karolina Staniak**†, **Agnieszka Bojko, and Ewa Sikora***

Laboratory of Molecular Bases of Aging, Nencki Institute of Experimental Biology, Polish Academy of Sciences, Warsaw, Poland

*Corresponding author: e-mail address: e.sikora@nencki.edu.pl

Contents

Abstract

Autophagy is an evolutionarily conserved process necessary to maintain cell homeostasis in response to various forms of stress such as nutrient deprivation and hypoxia as well as functioning to remove damaged molecules and organelles. The role of autophagy in cancer varies depending on the stage of cancer. Cancer therapeutics can also simultaneously evoke cancer cell senescence and ploidy increase. Both cancer cell senescence and polyploidization are reversible by depolyploidization giving rise to the progeny. Autophagy activation may be indispensable for cancer cell escape from senescence/polyploidy. As cancer cell polyploidy is proposed to be involved in cancer origin, the role of autophagy in polyploidization/depolyploidization of senescent cancer cells seems to be crucial. Accordingly, this review is an attempt to understand the complicated interrelationships between reversible cell senescence/polyploidy and autophagy.

† These authors equally participated in preparing the manuscript.

Advances in Cancer Research, Volume 150
ISSN 0065-230X
https://doi.org/10.1016/bs.acr.2021.01.004

1. Introduction

Our body's cells are constantly exposed to many different types of external and internal stress, including genotoxic forms. The response of a cell appears to depend on the type and severity of the insult. The cell's initial response to a stressful stimulus is to defend itself against the insult and to recover; that is, to maintain homeostasis and avoid cancerous transformation. However, if the noxious stimulus is not neutralized, the cell will not survive. The fact that the cell's survival critically depends on the ability to mount an appropriate response toward environmental or intracellular stress stimuli can explain why this reaction is highly conserved in evolution (Fulda, Gorman, Hori, & Samali, 2010). Indeed, cell responses that enhance the chance of survival can vary depending on cell condition and cell type, although several conserved mechanisms exist, among which are autophagy and cell senescence. Although these two processes are preserved both in normal and cancer cells, they can be differentially regulated in cells that either have limited (normal cells) and unlimited (cancer cells) proliferation capacity. In fact, somatic cells, with their restricted capacity to proliferate, were shaped by evolution for a limited period of body life, whereas cancer cells escaped this limitation and are generally more refractory to harmful conditions.

Recent studies have suggested that cell senescence and autophagy might act as critical mechanisms for the regulation of cancer development and progression as well as in determining the response of tumor cells to anticancer therapy. Elimination of tumor cells by commonly used anticancer drugs and ionizing radiation is mediated by triggering cell death, predominantly apoptosis, or necroptosis. However, tumors often evade cell death by engaging various regulatory pathways, thereby enhancing their survival potential. Autophagy modulation and cell senescence may be two of the primary mechanisms for cancer cell resistance to chemo/radiotherapy.

Autophagy is a process whereby cellular material is degraded in lysosome-dependent manner and recycled. According to Yim and Mizushima there are four kinds of autophagy: macroautophagy, chaperon-mediated autophagy (CMA), RN/DNautophagy and microautophagy (Yim & Mizushima, 2020). Microautophagy is the uptake of cytosolic material by invagination of lysosomal membrane (Oku & Sakai, 2018). CMA involves direct uptake of proteins into lysosomes via translocation complex consisting among others of lysosomal membrane protein 2A (LAMP-2A) and cytosolic chaperons (Kaushik & Cuervo, 2012). RN/DNautophagy is also a direct uptake of nucleic acids into lysosomes, however, in LAMP-2C dependent manner. The most frequently observed type of autophagy is macroautophagy (herein referred to as autophagy), which is characterized by the formation of autophagosomes, their fusion with lysosomes and cargo degradation (Mizushima & Komatsu, 2011). Thus, the specific stages of autophagy are induction, formation of the isolation membrane (phagophore), formation and maturation of the autophagosome and, finally fusion with the lysosome to form the autolysosome (Nakamura & Yoshimori, 2018). Autophagy is an evolutionarily conserved physiological process with a fundamental role during development, differentiation, and survival, which is essential for homeostasis in eukaryotic cells. Autophagy dysregulation is associated with a number of pathological conditions in humans, such as neurodegenerative diseases, heart disease and cancer. Active autophagy is considered to contribute to a healthy life span and longevity. Autophagy can be inhibited at any stage, leading to accumulation of undigested defective molecules and organelles, and thereby contributing to age-related disorders (Nakamura & Yoshimori, 2018).

Cell senescence is a process of cessation of proliferation, which can be caused by oncogene activation (oncogene-induced senescence—OIS), various forms of stress (stress-induced premature senescence—SIPS) or exhaustion of cell division potential due to telomere erosion (replicative senescence). Senescent cells are alive, metabolically active, but have disturbed proteostasis (Sabath et al., 2020). Senescent cells secrete a host of factors, including cytokines, growth factors and metalloproteinases, commonly described as a senescence-associated secretory phenotype (SASP) (Freund, Orjalo, Desprez, & Campisi, 2010). Senescent cells accumulate with age and are the main culprit of age-related diseases (Sikora, Bielak-Zmijewska, & Mosieniak, 2019). Thus, logically, senescent cells should have defects in autophagy. However, the relationship between senescence and autophagy is not straightforward (Gewirtz, 2013).

Recently, it has been documented that not only normal, but also cancer cells can undergo therapy-induced senescence (TIS) which is often associated with a ploidy increase (Mosieniak & Sikora, 2010; Saleh, Carpenter, Bloukh, & Gewirtz, 2020; Sikora, Czarnecka-Herok, Bojko, & Sunderland, 2020). Along with increased expression of anti-apoptotic proteins, this contributes to their resistance to anticancer therapy, because senescent/polyploid cancer cells can divide asymmetrically and produce viable progeny, thus contributing to cancer renewal. The heterogeneity of cancer cells arising from polyploidization/depolyploidization and the role of cell senescence and autophagy in these processes, all contribute to the complexity of the cancer cell life cycle (Erenpreisa & Cragg, 2013).

Control of cell size is critical for regulating nutrient distribution for the cell and for regulating organ size and function in multicellular organisms (Visconti, Della Monica, & Grieco, 2016). It has been very recently proposed that optimal cell function requires maintenance of a narrow range of the ratio of nucleus: cytoplasm (N/C) size. When the cell size increases, due to prolonged arrest in the G1 phase of the cell cycle, this ratio leads to oversizing of cytoplasm and, as it has been recently proved, contributes to senescence (Neurohr et al., 2019).

We can speculate that ploidy increase in cancer senescent cells is necessary to counteract an unfavorable N/C ratio. Subsequently, polyploid giant cancer cells (PGCCs) divide atypically, generating more fit descendants with a normal or near normal number of chromosomes. Although the role of autophagy in cancer development and cancer cell survival is highly complex, we can anticipate that active autophagy is indispensable for depolyploidization of senescent/polyploid cancer cells as suggested by Erenpreisa and Cragg (2013). Accordingly, in this review we will consider autophagy and cell senescence regulation through the prism of polyploidy of cancer cells.

2. Polyploidy in cancer development

The vast majority of cancer cells are aneuploid (the chromosome number is not a multiple of the diploid complement) or polyploid (the chromosome number is a multiple of the normal diploid complement). Around 90% of solid tumors and 75% of hematopoietic cancers have abnormal chromosome numbers. Karyotype analysis of cancer cell lines has revealed a wide range of chromosome numbers ranging from hypodiploid to hypertetraploid (Davoli & de Lange, 2011). Recent studies identified whole-genome doubling in nearly 30% of almost 10,000 prospectively sequenced advanced cancers (Bielski et al., 2018). Polyploidy is the result of cell fusion or endoreplication without cytokinesis. Endoreplication can proceed via endocycle or endomitosis (Fig. 1). The first means that cells can perform DNA synthesis without mitosis, thus producing polyploid mononuclear cells. The

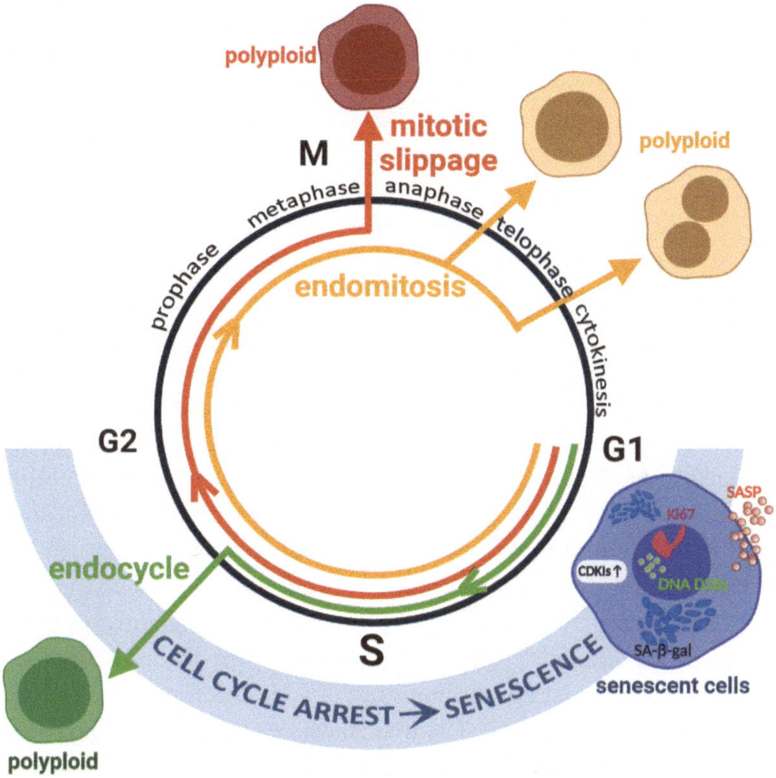

Fig. 1 The cell cycle in polyploids and senescent cells. Polyploidy is connected with endoreplication, i.e., the process of cell cycling proceeding with omission of phases of the cycle. Endoreplication includes: endocycle—multiple S-phases without intervening mitosis and cytokinesis, mitotic slippage—cell cycle progression to metaphase and skipping of the subsequent phases of mitosis and cytokinesis, endomitosis—cell cycling with bypassing the last phase of mitosis, telophase and cytokinesis (in early endomitosis) or cytokinesis (in late endomitosis). These types of endoreplication lead to formation of polyploid mononuclear cell, except for late endomitosis, which results in appearance of multinucleated polyploid cell. Cellular senescence is associated with cell cycle arrest in G1 or G2 phase. The senescent cell does not proliferate but also does not exit the cell cycle, as confirmed by the presence of Ki-67, a marker of cycling cells. The main markers of senescent cell are: increased activity of SA-β-gal, increased level of cyclin-dependent kinase inhibitors (CDKIs), DNA double strand breaks (DNA DSBs) and senescence-associated secretory phenotype (SASP). Scheme created with BioRender.com.

same effect occurs in the case of mitotic slippage or early endomitosis. Cells undergoing late endomitosis, by contrast, execute an abortive mitosis without cytokinesis producing multinucleated polyploid cells (Orr-Weaver, 2015). Both types of endoreplication can take place in cancer cells (Fox & Duronio, 2013).

Polyploidy has been observed for many years, both *in vitro* and in specimens such as a giant or monster cells (White-Gilbertson & Voelkel-Johnson, 2020). Now it is customary to describe them under a common name—polyploid giant cancer cells (PGCCs) (Amend et al., 2019).

Currently, increase in ploidy seems to be a suitable explanation for cancer origin and heterogeneity. Liu (2018, 2020) proposed the theory of a "life code" that unifies the human life cycle and the origin of human tumors. The main assumption of this theory is that early embryogenesis can be interpreted as a progressive increase in ploidy and the 32-cell morula can be considered as a multinucleated giant cell. Morphological transition from a morula to a blastocyst and formation of an inner cell mass gives rise to a new embryonic life. If the subsequent differentiation leads to full maturation, then normal life will occur, but if differentiation is blocked at any point along the continuum of primordial germ cell to embryonic maturation and to fetal organ maturation, a well-differentiated tumor will develop. Yet, undifferentiated tumors are derived from mature somatic cells through dedifferentiation by a reprogramming mechanism called the giant cell life cycle or the giant cell cycle. This mechanism can initiate "somatic embryogenesis" through an increase in ploidy similar to that which occurs in normal embryogenesis. This dedifferentiation mechanism leads to the formation of PGCCs, that is, cancer stem-like cells that mimic an embryo in the blastomeric stage. Thus, the increase in ploidy explains not only normal embryogenesis in the origin of well-differentiated tumors but also "somatic embryogenesis" in the development of undifferentiated tumors (Liu, 2018, 2020).

Generally, cancer cell polyploidization is associated with poor prognosis and is significantly associated with disease relapse (Dewhurst et al., 2014; Lundberg et al., 2013; Storchova & Kuffer, 2008). It has been proposed that tumor cells with an elevated genomic content are key players in the evolution of cancer (Coward & Harding, 2014). PGCCs can participate in cancer cell repopulation via depolyploidization, which generates viable and likely more fit descendants. Different mechanisms of such divisions have been proposed. Specifically, Erenpreisa's group observed that X-ray irradiation of several cancer cell lines led to repeated mitotic catastrophe, and, in consequence, to the formation of multinuclear cells that were able to survive. Those cells underwent multipolar cell divisions, giving rise to viable and proliferating cells with lower ploidy (Ianzini et al., 2009; Kalejs et al., 2006). Although, originally, mitotic catastrophe was characterized by an unscheduled or prolonged activation of mitotic drivers followed by cell death (Vitale, Galluzzi, Castedo, & Kroemer, 2011), it seems that mitotic catastrophe can precede mitotic slippage, leading to increased ploidy and survival (Salmina et al., 2020).

An alternative mechanism that enables a polyploid/multinuclear cell to proliferate was proposed by the group of Rajaraman. Based on a prolonged observation via live-cell imaging, polyploid/multinuclear cells, which can arise from normal cells exposed to carcinogens or cancer cells exposed to DNA damaging agents, undergo neither meiotic nor mitotic divisions, but a so-called neosis. Generally, neosis is characterized by cell budding. Rajaraman hypothesizes that the process of polyploidization and neosis can be repeated several times and give rise to a progeny called Raju cells, which temporally acquire features of cancer stem cells (Rajaraman, Guernsey, Rajaraman, & Rajaraman, 2006). In another report, successful purification and culture of PGCCs from human ovarian cancer cell lines and primary ovarian cancer, was documented. These cells generated regular-sized cancer cells quickly through budding or bursting. PGCCs expressed normal and cancer stem cell markers, divided asymmetrically and cycled slowly. A single PGCC formed cancer spheroids *in vitro* and generated tumors in immunodeficient mice. These PGCC derived tumors gained a mesenchymal phenotype with increased expression of cancer stem cell markers and became more resistant to treatment with cisplatin (Zhang, Mercado-Uribe, Xing, et al., 2014). However, two recent papers reported that mimicking hypoxia in ovarian, breast and colon cancer cells by treatment with $CoCl_2$ led to selection of giant polyploid cells with characteristics of stemness and the ability to divide by budding (Zhang, Mercado-Uribe, & Liu, 2014). Currently, it appears that amitotic divisions of PGCCs giving rise to mononuclear progeny has been accepted by the scientific community. There are several excellent reviews on this topic, which include information about the mechanisms of depolyploidization and readers are encouraged to refer to these for further details (Bharadwaj & Mandal, 2020; Liu, 2020; Mirzayans, Andrais, Scott, Wang, & Murray, 2013; Mirzayans & Murray, 2020; Mosieniak & Sikora, 2010; Saleh et al., 2019; Sikora et al., 2020; White-Gilbertson & Voelkel-Johnson, 2020).

3. Senescence/polyploidy in cancer cell repopulation after therapy

Cellular senescence is a state in which dividing cells cease to proliferate but remain alive, metabolically active and, very often, resistant to apoptosis (Rodier & Campisi, 2011). Senescent cells are characterized by the senescence-associated secretory phenotype (SASP). The SASP leads to the development of a low inflammatory state, which is conducive to organismal aging and age-related diseases (reviewed by, e.g., (Campisi, 2011; Gorgoulis et al., 2019; Sikora, Arendt, Bennett, & Narita, 2011)). On the other hand, cell senescence plays a beneficial role in embryonic growth and pattering (so-called programmed senescence) (Munoz-Espin et al., 2013; Storer et al., 2013) as well as proper wound healing (Krizhanovsky et al., 2008). The execution of cell senescence is correlated with the appearance of several cellular features such as cell cycle arrest accompanied by increased expression of cyclin-dependent kinase inhibitors (CDKIs), $p21^{Cip1/WAF1}$ and $p16^{INK4a}$, altered cellular morphology, increased activity of senescence associated-β-galactosidase (SA-β-gal), accumulation of DNA damage foci and chromatin alterations (senescence-associated heterochromatin foci, SAHF) (Bojko, Czarnecka-Herok, Charzynska, Dabrowski, & Sikora, 2019; Gorgoulis et al., 2019) (Fig. 1). Additionally, in contrast to what might have been anticipated, we have recently shown that senescent cancer cells are also positive for Ki-67, a cycling cell marker (Bojko et al., 2019; Sikora, Mosieniak, & Sliwinska, 2016). The Ki-67 protein is present during all active phases of the cell cycle (G1, S, G2, and mitosis), but is absent from resting (quiescent) cells (G0). Ki-67 can be also an indicator of endocycling cells.

It is believed that the primary positive attribute of cell senescence is the formation of a barrier to cancer development by preventing proliferation of altered, potentially harmful cells. Oncogene-induced senescence (OIS), which was first shown *in vitro* (Serrano, Lin, McCurrach, Beach, & Lowe, 1997) and then confirmed by several groups in different models of oncogene-induced cancer *in vivo* (Braig et al., 2005; Chen et al., 2005; Collado et al., 2005; Michaloglou et al., 2005), argues in favor of senescence as a cancer barrier, especially that senescent cells were found in benign but not malignant cancers (Kuilman, Michaloglou, Mooi, & Peeper, 2010). Moreover, it is believed that senescent cells (Prata, Ovsyannikova, Tchkonia, & Kirkland, 2018), like apoptotic cells (Devitt & Marshall, 2011), are cleared by the immune system, but when we get older the immune system became less efficient and there is accumulation of senescent cells in body (Ovadya & Krizhanovsky, 2014). Thus, senescent cells, by secreting numerous pro-inflammatory factors, can support tumorigenesis. Moreover, a growing body of evidence supports the contention that cells can escape from senescence (Saleh, Carpenter, et al., 2020; Saleh et al., 2019; Sikora et al., 2020), which contradicts the idea that senescent cells are barrier to cancer and that senescence is synonymous with an irreversible growth arrest (Gosselin et al., 2009; Walen, 2004, 2005).

Recently, using human and mice cells and a plethora of experimental approaches, Leikam et al. have shown that differentiated melanocytes, when driven into senescence by an oncogene (OIS), gave rise to highly aggressive tumor-initiating cells, which possessed stemness characteristics. Moreover, it was documented that senescent cells were polyploid (multinucleated) (Leikam et al., 2015). These studies provide strong support to the still not commonly accepted idea that senescence may be not a barrier, but quite to the contrary, a driving force to cancer development, and that polyploidy formation could be crucial in this process.

Both increased polyploidy and senescence characteristics have been observed not only during cancer development, but also after radio/chemotherapy. A new term for induced cancer cell senescence was coined, namely, therapy-induced senescence (TIS) (Nardella, Clohessy, Alimonti, & Pandolfi, 2011). In our recent papers, we thoroughly characterized a variety of senescent phenotypes of diverse cancer cell lines after treatment with different chemotherapeutics and found that senescence depends very much on the drug and cell type (Bojko et al., 2019; Was et al., 2018). Interestingly, for a short period of time, TIS was considered as beneficial for patients; however, it appeared very quickly that this could be quite the opposite. In fact, TIS can be a form of cancer resistance as it contributes to cancer reoccurrence. It was shown that patients with non-small cell lung cancer expressing markers of therapy-induced senescence had a significantly worse prognosis than those with no TIS-expressing tumors (Wang et al., 2013). We showed that both commonly used anticancer drugs, doxorubicin and a natural phenol, curcumin, induced senescence coupled with high polyploidy in colon cancer HCT116 cells (Mosieniak et al., 2012; Sliwinska et al., 2009) and that doxorubicin induced senescence/polyploidy in MDA-MB-231 breast cancer cells (Salmina et al., 2020). There is also a growing body of evidence, showing that TIS is not permanent, with rare cells being able to escape senescence (Saleh et al., 2019; Saleh, Bloukh, et al., 2020; Saleh, Carpenter, et al., 2020; Sikora et al., 2020). Moreover, it seems that post-senescent cells acquire the characteristics of more aggressive cancer (Milanovic et al., 2018).

It cannot be excluded that both increased ploidy and senescence are indispensable for cancer renewal. Other researchers have observed that spontaneously arising multinuclear cancer cells were positive for SA-β-gal and were responsible for the majority of clonogenic activity. These multinuclear cells were also more resistant to chemotherapeutic agents than mononuclear cells. Moreover, these polyploid cells could undergo asymmetric divisions (producing mononuclear cells) and self-renewal *in vitro* and *in vivo*; most importantly, a single giant cancer cell produced orthotopic subcutaneous tumors (composed mainly of mononuclear cells) that gave rise to spontaneous lung metastases in nude mice (Weihua, Lin, Ramoth, Fan, & Fidler, 2011). Sabisz and Skladanowski (2009) documented that non-small lung adenocarcinoma cells underwent polyploidization and senescence, but a small fraction of cells escaped senescence and started to re-grow. These cells possessed the characteristics of stem cells. Puig et al. (2008) documented that sorted senescent giant cells gave rise to small cell progeny. Moreover, the giant senescent cells were able to form tumors *in vivo*. Studies by Was et al. also showed the presence of cancer stem cell markers, CD24 and increase of side population, in HCT116 and SW480 colon cancer cells undergoing polyploidization upon senescence induction (Was et al., 2017, 2018). Altogether, it seems that, during cancer development and cancer treatment, the cancer cell response may include polyploidization associated with senescence characteristics and followed by depolyploidization. Thus, in this context, cell senescence clearly appears to be reversible.

4. Autophagy in cancer development and anticancer therapy

Autophagy is a quality control system removing damaged cell components and, in consequence, protecting cells from oxidative stress and genomic instability. Accordingly, impairment of the autophagy process leads to accumulation of harmful structures and may result in an environment conducive to mutation and cancer development (Fig. 2). In this context, autophagy can be viewed as a tumor-suppression mechanism. This concept originates from the period between 1990 and 2010, when the first studies of this phenomenon revealed mutation of autophagy genes in many tumors (Wilde, Tanson, Curry, & Martinez-Outschoorn, 2018). Thus, the monoallelic loss of the *BECN1* gene, encoding the Beclin-1 protein necessary for the formation of the phagophore, was found in human breast, prostate and ovarian tumors, whereas expression of Beclin-1 was reduced in human colon cancer, brain tumors, hepatocellular carcinomas and cervical cancers (Dalby, Tekedereli, Lopez-Berestein, & Ozpolat, 2010). Mutations and downregulation of other autophagy genes have been also identified in a variety of cancers (Coppola et al., 2008; Costa, Prak, Aldous, Gewinner, & Ketteler, 2016; Ionov, Nowak, Perucho, Markowitz, & Cowell, 2004). Similarly, studies with knockout mice confirmed an inverse correlation

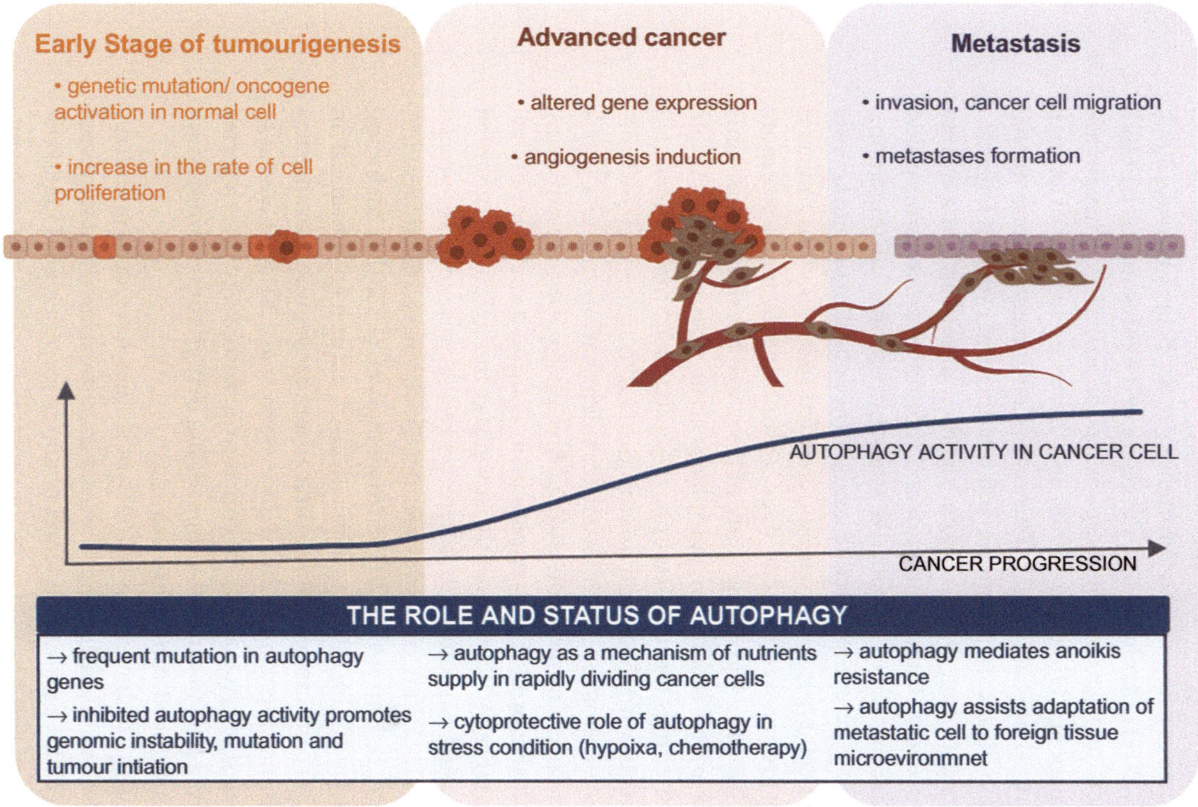

Fig. 2 The role and status of autophagy in subsequent stages of carcinogenesis. During early stage of tumorigenesis autophagy is frequently inactive, which contributes to genomic instability, mutations and formation of tumor initiating cells with increased rate of proliferation. In advanced cancer autophagy may have opposite role. In highly proliferating cancer cells with increased demand for nutrients autophagy acts as a mechanism of nutrients supply. Also, autophagy is cytoprotective in cancer cells exposed to stress factors, such as hypoxia, chemotherapy. Additionally, autophagy may promote metastasis by mediating anoikis resistance and adaptation of metastatic cells to new microenvironment. Fluctuations in autophagy activity are associated with cancer progression. Namely, inhibition of autophagy at early stage of carcinogenesis and its activation in advanced or metastatic cancer was observed in many tumors. In line with this, autophagy is defined as a process that plays a seemingly contradictory role in cancer by suppressing benign tumor progression but promoting advanced cancer growth. Scheme created with BioRender.com.

between autophagy functionality and cancer occurrence. *Becn1* heterozygous deletion or mosaic deletion of *Atg5* or *Atg7* genes encoding proteins indispensable for the early phase of autophagy, resulted in increased susceptibility to spontaneous malignancies (Mortensen et al., 2011; Qu et al., 2003; Takamura et al., 2011). Additionally, the decreased expression of BECN1 was associated with a malignant phenotype and poor prognosis of hepatocellular carcinoma patients (Ding et al., 2008) and shortened survival of ovarian carcinoma patients (Lin et al., 2013). This makes autophagy genes potential biomarkers for the diagnosis, prognosis, and treatment of cancer (Han et al., 2020; Luo, Huang, & Liu, 2020). Taken together, autophagy, as a quality control system, prevents tumor initiation and suppresses cancer progression by limiting genome mutations, oxidative stress and chronic inflammation.

Autophagy plays a cytoprotective and homeostasis maintaining role independently of cell type, origin and differentiation status. This indicates that after tumor progression to a late stage, autophagy reverts to a tumor-promotion mechanism. A growing body of literature has shown that some tumor cells are more reliant on autophagy than normal cells (Wilde et al., 2018), sometimes termed "autophagy-addicted." Some cancer cells reactivate/stimulate autophagy and use this process to survive in unfavorable environment condition, such as nutrient deprivation or stress (Lock et al., 2011; Yang & Kimmelman, 2011). Autophagy, besides supplying nutrients, should also protect cancer cells from damage caused by physical or chemical stress (Fig. 2). Consequently, autophagy induction was observed in cancer cells during exposition to different stress factors, such as hypoxia (Degenhardt et al., 2006) or toxic agents (Ciechomska, Gabrusiewicz, Szczepankiewicz, & Kaminska, 2013; Mosieniak et al., 2012) including chemotherapeutics (Fulda, 2017). Activation of autophagy occurred in response to various anticancer agents, such as temozolomide, dexamethasone, 6-thioguanine, and camptothecin (Sorokina et al., 2017) as well as to ionizing radiation (Gewirtz, 2020). In conclusion, autophagy plays a seemingly contradictory role in cancer by suppressing benign tumor progression but promoting advanced cancer growth.

Not surprisingly, an increasing number of studies have demonstrated a beneficial effect of autophagy modulation on cancer treatment and suggested that autophagy can be a promising target of anticancer therapy. Due to the dichotomous role of autophagy as a tumor suppressor and tumor promoter, both induction and inhibition of autophagy have been proposed as potential strategies. Targets, which can be modulated by small molecules, have been identified at each stage of the autophagy process. That, in theory, should allow for the adjustment of treatment to the expected effect on autophagy in cancer cells (Onorati, Dyczynski, Ojha, & Amaravadi, 2018). The strategy of autophagy induction has been developed primarily for cancer cells with a non-functional autophagy machinery, in which stimulation of this process may lead to cell death (Sharma, Le, Alotaibi, & Gewirtz, 2014; Bishop & Bradshaw, 2018). However, "autophagy-addicted" cancer cells with high autophagy activity are not able to cope with additional autophagic flux stimulation. Thus, over-induction of autophagy in these cells results in energy exhaustion and autophagic cell death (Bialik, Dasari, & Kimchi, 2018; Fulda & Kogel, 2015). Among many autophagy inducers, we can distinguish several groups with verified anticancer activity and potential therapeutic effect such as mTOR kinase inhibitors (Tian, Li, & Zhang, 2019), histone deacetylase inhibitors (Harrison et al., 2012) and inhibitors or antagonists of receptor tyrosine kinase (Kohli et al., 2013).

As previously mentioned, inhibition of a cytoprotective quality control system, such as autophagy, is associated with damage accumulation and, in consequence, cell death. The same is true for cancer cells. Furthermore, blockade of autophagy in RAS-driven cancers led to severe, lethal metabolic stress (Yang et al., 2011). Studies by Morgan et al. showed that metastatic bladder cancer cells were more sensitive to lysosomal inhibitors, such as chloroquine, than benign bladder cancer cells (Morgan et al., 2018). Taking into account the complexity of the autophagy process, it should be noted that autophagy can be inhibited at the early or late stage. Early inhibition prevents formation of autophagosome, but late inhibition leads to accumulation of autophagic vesicles. However, in consequence, both types of autophagy modulations may result in cancer cell death. Inhibitors of the initial phase of autophagy blocked autophagosome formation, which prevented induction of autophagy in chemotherapeutic-treated cancer cells and enhanced their apoptosis (Liu, Yang, Liu, & Wang, 2011). Agents blocking the late stage of autophagy prevented degradation of cargo in autophagic vesicles. These modulators exerted also high toxicity against cancer cells and significantly enhanced the efficiency of currently applied therapies. Indeed, one of them, hydroxychloroquine, is the clinically-approved autophagy inhibitor (Chude & Amaravadi, 2017).

Taken together, the effectiveness of autophagy modulation as an anticancer therapy is closely correlated with the stage and type of tumor, mutations, basal autophagy level, response to therapy (chemoresistance) (Amaravadi, Kimmelman, & Debnath, 2019; Perez-Hernandez et al., 2019; Santana-Codina, Mancias, & Kimmelman, 2017). Although a plethora of autophagy modulators have been exploited with variable success, there are currently no therapies, which focus exclusively on autophagy modulation. This highlights the concept that autophagy modulation, due to a strong tumor- and context-dependent effect, might be applied as a supporting strategy together with classical chemotherapy. Indeed, many reports have confirmed an increase in the effectiveness of autophagy-inducing anticancer drugs in combination with genetic or pharmacological autophagy inhibition (Marinkovic, Sprung, Buljubasic, & Novak, 2018). Only a minority of clinical trials are testing autophagy inhibitors as a monotherapy. The reason is that, thus far, specific and effective agents that could replace the combined therapy have not been identified. However, we have shown that newly synthesized tacrine-melatonin heterodimer, C10, simultaneously stimulated and blocked autophagy in MCF-7 breast cancer cells (Kucharewicz et al., 2018). Others have shown a similar effect using other molecules, such as AMDE-1 (Li et al., 2015) and ML-9 (Kondratskyi et al., 2014). These are promising autophagy-targeting agents, which could theoretically be utilized as stand-alone drugs instead of a combined therapy. However, further studies are necessary to identify and compare the benefits and side effects of simultaneous autophagy induction and inhibition by monotherapy and combining therapy.

As autophagy is equally important in cancer and normal cells, a critical question is whether therapy-evoked modulation of autophagy will be able to inhibit tumor growth without lethal effect on normal cells or will have similar deleterious consequences for both cell types. The long-term effects of autophagy-modulated therapy also remain unresolved. The involvement of autophagy as a mechanism of survival in tumor cell dormancy and metastasis has been demonstrated (Gewirtz, 2009; Sosa, Bragado, Debnath, & Aguirre-Ghiso, 2013). For example, dormancy-associated phenotypes are exhibited by breast cancer stem cells (BCSCs), which upregulate autophagy during metastatic dormancy as a means to survive in nutrient poor conditions and environmental stress (Flynn & Schiemann, 2019). Gewirtz's group showed that stable knockdown of autophagy (*Atg5* knockout) resulted in early escape from dormancy and recurrence of mouse mammary carcinoma treated with doxorubicin. Interestingly, *Atg5* knockout cells contained more polyploid cells upon treatment (Aqbi et al., 2018). Thus, not the cytotoxic effect occurring shortly after autophagy modulation, but rather the long-term influence of this modulation on the development of metastases may be crucial for therapy effectiveness (Fig. 2). As was shown by Barnard et al., autophagy modulation exerted different effect (cytostatic or not) on cancer cells dependent on the stage of metastatic disease (Barnard et al., 2016). Given that TIS, as the main outcome of therapy, could represent a mechanism of evasion from the therapeutic toxicity, understanding the role of autophagy in senescence development and escape from senescence may substantially improve efficiency of autophagy-directed therapy.

5. Relationship between autophagy and senescence

Autophagy is essential to maintain cell homeostasis, and is therefore continuously active at a basal level. Yet, similarly to senescence, autophagy may be stimulated in response to stress. Autophagy induced in response to stress may be a crucial pro-survival mechanism. Nevertheless, severe stress may induce autophagy-mediated cell death (Bialik et al., 2018). Since both cell senescence and autophagy enable cancer cells to avoid death these may be considered as tumor dormancy mechanisms, that can be followed by cancer cell escaping from senescence and cancer regrowth. The connection between autophagy and cell senescence is undeniable; however, the precise nature of this relationship is still unclear, especially given that autophagy may act as either a pro-senescence or an anti-senescence mechanism (Gewirtz, 2013). Some reports, including a seminal one performed on normal fibroblasts induced to senescence by oncogene activation (Young et al., 2009), demonstrated involvement of autophagy in the development of the senescence phenotype. Results from our laboratory have shown that in colon cancer HCT116 cells treated with a natural polyphenol, curcumin, senescence was accompanied by autophagy protein markers and vesicle accumulation. Inhibition of autophagy by ATG5 RNAi decreased the number of senescent cells induced by curcumin, but did not lead to increased cell death (Mosieniak et al., 2012). Autophagy attenuation induced senescence in glioblastoma (Gammoh et al., 2016) and primary human fibroblasts (Kang & Elledge, 2016), arguing for autophagy suppression of senescence in this system. Our recent study showed that tacrine-melatonin heterodimer, C10, induced senescence in 20% of MCF-7 breast cancer cell population by inducing and, simultaneously, blocking the autophagic flux (Kucharewicz et al., 2018). From these results, we conclude that autophagy blockade at the late stage, associated with accumulation of non-functional vesicles, promotes intracellular stress and results in senescence induction whereas autophagy inhibition at the early stage, e.g., by *ATG5* silencing, in stress condition prevents vesicle formation and development of the senescence phenotype.

Autophagy acts as a barrier to malignant transformation during so-called replication or telomere crisis, which leads to cell immortalization. In this process, senescence is an intermediate stage. Autophagy promotes senescence through degradation of the nuclear lamina (Lenain, Gusyatiner, Douma, van den Broek, & Peeper, 2015). This, in turn, leads to extrusion of chromatin fragments and portions of the nucleus, and subsequent generation of cytoplasmic chromatin fragments (CCFs), which are exposed to the DNA-sensing pathway, promoting the SASP. Senescence can be bypassed due to dysfunctional cell cycle check points and cells continue to divide with chromosomal fusions and subsequent breaks, which are initiated by the loss of telomere protection. Telomere dysfunction generates CCFs that activate the DNA-sensing cGAS–STING (cyclic GMP-AMP synthase-stimulator of interferon response cGAMP interactor 1) pathway, involved in a cell death during telomere crisis, thus preventing malignant transformation (Ambrosio & Majello, 2020). However, some cells can escape the crisis and became immortal. Interestingly, using several experimental models of telomere crisis, it was demonstrated that endoreplication and mitotic failure created tetraploid cells during crisis (Ambrosio & Majello, 2020). Telomere-driven tetraploidization promotes tumorigenic transformation of mouse cells, thus facilitating cancer development (Davoli & de Lange, 2012). Another study documented that autophagy may regulate SASP. Kang et al. showed that cellular senescence was prevented by autophagic degradation of transcription factor GATA4 (Kang et al., 2015), whereas GATA4 facilitates development of SASP. In contrast, Young et al. documented autophagy activation upon induction of senescence and showed that autophagy inhibition, by depletion of ATG5 or ATG7, attenuated SASP manifestations, namely, IL6 and IL8 secretion. Nevertheless, autophagy inhibition did not affect the extent of cellular senescence, but only delayed its development (Young et al., 2009).

An explanation of the complex relationship between autophagy and senescence has been recently proposed by Kwon, Kim, Jeoung, Kim, and Kang (2017) who claim that autophagy can be either anti- or pro-senescent, depending on the type of autophagy involved. Indeed, a specialized type of selective autophagy actively suppresses cellular senescence through the degradation of the SASP regulator, GATA4 (Kang et al., 2015), whereas general autophagy supports senescence transition through the TASCC (TOR-autophagy spatial coupling compartment). The key ingredients of this compartment are autophagosomes and the mTOR Complex 1 (mTORC1). In the proposed model, autophagic vesicles, by degrading aged organelles and proteins, generate a high flux of amino acids and other metabolites needed for biosynthesis. In turn, mTORC1 promotes the utilization of these metabolites for the synthesis of the SASP components (Narita et al., 2011). Indeed, this concept may help to explain the very complicated and not unequivocal relationship between cell senescence and autophagy.

6. Autophagy in polyploidization/depolyploidization and senescence escape

The cell cycle must be strictly controlled to avoid the replication of damaged and unwanted cells. The fidelity of the cell cycle is controlled by several checkpoints. Senescent cells are arrested either in the G1 or G2 phase of the cell cycle (Fig. 1) due to activity of two signaling pathways, namely, $p16^{INK4a}$/Rb and $p53/p21^{Cip1/WAF1}$. Thus, CDKIs such as $p16^{INK4a}$ and $p21^{Cip1/WAF1}$ are one of the most reliable markers of cell senescence. The cell cycle checkpoints are critical for the protection against uncontrolled cell division, which is the main feature of cancer development. There are a number of documented genetic lesions in the checkpoint genes, or in the cell cycle genes themselves, which either result directly in cancer or in a predisposition to certain cancer types (Cordon-Cardo, 1995). Impairment in the cell cycle check points is also a cause of polyploidy. However, it seems that this cannot be a sufficient explanation for such a widespread occurrence of PGCCs.

However, the question is whether autophagy actually plays a role in cancer cell polyploidization? Some indication lies in the connection between autophagy and the cell cycle. Regardless of controversies concerning the status of autophagy in different phases of the cell cycle, the conclusion that can be drawn from several studies is that a basal level of active autophagy is detected in all cell cycle phases, even during mitosis (Li, Ji, Wang, Liu, & Zhang, 2016; Liu, Xie, Nguyen, Ye, & McKeehan, 2009). Moreover, it seems that, besides the proteasomal system, selective autophagy can be indispensable for degradation of proteins involved in different phases of the cell cycle (Mathiassen, De Zio, & Cecconi, 2017; Zheng, He, Kitazato, & Wang, 2019). Furthermore, SQSTM1/p62-mediated selective autophagy has been clearly shown to orchestrate cytokinesis via restricting activation of RhoA, a small GTPase that regulates cell shape and completion of cytokinesis by F-actin reticulation at the midbody (Belden & Dunlap, 2008). The authors documented that compromised autophagy (irrespective of the defect: autophagosome formation, sequestration, or degradation) drives cytokinesis failure, multinucleation and aneuploidy.

Interestingly, polyploidization of megakaryocytes is considered to be an example of so-called developmental polyploidization. It has been documented that mTOR activity is required for megakaryocyte polyploidization (Ma et al., 2009) and that its inhibition by rapamycin is concomitant with autophagy activation, disturbed maturation and polyploidization of adult megakaryocytes, and with a reduced number of newly formed platelets (Liu, Italiano, et al., 2011). Interestingly, megakaryopoesis occurring *in vitro* was accompanied by the appearance of senescence markers such as increased $p21^{Cip1/WAF1}$ expression and activity of SA-β-gal in mature polyploid megakaryocytes. Also, mature megakaryocytes isolated from adult donors were SA-β-gal-positive (Besancenot et al., 2010). In contrast, in the process of normal cardiomyocyte multinucleation, mTOR inhibition and mitophagy enhancement were observed (Jiang et al., 2020).

As mentioned previously, PGCCs facilitate cancer recurrence. Some studies indicate that polyploidization not only facilitates resumption of cancer cell proliferation but, in senescent cells, seems to be a driving force behind atypical cell divisions. However, what is the trigger that switches the state of cell senescence/polyploidy to cell division? As already mentioned, the nucleus: cytoplasm (N/C) ratio contributes to a functional decline in senescent cells (Neurohr et al., 2019). The authors showed that cells in the G1 phase of the cell cycle increased their volume. Additionally, when the cell cycle was blocked, e.g., after chemical treatment, cells were continuing to increase in size. The increased size and cell volume resulted in impairment of gene induction, insufficient RNA and protein synthesis and, in consequence, cytoplasm dilution. Such changes severely affected cell functionality and contributed to cell senescence. In line with this finding, we speculate that senescent cells (at least cancer senescent cells) are forced to change the N/C ratio for their further reprogramming and division, which can be achieved by polyploidization. As mentioned previously, autophagy plays an important role in cell cycle progression and polyploidization. Resumption of proliferation of polyploid/senescent cells requires reorganization of the cell interior. Therefore, autophagy seems likely to be involved in senescence escape, but thus far the published literature concerning this subject is rather scarce. Nonetheless, as was suggested by Erenpreisa et al., autophagy of whole sub-nuclei in multi-nucleated cells appears to favor depolyploidization, mitigation of aneuploidy with its adverse effects, thereby promoting the survival fitness of descendants as well as treatment resistance (Erenpreisa, Huna, Salmina, Jackson, & Cragg, 2012).

Our results confirm the concept of autophagy involvement in cancer recurrence (Bojko et al., 2020). We showed that induction of senescence by DNA damaging agent (doxorubicin) increased cell ploidy. Polyploidization was accompanied by transient inactivation of mTOR (Fig. 3A), which should result in autophagy induction. However, accumulation of autophagy markers, namely, LC3B and SQSTM1/p62, and lack of enhancement of autophagic vesicle formation and degradation (low autophagic flux) indicated unsuccessful, "abortive autophagy" in senescent cancer cells. Moreover, we observed that, regardless of the basal autophagy status, parental cells underwent drug-induced senescence/polyploidization accompanied by almost non-functional autophagy and generated progeny with elevated and functional autophagic flux (measured by autophagic index) (Fig. 3B). The autophagic index in descendants was similar to or higher than in parental, non-senescent cells. However, we cannot exclude the possibility that senescent cells activate autophagy just prior to division. Thus, we speculate that functional autophagy may be indispensable for senescence/polyploidy escape and repopulation of cancer cells.

Other experimental results support this claim. Autophagy blockade with Bafilomycin A1 in senescent cancer HCT116 cells resulted in an increased number of polyploid cells and a substantial delay in senescence escape in comparison with untreated senescent cells (21 days compared to 14) (Was et al., 2017). Nevertheless, pulse treatment of senescent cells with Bafilomycin A1 increased their migration and invasion and led to formation of larger tumors upon injection into NOD/SCID mice. The authors concluded that the pulse of Bafilomycin A1 postponed short-term *in vitro* cancer cell repopulation but, in the long-term, such treatment resulted in an increased number of polyploid cells, reactivation of autophagy in a subpopulation of HCT116 cells and increased proliferation. Aqbi et al. showed clearly that permanent autophagy inhibition evoked by ATG5 knockdown, but not the pulse treatment with an autophagy inhibitor, chloroquine, led to accumulation of adriamycin-induced polyploid cells and significantly earlier development of lung metastasis (Aqbi et al., 2018). The authors suggested that chloroquine prolonged tumor dormancy whereas stable knockdown of autophagy resulted in early escape from dormancy and cancer recurrence. Chiticova et al. demonstrated that mTOR was downregulated in immortalized rat embryonic fibroblasts after irradiation-induced (IR) polyploidization and senescence, which suggests activation of autophagy (Chitikova et al., 2014). In the authors' opinion, activated autophagy is indispensable for the reversion from IR-induced senescence, which

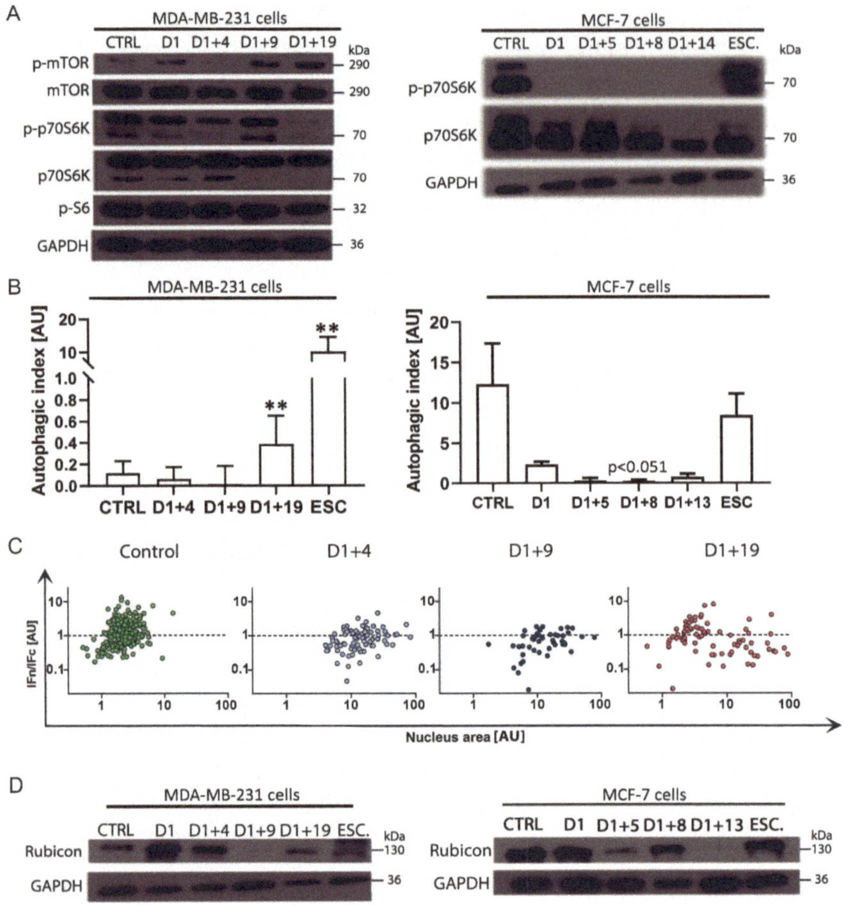

Fig. 3 The status of autophagy-related signaling pathways in senescence escape and breast cancer repopulation. (A) Representative western blots showing protein level of p-mTOR (Ser2448), mTOR, p-p70S6K, p70S6K, p-S6 and GAPDH in MDA-MB-231 (left panel) and MCF-7 cells (right panel). (B) Quantitative analysis of autophagic index based on LC3B protein levels in untreated and chloroquine-treated MDA-MB-231 (left panel) and MCF-7 cells (right panel). Bars: mean value, error bars: SEM, $N=4$. Statistical significance (in relation to control): **$0.001 < P < 0.01$; (C) Quantitative analysis of the intensity of TFEB fluorescence in nucleus/cytoplasm (IFn/IFc) vs nucleus area in MDA-MB-231 cells. (D) Representative western blots showing protein level of Rubicon in MDA-MB-231 (left panel) and MCF-7 cells (right panel). ESC—escapers. *All figures, except the A right panel, are copied and modified from Bojko, A., Staniak, K., Czarnecka-Herok, J., Sunderland, P., Dudkowska, M., et al. (2020). Improved autophagic flux in escapers from doxorubicin-induced senescence/polyploidy of breast cancer cells.* International Journal of Molecular Sciences. 21, *6084.*

is manifested by the appearance of SA-β-gal-negative cells of near normal size and ploidy that exhibit high proliferative potential and restoration of the cell population.

However, Gewirtz's group (Gewirtz, 2020) questioned the role of autophagy in senescence escape and cancer recurrence. Thus, some reports revealed that autophagy inhibition attenuated senescent cancer regrowth dependent on the p53 status of cancer cells (Sharma, Yao, et al., 2014; Xu et al., 2018; Patel et al., 2020). The authors postulated that p53wt and p53mutant cells possess different autophagy status, which determines the effect of autophagy inhibition on the escape from senescence induced by DNA damage treatment. Both p53wt and p53mutant cells escaped from senescence with similar dynamics. However, autophagy inhibition attenuated senescent cancer cell regrowth with respect to p53 status. Namely, blockade of autophagy ameliorated cancer repopulation only in p53wt (Xu et al., 2018) or only in p53mutant cells (Sharma, Yao, et al., 2014; Patel et al., 2020). In line with this, Jakhar et al. showed that sensitivity to combinatorial treatment with an anticancer drug (paclitaxel) and autophagy inhibitor, chloroquine, was dependent on p53 status (Jakhar et al., 2018). In this study postmitotic slippage, leading to tetraploidy of cancer cells, was dependent on autophagy induction. Precisely, inhibition by chloroquine or silencing of an autophagy-related gene, ATG5, in cancer cells treated with an antimitotic drug, led to bypass of the G1 arrest/senescence and an increase in apoptosis. Importantly, chloroquine attenuated the growth of xenograft tumors generated by subcutaneous injection of p53wt HCT116 cells, but not p53 null HCT116 cells, in nude mice. In contrast, Saleh et al. demonstrated that in both, p53wt and p53 null, cancer senescent cells, autophagy inhibition by pulse treatment with Bafilomycin A1, chloroquine or ATG5 silencing did not reduce cancer cell repopulation (Saleh, Tyutyunyk-Massey, et al., 2020). However, in those studies, autophagy was not inhibited in senescent cells but before and during cell exposition to DNA damage treatment, which may be the cause of a different conclusion concerning the role of autophagy in senescence escape.

The reports on the role of autophagy in cancers exposed to therapy are inconsistent also due to different interpretation of the results concerning the autophagy process. Properly functioning autophagy can be described as a multi-stage phenomenon. Since autophagy is a highly dynamic process, for proper estimation of autophagy efficiency, it is crucial to measure the autophagic flux, which determines the degradation activity (Mizushima, Ohsumi, & Yoshimori, 2002). Therefore, the level of autophagy markers does not always equal autophagy effectiveness. We have shown that the increased level of autophagy proteins and accumulation of vesicles in cells treated with anticancer drugs may be the result of both autophagy induction and blockade or even a simultaneous occurrence of both processes (Kucharewicz et al., 2018). Thus, we propose, as a rule, to determine autophagy activity by estimating the autophagic index or the rate of formation and degradation of autophagic vesicles, and not only by measuring autophagy protein levels (Bojko et al., 2020).

Another parameter which can lead to different interpretation of the role of autophagy in cancer treatment is the type of autophagy modulator and the treatment scheduled. Pulse treatment with autophagy inhibitor, attenuated autophagy activity in cells soon after treatment whereas long-term analysis revealed autophagy reactivation in the cells (Li et al., 2013). A similar scenario was noticed by Was et al. who observed short-term attenuation of senescence escape but, in a long-term, stimulation of Bafilomycin A1-treated cancer cell repopulation (Was et al., 2017). In contrast, Saleh et al. did not find any effect of pulse autophagy inhibition on senescent cancer recurrence (Saleh, Tyutyunyk-Massey, et al., 2020). In other studies, a prolonged inhibition of autophagy by ATG gene silencing or long-term treatment with an inhibitor (Aqbi et al., 2018; Jakhar et al., 2018) usually exerted a higher impact on cancer repopulation than pulse autophagy blockade (Was et al., 2017; Patel et al., 2020; Saleh, Tyutyunyk-Massey, et al., 2020). Currently, it is postulated that prevention of cancer recurrence by autophagy inhibition depends on the type of autophagy that is inhibited (Aqbi et al., 2018; Gewirtz, 2016, 2020). As was mentioned previously, cancer cells are characterized by different basal autophagy status. Thus, inhibition of basal autophagy by ATG gene silencing may have a completely different effect than blockade of the autophagic flux during cell exposition to stress factors. Moreover, even in tumors with almost non-functional basal autophagy, transient activation of cytoprotective autophagy was observed in response to stress (Gewirtz, 2020).

7. Mechanistic view on autophagy activation in the escape from senescent cancer cell polyploidy

An increasing number of reports have argued for the involvement of autophagy in cell reorganization processes during senescence escape. It has been proposed by Erenpreisa et al. that autophagy is required to fuel the process of senescence/polyploidy escape (Erenpreisa et al., 2012). This group postulated that autophagy was responsible for elimination of micronuclei with DNA damage in senescent/polyploid cells (Erenpreisa, Salmina, & Cragg, 2017; Erenpreisa et al., 2011). Upon senescence and polyploidization of cancer cells, DNA damage accumulation was evident, but the mechanism of repair seemed to be insufficient (Bojko et al., 2019). In our own study, the progeny of senescent/polyploid cells were characterized by a reduced number of DNA damage foci and by fully functional autophagy (Bojko et al., 2020). However, we did not observe the expected degradation of micronuclei with accumulated damaged DNA by autophagy in PGCC mother cells even though, in theory, small fragments of cytoplasmic damaged DNA (CCFs) should activate the autophagy process; indeed, we have observed such fragments in cells induced to senescence/polyploidy (Bojko et al., 2020; Salmina et al., 2020). Previously, it has been shown that free cytoplasmic DNA activated the cGas/STING pathway (Dunphy et al., 2018). Recently, Gui et al. have described that free DNA in the cytoplasm activated the cGas/STING pathway, leading to LC3 lipidation and autophagosome biogenesis independently from ULK1 (Gui et al., 2019). A similar mechanism of autophagy induction can occur upon polyploidization (Fig. 4A).

Autophagy is regulated by mTOR- and AMPK-dependent signaling pathways. These are modulated in response to nutrient and energy deficiency, respectively. mTOR is a component of two protein complexes: mTORC1 and mTORC2. High lysosomal amino acid concentration keeps the mTORC1 complex activated in a manner dependent on the lysosomal proton pump v-ATPase complex (Kawabata & Yoshimori, 2020). When mTOR is activated, mTORC1 is bound to the lysosomal surface and catalyzes phosphorylation of a variety of substrates, such as ULK1 and transcription factor EB (TFEB), among others, which leads to autophagy inhibition. Conversely, mTOR inhibition results in autophagy induction. In contrast, AMPK-mediated phosphorylation of ULK1 induces autophagy. Thus, mTORC1 and AMPK act as antagonists in regulating ULK1 activity and, in consequence, autophagy. AMPK further influences mTOR activity via phosphorylation of TSC2, leading to mTORC1 inhibition. Both mTOR and AMPK kinase signaling are regulated by stress factors, which makes them potential candidates for the autophagy "switch on" mechanism during escape from polyploidy/senescence (Grasso, Renna, & Vaccaro, 2018) (Fig. 4B). Due to a decreased level of the phosphorylated mTORC1

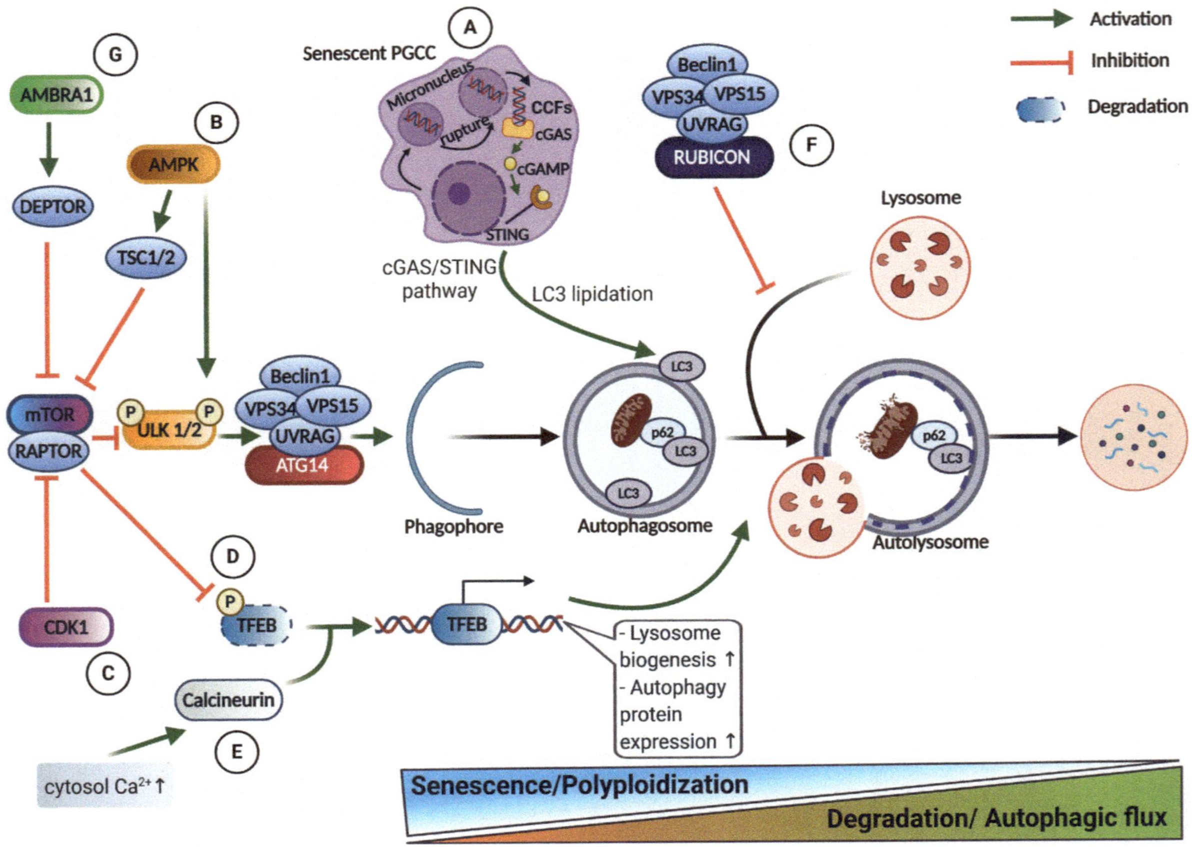

Fig. 4 See figure legend on opposite page.

substrate, S6K, autophagy induction in irradiation-induced polyploid/senescent cells was suggested to occur through an AMPK/mTOR dependent mechanism. Even though this was not investigated, the authors assumed p53- and ATM-dependent activation of AMPK and complex TSC1/2 (Chitikova et al., 2014). In contrast, despite mTORC1 activity fluctuation, we did not observe autophagy upregulation in senescent/polyploid cells, which was also confirmed by low autophagic flux (Bojko et al., 2020) (Fig. 3B). Interestingly, in MDA-MB-231 cells, when the daughter cells appeared on 19th day of culture in drug free medium, after pulse treatment (24 h incubation) with doxorubicin, the level of phosphorylated mTOR was still pronounced, contrary to the reduced level of its substrate, namely, phosphorylated p70S6K (Fig. 3A). We concluded that despite mTOR phosphorylation the mTORC1 complex was not active,

Fig. 4 Possible mechanisms of autophagy activation/induction during senescence escape. Autophagy is a multistep process, with specific stages: induction, formation of the isolation membrane (phagophore), formation and maturation of the autophagosome and, finally autolysosome formation by autophagosome fusion with lysosome. The endpoint of autophagy is cargo degradation and its recycling. Autophagy activation during cancer repopulation can be caused by several mechanisms: (A) cGAS/STING Pathway: due to polyploidization cytoplasmic chromatin fragments (CCFs) accumulate in cytoplasm. They are recognized by cGAS/STING, which leads to LC3 lipidation and autophagosome biogenesis independently from ULK-1. (B) AMPK Pathway: AMPK by phosphorylation activates ULK1 and inhibits mTORC1 via TSC2 phosphorylation. Both processes lead to autophagy induction. (C) CDK1 Pathway: some proteins involved in cell cycle progression regulate autophagy. CDK1 phosphorylates RAPTOR, a member of mTORC1 complex, leading to mTORC1 inhibition, what should induce autophagy. (D) mTOR/TFEB Pathway: mTOR dependent phosphorylation of TFEB prevents its translocation to nucleus leads to its degradation in the cytoplasm. This ultimately results in downregulation of lysosomal biogenesis. (E) Calcineurin/TFEB Pathway: TFEB is dephosphorylated by calcineurin in Ca^{2+}-dependent manner. Dephosphorylated TFEB is translocated then into nucleus where it acts as a transcription factor. This regulation is mTOR independent. (F) Rubicon: Rubicon blocks autophagosome and lysosome fusion, thus acts like endogenous autophagy inhibitor. Rubicon level is elevated during senescence and decreases during escape, what favor autophagy activation. (G) Ambra1: Ambra1, endogenous autophagy inductor, prevents DEPTOR, a mTORC1 inhibitor, degradation. It results in mTOR inhibition and autophagy activation. Abbreviations: PGCC: polyploid giant cancer cell; P: phosphorylation; CCFs: cytoplasmic chromatin fragments. Scheme created with BioRender.com.

suggesting mTOR-dependent autophagy activation via a mechanism other than mTOR phosphorylation at Ser 2488. However, in MCF-7 cells p70S6K phosphorylation and, in turn, mTORC1 activity, decreased during senescence induction and escape, whereas in the MCF-7 escapers, the p70S6K phosphorylation was restored. Autophagy regulation in this case seems to be more complex.

It has already been mentioned that autophagy may affect cell cycle but this relation is bilateral in that select proteins involved in cell cycle progression can likewise regulate autophagy. CDK1 was found to phosphorylate several proteins involved in autophagy induction and lysosome biogenesis (Li et al., 2020; Odle et al., 2020). In mitosis, CDK1-dependent phosphorylation of RAPTOR, a member of the mTORC1 complex, led to mTORC1 inhibition that should induce autophagy. However, CDK1 by phosphorylation downregulated also ULK1, ATG13, ATG14 and TFEB, which, in consequence, led to autophagy repression in mitosis. It could be suggested that CDK1-dependent downregulation of mTORC1, when is not accompanied by CDK1-dependent inhibition of autophagy proteins, may lead to autophagy activation. Thus, in some scenarios, CDK1 may be a potent autophagy activator in senescence escaping cancer cells (Fig. 4C).

The mTOR-dependent signaling pathway involved in autophagy activation could be also regulated at a level different from mTORC1. TFEB, a key regulator of lysosome biogenesis, is downregulated by mTOR-catalyzed phosphorylation. Its translocation to the nucleus resulted in the upregulation of lysosomal biogenesis, whereas TFEB phosphorylation by mTORC1 prevented its translocation and led to TFEB degradation in the cytoplasm (Napolitano & Ballabio, 2016; Puertollano, Ferguson, Brugarolas, & Ballabio, 2018) (Fig. 4D). Our results showed that TFEB translocation into the cell nucleus correlated with the appearance of progeny in an mTOR-dependent manner (Bojko et al., 2020) (Fig. 3C).

However, TFEB can also be regulated in an mTOR independent way. TFEB is dephosphorylated by calcineurin in a Ca^{2+}-dependent manner (Medina et al., 2015) (Fig. 4E). Dephosphorylated TFEB is translocated into the nucleus where it acts as a transcription factor. Ikari et al. showed that during starvation cytosolic Ca^{2+} increased and that this resulted in TFEB activation by calcineurin (Ikari et al., 2020). The elevated level of Ca^{2+} in cytosol occurs in response to ER or mitochondrial stress (Bahar, Kim, & Yoon, 2016). Accordingly, under ER stress, TFEB was activated and translocated to the nucleus in a process that was dependent on calcineurin and independent of mTOR (Martina, Diab, Brady, & Puertollano, 2016).

Apart from autophagy activators, at least one endogenous cellular autophagy inhibitor, Rubicon, should be mentioned. Rubicon inhibits autophagy by several mechanisms; among these, the inhibition of autophagosome and lysosome fusion is the best described (Matsunaga et al., 2009). Elevation of Rubicon was found to be correlated with cellular senescence and aging (Nakamura et al., 2019). In cells escaping the senescence/polyploidy state, we have observed a decreased level of Rubicon (Bojko et al., 2020) (Fig. 3D). This finding suggests that the inhibition of Rubicon may be a possible autophagy "switch on" mechanism (Fig. 4F).

Another intracellular autophagy regulator is the Activating Molecule in Beclin-1-Regulated Autophagy (Ambra1) that is an autophagy activator. Ambra1 coordinates the cell response upon a variety of cellular stresses (Cianfanelli et al., 2015) and is involved in such crucial physiological events as cell metabolism, cell death and division. mTORC1 can inactivate Ambra1 by phosphorylation, whereas during stress, Ambra1 prevents degradation of DEP-domain containing mTOR interacting protein (DEPTOR), a mTORC1 inhibitor. This results in mTOR inhibition and autophagy activation. Independently of mTOR inhibition, Ambra1 can stabilize ULK1, which, in turn, favors autophagy induction. The multitarget activation of autophagy by Ambra1 provides grounds for consideration of this protein as a possible autophagy activator in senescence escaping cancer cells (Fig. 4G). Nonetheless, Ambra1 involvement in autophagy activation in cancer cells escaping from senescence/polyploidy needs to be further elucidated.

Summing up, we conclude that the mechanism of autophagy activation in cancer cells escaping senescence/polyploidy is complex and cancer cell specific. Due to the frequent autophagy protein mutations in cancers, a variety of signaling pathways may be triggered to overcome this burden. However, regardless of the cancer cell context, the aim remains the same—to survive and reassume cell proliferation.

8. Conclusions

Autophagy is a self-degradative process that appears to play a central role in balancing the sources of energy at critical times in development and in response to stress. Autophagy also plays a housekeeping role by removing misfolded or aggregated proteins, clearing damaged organelles, including mitochondria and lysosomes, and eliminating intracellular pathogens. Thus, autophagy is generally considered as a survival mechanism (Glick, Barth, & Macleod, 2010). However, during an early stage of tumorigenesis, autophagy is frequently inhibited. Autophagy inhibition favors genomic instability, mutations and formation of tumor initiating cells with increased proliferation rates. In advanced cancer, autophagy appears to play primarily opposite role. Since highly proliferating cancer cells have an increased demand for nutrients, autophagy may act as a mechanism for supplying them. Autophagy may also represent a cytoprotective process for cancer cells in case of exposition to stress factors, such as hypoxia (Liu et al., 2010), chemotherapy (Ciechomska et al., 2013) or radiotherapy (Das, Joshi, Parashiva, & Rao, 2020). Additionally, autophagy may promote metastasis by mediating anoikis resistance and adaptation of metastatic cells to a new microenvironment (Sosa et al., 2013).

Senescence of cancer cells has characteristics typical also for the senescence of normal cells. Senescent cells are permanently or temporally arrested in the cell cycle. They have unrepaired DNA lesions, damaged mitochondria, increased number of vesicles, impaired proteostasis with very characteristic accumulation of undigested lipofuscin, which clogs the lysosomes (Jung, Bader, & Grune, 2007). Impaired selective autophagy of GATA4, involved in the development of SASP, was demonstrated in senescent cells (Kang & Elledge, 2016). Generally, these are strong arguments for dysfunctional autophagy in an established senescence phenotype. However, this does not exclude the involvement of active autophagy in the development of the senescence phenotype, during which the cell interior is rebuilt. Moreover, regarding senescence of cancer cells, the situation can be even more complicated if we consider senescence of benign tumors, with already impaired autophagy, and that of developed cancers, which usually have active autophagy. On the other hand, we have documented that senescence can be evoked in cancer cells regardless of the level of basal autophagy. Thus, according to our results, we speculate that development of senescence phenotype does not rely on autophagy, but senescence is frequently associated with diminished autophagic flux (Bojko et al., 2020).

It cannot be denied that autophagy regulation changes during cell senescence/polyploidization and senescence escape/depolyploidization (Erenpreisa & Cragg, 2013; Mirzayans, Andrais, & Murray, 2018; Mosieniak & Sikora, 2010). Therefore, autophagy regulation should be added to senescent/polyploid cancer cell characteristics. Accordingly, we speculate that, despite the basal level of autophagy in cancer cells, these cells undergo senescence and they increase ploidy upon treatment with chemotherapeutic drugs, i.e., in response to stress. Autophagy inhibition favors ploidy increase (Aqbi et al., 2018). Senescent/polyploid cells, which are really giant, multinucleated and full of waste, including fragments of damaged DNA and impaired organelles, cannot survive for long without functional autophagy. It is likely that the majority of these cells die. There are, however, very rare escapers, and to release these escapers, PGCCs need to activate autophagy.

In conclusion, taking into consideration the recent reports on senescence, polyploidy and autophagy involvement in cancer relapse, including our own results, we hypothesize that autophagy is crucial for senescent/polyploid cancer cells division, as it has been previously proposed by Erenpreisa and Cragg (2013). Moreover, polyploidy formation and autophagy are intertwined and autophagy modulation may stimulate or affect polyploidization and, in consequence, modulate the escape from the tumor dormancy state/senescence. Fig. 4 presents the proposed mechanisms of autophagy regulation involved in the escape from polyploidy/senescence.

We do not know, yet, what is the role of cell senescence and autophagy modulation in normal life cycle at the very early stage of embryogenesis, however, their role in cancer cell life cannot be overestimated. Cell senescence is considered as a cancer barrier (Campisi, Andersen, Kapahi, & Melov, 2011). The question is why and how this barrier is broken in PGGCs? Just like "cell crisis" leads to cell death of the majority of senescent cells and the escape of rare immortal cells, also polyploid/senescent cancer cells can undergo "crisis" that leads to the delivery of progeny. We speculate that, to complete this program, active cytoprotective autophagy is necessary. Fig. 5 shows the life cycle of cancer cells with the emphasis on the role of autophagy in this life cycle.

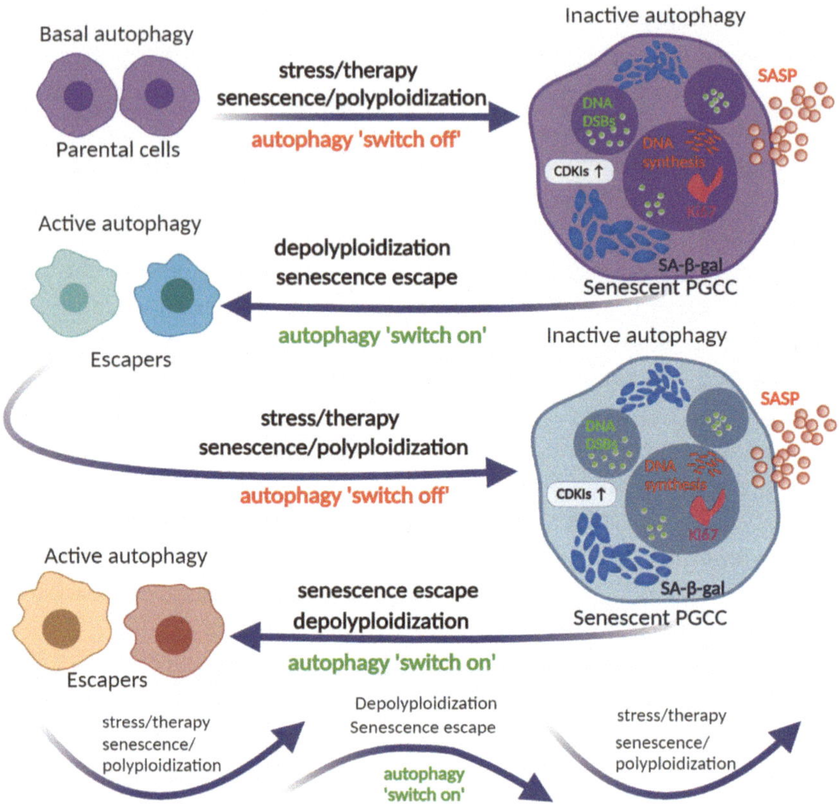

Fig. 5 Cancer life cycle. Stress or treatment (chemotherapy or irradiation) causes perturbances in autophagy (autophagy "switch off"), leading to senescence and emergence of Polyploid Giant Cancer Cell (PGCC). Those senescent PGCCs are characterized by both senescence and polyploidization markers as well as inactive autophagy. Due to autophagy "switch on" PGCCs escape senescence and depolyplodize giving viable progeny. Escaper cells are different than parental cells in terms of: autophagy, aggressiveness, stemness, SASP secretion and drug sensitivity. Escaper cells can undergo subsequent rounds of autophagy switch off/on and senescence/polyplodization induction and escape each time giving a different kind of progeny (escapers). Definitions/ Abbreviations: Parental cells: initial cancer cells, before treatment. Senescent PGCC: Senescent Polyploid Giant Cancer Cell. Escapers: descendants of senescent polyploid giant cancer cells. Basal autophagy: status of the autophagic flux in untreated cells. Inactive autophagy: significant attenuation of autophagic flux, resulting in diminished ability to degrade cargo within autolysosomes. Active autophagy: unblocked autophagic flux, resulting in increased ability to degrade cargo within autolysosomes. Markers of senescence: cell enlargement, DNA double strand breaks (DNA DSBs), cyclin dependent kinase inhibitors (CDKIs), e.g., p16INK4a, p21 Cip1/WAF1, Senescence Associated Secretory Phenotype (SASP), activity of Senescence-Associated β-galactosidase (SA-β-gal). Markers of polyploidization: DNA synthesis, Ki-67, marker of cycling cells. Scheme created with BioRender.com.

Acknowledgment

This work was supported by the National Science Centre, grant UMO-2015/17/B/NZ3/03531 for E.S.

References

Amaravadi, R. K., Kimmelman, A. C., & Debnath, J. (2019). Targeting autophagy in cancer: Recent advances and future directions. *Cancer Discovery*, *9*, 1167–1181.

Ambrosio, S., & Majello, B. (2020). Autophagy roles in genome maintenance. *Cancers*, *12*, 1793.

Amend, S. R., Torga, G., Lin, K. C., Kostecka, L. G., de Marzo, A., Austin, R. H., et al. (2019). Polyploid giant cancer cells: Unrecognized actuators of tumorigenesis, metastasis, and resistance. *The Prostate*, *79*, 1489–1497.

Aqbi, H. F., Tyutyunyk-Massey, L., Keim, R. C., Butler, S. E., Thekkudan, T., Joshi, S., et al. (2018). Autophagy-deficient breast cancer shows early tumor recurrence and escape from dormancy. *Oncotarget*, *9*, 22113–22122.

Bahar, E., Kim, H., & Yoon, H. (2016). ER stress-mediated signaling: Action potential and Ca(2+) as key players. *International Journal of Molecular Sciences*, *17*, 1558.

Barnard, R. A., Regan, D. P., Hansen, R. J., Maycotte, P., Thorburn, A., & Gustafson, D. L. (2016). Autophagy inhibition delays early but not late-stage metastatic disease. *The Journal of Pharmacology and Experimental Therapeutics*, *358*, 282–293.

Belden, W. J., & Dunlap, J. C. (2008). SIRT1 is a circadian deacetylase for core clock components. *Cell*, *134*, 212–214.

Besancenot, R., Chaligne, R., Tonetti, C., Pasquier, F., Marty, C., Lecluse, Y., et al. (2010). A senescence-like cell-cycle arrest occurs during megakaryocytic maturation: Implications for physiological and pathological megakaryocytic proliferation. *PLoS Biology*, *8*, e1000476.

Bharadwaj, D., & Mandal, M. (2020). Senescence in polyploid giant cancer cells: A road that leads to chemoresistance. *Cytokine & Growth Factor Reviews*, *52*, 68–75.

Bialik, S., Dasari, S. K., & Kimchi, A. (2018). Autophagy-dependent cell death—Where, how and why a cell eats itself to death. *Journal of Cell Science*, *131*, jcs215152.

Bielski, C. M., Zehir, A., Penson, A. V., Donoghue, M. T. A., Chatila, W., Armenia, J., et al. (2018). Genome doubling shapes the evolution and prognosis of advanced cancers. *Nature Genetics*, *50*, 1189–1195.

Bishop, E., & Bradshaw, T. D. (2018). Autophagy modulation: A prudent approach in cancer treatment? *Cancer Chemotherapy and Pharmacology*, *82*, 913–922.

Bojko, A., Czarnecka-Herok, J., Charzynska, A., Dabrowski, M., & Sikora, E. (2019). Diversity of the senescence phenotype of cancer cells treated with chemotherapeutic agents. *Cells*, *8*, 1501.

Bojko, A., Staniak, K., Czarnecka-Herok, J., Sunderland, P., Dudkowska, M., Sliwinska, M. A., et al. (2020). Improved autophagic flux in escapers from doxorubicin-induced senescence/polyploidy of breast cancer cells. *International Journal of Molecular Sciences*, *21*, 6084.

Braig, M., Lee, S., Loddenkemper, C., Rudolph, C., Peters, A. H., Schlegelberger, B., et al. (2005). Oncogene-induced senescence as an initial barrier in lymphoma development. *Nature*, *436*, 660–665.

Campisi, J. (2011). Cellular senescence: Putting the paradoxes in perspective. *Current Opinion in Genetics & Development*, *21*, 107–112.

Campisi, J., Andersen, J. K., Kapahi, P., & Melov, S. (2011). Cellular senescence: A link between cancer and age-related degenerative disease? *Seminars in Cancer Biology*, *21*, 354–359.

Chen, Z., Trotman, L. C., Shaffer, D., Lin, H. K., Dotan, Z. A., Niki, M., et al. (2005). Crucial role of p53-dependent cellular senescence in suppression of Pten-deficient tumorigenesis. *Nature, 436*, 725–730.
Chitikova, Z. V., Gordeev, S. A., Bykova, T. V., Zubova, S. G., Pospelov, V. A., & Pospelova, T. V. (2014). Sustained activation of DNA damage response in irradiated apoptosis-resistant cells induces reversible senescence associated with mTOR down-regulation and expression of stem cell markers. In *Cell cycle* (pp. 1424–1439). Taylor & Francis Online. United States.
Chude, C. I., & Amaravadi, R. K. (2017). Targeting autophagy in cancer: Update on clinical trials and novel inhibitors. *International Journal of Molecular Sciences, 18*.
Cianfanelli, V., De Zio, D., Di Bartolomeo, S., Nazio, F., Strappazzon, F., & Cecconi, F. (2015). Ambra1 at a glance. *Journal of Cell Science, 128*, 2003–2008.
Ciechomska, I. A., Gabrusiewicz, K., Szczepankiewicz, A. A., & Kaminska, B. (2013). Endoplasmic reticulum stress triggers autophagy in malignant glioma cells undergoing cyclosporine a-induced cell death. *Oncogene, 32*, 1518–1529.
Collado, M., Gil, J., Efeyan, A., Guerra, C., Schuhmacher, A. J., Barradas, M., et al. (2005). Tumour biology: Senescence in premalignant tumours. *Nature, 436*, 642.
Coppola, D., Khalil, F., Eschrich, S. A., Boulware, D., Yeatman, T., & Wang, H. G. (2008). Down-regulation of Bax-interacting factor-1 in colorectal adenocarcinoma. *Cancer, 113*, 2665–2670.
Cordon-Cardo, C. (1995). Mutations of cell cycle regulators. Biological and clinical implications for human neoplasia. *The American Journal of Pathology, 147*, 545–560.
Costa, J. R., Prak, K., Aldous, S., Gewinner, C. A., & Ketteler, R. (2016). Autophagy gene expression profiling identifies a defective microtubule-associated protein light chain 3A mutant in cancer. *Oncotarget, 7*, 41203–41216.
Coward, J., & Harding, A. (2014). Size does matter: Why polyploid tumor cells are critical drug targets in the war on cancer. *Frontiers in Oncology, 4*, 123.
Dalby, K. N., Tekedereli, I., Lopez-Berestein, G., & Ozpolat, B. (2010). Targeting the prodeath and prosurvival functions of autophagy as novel therapeutic strategies in cancer. *Autophagy, 6*, 322–329.
Das, S., Joshi, M. B., Parashiva, G. K., & Rao, S. B. S. (2020). Stimulation of cytoprotective autophagy and components of mitochondrial biogenesis / proteostasis in response to ionizing radiation as a credible pro-survival strategy. *Free Radical Biology & Medicine, 152*, 715–727.
Davoli, T., & de Lange, T. (2011). The causes and consequences of polyploidy in normal development and cancer. *Annual Review of Cell and Developmental Biology, 27*, 585–610.
Davoli, T., & de Lange, T. (2012). Telomere-driven tetraploidization occurs in human cells undergoing crisis and promotes transformation of mouse cells. *Cancer Cell, 21*, 765–776.
Degenhardt, K., Mathew, R., Beaudoin, B., Bray, K., Anderson, D., Chen, G., et al. (2006). Autophagy promotes tumor cell survival and restricts necrosis, inflammation, and tumorigenesis. *Cancer Cell, 10*, 51–64.
Devitt, A., & Marshall, L. J. (2011). The innate immune system and the clearance of apoptotic cells. *Journal of Leukocyte Biology, 90*, 447–457.
Dewhurst, S. M., McGranahan, N., Burrell, R. A., Rowan, A. J., Gronroos, E., Endesfelder, D., et al. (2014). Tolerance of whole-genome doubling propagates chromosomal instability and accelerates cancer genome evolution. *Cancer Discovery, 4*, 175–185.
Ding, Z. B., Shi, Y. H., Zhou, J., Qiu, S. J., Xu, Y., Dai, Z., et al. (2008). Association of autophagy defect with a malignant phenotype and poor prognosis of hepatocellular carcinoma. *Cancer Research, 68*, 9167–9175.

Dunphy, G., Flannery, S. M., Almine, J. F., Connolly, D. J., Paulus, C., Jonsson, K. L., et al. (2018). Non-canonical activation of the DNA sensing adaptor STING by ATM and IFI16 mediates NF-kappaB signaling after nuclear DNA damage. *Molecular Cell*, *71*, 745–760 e745.

Erenpreisa, J., & Cragg, M. S. (2013). Three steps to the immortality of cancer cells: Senescence, polyploidy and self-renewal. *Cancer Cell International*, *13*, 92.

Erenpreisa, J., Huna, A., Salmina, K., Jackson, T. R., & Cragg, M. S. (2012). Macroautophagy-aided elimination of chromatin: Sorting of waste, sorting of fate? *Autophagy*, *8*, 1877–1881.

Erenpreisa, J., Salmina, K., & Cragg, M. S. (2017). Accelerated senescence of cancer stem cells: A failure to thrive or a route to survival? In *Senescence—Physiology or pathology* IntechOpen.

Erenpreisa, J., Salmina, K., Huna, A., Kosmacek, E. A., Cragg, M. S., Ianzini, F., et al. (2011). Polyploid tumour cells elicit paradiploid progeny through depolyploidizing divisions and regulated autophagic degradation. *Cell Biology International*, *35*, 687–695.

Flynn, A. B., & Schiemann, W. P. (2019). Autophagy in breast cancer metastatic dormancy: Tumor suppressing or tumor promoting functions? *Journal of Cancer Metastasis and Treatment*, *5*, 43.

Fox, D. T., & Duronio, R. J. (2013). Endoreplication and polyploidy: Insights into development and disease. *Development*, *140*, 3–12.

Freund, A., Orjalo, A. V., Desprez, P. Y., & Campisi, J. (2010). Inflammatory networks during cellular senescence: Causes and consequences. *Trends in Molecular Medicine*, *16*, 238–246.

Fulda, S. (2017). Autophagy in cancer therapy. *Frontiers in Oncology*, 7, 128.

Fulda, S., Gorman, A. M., Hori, O., & Samali, A. (2010). Cellular stress responses: Cell survival and cell death. *International Journal of Cell Biology*, *2010*, 214074.

Fulda, S., & Kogel, D. (2015). Cell death by autophagy: Emerging molecular mechanisms and implications for cancer therapy. *Oncogene*, *34*, 5105–5113.

Gammoh, N., Fraser, J., Puente, C., Syred, H. M., Kang, H., Ozawa, T., et al. (2016). Suppression of autophagy impedes glioblastoma development and induces senescence. *Autophagy*, *12*, 1431–1439.

Gewirtz, D. A. (2009). Autophagy, senescence and tumor dormancy in cancer therapy. *Autophagy*, *5*, 1232–1234.

Gewirtz, D. A. (2013). Autophagy and senescence: A partnership in search of definition. *Autophagy*, *9*, 808–812.

Gewirtz, D. A. (2016). The challenge of developing autophagy inhibition as a therapeutic strategy. *Cancer Research*, *76*, 5610–5614.

Gewirtz, D. A. (2020). The switch between protective and nonprotective autophagy; implications for autophagy inhibition as a therapeutic strategy in cancer. *Biology*, *9*, 12.

Glick, D., Barth, S., & Macleod, K. F. (2010). Autophagy: Cellular and molecular mechanisms. *The Journal of Pathology*, *221*, 3–12.

Gorgoulis, V., Adams, P. D., Alimonti, A., Bennett, D. C., Bischof, O., Bishop, C., et al. (2019). Cellular senescence: Defining a path forward. *Cell*, *179*, 813–827.

Gosselin, K., Martien, S., Pourtier, A., Vercamer, C., Ostoich, P., Morat, L., et al. (2009). Senescence-associated oxidative DNA damage promotes the generation of neoplastic cells. *Cancer Research*, *69*, 7917–7925.

Grasso, D., Renna, F. J., & Vaccaro, M. I. (2018). Initial steps in mammalian autophagosome biogenesis. *Frontiers in Cell and Developmental Biology*, *6*, 146.

Gui, X., Yang, H., Li, T., Tan, X., Shi, P., Li, M., et al. (2019). Autophagy induction via STING trafficking is a primordial function of the cGAS pathway. *Nature*, *567*, 262–266.

Han, B., Zhang, H., Zhu, Y., Han, X., Wang, Z., Gao, Z., et al. (2020). Subtype-specific risk models for accurately predicting the prognosis of breast cancer using differentially expressed autophagy-related genes. *Aging, 12*, 13318–13337.

Harrison, S. J., Bishton, M., Bates, S. E., Grant, S., Piekarz, R. L., Johnstone, R. W., et al. (2012). A focus on the preclinical development and clinical status of the histone deacetylase inhibitor, romidepsin (depsipeptide, Istodax((R))). *Epigenomics, 4*, 571–589.

Ianzini, F., Kosmacek, E. A., Nelson, E. S., Napoli, E., Erenpreisa, J., Kalejs, M., et al. (2009). Activation of meiosis-specific genes is associated with depolyploidization of human tumor cells following radiation-induced mitotic catastrophe. *Cancer Research, 69*, 2296–2304.

Ikari, S., Lu, S. L., Hao, F., Imai, K., Araki, Y., Yamamoto, Y. H., et al. (2020). Starvation-induced autophagy via calcium-dependent TFEB dephosphorylation is suppressed by Shigyakusan. *PLoS One, 15*, e0230156.

Ionov, Y., Nowak, N., Perucho, M., Markowitz, S., & Cowell, J. K. (2004). Manipulation of nonsense mediated decay identifies gene mutations in colon cancer Cells with microsatellite instability. *Oncogene, 23*, 639–645.

Jakhar, R., Luijten, M. N. H., Wong, A. X. F., Cheng, B., Guo, K., Neo, S. P., et al. (2018). Autophagy governs protumorigenic effects of mitotic slippage-induced senescence. *Molecular Cancer Research, 16*, 1625–1640.

Jiang, Y. H., Wang, H. L., Peng, J., Zhu, Y., Zhang, H. G., Tang, F. Q., et al. (2020). Multinucleated polyploid cardiomyocytes undergo an enhanced adaptability to hypoxia via mitophagy. *Journal of Molecular and Cellular Cardiology, 138*, 115–135.

Jung, T., Bader, N., & Grune, T. (2007). Lipofuscin: Formation, distribution, and metabolic consequences. *Annals of the New York Academy of Sciences, 1119*, 97–111.

Kalejs, M., Ivanov, A., Plakhins, G., Cragg, M. S., Emzinsh, D., Illidge, T. M., et al. (2006). Upregulation of meiosis-specific genes in lymphoma cell lines following genotoxic insult and induction of mitotic catastrophe. *BMC Cancer, 6*, 6.

Kang, C., & Elledge, S. J. (2016). How autophagy both activates and inhibits cellular senescence. *Autophagy, 12*, 898–899.

Kang, C., Xu, Q., Martin, T. D., Li, M. Z., Demaria, M., Aron, L., et al. (2015). The DNA damage response induces inflammation and senescence by inhibiting autophagy of GATA4. *Science, 349*, aaa5612.

Kaushik, S., & Cuervo, A. M. (2012). Chaperone-mediated autophagy: A unique way to enter the lysosome world. *Trends in Cell Biology, 22*, 407–417.

Kawabata, T., & Yoshimori, T. (2020). Autophagosome biogenesis and human health. *Cell Discovery, 6*, 33.

Kohli, L., Kaza, N., Coric, T., Byer, S. J., Brossier, N. M., Klocke, B. J., et al. (2013). 4-Hydroxytamoxifen induces autophagic death through K-Ras degradation. *Cancer Research, 73*, 4395–4405.

Kondratskyi, A., Yassine, M., Slomianny, C., Kondratska, K., Gordienko, D., Dewailly, E., et al. (2014). Identification of ML-9 as a lysosomotropic agent targeting autophagy and cell death. *Cell Death & Disease, 5*, e1193.

Krizhanovsky, V., Yon, M., Dickins, R. A., Hearn, S., Simon, J., Miething, C., et al. (2008). Senescence of activated stellate cells limits liver fibrosis. *Cell, 134*, 657–667.

Kucharewicz, K., Dudkowska, M., Zawadzka, A., Ogrodnik, M., Szczepankiewicz, A. A., Czarnocki, Z., et al. (2018). Simultaneous induction and blockade of autophagy by a single agent. *Cell Death & Disease, 9*, 353.

Kuilman, T., Michaloglou, C., Mooi, W. J., & Peeper, D. S. (2010). The essence of senescence. *Genes & Development, 24*, 2463–2479.

Kwon, Y., Kim, J. W., Jeoung, J. A., Kim, M. S., & Kang, C. (2017). Autophagy is pro-senescence when seen in close-up, but anti-senescence in long-shot. *Molecules and Cells, 40*, 607–612.

Leikam, C., Hufnagel, A. L., Otto, C., Murphy, D. J., Muhling, B., Kneitz, S., et al. (2015). In vitro evidence for senescent multinucleated melanocytes as a source for tumor-initiating cells. *Cell Death & Disease*, *6*, e1711.

Lenain, C., Gusyatiner, O., Douma, S., van den Broek, B., & Peeper, D. S. (2015). Autophagy-mediated degradation of nuclear envelope proteins during oncogene-induced senescence. *Carcinogenesis*, *36*, 1263–1274.

Li, Z., Ji, X., Wang, D., Liu, J., & Zhang, X. (2016). Autophagic flux is highly active in early mitosis and differentially regulated throughout the cell cycle. *Oncotarget*, *7*, 39705–39718.

Li, M., Khambu, B., Zhang, H., Kang, J. H., Chen, X., Chen, D., et al. (2013). Suppression of lysosome function induces autophagy via a feedback down-regulation of MTOR complex 1 (MTORC1) activity. *Journal of Biological Chemistry*, *288*, 35769–35780.

Li, Z., Tian, X., Ji, X., Wang, J., Chen, H., Wang, D., et al. (2020). ULK1-ATG13 and their mitotic phospho-regulation by CDK1 connect autophagy to cell cycle. *PLoS Biology*, *18*, e3000288.

Li, M., Yang, Z., Vollmer, L. L., Gao, Y., Fu, Y., Liu, C., et al. (2015). AMDE-1 is a dual function chemical for autophagy activation and inhibition. *PLoS One*, *10*, e0122083.

Lin, H. X., Qiu, H. J., Zeng, F., Rao, H. L., Yang, G. F., Kung, H. F., et al. (2013). Decreased expression of Beclin 1 correlates closely with Bcl-xL expression and poor prognosis of ovarian carcinoma. *PLoS One*, *8*, e60516.

Liu, J. (2018). The dualistic origin of human tumors. *Seminars in Cancer Biology*, *53*, 1–16.

Liu, J. (2020). The "life code": A theory that unifies the human life cycle and the origin of human tumors. *Seminars in Cancer Biology*, *60*, 380–397.

Liu, Z. J., Italiano, J., Jr., Ferrer-Marin, F., Gutti, R., Bailey, M., Poterjoy, B., et al. (2011). Developmental differences in megakaryocytopoiesis are associated with up-regulated TPO signaling through mTOR and elevated GATA-1 levels in neonatal megakaryocytes. *Blood*, *117*, 4106–4117.

Liu, X. W., Su, Y., Zhu, H., Cao, J., Ding, W. J., Zhao, Y. C., et al. (2010). HIF-1alpha-dependent autophagy protects HeLa cells from fenretinide (4-HPR)-induced apoptosis in hypoxia. *Pharmacological Research*, *62*, 416–425.

Liu, L., Xie, R., Nguyen, S., Ye, M., & McKeehan, W. L. (2009). Robust autophagy/mitophagy persists during mitosis. *Cell Cycle*, *8*, 1616–1620.

Liu, D., Yang, Y., Liu, Q., & Wang, J. (2011). Inhibition of autophagy by 3-MA potentiates cisplatin-induced apoptosis in esophageal squamous cell carcinoma cells. *Medical Oncology*, *28*, 105–111.

Lock, R., Roy, S., Kenific, C. M., Su, J. S., Salas, E., Ronen, S. M., et al. (2011). Autophagy facilitates glycolysis during Ras-mediated oncogenic transformation. *Molecular Biology of the Cell*, *22*, 165–178.

Lundberg, G., Jin, Y., Sehic, D., Ora, I., Versteeg, R., & Gisselsson, D. (2013). Intratumour diversity of chromosome copy numbers in neuroblastoma mediated by on-going chromosome loss from a polyploid state. *PLoS One*, *8*, e59268.

Luo, M. S., Huang, G. J., & Liu, H. B. (2020). An autophagy-related model of 4 key genes for predicting prognosis of patients with laryngeal cancer. *Medicine*, *99*, e21163.

Ma, D., Yu, H., Lin, D., Sun, Y., Liu, L., Liu, Y., et al. (2009). S6K1 is involved in polyploidization through its phosphorylation at Thr421/Ser424. *Journal of Cellular Physiology*, *219*, 31–44.

Marinkovic, M., Sprung, M., Buljubasic, M., & Novak, I. (2018). Autophagy modulation in cancer: Current knowledge on action and therapy. *Oxidative Medicine and Cellular Longevity*, *2018*, 8023821.

Martina, J. A., Diab, H. I., Brady, O. A., & Puertollano, R. (2016). TFEB and TFE3 are novel components of the integrated stress response. *The EMBO Journal*, *35*, 479–495.

Mathiassen, S. G., De Zio, D., & Cecconi, F. (2017). Autophagy and the cell cycle: A complex landscape. *Frontiers in Oncology*, *7*, 51.

Matsunaga, K., Saitoh, T., Tabata, K., Omori, H., Satoh, T., Kurotori, N., et al. (2009). Two Beclin 1-binding proteins, Atg14L and Rubicon, reciprocally regulate autophagy at different stages. *Nature Cell Biology*, *11*, 385–396.

Medina, D. L., Di Paola, S., Peluso, I., Armani, A., De Stefani, D., Venditti, R., et al. (2015). Lysosomal calcium signalling regulates autophagy through calcineurin and TFEB. *Nature Cell Biology*, *17*, 288–299.

Michaloglou, C., Vredeveld, L. C., Soengas, M. S., Denoyelle, C., Kuilman, T., van der Horst, C. M., et al. (2005). BRAFE600-associated senescence-like cell cycle arrest of human naevi. *Nature*, *436*, 720–724.

Milanovic, M., Fan, D. N. Y., Belenki, D., Dabritz, J. H. M., Zhao, Z., Yu, Y., et al. (2018). Senescence-associated reprogramming promotes cancer stemness. *Nature*, *553*, 96–100.

Mirzayans, R., Andrais, B., & Murray, D. (2018). Roles of polyploid/multinucleated giant cancer cells in metastasis and disease relapse following anticancer treatment. *Cancers*, *10*, 118.

Mirzayans, R., Andrais, B., Scott, A., Wang, Y. W., & Murray, D. (2013). Ionizing radiation-induced responses in human cells with differing TP53 status. *International Journal of Molecular Sciences*, *14*, 22409–22435.

Mirzayans, R., & Murray, D. (2020). Intratumor heterogeneity and therapy resistance: Contributions of dormancy, apoptosis reversal (anastasis) and cell fusion to disease recurrence. *International Journal of Molecular Sciences*, *21*, 1308.

Mizushima, N., & Komatsu, M. (2011). Autophagy: Renovation of cells and tissues. *Cell*, *147*, 728–741.

Mizushima, N., Ohsumi, Y., & Yoshimori, T. (2002). Autophagosome formation in mammalian cells. *Cell Structure and Function*, *27*, 421–429.

Morgan, M. J., Fitzwalter, B. E., Owens, C. R., Powers, R. K., Sottnik, J. L., Gamez, G., et al. (2018). Metastatic cells are preferentially vulnerable to lysosomal inhibition. *Proceedings of the National Academy of Sciences of the United States of America*, *115*, E8479–E8488.

Mortensen, M., Soilleux, E. J., Djordjevic, G., Tripp, R., Lutteropp, M., Sadighi-Akha, E., et al. (2011). The autophagy protein Atg7 is essential for hematopoietic stem cell maintenance. *The Journal of Experimental Medicine*, *208*, 455–467.

Mosieniak, G., Adamowicz, M., Alster, O., Jaskowiak, H., Szczepankiewicz, A. A., Wilczynski, G. M., et al. (2012). Curcumin induces permanent growth arrest of human colon cancer cells: Link between senescence and autophagy. *Mechanisms of Ageing and Development*, *133*, 444–455.

Mosieniak, G., & Sikora, E. (2010). Polyploidy: The link between senescence and cancer. *Current Pharmaceutical Design*, *16*, 734–740.

Munoz-Espin, D., Canamero, M., Maraver, A., Gomez-Lopez, G., Contreras, J., Murillo-Cuesta, S., et al. (2013). Programmed cell senescence during mammalian embryonic development. *Cell*, *155*, 1104–1118.

Nakamura, S., Oba, M., Suzuki, M., Takahashi, A., Yamamuro, T., Fujiwara, M., et al. (2019). Suppression of autophagic activity by Rubicon is a signature of aging. *Nature Communications*, *10*, 847.

Nakamura, S., & Yoshimori, T. (2018). Autophagy and longevity. *Molecules and Cells*, *41*, 65–72.

Napolitano, G., & Ballabio, A. (2016). TFEB at a glance. *Journal of Cell Science*, *129*, 2475–2481.

Nardella, C., Clohessy, J. G., Alimonti, A., & Pandolfi, P. P. (2011). Pro-senescence therapy for cancer treatment. *Nature Reviews. Cancer*, *11*, 503–511.

Narita, M., Young, A. R., Arakawa, S., Samarajiwa, S. A., Nakashima, T., Yoshida, S., et al. (2011). Spatial coupling of mTOR and autophagy augments secretory phenotypes. *Science*, *332*, 966–970.

Neurohr, G. E., Terry, R. L., Lengefeld, J., Bonney, M., Brittingham, G. P., Moretto, F., et al. (2019). Excessive cell growth causes cytoplasm dilution and contributes to senescence. *Cell*, *176*, 1083–1097 e1018.

Odle, R. I., Walker, S. A., Oxley, D., Kidger, A. M., Balmanno, K., Gilley, R., et al. (2020). An mTORC1-to-CDK1 switch maintains autophagy suppression during mitosis. *Molecular Cell*, 77, 228–240 e227.

Oku, M., & Sakai, Y. (2018). Three distinct types of microautophagy based on membrane dynamics and molecular machineries. *BioEssays: News and Reviews in Molecular, Cellular and Developmental Biology*, *40*, e1800008.

Onorati, A. V., Dyczynski, M., Ojha, R., & Amaravadi, R. K. (2018). Targeting autophagy in cancer. *Cancer*, *124*, 3307–3318.

Orr-Weaver, T. L. (2015). When bigger is better: The role of polyploidy in organogenesis. *Trends in Genetics*, *31*, 307–315.

Ovadya, Y., & Krizhanovsky, V. (2014). Senescent cells: SASPected drivers of age-related pathologies. *Biogerontology*, *15*, 627–642.

Patel, N. H., Xu, J., Saleh, T., Wu, Y., Lima, S., & Gewirtz, D. A. (2020). Influence of nonprotective autophagy and the autophagic switch on sensitivity to cisplatin in non-small cell lung cancer cells. *Biochemical Pharmacology*, *175*, 113896.

Perez-Hernandez, M., Arias, A., Martinez-Garcia, D., Perez-Tomas, R., Quesada, R., & Soto-Cerrato, V. (2019). Targeting autophagy for cancer treatment and tumor chemosensitization. *Cancers*, *11*, 1599.

Prata, L., Ovsyannikova, I. G., Tchkonia, T., & Kirkland, J. L. (2018). Senescent cell clearance by the immune system: Emerging therapeutic opportunities. *Seminars in Immunology*, *40*, 101275.

Puertollano, R., Ferguson, S. M., Brugarolas, J., & Ballabio, A. (2018). The complex relationship between TFEB transcription factor phosphorylation and subcellular localization. *The EMBO Journal*, *37*, e98804.

Puig, P. E., Guilly, M. N., Bouchot, A., Droin, N., Cathelin, D., Bouyer, F., et al. (2008). Tumor cells can escape DNA-damaging cisplatin through DNA endoreduplication and reversible polyploidy. *Cell Biology International*, *32*, 1031–1043.

Qu, X., Yu, J., Bhagat, G., Furuya, N., Hibshoosh, H., Troxel, A., et al. (2003). Promotion of tumorigenesis by heterozygous disruption of the beclin 1 autophagy gene. *The Journal of Clinical Investigation*, *112*, 1809–1820.

Rajaraman, R., Guernsey, D. L., Rajaraman, M. M., & Rajaraman, S. R. (2006). Stem cells, senescence, neosis and self-renewal in cancer. *Cancer Cell International*, *6*, 25.

Rodier, F., & Campisi, J. (2011). Four faces of cellular senescence. *The Journal of Cell Biology*, *192*, 547–556.

Sabath, N., Levy-Adam, F., Younis, A., Rozales, K., Meller, A., Hadar, S., et al. (2020). Cellular proteostasis decline in human senescence. *Proceedings of the National Academy of Sciences of the United States of America*, *117*, 31902–31913.

Sabisz, M., & Skladanowski, A. (2009). Cancer stem cells and escape from drug-induced premature senescence in human lung tumor cells: Implications for drug resistance and in vitro drug screening models. *Cell Cycle*, *8*, 3208–3217.

Saleh, T., Bloukh, S., Carpenter, V. J., Alwohoush, E., Bakeer, J., Darwish, S., et al. (2020). Therapy-induced senescence: An "old" friend becomes the Enemy. *Cancers*, *12*, 822.

Saleh, T., Carpenter, V. J., Bloukh, S., & Gewirtz, D. A. (2020). Targeting tumor cell senescence and polyploidy as potential therapeutic strategies. *Seminars in Cancer Biology*.

Saleh, T., Tyutyunyk-Massey, L., Murray, G. F., Alotaibi, M. R., Kawale, A. S., Elsayed, Z., et al. (2019). Tumor cell escape from therapy-induced senescence. *Biochemical Pharmacology, 162*, 202–212.

Saleh, T., Tyutyunyk-Massey, L., Patel, N. H., Cudjoe, E. K., Jr., Alotaibi, M., & Gewirtz, D. A. (2020). Studies of non-protective autophagy provide evidence that recovery from therapy-induced senescence is independent of early autophagy. *International Journal of Molecular Sciences, 21*, 1427.

Salmina, K., Bojko, A., Inashkina, I., Staniak, K., Dudkowska, M., Podlesniy, P., et al. (2020). "Mitotic slippage" and extranuclear DNA in cancer chemoresistance: A focus on telomeres. *International Journal of Molecular Sciences, 21*, 2779.

Santana-Codina, N., Mancias, J. D., & Kimmelman, A. C. (2017). The role of autophagy in cancer. *Annual Review of Cancer Biology, 1*, 19–39.

Serrano, M., Lin, A. W., McCurrach, M. E., Beach, D., & Lowe, S. W. (1997). Oncogenic ras provokes premature cell senescence associated with accumulation of p53 and p16INK4a. *Cell, 88*, 593–602.

Sharma, K., Le, N., Alotaibi, M., & Gewirtz, D. A. (2014). Cytotoxic autophagy in cancer therapy. *International Journal of Molecular Sciences, 15*, 10034–10051.

Sharma, S., Yao, H. P., Zhou, Y. Q., Zhou, J., Zhang, R., & Wang, M. H. (2014). Prevention of BMS-777607-induced polyploidy/senescence by mTOR inhibitor AZD8055 sensitizes breast cancer cells to cytotoxic chemotherapeutics. *Molecular Oncology, 8*, 469–482.

Sikora, E., Arendt, T., Bennett, M., & Narita, M. (2011). Impact of cellular senescence signature on ageing research. *Ageing Research Reviews, 10*, 146–152.

Sikora, E., Bielak-Zmijewska, A., & Mosieniak, G. (2019). Targeting normal and cancer senescent cells as a strategy of senotherapy. *Ageing Research Reviews, 55*, 100941.

Sikora, E., Czarnecka-Herok, J., Bojko, A., & Sunderland, P. (2020). Therapy-induced polyploidization and senescence: Coincidence or interconnection? *Seminars in Cancer Biology*. https://doi.org/10.1016/j.semcancer.2020.11.015.

Sikora, E., Mosieniak, G., & Sliwinska, M. A. (2016). Morphological and functional characteristic of senescent cancer cells. *Current Drug Targets, 17*, 377–387.

Sliwinska, M. A., Mosieniak, G., Wolanin, K., Babik, A., Piwocka, K., Magalska, A., et al. (2009). Induction of senescence with doxorubicin leads to increased genomic instability of HCT116 cells. *Mechanisms of Ageing and Development, 130*, 24–32.

Sorokina, I. V., Denisenko, T. V., Imreh, G., Tyurin-Kuzmin, P. A., Kaminskyy, V. O., Gogvadze, V., et al. (2017). Involvement of autophagy in the outcome of mitotic catastrophe. *Scientific Reports, 7*, 14571.

Sosa, M. S., Bragado, P., Debnath, J., & Aguirre-Ghiso, J. A. (2013). Regulation of tumor cell dormancy by tissue microenvironments and autophagy. *Advances in Experimental Medicine and Biology, 734*, 73–89.

Storchova, Z., & Kuffer, C. (2008). The consequences of tetraploidy and aneuploidy. *Journal of Cell Science, 121*, 3859–3866.

Storer, M., Mas, A., Robert-Moreno, A., Pecoraro, M., Ortells, M. C., Di Giacomo, V., et al. (2013). Senescence is a developmental mechanism that contributes to embryonic growth and patterning. *Cell, 155*, 1119–1130.

Takamura, A., Komatsu, M., Hara, T., Sakamoto, A., Kishi, C., Waguri, S., et al. (2011). Autophagy-deficient mice develop multiple liver tumors. *Genes & Development, 25*, 795–800.

Tian, T., Li, X., & Zhang, J. (2019). mTOR signaling in cancer and mTOR inhibitors in solid tumor targeting therapy. *International Journal of Molecular Sciences, 20*, 755.

Visconti, R., Della Monica, R., & Grieco, D. (2016). Cell cycle checkpoint in cancer: A therapeutically targetable double-edged sword. *Journal of Experimental & Clinical Cancer Research, 35*, 153.

Vitale, I., Galluzzi, L., Castedo, M., & Kroemer, G. (2011). Mitotic catastrophe: A mechanism for avoiding genomic instability. *Nature Reviews. Molecular Cell Biology, 12*, 385–392.

Walen, K. H. (2004). Spontaneous cell transformation: Karyoplasts derived from multinucleated cells produce new cell growth in senescent human epithelial cell cultures. *In Vitro Cellular & Developmental Biology. Animal, 40*, 150–158.

Walen, K. H. (2005). Budded karyoplasts from multinucleated fibroblast cells contain centrosomes and change their morphology to mitotic cells. *Cell Biology International, 29*, 1057–1065.

Wang, Q., Wu, P. C., Dong, D. Z., Ivanova, I., Chu, E., Zeliadt, S., et al. (2013). Polyploidy road to therapy-induced cellular senescence and escape. *International Journal of Cancer, 132*, 1505–1515.

Was, H., Barszcz, K., Czarnecka, J., Kowalczyk, A., Bernas, T., Uzarowska, E., et al. (2017). Bafilomycin A1 triggers proliferative potential of senescent cancer cells in vitro and in NOD/SCID mice. *Oncotarget, 8*, 9303–9322.

Was, H., Czarnecka, J., Kominek, A., Barszcz, K., Bernas, T., Piwocka, K., et al. (2018). Some chemotherapeutics-treated colon cancer cells display a specific phenotype being a combination of stem-like and senescent cell features. *Cancer Biology & Therapy, 19*, 63–75.

Weihua, Z., Lin, Q., Ramoth, A. J., Fan, D., & Fidler, I. J. (2011). Formation of solid tumors by a single multinucleated cancer cell. *Cancer, 117*, 4092–4099.

White-Gilbertson, S., & Voelkel-Johnson, C. (2020). Giants and monsters: Unexpected characters in the story of cancer recurrence. *Advances in Cancer Research, 148*, 201–232.

Wilde, L., Tanson, K., Curry, J., & Martinez-Outschoorn, U. (2018). Autophagy in cancer: A complex relationship. *The Biochemical Journal, 475*, 1939–1954.

Xu, M., Pirtskhalava, T., Farr, J. N., Weigand, B. M., Palmer, A. K., Weivoda, M. M., et al. (2018). Senolytics improve physical function and increase lifespan in old age. *Nature Medicine, 24*, 1246–1256.

Yang, S., & Kimmelman, A. C. (2011). A critical role for autophagy in pancreatic cancer. *Autophagy, 7*, 912–913.

Yang, S., Wang, X., Contino, G., Liesa, M., Sahin, E., Ying, H., et al. (2011). Pancreatic cancers require autophagy for tumor growth. *Genes & Development, 25*, 717–729.

Yim, W. W., & Mizushima, N. (2020). Lysosome biology in autophagy. *Cell Discovery, 6*, 6.

Young, A. R., Narita, M., Ferreira, M., Kirschner, K., Sadaie, M., Darot, J. F., et al. (2009). Autophagy mediates the mitotic senescence transition. *Genes & Development, 23*, 798–803.

Zhang, S., Mercado-Uribe, I., & Liu, J. (2014). Tumor stroma and differentiated cancer cells can be originated directly from polyploid giant cancer cells induced by paclitaxel. *International Journal of Cancer, 134*, 508–518.

Zhang, S., Mercado-Uribe, I., Xing, Z., Sun, B., Kuang, J., & Liu, J. (2014). Generation of cancer stem-like cells through the formation of polyploid giant cancer cells. *Oncogene, 33*, 116–128.

Zheng, K., He, Z., Kitazato, K., & Wang, Y. (2019). Selective autophagy regulates cell cycle in cancer therapy. *Theranostics, 9*, 104–125.

CHAPTER SIX

Autophagy and PTEN in DNA damage-induced senescence

Arishya Sharma[a,*] and Alexandru Almasan[a,b,c,*]
[a]Department of Cancer Biology, Lerner Research Institute, Cleveland Clinic, Cleveland, OH, United States
[b]Department of Radiation Oncology, Taussig Cancer Institute, Cleveland Clinic, Cleveland, OH, United States
[c]Case Comprehensive Cancer Center, Case Western Reserve University School of Medicine, Cleveland, OH, United States
*Corresponding authors: e-mail address: sharmaa2@ccf.org; almasaa@ccf.org

Contents

Abstract

The use of DNA-damaging agents such as radiotherapy and chemotherapy has been a mainstay treatment protocol for many cancers, including lung and prostate. Recently, FDA approval of inhibitors of DNA repair, and targeting innate immunity to enhance the efficacy of DNA-damaging agents have gained much attention. Yet, inherent or

Advances in Cancer Research, Volume 150
ISSN 0065-230X
https://doi.org/10.1016/bs.acr.2021.01.006

acquired resistance against DNA-damaging therapies persists as a fundamental drawback. While cancer eradication by causing cancer cell death through induction of apoptosis is the ultimate goal of anti-cancer treatments, autophagy and senescence are two major cellular responses induced by clinically tolerable doses of DNA-damaging therapies. Unlike apoptosis, autophagy and senescence can act as both pro-tumorigenic as well as tumor suppressive mechanisms. DNA damage-induced senescence is associated with a pro-inflammatory secretory phenotype, which contributes to reshaping the tumor- immune microenvironment. Moreover, PTEN (phosphatase and tensin homolog) is a tumor supressor deleted in many tumors, and has been implicated in both senescence and autophagy. This review presents an overview of the literature on the regulation and consequences of DNA damage- induced senescence in cancer cells, with a specific focus on autophagy and PTEN. Both autophagy and senescence occur concurrently in the same cells in response to DNA damaging agents. However, a deterministic relationship between these fundamental processes has been controversial. We present experimental evidence obtained with tumor cells, with a prime focus on two models of cancer, prostate and lung. A better understanding of mechanisms associated with DNA damage-induced cellular senescence is central to fully exploit the potential of DNA-damaging agents against cancer.

1. Introduction

DNA damage can be produced by endogenous, environmental, and clinical therapy stresses, including DNA damage produced as a result of radio- or chemo-therapy (Ciccia & Elledge, 2010). Among the various types of DNA damage, DNA double strand breaks (DSBs) that are generated by direct ionization of DNA bases or indirectly by reaction with reactive oxygen species produced by DNA damaging agents such as radiation, are the most lethal, and will be a prime focus of this review. When mammalian cells encounter DNA damage, they activate an evolutionally conserved pathway, the DNA-damage response, which senses the damage, followed by signal transduction through a cascade of post-translational modifications, and ultimately leading to DNA damage repair in order to maintain genomic integrity. Ataxia-telangiectasia mutated (ATM) is the key kinase involved in the DNA-damage response. DSBs recruit ATM to the damaged DNA sites, where it phosphorylates histone H2AX. H2AX acts as a signal for recruiting the DSB repair-associated factors (Ciccia & Elledge, 2010; Rogakou, Pilch, Orr, Ivanova, & Bonner, 1998).

In addition to H2AX, ATM also phosphorylates CHK1 and CHK2, and leads to subsequent activation of the p53/p21^{WAF1}-dependent cell cycle checkpoint, to allow time to repair the DNA damage, as indicated in Fig. 1 (Smith, Southgate, Tweddle, & Curtin, 2020). In addition, autophagy

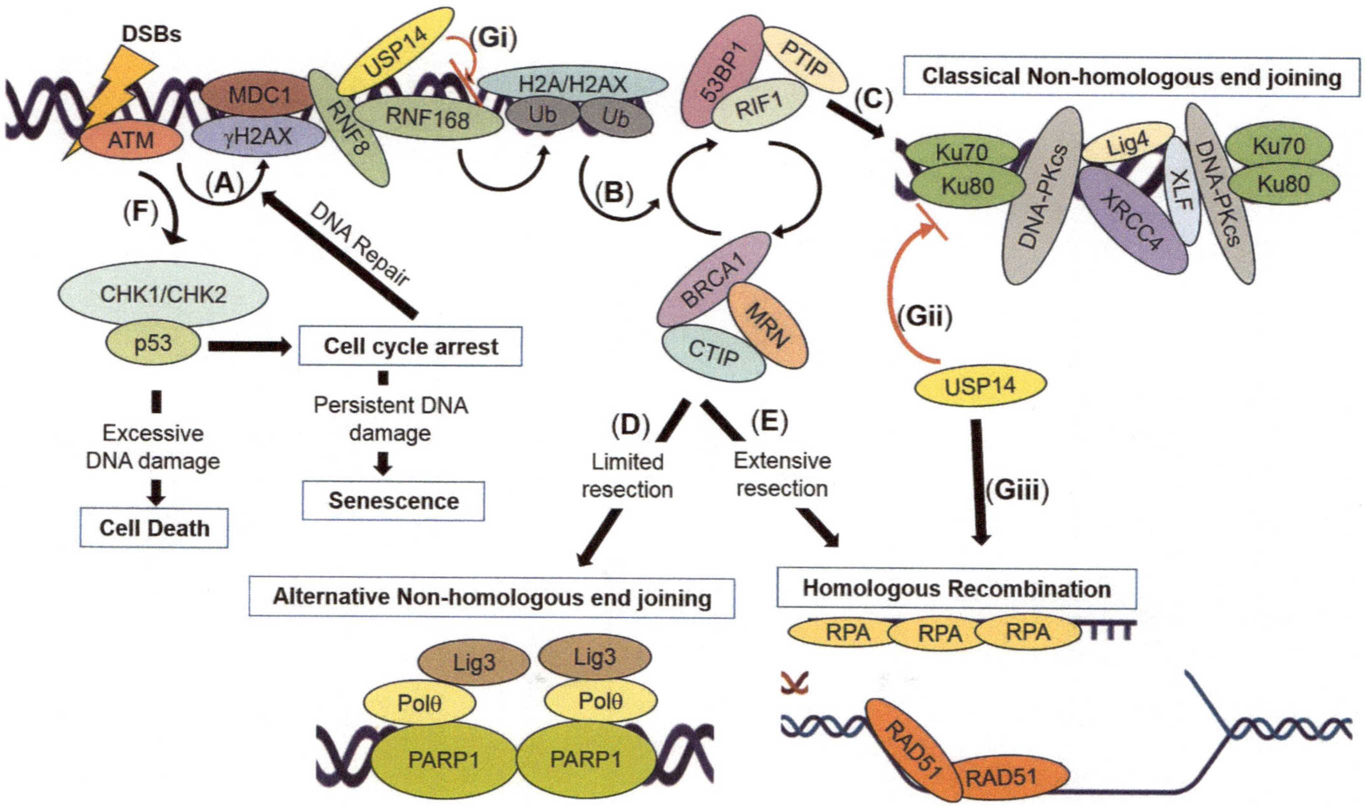

Fig. 1 See figure legend on next page.

is activated to provide the basic bochemical components necessary to carry out essential biological functions (Mizushima, 2007). Cell cycle arrest, DNA repair, and autophagy provide the immediate responses to DNA damage. If the DNA damage is too severe and/or persistent, cells can undergo senescence, permanent cell cycle arrest, or cell death, through apoptosis or mitotic catastrophe (Gewirtz, Alotaibi, Yakovlev, & Povirk, 2016; Patel, Sohal, Manjili, Harrell, & Gewirtz, 2020). Thus, both DNA repair and autophagy are primarily acute, transient events that preceed senescence or cell death, biological processes by which cells are removed from the cell population through a cytostatic or cytotoxic process. Autophagic cell death can also occur when the requirement of autophagy machinery can be direclty established (Denton & Kumar, 2019; Sharma et al., 2013).

Senescent cells, in addition to growth arrest, demonstrate global gene expression alterations, associated with the secretion of inflammatory cytokines, immune modulators, growth factors, and proteases collectively termed the senescence-associated secretory phenotype (SASP). The SASP can be pro tumorigenic, or can act as a tumor suppressive mechanism by reinforcing senescence, immune recognition of the tumor cells or tissue repair (Coppe, Desprez, Krtolica, & Campisi, 2010; Kuilman & Peeper, 2009). Unlike apoptosis, cellular senescence can be either tumor suppressive or oncogenic, depending upon the patho-physiological context and tumor microenvironment. In cancer therapy the net outcome will be a stable tumor or

Fig. 1 Schematic diagram of the cellular response to DNA double-strand breaks. (A) DNA double stand break (DSB) induction leads to activation of a signal transduction cascade that involves activation of ATM that phosphorylates H2AX and MDC1. These early participants recruit the ubiquitin ligases RNF8 and RNF168 to ubiquitylate histones H2A/H2AX. (B) This ubiquitination is important determinant for DSB repair pathway choice. Recruitment of 53BP1 and anti-resection factors RIF1 and PTIP promotes non-homologous end joining (NHEJ), and that of BRCA1/CtIP/MRN promotes homologous recombination (HR). (C) NHEJ depends on the binding of NHEJ-core complex, comprising of the Ku70/Ku80 heterodimer, DNA–PKcs, XRCC4, XLF, and Lig4. (D) If resection is limited, and NHEJ is non-functional alternative-NHEJ (alt-NHEJ) is active, involving PARP1, Polθ, and Lig3. (E) HR is the pathway choice when a homologous sister chromatid is available, and it requires excessive resection of the damaged DNA to generate a single-stranded DNA overhang that is coated by RPA, which is replaced by RAD51, which then invades the sister-chromatid to facilitate homologous template-directed repair. (F) ATM also activates p53-dependent signal transduction that arrests cell cycle to facilitate DNA repair; if the damage is excessive it leads to apoptosis or senescence. (G) USP14 is a deubiquitinase enzyme that regulates DSB at various levels: (i) by antagonizing RNF168-dependent ubiquitination signaling; (ii) by inhibition of Ku70 recruitment to damage sites, and (iii) by inhibition of HR.

a tumor that is shrinking and that may be completely erradicated, depending on the extent of damage and the overall interaction among the various cellular response pathways, including the DNA-damage response, autophagy, and senescence.

2. DNA damage response

DNA damaging anticancer therapy eradicates tumors by inducing lethal DSBs in cells. In tumor cells, DSB repair pathways contribute to resistance against the treatment. Two major pathways are employed by eukaryotic cells for the repair of DSBs, non-homologous end joining (NHEJ) and homologous recombination (HR), as described in Fig. 1. NHEJ is active throughout the cell cycle and is, therefore, the major pathway choice responsible for DSB repair (Davis & Chen, 2013). In contrast, HR depends on the presence of a sister chromatid as a template and is, therefore, restricted to the late S- and G2-phases of the cell cycle (San Filippo, Sung, & Klein, 2008). Thus, an appropriate DNA repair pathway choice is tightly regulated through the cell cycle of both normal and cancer cells to maintain cellular genomic stability. Therefore, disruption of the balance between these two pathways is an attractive strategy to sensitize cancer cells to DNA damaging therapy (Sharma & Almasan, 2020). Ubiquitination of histone H2A and its variant H2AX by E3 ligases RNF8 and RNF168 plays an important part in DNA repair pathway choice by recruiting 53BP1 to DSBs. 53BP1, together with its partner proteins RIF1 (Rap1-interacting factor 1) and PAX transcription activation domain interacting protein (PTIP), inhibits Breast Cancer gene 1 (BRCA1)-CTBP interacting protein (CtIP) complex-dependent DSB end resection (Panier & Boulton, 2014). This promotes a rapid NHEJ of the DSB ends and inhibits the HR pathway (Fig. 1).

Classical NHEJ involves sensing and then binding of the Ku70/Ku80 heterodimer to DNA DSBs (Fig. 1), with subsequent recruitment of DNA-dependent protein kinase catalytic subunit (DNA-PKcs) and end-processing factors leading to repair by the DNA ligase IV/X-ray repair cross-complementing protein 4 (XRCC4)/XRCC4-like factor (XLF) complex (Davis & Chen, 2013). The CtIP complex-dependent DNA end resection is an initial critical step in HR repair as it generates a 3′ single-stranded DNA (Limbo et al., 2007). The 3′ single-stranded DNA tails are subsequently coated by RPA, composed of three subunits, RPA70, RPA32, and RPA14 that are often used as a marker for DNA end processing (Raderschall, Golub, & Haaf, 1999). Rad51 then replaces RPA and forms extended filaments on the single-stranded DNA. These nucleoprotein filaments guide the single-stranded DNA to pair with the homologous template DNA (Fig. 1).

In addition to classical NHEJ, alternative-NHEJ (Alt-NHEJ) also operates in mammalian cells as a back-up mechanism (Fig. 1). Alt-NHEJ is kinetically less efficient than NHEJ, and can cause larger deletions than classical NHEJ, and thus be responsible for translocations and chromosomal rearrangements. Indeed, alt-NHEJ is recognized as a major source of genomic instability. A subtype of Alt-NHEJ utilizes microhomologies at DNA ends, and thus it is called microhomology-mediated end joining (MMEJ) (Boulton & Jackson, 1996; Chiruvella, Liang, & Wilson, 2013; Ma, Kim, Haber, & Lee, 2003). MMEJ is known to be crucial for HR-deficient cancers (Ceccaldi et al., 2015). The understanding of the molecular mechanisms regulating Alt-NHEJ is still an active research area; it relies on different proteins, including factors associated with HR, such as PARP1, MRN, and CtIP (Patterson-Fortin & D'Andrea, 2020). PARP-1 binds to DSBs ends and recruits MRN and CtIP, which promote end resection. Resection generates single-stranded DNA ends, which are coated by 5-hydroxymethylcytosine-binding protein (HMCES), followed by end-bridging and alignment of microhomologies mediated by a PARP1, MRN, and POLQ. Ultimately, nonhomologous 3′ tails are processed by ERCC1/XPF, FEN1, and yet unidentified endonucleases, POLQ fills any gaps, and LIG3/XRCC1 ligate the ends together (Patterson-Fortin & D'Andrea, 2020). MMEJ is increasingly being recognized as more than merely a back-up DSB repair mechanism (Patterson-Fortin & D'Andrea, 2020). Thus, understanding the molecular mechanisms of MMEJ and its role in DSB repair is a major research interest in the field.

3. Cellular senescence

Cellular senescence has been defined as a highly stable cell cycle arrest that is mounted in response to different stresses, including those induced by a persistent DNA damage response (Gorgoulis et al., 2019; Herranz & Gil, 2018). The key signaling players in the process, as shown in Fig. 2, include: p53, $p21^{WAF1}$, $p16^{INK4a}$, retinoblastoma (Rb), and formation of Rb-E2F complexes (Chang et al., 2002; Macip et al., 2002). Persistent activation of the DNA-damage response induces expression of the cyclin-dependent kinase (CDK) inhibitors $p16^{INK4a}$ that inhibits the cyclin D-CDK4/6

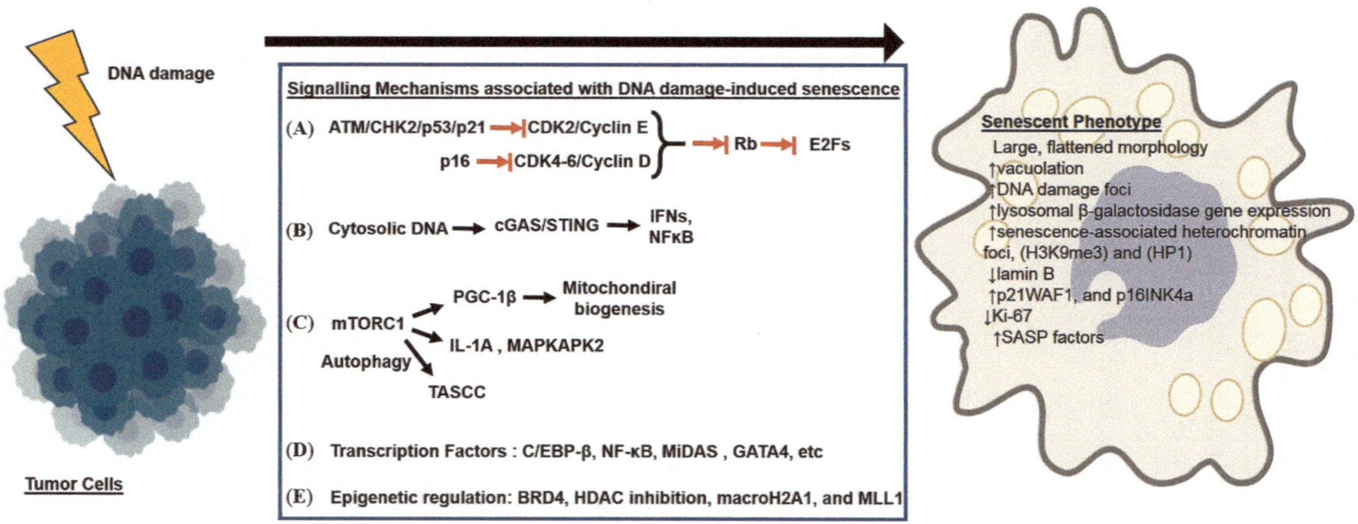

Fig. 2 Signaling and morphological features associated with DNA damage-induced senescence. Exposure of tumor cells to DNA damage activates multiple signal transduction pathways that lead to a senescence phenotype. The important signaling mechanisms include: (A) ATM/p53/p21- and p16-dependent inhibition of Rb. (B) Activation of cytosolic DNA sensing cGAS/STING pathway that leads to IFN response and NFκB activation. (C) mTORC1 activation, leading to mitochondrial biogenesis through PGC1β; or by inducing translation of IL-1A or MAPKAP2, or by coordinating with autophagy through specialized TASCC structures. (D) Upregulation of senescence-promoting transcription factors. (E) Epigenetic alterations.

complexes, and p21^{WAF1} that inhibits cyclin *E*-CDK2 activity. Both of these CDK inhibitors induce senescence by acting on the Rb protein. Rb-mediated heterochromatin formation and silencing of E2F target genes plays a critical role in activating cellular senescence (Narita et al., 2003). The Rb family of pocket proteins also includes p130 and p107 (Dick & Rubin, 2013). These proteins bind to the E2F family of transcription factors to control gene expression. These include E2F1 and E2F4 that play an important role in the outcome of the DNA damage response, particularly on cell cycle control (Almasan et al., 1995; Crosby, Jacobberger, Gupta, Macklis, & Almasan, 2007; DuPree, Mazumder, & Almasan, 2004; Plesca, Crosby, Gupta, & Almasan, 2007).

Some of the typical morphological features of senescent cells, as shown in Fig. 2, include: (i) cell enlargement, flattened shape and vacuolization (Mikula-Pietrasik, Niklas, Uruski, Tykarski, & Ksiazek, 2020), (ii) DNA damage foci, particularly 53BP1 and γH2AX(Rouleau, Patel, Hendzel, Kaufmann, & Poirier, 2010), (iii) upregulation of the lysosomal β-galactosidase gene expression, as measured by senescence-associated β-galactosidase (SA-β-gal) staining (Dimri et al., 1995), (iv) senescence-associated heterochromatin foci, including trimethylated histone H3 lysine 9 (H3K9me3) and heterochromatin protein 1 (HP1) (Aird & Zhang, 2013), (v) downregulation of the nuclear lamina protein, lamin B (Shimi et al., 2011), and (vi) elevated p21^{WAF1}, p16^{INK4a} (Calio et al., 2015) and negative proliferative markers, particularly Ki-67 (Lawless et al., 2010) and SASP (Coppe et al., 2010; Kuilman & Peeper, 2009).

Senescence is of three main types: replicative senescence, associated with telomere attrition, oncogene-induced senescence and therapy-induced premature senescence (Mikula-Pietrasik et al., 2020). Replicative senescence in human fibroblasts is a result of telomere shortening over cell divisions to a degree at which chromosome ends become exposed and mount a chronic DNA damage response sensed as a DSB. Similarly, in tumor cells oncogene-induced senescence leads to a highly proliferative stage that also induces a DNA-damage response because of abnormal usage of DNA replication forks that leads to DNA damage. Oncogene-induced senescence is part of the tumorigenesis barrier imposed by DNA damage checkpoints (Bartkova et al., 2005; Bartkova et al., 2006). Senescence thus provides an effective anticancer mechanism that prevents replication of pre-neoplastic cells. PTEN (phosphatase and tensin homologue)-loss-induced cellular senescence was first reported to be a novel type of cellular senescence that does not induce DNA damage (Alimonti et al., 2010; Trotman et al., 2006).

3.1 Therapy-induced senescence

Therapy-induced senescence emerges as a result of exposure to a number of cancer therapeutics, including DNA-damaging agents (Saleh Bloukh, et al., 2020). Thus, therapy-induced senescence is an important determinant of response to anti-cancer therapy. In general, high doses of DNA damaging agents induce apoptosis and lower doses induce senescence (Y. Wang, Boerma, & Zhou, 2016). Radiotherapy, a prototypical DSB-inducing agent, is a highly effective treatment modality for local control of many, if not most cancer histologies. It has been observed in experimental lung cancer cell lines, with both 0.5–10 Gy (low dose) and > 10Gy (high dose) radiation inducing senescence, whereas apoptosis is mainly observed only at high doses. This response is primarily a function of the amount of DNA damage and the resulting DNA-damage response. A low dose radiation was shown to induce senesence in the H460 cell line (Wang, Boerma, & Zhou, 2016). Similarly, a 6 Gy dose of radiation induced senescence in A549 and H460 cells in a p53-dependent manner (Xu et al., 2018). Prostate cancer cells underwent apoptosis when treated with 250 nM of doxorubicin (Ewald, Peters, Desotelle, Hoffmann, & Jarrard, 2009), however, when they were treated with a 10 times lower drug concentration, cellular senescence was observed (Schwarze, Fu, Desotelle, Kenowski, & Jarrard, 2005). In our findings in the C4–2 and PC3 prostate cancer cells, the lowest dose rate of radiation directly correlated with more extensive DNA damage, resulting in fewer surviving colonies and more γH2AX and 53BP1 DNA damage foci compared to the highest dose rate. Growth control was indeed associated with increased senescence (Chatterjee et al., 2013). These apparently contradicting observations, *i.e.,* low dose rate resulting in more extensive damage compared to high dose rate can be explained by difference in repair kinetics, and are in agreement with previous report where the low dose radiation in human mesenchymal stem cells has resulted in slower repair kinetics as indicate by γH2AX and pATM foci than the high dose radiation (Ulyanenko et al., 2019). It has also been reported that low dose-rate exposures may not be capable of activating ATM-mediated DDR resulting in an increased cell death (Collis et al., 2004). A delayed resolution of DNA double strand breaks has been associated with a higher frequency of mis-joining events, resulting in chromosomal aberrations, as unrepaired, free DNA double strand breaks get more time and opportunity for exchange (Schmid, Zlobinskaya, & Multhoff, 2012; Varga & Aplan, 2005). This is consistent with the reports in which low dose-rate has shown an increased incidence of chromosomal aberrations or mutations and increased cytotoxicity (Amundson & Chen, 1996; Mitchell, Folkard, & Joiner, 2002; Stevens, Bradley, Goodhead, & Hill, 2014). A recent study has attributed an inverted dose-rate effect to changes in the S-phase fraction of cells (Matsuya et al., 2018).

Among chemotherapeutics, Poly (ADP-ribose) polymerase (PARP) inhibitors represent a new class of agents, recently approved for use for prostate cancer therapy (Albert et al., 2007; de Bono et al., 2020; Marshall & Antonarakis, 2020; Sokolova, Yu, & Cheng, 2020). PARP inhibitors have gained considerable interest because of their reported *"synthetic lethality"* (Bryant et al., 2005) in tumor models with defective BRCA1 or BRCA2, which, as a consequence have defective homologous recombination (Brenner et al., 2011) and are thus amenable to use of PARP inhibitors as a monotherapy (Lord, Tutt, & Ashworth, 2015). Radiation induces PARP activity and its inhibition enhances cell death and improves tumor growth delay in irradiated lung cancer models (Albert et al., 2007). Addition of the PARP inhibitor rucaparib significantly increased radiosensitization in prostate cancer cell lines. PARP inhibition led to radiosensitization, irrespective of the p53 status (Chatterjee et al., 2013). Moreover, the combination treatment mounted a robut senescence response that was absent in DU145 cells (Chatterjee et al., 2013) that have a functional *PTEN* allele and a truncated, non-functional Rb (Bookstein, Shew, Chen, Scully, & Lee, 1990). The low dose-rate doses used mimicked those employed in clinical practice for prostate cancer brachytherapy (Nag, 2000) that may employ molecular mechanisms that promote enhanced cancer cell radiosensitivity through enhanced senescence. Similar to our studies, PC3 but not DU145 cells and tumors were shown to contain abundant senescent cells displaying persistent DNA damage foci following the radiation and PARP inhibitor combination treatment (Barreto-Andrade et al., 2011). Radiosensitization by PARP inhibition has been reported also in DNA repair proficient and deficient colon carcinoma tumor cells, with proliferative recovery observed in the senescent cells (Alotaibi et al., 2016).

DNA damaging therapy-induced senescence in cancer cells is a mechanism that allows cells to transiently arrest cell growth in an attempt to repair genomic damage. Subsequently, cells may be eradicated through apoptotic cell death if the extent of damage is extensive or, if the damage is repaired, cells may continue to proliferate. While senescence has been considered a durable form of cell cycle arrest, some senescent cells may recover and start proliferating again and thus share similar properties with dormant cells (Saleh, Tyutyunyk-Massey, & Gewirtz, 2019) as well as stem cells (De Angelis, Francescangeli, La Torre, & Zeuner, 2019), leading to disease recurrence in a more aggressive state (Demaria et al., 2017; Saleh, Bloukh, et al., 2020). As mentioned above, senescent cells also secrete proinflammatory cytokines and chemokines, collectively referred to as the SASP. SASP factors can alter the tissue microenvironment in an autocrine or paracrine manner (Malaquin, Martinez, & Rodier, 2016; Ritschka et al., 2017), leading to tumor suppression by reinforcing senescence to prevent the growth of damaged cells. Transcriptional regulators of the SASP include, CCAAT/enhancer-binding protein β (C/EBP-β), NF-κB, mitochondrial dysfunction-associated senescence (MiDAS) and GATA4 (Gluck & Ablasser, 2019; Kang et al., 2015; Kuilman et al., 2008; Ohanna et al., 2011; Wiley et al., 2016). p38MAPK was shown to induce SASP by acting upstream of NF-κb (Borodkina, Deryabin, Giukova, & Nikolsky, 2018). mTOR also regulates SASP at the post-transcriptional level (Fig. 2) by inducing PGC-1β-dependent mitochondrial biogenesis (Correia-Melo et al., 2016), translation of IL-1A (Laberge et al., 2015) and MAP kinase-activated protein kinase 2 (MAPKAPK2) (Herranz et al., 2015). Further, BRD4, SIRT1, macroH2A1, and MLL1 have been shown to epigenetically regulate SASP (Chen et al., 2015; Hayakawa et al., 2015).

3.2 cGAS/STING and SASP

Recently, the DNA-sensing receptor Cyclic GMP-AMP Synthase (cGAS) and its downstream signaling effector STimulator of INterferon Genes (STING) pathway has been shown to be an important regulator of the SASP and senescence in response to senescence-inducing stresses, including genotoxic stress, shown in Fig. 2 (Gluck et al., 2017; Li & Chen, 2018; Yang, Wang, Ren, Chen, & Chen, 2017). Upon genotoxic insult, cGAS seems to follow a biphasic pattern (Gluck & Ablasser, 2019). Exposure of cells to DNA damaging agents causes aberrant increase in cytosolic DNA, such as through rupture of micronuclei, or leakage of mitochondrial DNA (mtDNA), which causes cGAS/STING pathway activation and subsequently secretion of type 1 interferons and other cytokines (Fenech et al., 2011; Gluck & Ablasser, 2019; Harding et al., 2017; Mackenzie et al., 2017; Yamazaki et al., 2020). In certain contexts, the cGAS/STING-mediated-secretome, including several families of soluble and insoluble factors, particularly NF-κB-dependent cytokines and chemokines, leads to activation of senescence by acting in an autocrine as well as paracrine manner (Moiseeva, Mallette, Mukhopadhyay, Moores, & Ferbeyre, 2006; Yu et al., 2015). Importantly, activation of senescence coincides with further morphological changes in the nucleus, including the downregulation of Lamin B1, resulting in protrusion of chromatin from the nucleus that gives rise to cytosolic chromatin fragments. These chromatin fragments further serve as a stimulus for activation of cGAS in senescent cells (Ivanov et al., 2013). Moreover, DNA damage also inhibits members of the E2F/DP family of transcription factors of cellular nucleases TREX1 and DNase II, that results in aberrant accumulation of self-DNA in the cytosol of senescent cells (Takahashi et al., 2018).

Through SASP, senescent cells shape an inflammatory/immune microenvironment and recruit immune cells to target the senescent cells (Wan, Jiang, & Hao, 2020). Previously, senescent liver cells were shown to attract natural killer (NK) lymphocytes and neutrophils in a SASP-dependent manner (Xue et al., 2007). Further, it was reported that the cGAS–STING pathway is essential for the tumor-suppressive effect of SASP using an NRasV12 liver cancer model. It was shown that the STING null mice had attenuated immune cell infiltration, reduced clearance of senescent cells at the early stage, and increased tumor growth at later stages (Dou et al., 2017). However, the SASP is not always tumor suppressive and it can have dichotomous effects. Thus, it may promote immune-stimulation and inhibit tumorigenesis, and may also drive tumorigenesis. Senescent fibroblasts secreting SASP factors, such as IL-6 and IL-8, have been shown to induce proliferation and invasiveness of epithelial cells (Coppe et al., 2008; Krtolica, Parrinello, Lockett, Desprez, & Campisi, 2001). Similarly, other SASP factors have additional tumorigenic effects: (i) the matrix metalloprotease increases growth factor availability (Liu & Hornsby, 2007), and (ii) the vasculature endothelium growth factor (VEGF) promotes tumor angiogenesis (Coppe, Kauser, Campisi, & Beausejour, 2006). Given the critical roles of cGAS–STING pathway in cellular senescence, such dichotomous roles of the cGAS/STING pathway should be carefully investigated in future studies.

4. Autophagy

Autophagy is an intracellular degradation process that delivers cytoplasmic components to the lysosome. Autophagy can enhance cellular survival during nutrient starvation and other forms of cellular stress, such as DNA damage (Kroemer, Marino, & Levine, 2010). Autophagy signaling is upregulated in advanced cancers (Burdelski et al., 2015; Chen et al., 2015). A vast body of literature demonstrates detrimental effects of targeting autophagy on tumor cell survival in support of a prosurvival role of autophagy in response to stress. Thus, autophagy-associated proteins have gained considerable attention as potential therapeutic anti-cancer targets, either alone or in combination with other conventional therapies, including radiotherapy. However, such extensive preclinical research has not been yet successfully translated into the clinic for use in cancer patients. This may be attributed to lack of availability of specfic autophagy inhibitors, but also to the fact that functional autophagy may be required for inducing immune targeting of cancer cells by several conventional anticancer therapies (Michaud et al., 2011; Rao et al., 2014). We will only discuss in more detail aspects of protective autophagy, when its inhibition leads to sensitization of tumor cells. Aspects of non-protective autophagy have been reviewed recently (Saleh, Tyutyunyk-Massey, et al., 2020).

Autophagy signaling is regulated by autophagy-related genes (ATGs), which are evolutionarily conserved (Klionsky, 2007; Maiuri, Zalckvar, Kimchi, & Kroemer, 2007). The most widely studied autophagy protein is the microtubule associated protein 1 light chain 3 beta (LC3b; referred as LC3) also known as ATG8. Upon induction of autophagy, pro-LC3 gets cleaved at the C-terminus to expose Gly 120 (LC3 I) by a Cysteine protease ATG4b and is further lipidated to its phosphatidylethanolamine-conjugated form (LC3 II), which can then be localized to autophagosomal membranes (Behrends, Sowa, Gygi, & Harper, 2010; Kabeya et al., 2000; Tanida, Sou, Minematsu-Ikeguchi, Ueno, & Kominami, 2006). Other ATGs involved in LC3 modifications are ATG3 and ATG7 that act as E1 and E2 ubiquitin ligases for LC3 (Tanida et al., 2006). The serine/threonine-protein kinase ULK1, the mammalian homologue of yeast ATG1, complexes with the ATG13/FIP200 kinase to mediate mTOR inhibition and autophagy initiation (Jung et al., 2009; Mizushima, 2010). The ubiquitin-binding receptor, Sequestosome 1 (SQSTM1)/p62 participates in autophagic protein degradation through interaction with ubiquitinated protein aggregates that it targets to autophagosomes (Moscat & Diaz-Meco, 2009).

In the context of DNA damage, autophagy is known to regulate genomic stability either indirectly such as through modulation of ROS (Hewitt & Korolchuk, 2017; Karantza-Wadsworth et al., 2007; Mathew et al., 2007) or through selective degradation of nuclear components (Dou et al., 2015; Park et al., 2009). We and others have shown that autophagy directly targets DSB repair-associated proteins (Graeme Hewitt et al., 2016; Sharma et al., 2018; Wang, Zhang, et al., 2016).

4.1 USP14: A link between DNA repair and autophagy

Our studies have identified USP14, a proteasome associated deubiquitinase, as a substrate of autophagy, and a novel regulator of the NHEJ DSB repair pathway (Fig. 1). USP14 was identified through mass spectrometry in an ubiquitin-binding based activity assay, as one of the deubiquitinases that showed increased activity in autophagy-deficient cells. USP14 was recruited to DSB sites, as shown by co-localization analysis with γH2AX. USP14 protein stability and nuclear foci formation increased in autophagy-deficient cells. Moreover, USP14 directly interacted with LC3 and SQSTM1 that led us to conclude that USP14 is a substrate of autophagy. Importantly, USP14 was found to directly interact with, and antagonize RNF168-dependent ubiquitination signaling (Sharma et al., 2018). These and subsequent studies identified USP14 as a critical determinant of DSB repair, and a critical link between autophagy and DSB repair (Sharma et al., 2018; Sharma et al., 2020; Sharma & Almasan, 2020).

4.2 Autophagy and senescence

Senescence often co-exists with autophagy, in response to various stresses, including DNA damage (Huang et al., 2014). While a direct relationship between them is still not fully understood, the two pathways do share multiple common upstream regulators of overlapping signaling mechanisms, such as ROS, ATM, p53, mTOR, and SASP (as shown in Fig. 3, and discussed below).

Deacetylation of lysosome-associated membrane protein 1 (LAMP1) has been one mechanism reported to account for senescence regulation. It drives lipophagy-dependent generation of free fatty acids by Abrus agglutinin to promote senescence through autophagy in prostate cancer (Panda et al., 2020). Our data in prostate cancer cell lines shows a similar association of

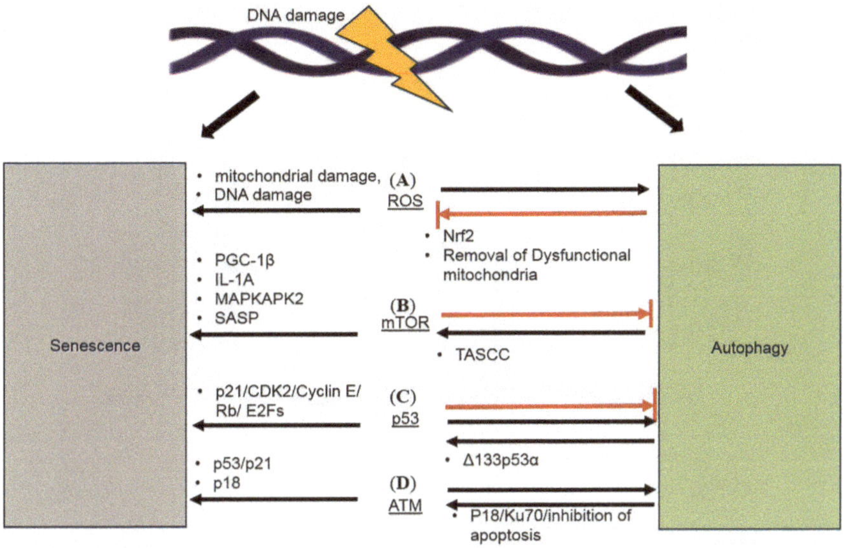

Fig. 3 Overlapping signaling mechanisms between autophagy and senescence. Both autophagy and senescence coexist in response to DNA damaging treatment. (A) ROS leads to mitochondrial and DNA damage, both of which lead to senescence. Reactive oxygen species (ROS)-generating signals simultaneously activate autophagy, which alleviates ROS by removal of damaged mitochondria or by stabilization of Nrf2. (B) mTOR positively regulates senescence by multiple mechanisms (refer to text), mTOR is a negative regulator of autophagy. Autophagy is essential to maintain mTOR activation and ongoing mass synthesis of chemokines and cytokines during SASP through TASCC by providing amino acids. (C) p53 is a central to induction of senescence. p53 Can both activate and inhibit autophagy, depending on cellular context. Autophagy targets Δ133p53α and positively regulates p53. (D) ATM induces senescence by multiple mechanisms (refer to text). ATM simulates autophagy *via* ULK1 activation. Autophagy, in turn, stimulates senescence by antagonizing apoptosis.

DNA-damage-induced senescence with autophagy. Irradiation or PARP inhibitor treatment, alone or in combination in PC3 and C4–2 cells, showed an increased percentage of senescent, β-gal positive cells. Autophagy inhibition in these cell lines by either stable (shRNA) or inducible (doxycycline) inactivation of ATG7 led to a dramatic reduction in the number of β-gal positive cells after treatment (Fig. 4A). In inducible shATG7 LNCaP cells, doxycycline treatment led to decreased number of senescent cells following radiation. In cells without doxycycline treatment, after 6 days following 4 Gy radiation ~40% increase in the number of β-gal positive cells was observed compared to the cells without radiation (Fig. 4B), indicating that these cells are capable of undergoing a normal process of senescence (Chatterjee, 2014).

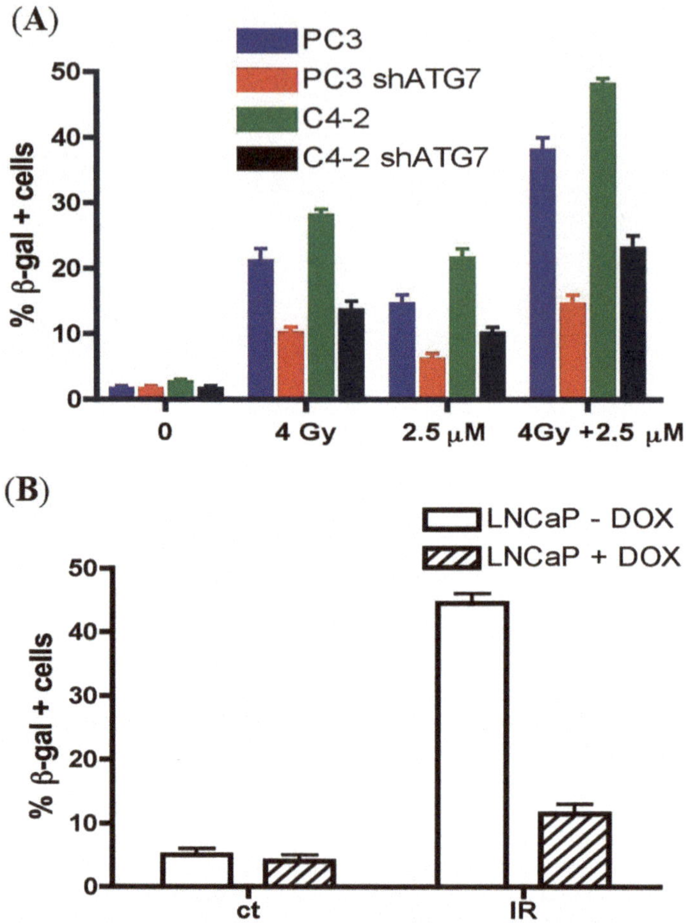

Fig. 4 Autophagy inhibition leads to reduce senescence. (A) Prostate cancer PC3 and C4–2 parental cells rendered ATG7 deficient by stable ATG7 shRNA expression showed β-gal positive cells following 4 Gy ionizing radiation and 2.5 μM rucaparib for PARP inhibition. (B) Doxycycline-inducible shATG7 LNCaP cells displayed β-gal staining following radiation in the presence or absence of doxycycline (Chatterjee, 2014).

4.3 ROS

In addition to direct DNA damage, ionizing radiation and many therapeutic agents can act indirectly by producing ROS (Mitra et al., 2019). Autophagy is a cellular mechanism that is activated to alleviate the ROS-mediated stress by removing damaged organelles and proteins (Azad, Chen, & Gibson, 2009). Thus, dysregulation of autophagy generates mitochondrial dysfunction, which in turn leads to cell senescence.

High ROS levels also mediate p53 activation that is an important regulator of autophagy (Liu, Chen, & St Clair, 2008; Macip et al., 2003; Tasdemir et al., 2008). Upregulation of ULK1 expression in prostate cancer PC-3 cells following tumor protein p53 transfection was another mechanism proposed. Examination of the association between autophagy and p53 using RT-PCR and western blot analysis has found increased levels of ULK1 and reduced cell viability following enhanced transfection of p53 (Wang et al., 2016). Moreover, resveratrol was shown to enhance radiosensitivity in non-small cell lung cancer (NSCLC) cell lines, including A549 and H460, through a ROS-dependent induction of senesence (Luo, Wang, et al., 2013). Further, low-dose ionizing radiation led to elevation of ROS which was shown to induce resistance to a subsequently administered high dose radiation in the A549 cell line. ROS production in this case was shown to induce senesence in an autophagy-dependent manner, and involved NRF2 and its downstream target heme-oxygenase-1 (N. Chen, Wu, Yuan, & Wang, 2015).

4.4 p53

p53-Mediated transcription of $p21^{WAF1}$ and $p16^{INK4a}$ that regulate interaction between cyclins and cyclin-dependent kinases, plays an essential role in establishing and maintaining cell cycle control, senescence, and apoptosis (Artuso, Esteve, Bresil, Vuillaume, & Hall, 1995). Pharmacological activation of senescence by activation of p53 in H1299 cells using Nutlin-3a was shown to induce radiosensitization (Luo, Yount, et al., 2013). Further, using isogenic p53 wild type *vs* p53-null cells, p53 was shown to increase the induction of senesence and radiosensitivity in the H460 cell line (Xu et al., 2018). The role of p53-dependent senescence induction in suppression of prostate cancer growth was demonstrated in response to the cardiotonic compound bufalin in androgen receptor-positive and p53 wild-type human LNCaP prostate cancer cells (Zhang et al., 2018).

Wild-type p53 was essential to induce senescence and radiosensitization by MEK inhibition in KRAS-LKB1–mutated cells (Wang et al., 2011). Moreover, activation of AMPK-dependent autophagy in wild type LKB1 expressing KRAS-mutant cells rescued the cells from radiation-induced senescence and promoted radioresistance. Importantly, inhibition of EGFR led to radiosensitization in NSCLC cell lines by inducing senescence. However, in 45% of the NSCLC lines tested, senescence was shown to occur independently of p53, in a $p16^{INK4a}$-dependent manner. This effect of EGFR inhibition on senescence relied on incresed DNA damage and disruption of the MEK-ERK pathway (Wang et al., 2011). In another study, radioresistance in KRAS-mutant, the most commonly mutated gene in NSCLC, could be overcome by EGFR inhibition. In this report, EGFR–PKCα-was shown to inhibit senescence and promote radioresistance by maintaining condensed chromatin, making it less vulnerable to ionizing radiation-mediated free radical-induced DNA damage. Importantly, co-targeting EGFR and autophagy converted radiation-induced senescence response to a more desirable apoptotic response in the KRAS-null setting (Wang et al., 2014).

4.5 ATM

We have studied the role of chronic expression of an apoptosis inducer, p18-Cyclin E in cell survival through regulation of DNA damage-induced autophagy and senescence. p18-Cyclin E, which is produced as a result of caspase-dependent cleavage of cyclin E in response to irradiation (Mazumder et al., 2002), acts independent of the Cyclin-dependent kinase 2 (Cdk2) as it no longer interacts with it (Mazumder, Plesca, & Almasan, 2007). Once p18-Cyclin E is generated, it interacts with Ku70 and dissociates the Bax-Ku70 complex, and thus provides the first step of Bax activation leading to apoptosis (Mazumder, Plesca, Kinter, & Almasan, 2007). In a subsequent study, we were able to delineate how acute expression of p18-Cyclin E induced apoptosis while its chronic expression facilitated senescence associated with autophagy. Autophagic signaling has been reported to be initiated by activation of AMPK and ULK1 kinase complexes (Jung et al., 2009; Mizushima, 2010). Examining the link between DNA damage and autophagy, we found that DNA damage caused by ionizing radiation led to phosphorylation of ATM at Ser1981, primarily in the nuclei of parental cells. However, in p18-Cyclin *E*-expressing cells, ATM phosphorylation was predominantly in the cytoplasm and not the nucleus, which directly led to increased AMPK and ULK1 activation. This activated autophagy to mediate cell survival, as shown by reduced long-term clonogenic survival in shATG7-expressing cells following radiation in p18-Cyclin *E*-expressing compared to the parental cells expressing non-target shRNA control. β-Gal staining and senescence-associated heterochromatin foci, which were marked with HP1γ, were increased in number and size in p18-Cyclin E-expressing cells following irradiation, but were reduced instead in shATG7-expressing cells. Thus, DNA damage-induced autophagy mediates senescence, which is enhanced in p18-Cyclin E-expressing cells, resulting in decreased long-term clonogenic survival, whereas apoptosis being responsible for reduced cell survival in autophagy-compromised cells.

Following irradiation, p18-Cyclin E and Ku70 co-localized in the cytoplasm, where they were then degraded in a coordinated manner by autophagy (Singh, Matsuyama, Drazba, & Almasan, 2012), which in turn was associated with induction of senescence and reduced cell proliferation. These findings provide more insights into the mechanism of senescence induction since other reports indicate that reduced cytosolic levels of Ku70 are correlated with the senescence phenotype (Ju et al., 2006; Seluanov, Danek, Hause, & Gorbunova, 2007). Thus, autophagy acts as an adaptive mechanism mediating cell survival upon chronic apoptotic stress, in this case provided by stable expression of p18-Cyclin E. Since autophagy can mediate cell survival, inhibiting autophagy can sensitize cells to DNA damage-induced cell death. Whether there was a strict deterministic relationship between autophagy and senescence at the cellular level was not examined.

4.6 mTOR

mTOR is an important upstream regulator of both senescence and autophagy. mTOR-dependent positive regulation of senescence has been described above. mTOR negatively regulates autophagy by directly targeting multiple compenents of the autophagy pathway, including ULK1, Atg13. Atg14, AMBRA1 and NRBF2, p300, WIPI2, UVRAG and Pacer (Dossou & Basu, 2019). The SASP outcome may be dependent on the co-operative action of the mammalian target of rapamycin (mTOR) and the autophagy regulatory complexes. A specialized compartment designated as the TOR-autophagy spatial coupling compartment (TASCC) was shown to be localized on the lysosome. Such proximity coupling of the cellular catabolic and anabolic machinery could facilitate the mass synthesis of these secretory proteins (Narita et al., 2011).

Thus, autophagy provides amino acids for mTOR activation, which was essential for the senescence phenotype by stimulating the massive synthesis of SASP factors.

5. PTEN loss induces senescence

PTEN is a tumor suppressor that antagonizes the PI3K/AKT survival pathway by its phospholipid 3-phosphatase activity, which, in turn, regulates proliferation, migration, and apoptosis (Trotman et al., 2006). Loss of *PTEN* gene leads to subsequent activation of the serine/threonine kinase AKT that is critical for prostate cancer (Majumder & Sellers, 2005). Loss of PTEN function significantly correlates with poor survival, resistance to chemotherapy, and advanced disease in many human cancers, including that of prostate (Mithal et al., 2014). Moreover, aberrant constitutive activation of PI3K/AKT signaling that is largely a consequence of *PTEN* inactivation, has been identified in many cancers, including about 40% of early and 70–100% of advanced prostate cancer (Taylor et al., 2010).

Loss of PTEN induces cellular senescence as a failsafe mechanism to defend against tumorigenesis. A p53-dependent cellular response was reported to suppress PTEN-deficient tumorigenesis in mice (Chen et al., 2005). Conditional inactivation of *TP53* in the mouse prostate failed to produce a tumor phenotype, whereas complete *PTEN* inactivation triggerd non-lethal invasive prostate cancer after a long latency. Combined inactivation of *PTEN* and *TP53* elicited invasive prostate cancer as early as 2 weeks after puberty and was invariably lethal by 7 months of age. Acute *PTEN* inactivation induced growth arrest through the p53-dependent cellular senescence pathway both *in vitro* and *in vivo*, which was rescued by combined loss of *TP53* (Chen et al., 2005). Cellular senescence can thus restrict tumorigenesis in mice *in vivo* by cooperative tumor suppression in which p53 is an essential failsafe protein of PTEN-deficient tumors (Chen et al., 2005). More recently, genetic ablation of *PTEN* in prostatic epithelial cells induced replication stress and senescence *in vivo* (Parisotto et al., 2018). This study reveals that *PTEN* ablation in prostatic luminal epithelial cells of adult mice stimulates prostatic epithelial cell proliferation, leading to prostatic intraepithelial neoplasia (PIN). These proliferating PTEN-deficient cells undergo replication stress and mount a DNA damage response leading to p53 stabilization, which is delayed by Mdm2-mediated p53 down-regulation. The DNA damage response mounted in these *PTEN*-deficient mouse cells thus resembles what is characteristic to oncogene-induced senescence (Parisotto et al., 2018; Parisotto, Grelet, El Bizri, & Metzger, 2019).

5.1 PI3K/Akt/mTOR pathway

Multiple studies have demonstrated a role of PTEN in promoting cellular senescence mediated by inhibition of the PI3K/Akt/mTOR pathway (Jung et al., 2019; Lee et al., 2011). Wild-type PTEN overexpression was shown to increase radiosensitization in H1299 cells, which was accompanied by G2/M cell cycle arrest as a consequence of $p21^{WAF1}$ leading to CDK1 inactivation (Park et al., 2008). Moreover, PTEN knockdown in endogenous wild-type PTEN-containing H460 cells failed to induce cell-cycle arrest in response to ionizing radiation (Jung, Kang, Kim, & Kim, 2010). These findings suggest a role of the PTEN-PI3K/AKT-p21-CDK1 pathway in regulating the outcome of irradiation in NSCLC cells. In addition, mTORC1 is an important regulator of both autophagy and senescence. Inhibition of mTOR led to radiosensitization in multiple cancer cell lines, including those of lung cancer by induction of senescence in an autophagy-dependent manner (Nam et al., 2013).

Downregulation of PTEN mediated premature senescence induced by the radiomimetic bleomycin in lung cancer cells by suppressing autophagy and thus decreasing tumorigenesis. PTEN depletion in A549 cells, followed by stimulation with bleomycin for 5 days led to activation of key downstream PI3K/Akt/mTOR pathway components. Importantly, silencing PTEN reduced autophagy and accelerated senescence of A549 cells. Autophagy levels were also increased and senescence markers were reduced after inhibiting mTOR. This study concludes that downregulation of PTEN mediates bleomycin-induced premature senescence in lung cancer cells by suppressing autophagy through the PI3K/Akt/mTOR pathway (Sai, Qin, Wu, Zhao, & Bian, 2020). A recent report showed that mTOR kinase, a major component of mTORC1 and mTORC2, leads to PTEN-loss-induced cellular senescence by phosphorylating p53 (Jung et al., 2019). It thus acts as a critical relay molecule downstream of PI3K/AKT and upstream of p53 in PTEN-induced senescence. mTOR kinase binds p53 and phosphorylates it on Ser15. mTORC1 and mTORC2 compete with MDM2, an E3 ubiquiting ligase for p53, and thus increase the stability of p53 to induce cellular senescence through accumulation of the cell cycle and CDK-inhibitor $p21^{WAF1}$ (Jung et al., 2019). In PTEN-knockout mouse embryonic fibroblasts, PTEN deficiency also induced mTORC1 and mTORC2 to bind to p53 instead of MDM2, leading to cellular senescence. This study concludes that mTOR plays a critical role in switching cells from proliferation to senescence signaling *via* a direct link between the growth-promoting activity of AKT and the growth-suppressing activity of p53 (Jung et al., 2019).

5.2 PARP inhibition

Similar to BRCA1/2-defective tumor cells, *PTEN*-null prostate cancer cells have been reported to be sensitive to PARP inhibition. Ionizing radiation and other DNA-damaging agents are known to induce PARP1 expression and levels of PARP are higher in these tumors (Schiewer et al., 2012; Zaremba et al., 2011). Senescence is known to represent an important response to both radiation and PARP inhibitors that impacts on prostate cancer cell proliferation and ultimately clonogenic survival. *PTEN* null PC3, LNCaP, and C4–2 cell lines showed a radiation and PARP inhibitor dose-dependent increase in β-gal-positive cells (Chatterjee et al., 2013). In contrast, DU145 cells that have a wild-type *PTEN* allele showed almost no senescent cells even at the highest doses of treatment used either with radiation or rucaparib. The combination index for the senescence β-gal staining indicated a moderate-strong synergy in PC3, LNCaP, and C4–2 cells. The senescence characteristics were sustained up to at least 12 days, with a slight increase in the number of β-gal-positive cells. Importantly, these data indicate that senescence contributes to decreased survival of *PTEN*-deficient cells and that the extent of senescence correlates with loss of clonogenic cell survival (Chatterjee et al., 2013).

5.3 TMPRSS2-ERG

The fusion between *TMPRSS2*, an androgen-regulated oncogene, and an ETS transcription factor estrogen-regulated gene, ERG, generated by an interstitial deletion on chromosome 21 or by reciprocal translocation is the most common prostate cancer fusion gene. It is present in ~50% of early stage prostate cancer (Rubin, Maher, & Chinnaiyan, 2011). VCaP cells that express *TMPRSS2-ERG* endogenously were radiosensitized by rucaparib and acquired β-gal-positive cells following radiation and PARP inhibition (Chatterjee et al., 2013). These cells are dependent on *TMPRSS2-ERG* as they stop proliferating when it is depleted by siRNA. Therefore, PC3 cells that lack *TMPRSS2-ERG* were transfected with *TMPRSS2-ERG*. Its stable expression did not have a significant effect on radiosensitivity. However, when radiation was administered together with rucaparib, the number of colonies was reduced significantly (Chatterjee et al., 2013). This effect was comparable to what was obtained in PC3 cells treated with the

DNA-PKcs inhibitor NU7441. NU7441 alone did not induce senescence in either cell line. The NU7441 and rucaparib combination further radiosensitized these cells, to an extent comparable to cells expressing the *TMPRSS2-ERG* fusion gene treated with rucaparib. In combination with radiation, rucaparib treatment significantly increased the number of senescent cells in *TMPRSS2-ERG*-expressing PC3 cells. These findings indicate that DNA-PKcs activity is critical for avoiding senescence, as indicated by the increased number of senescent cells in *TMPRSS2-ERG*-expressing cells or when DNA-PKcs is inhibited (Chatterjee et al., 2015). Similarly, in a recent study "PTEN low" breast cancer basal-like cells were highly sensitive to the WEE1 inhibitor, AZD1775, and could not recover following AZD1775 removal, while "PTEN high" cells recovered. Deletion of DNA-PK or PTEN sensitized recovering cells to AZD1775. This finding was linked to induction of apoptosis as well as an increased number of β-gal-positive cells (Brunner et al., 2020).

5.4 AKT

AKT overactivation was reported to suppress DNA repair also through the p70S6 kinase-dependent downregulation of MRE11 (Piscitello et al., 2018). In response to radiation treatment, PTEN deletion in HCT116 and DLD1 colon carcinoma cells led to suppression of CHK1 and CHK2 activation, impaired G2 checkpoint, and radiosensitization. These defects were associated with reduced expression of components of the MRE11/RAD50/NBS1 DNA damage response complex (MRE11, RAD50, and NBS1). Consistent with a reduction of this DNA repair complex function, PTEN-deficient cells failed to resect DSBs efficiently after irradiation and showed diminished DNA repair proficiency through homologous recombination. MRE11 was highly unstable in PTEN-deficient cells, but its stability could be restored by inhibiting mTORC1 or p70S6 kinase, downstream kinases whose activities are stimulated by AKT, or by mutating a residue in MRE11 that is phosphorylated by p70S6 kinase. In primary human fibroblasts, activated AKT suppressed this DNA repair complex expression to enhance RAS-induced DNA damage and thus reinforce oncogene-induced senescence (Piscitello et al., 2018).

In our studies, we found that AKT inhibition, similar to USP14 block, rescued the activity of NHEJ- DNA damage and repair proteins in autophagy- and PTEN-deficient prostate cancer cells. These findings reveal a novel PTEN/Akt-dependent regulation of NHEJ by USP14 (Sharma et al., 2020). A role for USP14 in senescence has not yet been examined. However, USP14 shares many properties with the USP17 deubiquitinase. It has been reported that inhibition of USP7 activity selectively eliminates senescent cells, in part through restoration of p53 activity (He et al., 2020).

5.5 PTEN regulation

Besides PTEN complete inactivation, its partial inactivation by transcriptional, translational, and post-translational control may also play a role, particularly in lung cancer (Gkountakos et al., 2019). Thus, its partial inactivation may be accomplished by noncoding RNAs that can also play a role in drug resistance and how they affect it. A negative regulation of PTEN by micro-RNA (miR)-221 and its association with drug resistance and cellular senescence was reported in lung cancer cells. The expression levels of miR-221 in different lung cancer cell lines (H226, H1299, and A549) were measured, with miR-221 levels found increased most in cisplatin-resistant A549 cells. Cell viability and senescence were established in H1299 and A549 cells transfected to modulate (up or down) miR-221 expression. Anti-miR-221 inhibited their proliferation and induced senescence (Wang, Zhu, Xu, Qian, & Zheng, 2018). Other miRs have been also reported to regulate PTEN in lung cancer, including miR-21, miR-29b, miR-92a, and miR-494 [reviewed in (Gkountakos et al., 2019)].

Foxo3a is a transcription factor controlled by AKT that has been reported to be regulated by autophagy (Fitzwalter et al., 2018). We have shown that PTEN expression is directly regulated by FOXO3a, with FOXO3a binding the *PTEN* promoter. Therapy resistant cells (in this case to an inhibitor of Bruton's tyrosine kinase, ibrutinib) had lower levels of FOXO3a and PTEN leading to increased Akt activation in leukemic cells. (Kapoor et al., 2019).

6. Concluding remarks

Senescence is a physiologically relevant phenomenon associated with DNA damaging therapy, that critically influences the overall clinical response. However, we still lack a complete understanding of the process, such as whether DNA damage *per se* causes senescence. Further, the cGAS/STING signaling has been identified as a crucial link between SASP and DNA damage. Rigorous investigation of the dependence of this relationship on cellular/genetic contexts, and on the stress- and tumor-type would allow to better understand the potential of combining DNA-damaging therapies with immune checkpoint inhibitors. In the context of radiation treatment, autophagy and senescence often co-exist. Moreover, the two pathways share common key regulators. However, whether a direct causal relationship exists between the two pathways is still not clear. Moreover, similar to DNA damage response and autophagy, therapy-induced senescence acts as a double-edged sword in the treatment of cancer. On one hand, senescence is a safe means to control cancer growth and attract immune cells to attack cancer cells at tolerable doses of DNA damaging agents. On the other hand, senescence-induced proinflammatory tumor micro-environment can contribute to increased invasiveness of cancer cells. Thus, a better understanding of the mechanisms associated with activation of senescence, careful examination of the factors or contexts enhancing tumor suppressive functions of senescence, and ultimately, well-designed pre-clinical and clinical studies would be needed to facilitate the development of effective strategies to eradicate senescent cancer cells.

Acknowledgments

The authors would like to thank for support the National Institutes of Health (NIH, CA184137), Department of Defense (DOD, W81XWH-20-1-0462), and Velosano Bike Ride.

Conflicts of interest

The authors declare no conflict of interest.

References

Aird, K. M., & Zhang, R. (2013). Detection of senescence-associated heterochromatin foci (SAHF). *Methods in Molecular Biology*, *965*, 185–196. https://doi.org/10.1007/978-1-62703-239-1_12.

Albert, J. M., Cao, C., Kim, K. W., Willey, C. D., Geng, L., Xiao, D., et al. (2007). Inhibition of poly(ADP-ribose) polymerase enhances cell death and improves tumor growth delay in irradiated lung cancer models. *Clinical Cancer Research*, *13*(10), 3033–3042. 13/10/3033 [pii]. https://doi.org/10.1158/1078-0432.CCR-06-2872.

Alimonti, A., Nardella, C., Chen, Z., Clohessy, J. G., Carracedo, A., Trotman, L. C., et al. (2010). A novel type of cellular senescence that can be enhanced in mouse models and human tumor xenografts to suppress prostate tumorigenesis. *The Journal of Clinical Investigation*, *120*(3), 681–693. https://doi.org/10.1172/JCI40535.

Almasan, A., Yin, Y., Kelly, R. E., Lee, E. Y., Bradley, A., Li, W., et al. (1995). Deficiency of retinoblastoma protein leads to inappropriate S-phase entry, activation of E2F-responsive genes, and apoptosis. *Proceedings of the National Academy of Sciences of the United States of America*, *92*(12), 5436–5440. https://doi.org/10.1073/pnas.92.12.5436.

Alotaibi, M., Sharma, K., Saleh, T., Povirk, L. F., Hendrickson, E. A., & Gewirtz, D. A. (2016). Radiosensitization by PARP inhibition in DNA repair proficient and deficient tumor cells: Proliferative recovery in senescent cells. *Radiation Research*, *185*(3), 229–245. https://doi.org/10.1667/RR14202.1.

Amundson, S. A., & Chen, D. J. (1996). Inverse dose-rate effect for mutation induction by gamma-rays in human lymphoblasts. *International Journal of Radiation Biology*, *69*(5), 555–563. https://doi.org/10.1080/095530096145562.

Artuso, M., Esteve, A., Bresil, H., Vuillaume, M., & Hall, J. (1995). The role of the Ataxia telangiectasia gene in the p53, WAF1/CIP1(p21)- and GADD45-mediated response to DNA damage produced by ionising radiation. *Oncogene*, *11*(8), 1427–1435.

Azad, M. B., Chen, Y., & Gibson, S. B. (2009). Regulation of autophagy by reactive oxygen species (ROS): Implications for cancer progression and treatment. *Antioxidants & Redox Signaling*, *11*(4), 777–790. https://doi.org/10.1089/ARS.2008.2270.

Barreto-Andrade, J. C., Efimova, E. V., Mauceri, H. J., Beckett, M. A., Sutton, H. G., Darga, T. E., et al. (2011). Response of human prostate cancer cells and tumors to combining PARP inhibition with ionizing radiation. *Molecular Cancer Therapeutics*, *10*(7), 1185–1193. 1535–7163.MCT-11-0061 [pii]. https://doi.org/10.1158/1535-7163.MCT-11-0061.

Bartkova, J., Horejsi, Z., Koed, K., Kramer, A., Tort, F., Zieger, K., et al. (2005). DNA damage response as a candidate anti-cancer barrier in early human tumorigenesis. *Nature*, *434*(7035), 864–870. https://doi.org/10.1038/nature03482.

Bartkova, J., Rezaei, N., Liontos, M., Karakaidos, P., Kletsas, D., Issaeva, N., et al. (2006). Oncogene-induced senescence is part of the tumorigenesis barrier imposed by DNA damage checkpoints. *Nature*, *444*(7119), 633–637. https://doi.org/10.1038/nature 05268.

Behrends, C., Sowa, M. E., Gygi, S. P., & Harper, J. W. (2010). Network organization of the human autophagy system. *Nature*, *466*(7302), 68–76. nature09204 [pii]. https://doi.org/10.1038/nature09204.

Bookstein, R., Shew, J. Y., Chen, P. L., Scully, P., & Lee, W. H. (1990). Suppression of tumorigenicity of human prostate carcinoma cells by replacing a mutated RB gene. *Science*, *247*(4943), 712–715. https://doi.org/10.1126/science.2300823.

Borodkina, A. V., Deryabin, P. I., Giukova, A. A., & Nikolsky, N. N. (2018). "Social life" of senescent cells: What is SASP and why study it? *Acta Naturae*, *10*(1), 4–14. PMID: 29713514.

Boulton, S. J., & Jackson, S. P. (1996). Saccharomyces cerevisiae Ku70 potentiates illegitimate DNA double-strand break repair and serves as a barrier to error-prone DNA repair pathways. *The EMBO Journal*, *15*(18), 5093–5103. PMID: 8890183.

Brenner, J. C., Ateeq, B., Li, Y., Yocum, A. K., Cao, Q., Asangani, I. A., et al. (2011). Mechanistic rationale for inhibition of poly(ADP-ribose) polymerase in ETS gene fusion-positive prostate cancer. *Cancer Cell*, *19*(5), 664–678. S1535-6108(11)00156-5 [pii]. https://doi.org/10.1016/j.ccr.2011.04.010.

Brunner, A., Suryo Rahmanto, A., Johansson, H., Franco, M., Viiliainen, J., Gazi, M., et al. (2020). PTEN and DNA-PK determine sensitivity and recovery in response to WEE1 inhibition in human breast cancer. *eLife*, *9*. https://doi.org/10.7554/eLife.57894.

Bryant, H. E., Schultz, N., Thomas, H. D., Parker, K. M., Flower, D., Lopez, E., et al. (2005). Specific killing of BRCA2-deficient tumours with inhibitors of poly(ADP-ribose) polymerase. *Nature*, *434*(7035), 913–917. nature03443 [pii]. https://doi.org/10.1038/nature03443.

Burdelski, C., Reiswich, V., Hube-Magg, C., Kluth, M., Minner, S., Koop, C., et al. (2015). Cytoplasmic accumulation of Sequestosome 1 (p62) is a predictor of biochemical recurrence, rapid tumor cell proliferation, and genomic instability in prostate cancer. *Clinical Cancer Research*, *21*(15), 3471–3479. https://doi.org/10.1158/1078-0432.ccr-14-0620.

Calio, A., Zamo, A., Ponzoni, M., Zanolin, M. E., Ferreri, A. J., Pedron, S., et al. (2015). Cellular senescence markers p16INK4a and p21CIP1/WAF are predictors of Hodgkin lymphoma outcome. *Clinical Cancer Research*, *21*(22), 5164–5172. https://doi.org/10.1158/1078-0432.CCR-15-0508.

Ceccaldi, R., Liu, J. C., Amunugama, R., Hajdu, I., Primack, B., Petalcorin, M. I., et al. (2015). Homologous-recombination-deficient tumours are dependent on Pol theta-mediated repair. *Nature, 518*(7538), 258–262. https://doi.org/10.1038/nature14184.

Chang, B. D., Swift, M. E., Shen, M., Fang, J., Broude, E. V., & Roninson, I. B. (2002). Molecular determinants of terminal growth arrest induced in tumor cells by a chemotherapeutic agent. *Proceedings of the National Academy of Sciences of the United States of America, 99*(1), 389–394. https://doi.org/10.1073/pnas.012602599.

Chatterjee, P. (2014). *DNA repair defects in prostate cancer cells in response to radiation*. PhD Kent State University. Retrieved from https://etd.ohiolink.edu/.

Chatterjee, P., Choudhary, G. S., Alswillah, T., Xiong, X., Heston, W. D., Magi-Galluzzi, C., et al. (2015). The TMPRSS2-ERG gene fusion blocks XRCC4-mediated nonhomologous end-joining repair and radiosensitizes prostate cancer cells to PARP inhibition. *Molecular Cancer Therapeutics, 14*(8), 1896–1906. https://doi.org/10.1158/1535-7163.MCT-14-0865.

Chatterjee, P., Choudhary, G. S., Sharma, A., Singh, K., Heston, W. D., Ciezki, J., et al. (2013). PARP inhibition sensitizes to low dose-rate radiation TMPRSS2-ERG fusion gene-expressing and PTEN-deficient prostate cancer cells. *PLoS One, 8*(4), e60408. https://doi.org/10.1371/journal.pone.0060408.

Chen, H., Ruiz, P. D., McKimpson, W. M., Novikov, L., Kitsis, R. N., & Gamble, M. J. (2015). MacroH2A1 and ATM play opposing roles in paracrine senescence and the senescence-associated secretory phenotype. *Molecular Cell, 59*(5), 719–731. https://doi.org/10.1016/j.molcel.2015.07.011.

Chen, Z., Trotman, L. C., Shaffer, D., Lin, H. K., Dotan, Z. A., Niki, M., et al. (2005). Crucial role of p53-dependent cellular senescence in suppression of Pten-deficient tumorigenesis. *Nature, 436*(7051), 725–730. https://doi.org/10.1038/nature03918.

Chen, S., Wang, C., Sun, L., Wang, D. L., Chen, L., Huang, Z., et al. (2015). RAD6 promotes homologous recombination repair by activating the autophagy-mediated degradation of heterochromatin protein HP1. *Molecular and Cellular Biology, 35*(2), 406–416. https://doi.org/10.1128/mcb.01044-14.

Chen, N., Wu, L., Yuan, H., & Wang, J. (2015). ROS/autophagy/Nrf2 pathway mediated low-dose radiation induced radio-resistance in human lung adenocarcinoma A549 cell. *International Journal of Biological Sciences, 11*(7), 833–844. https://doi.org/10.7150/ijbs.10564.

Chiruvella, K. K., Liang, Z., & Wilson, T. E. (2013). Repair of double-strand breaks by end joining. *Cold Spring Harbor Perspectives in Biology, 5*(5), a012757. https://doi.org/10.1101/cshperspect.a012757.

Ciccia, A., & Elledge, S. J. (2010). The DNA damage response: Making it safe to play with knives. *Molecular Cell, 40*(2), 179–204. https://doi.org/10.1016/j.molcel.2010.09.019.

Collis, S. J., Schwaninger, J. M., Ntambi, A. J., Keller, T. W., Nelson, W. G., Dillehay, L. E., et al. (2004). Evasion of early cellular response mechanisms following low level radiation-induced DNA damage. *The Journal of Biological Chemistry, 279*(48), 49624–49632. https://doi.org/10.1074/jbc.M409600200.

Coppe, J. P., Desprez, P. Y., Krtolica, A., & Campisi, J. (2010). The senescence-associated secretory phenotype: The dark side of tumor suppression. *Annual Review of Pathology, 5*, 99–118. https://doi.org/10.1146/annurev-pathol-121808-102144.

Coppe, J. P., Kauser, K., Campisi, J., & Beausejour, C. M. (2006). Secretion of vascular endothelial growth factor by primary human fibroblasts at senescence. *The Journal of Biological Chemistry, 281*(40), 29568–29574. https://doi.org/10.1074/jbc.M603307200.

Coppe, J. P., Patil, C. K., Rodier, F., Sun, Y., Munoz, D. P., Goldstein, J., et al. (2008). Senescence-associated secretory phenotypes reveal cell-nonautonomous functions of oncogenic RAS and the p53 tumor suppressor. *PLoS Biology, 6*(12), 2853–2868. https://doi.org/10.1371/journal.pbio.0060301.

Correia-Melo, C., Marques, F. D., Anderson, R., Hewitt, G., Hewitt, R., Cole, J., et al. (2016). Mitochondria are required for pro-ageing features of the senescent phenotype. *EMBO Journal*, *35*(7), 724–742. https://doi.org/10.15252/embj.201592862.

Crosby, M. E., Jacobberger, J., Gupta, D., Macklis, R. M., & Almasan, A. (2007). E2F4 regulates a stable G2 arrest response to genotoxic stress in prostate carcinoma. *Oncogene*, *26*(13), 1897–1909. https://doi.org/10.1038/sj.onc.1209998.

Davis, A. J., & Chen, D. J. (2013). DNA double strand break repair via non-homologous end-joining. *Translational Cancer Research*, *2*(3), 130–143. https://doi.org/10.3978/j.issn.2218-676X.2013.04.02.

De Angelis, M. L., Francescangeli, F., La Torre, F., & Zeuner, A. (2019). Stem cell plasticity and dormancy in the development of cancer therapy resistance. *Frontiers in Oncology*, *9*, 626. https://doi.org/10.3389/fonc.2019.00626.

de Bono, J., Mateo, J., Fizazi, K., Saad, F., Shore, N., Sandhu, S., et al. (2020). Olaparib for metastatic castration-resistant prostate cancer. *The New England Journal of Medicine*, *382*(22), 2091–2102. https://doi.org/10.1056/NEJMoa1911440.

Demaria, M., O'Leary, M. N., Chang, J., Shao, L., Liu, S., Alimirah, F., et al. (2017). Cellular senescence promotes adverse effects of chemotherapy and cancer relapse. *Cancer Discovery*, 7(2), 165–176. https://doi.org/10.1158/2159-8290.CD-16-0241.

Denton, D., & Kumar, S. (2019). Autophagy-dependent cell death. *Cell Death and Differentiation*, *26*(4), 605–616. https://doi.org/10.1038/s41418-018-0252-y.

Dick, F. A., & Rubin, S. M. (2013). Molecular mechanisms underlying RB protein function. *Nature Reviews. Molecular Cell Biology*, *14*(5), 297–306. https://doi.org/10.1038/nrm3567.

Dimri, G. P., Lee, X., Basile, G., Acosta, M., Scott, G., Roskelley, C., et al. (1995). A biomarker that identifies senescent human cells in culture and in aging skin in vivo. *Proceedings of the National Academy of Sciences of the United States of America*, *92*(20), 9363–9367. https://doi.org/10.1073/pnas.92.20.9363.

Dossou, A. S., & Basu, A. (2019). The emerging roles of mTORC1 in macromanaging autophagy. *Cancers (Basel)*, *11*(10), 1–17. https://doi.org/10.3390/cancers11101422.

Dou, Z., Ghosh, K., Vizioli, M. G., Zhu, J., Sen, P., Wangensteen, K. J., et al. (2017). Cytoplasmic chromatin triggers inflammation in senescence and cancer. *Nature*, *550*(7676), 402–406. https://doi.org/10.1038/nature24050.

Dou, Z., Xu, C., Donahue, G., Shimi, T., Pan, J.-A., Zhu, J., et al. (2015). Autophagy mediates degradation of nuclear lamina. *Nature*, *527*, 105. https://doi.org/10.1038/nature15548.

DuPree, E. L., Mazumder, S., & Almasan, A. (2004). Genotoxic stress induces expression of E2F4, leading to its association with p130 in prostate carcinoma cells. *Cancer Research*, *64*(13), 4390–4393. https://doi.org/10.1158/0008-5472.CAN-03-3695.

Ewald, J. A., Peters, N., Desotelle, J. A., Hoffmann, F. M., & Jarrard, D. F. (2009). A high-throughput method to identify novel senescence-inducing compounds. *Journal of Biomolecular Screening*, *14*(7), 853–858. https://doi.org/10.1177/1087057109340314.

Fenech, M., Kirsch-Volders, M., Natarajan, A. T., Surralles, J., Crott, J. W., Parry, J., et al. (2011). Molecular mechanisms of micronucleus, nucleoplasmic bridge and nuclear bud formation in mammalian and human cells. *Mutagenesis*, *26*(1), 125–132. https://doi.org/10.1093/mutage/geq052.

Fitzwalter, B. E., Towers, C. G., Sullivan, K. D., Andrysik, Z., Hoh, M., Ludwig, M., et al. (2018). Autophagy inhibition mediates apoptosis sensitization in cancer therapy by relieving FOXO3a turnover. *Developmental Cell*, *44*(5), 555–565. e553. https://doi.org/10.1016/j.devcel.2018.02.014.

Gewirtz, D. A., Alotaibi, M., Yakovlev, V. A., & Povirk, L. F. (2016). Tumor cell recovery from senescence induced by radiation with PARP inhibition. *Radiation Research*, *186*(4), 327–332. https://doi.org/10.1667/RR14437.1.

Gkountakos, A., Sartori, G., Falcone, I., Piro, G., Ciuffreda, L., Carbone, C., et al. (2019). PTEN in lung cancer: Dealing with the problem, building on new knowledge and turning the game around. *Cancers (Basel)*, *11*(8), 1141. https://doi.org/10.3390/cancers11081141.

Gluck, S., & Ablasser, A. (2019). Innate immunosensing of DNA in cellular senescence. *Current Opinion in Immunology*, *56*, 31–36. https://doi.org/10.1016/j.coi.2018.09.013.

Gluck, S., Guey, B., Gulen, M. F., Wolter, K., Kang, T. W., Schmacke, N. A., et al. (2017). Innate immune sensing of cytosolic chromatin fragments through cGAS promotes senescence. *Nature Cell Biology*, *19*(9), 1061–1070. https://doi.org/10.1038/ncb3586.

Gorgoulis, V., Adams, P. D., Alimonti, A., Bennett, D. C., Bischof, O., Bishop, C., et al. (2019). Cellular senescence: Defining a path forward. *Cell*, *179*(4), 813–827. https://doi.org/10.1016/j.cell.2019.10.005.

Harding, S. M., Benci, J. L., Irianto, J., Discher, D. E., Minn, A. J., & Greenberg, R. A. (2017). Mitotic progression following DNA damage enables pattern recognition within micronuclei. *Nature*, *548*(7668), 466–470. https://doi.org/10.1038/nature23470.

Hayakawa, T., Iwai, M., Aoki, S., Takimoto, K., Maruyama, M., Maruyama, W., et al. (2015). SIRT1 suppresses the senescence-associated secretory phenotype through epigenetic gene regulation. *PLoS One*, *10*(1), e0116480. https://doi.org/10.1371/journal.pone.0116480.

He, Y., Li, W., Lv, D., Zhang, X., Zhang, X., Ortiz, Y. T., et al. (2020). Inhibition of USP7 activity selectively eliminates senescent cells in part via restoration of p53 activity. *Aging Cell*, *19*(3), e13117. https://doi.org/10.1111/acel.13117.

Herranz, N., Gallage, S., Mellone, M., Wuestefeld, T., Klotz, S., Hanley, C. J., et al. (2015). mTOR regulates MAPKAPK2 translation to control the senescence-associated secretory phenotype. *Nature Cell Biology*, *17*(9), 1205–1217. https://doi.org/10.1038/ncb3225.

Herranz, N., & Gil, J. (2018). Mechanisms and functions of cellular senescence. *The Journal of Clinical Investigation*, *128*(4), 1238–1246. https://doi.org/10.1172/JCI95148.

Hewitt, G., Carroll, B., Sarallah, R., Correia-Melo, C., Ogrodnik, M., Nelson, G., et al. (2016). SQSTM1/p62 mediates crosstalk between autophagy and the UPS in DNA repair. *Autophagy*, *12*(10), 1917–1930. https://doi.org/10.1080/15548627.2016.1210368.

Hewitt, G., & Korolchuk, V. I. (2017). Repair, reuse, recycle: The expanding role of autophagy in genome maintenance. *Trends in Cell Biology*, *27*(5), 340–351. https://doi.org/10.1016/j.tcb.2016.11.011.

Huang, Y. H., Yang, P. M., Chuah, Q. Y., Lee, Y. J., Hsieh, Y. F., Peng, C. W., et al. (2014). Autophagy promotes radiation-induced senescence but inhibits bystander effects in human breast cancer cells. *Autophagy*, *10*(7), 1212–1228. https://doi.org/10.4161/auto.28772.

Ivanov, A., Pawlikowski, J., Manoharan, I., van Tuyn, J., Nelson, D. M., Rai, T. S., et al. (2013). Lysosome-mediated processing of chromatin in senescence. *The Journal of Cell Biology*, *202*(1), 129–143. https://doi.org/10.1083/jcb.201212110.

Ju, Y. J., Lee, K. H., Park, J. E., Yi, Y. S., Yun, M. Y., Ham, Y. H., et al. (2006). Decreased expression of DNA repair proteins Ku70 and Mre11 is associated with aging and may contribute to the cellular senescence. *Experimental & Molecular Medicine*, *38*(6), 686–693. doi:2006123012 [pii].

Jung, S. H., Hwang, H. J., Kang, D., Park, H. A., Lee, H. C., Jeong, D., et al. (2019). mTOR kinase leads to PTEN-loss-induced cellular senescence by phosphorylating p53. *Oncogene*, *38*(10), 1639–1650. https://doi.org/10.1038/s41388-018-0521-8.

Jung, C. H., Jun, C. B., Ro, S. H., Kim, Y. M., Otto, N. M., Cao, J., et al. (2009). ULK-Atg13-FIP200 complexes mediate mTOR signaling to the autophagy machinery. *Molecular Biology of the Cell*, *20*(7), 1992–2003. E08-12-1249 [pii]. https://doi.org/10.1091/mbc.E08-12-1249.

Jung, I. L., Kang, H. J., Kim, K. C., & Kim, I. G. (2010). PTEN/pAkt/p53 signaling pathway correlates with the radioresponse of non-small cell lung cancer. *International Journal of Molecular Medicine, 25*(4), 517–523. https://doi.org/10.3892/ijmm_00000372.

Kabeya, Y., Mizushima, N., Ueno, T., Yamamoto, A., Kirisako, T., Noda, T., et al. (2000). LC3, a mammalian homologue of yeast Apg8p, is localized in autophagosome membranes after processing. *The EMBO Journal, 19*(21), 5720–5728. https://doi.org/10.1093/emboj/19.21.5720.

Kang, C., Xu, Q., Martin, T. D., Li, M. Z., Demaria, M., Aron, L., et al. (2015). The DNA damage response induces inflammation and senescence by inhibiting autophagy of GATA4. *Science, 349*(6255), aaa5612. https://doi.org/10.1126/science.aaa5612.

Kapoor, I., Li, Y., Sharma, A., Zhu, H., Bodo, J., Xu, W., et al. (2019). Resistance to BTK inhibition by ibrutinib can be overcome by preventing FOXO3a nuclear export and PI3K/AKT activation in B-cell lymphoid malignancies. *Cell Death & Disease, 10*(12), 924. https://doi.org/10.1038/s41419-019-2158-0.

Karantza-Wadsworth, V., Patel, S., Kravchuk, O., Chen, G., Mathew, R., Jin, S., et al. (2007). Autophagy mitigates metabolic stress and genome damage in mammary tumorigenesis. *Genes & Development, 21*(13), 1621–1635. https://doi.org/10.1101/gad.1565707.

Klionsky, D. J. (2007). Autophagy: From phenomenology to molecular understanding in less than a decade. *Nature Reviews. Molecular Cell Biology, 8*(11), 931–937. nrm2245 [pii]. https://doi.org/10.1038/nrm2245.

Kroemer, G., Marino, G., & Levine, B. (2010). Autophagy and the integrated stress response. *Molecular Cell, 40*(2), 280–293. S1097-2765(10)00751-3 [pii]. https://doi.org/10.1016/j.molcel.2010.09.023.

Krtolica, A., Parrinello, S., Lockett, S., Desprez, P. Y., & Campisi, J. (2001). Senescent fibroblasts promote epithelial cell growth and tumorigenesis: A link between cancer and aging. *Proceedings of the National Academy of Sciences of the United States of America, 98*(21), 12072–12077. https://doi.org/10.1073/pnas.211053698.

Kuilman, T., Michaloglou, C., Vredeveld, L. C., Douma, S., van Doorn, R., Desmet, C. J., et al. (2008). Oncogene-induced senescence relayed by an interleukin-dependent inflammatory network. *Cell, 133*(6), 1019–1031. https://doi.org/10.1016/j.cell.2008.03.039.

Kuilman, T., & Peeper, D. S. (2009). Senescence-messaging secretome: SMS-ing cellular stress. *Nature Reviews. Cancer, 9*(2), 81–94. https://doi.org/10.1038/nrc2560.

Laberge, R. M., Sun, Y., Orjalo, A. V., Patil, C. K., Freund, A., Zhou, L., et al. (2015). MTOR regulates the pro-tumorigenic senescence-associated secretory phenotype by promoting IL1A translation. *Nature Cell Biology, 17*(8), 1049–1061. https://doi.org/10.1038/ncb3195.

Lawless, C., Wang, C., Jurk, D., Merz, A., Zglinicki, T., & Passos, J. F. (2010). Quantitative assessment of markers for cell senescence. *Experimental Gerontology, 45*(10), 772–778. https://doi.org/10.1016/j.exger.2010.01.018.

Lee, J. J., Kim, B. C., Park, M. J., Lee, Y. S., Kim, Y. N., Lee, B. L., et al. (2011). PTEN status switches cell fate between premature senescence and apoptosis in glioma exposed to ionizing radiation. *Cell Death and Differentiation, 18*(4), 666–677. https://doi.org/10.1038/cdd.2010.139.

Li, T., & Chen, Z. J. (2018). The cGAS-cGAMP-STING pathway connects DNA damage to inflammation, senescence, and cancer. *The Journal of Experimental Medicine, 215*(5), 1287–1299. https://doi.org/10.1084/jem.20180139.

Limbo, O., Chahwan, C., Yamada, Y., de Bruin, R. A., Wittenberg, C., & Russell, P. (2007). Ctp1 is a cell-cycle-regulated protein that functions with Mre11 complex to control double-strand break repair by homologous recombination. *Molecular Cell, 28*(1), 134–146. https://doi.org/10.1016/j.molcel.2007.09.009.

Liu, B., Chen, Y., & St Clair, D. K. (2008). ROS and p53: A versatile partnership. *Free Radical Biology & Medicine*, *44*(8), 1529–1535. https://doi.org/10.1016/j.freeradbiomed.2008.01.011.

Liu, D., & Hornsby, P. J. (2007). Senescent human fibroblasts increase the early growth of xenograft tumors via matrix metalloproteinase secretion. *Cancer Research*, *67*(7), 3117–3126. https://doi.org/10.1158/0008-5472.CAN-06-3452.

Lord, C. J., Tutt, A. N., & Ashworth, A. (2015). Synthetic lethality and cancer therapy: Lessons learned from the development of PARP inhibitors. *Annual Review of Medicine*, *66*, 455–470. https://doi.org/10.1146/annurev-med-050913-022545.

Luo, H., Wang, L., Schulte, B. A., Yang, A., Tang, S., & Wang, G. Y. (2013). Resveratrol enhances ionizing radiation-induced premature senescence in lung cancer cells. *International Journal of Oncology*, *43*(6), 1999–2006. https://doi.org/10.3892/ijo.2013.2141.

Luo, H., Yount, C., Lang, H., Yang, A., Riemer, E. C., Lyons, K., et al. (2013). Activation of p53 with Nutlin-3a radiosensitizes lung cancer cells via enhancing radiation-induced premature senescence. *Lung Cancer*, *81*(2), 167–173. https://doi.org/10.1016/j.lungcan.2013.04.017.

Ma, J. L., Kim, E. M., Haber, J. E., & Lee, S. E. (2003). Yeast Mre11 and Rad1 proteins define a Ku-independent mechanism to repair double-strand breaks lacking overlapping end sequences. *Molecular and Cellular Biology*, *23*(23), 8820–8828. https://doi.org/10.1128/mcb.23.23.8820-8828.2003.

Macip, S., Igarashi, M., Berggren, P., Yu, J., Lee, S. W., & Aaronson, S. A. (2003). Influence of induced reactive oxygen species in p53-mediated cell fate decisions. *Molecular and Cellular Biology*, *23*(23), 8576–8585. https://doi.org/10.1128/mcb.23.23.8576-8585.2003.

Macip, S., Igarashi, M., Fang, L., Chen, A., Pan, Z. Q., Lee, S. W., et al. (2002). Inhibition of p21-mediated ROS accumulation can rescue p21-induced senescence. *The EMBO Journal*, *21*(9), 2180–2188. https://doi.org/10.1093/emboj/21.9.2180.

Mackenzie, K. J., Carroll, P., Martin, C. A., Murina, O., Fluteau, A., Simpson, D. J., et al. (2017). cGAS surveillance of micronuclei links genome instability to innate immunity. *Nature*, *548*(7668), 461–465. https://doi.org/10.1038/nature23449.

Maiuri, M. C., Zalckvar, E., Kimchi, A., & Kroemer, G. (2007). Self-eating and self-killing: Crosstalk between autophagy and apoptosis. *Nature Reviews. Molecular Cell Biology*, *8*(9), 741–752. nrm2239 [pii]. https://doi.org/10.1038/nrm2239.

Majumder, P. K., & Sellers, W. R. (2005). Akt-regulated pathways in prostate cancer. *Oncogene*, *24*(50), 7465–7474. https://doi.org/10.1038/sj.onc.1209096.

Malaquin, N., Martinez, A., & Rodier, F. (2016). Keeping the senescence secretome under control: Molecular reins on the senescence-associated secretory phenotype. *Experimental Gerontology*, *82*, 39–49. https://doi.org/10.1016/j.exger.2016.05.010.

Marshall, C. H., & Antonarakis, E. S. (2020). Therapeutic targeting of the DNA damage response in prostate cancer. *Current Opinion in Oncology*, *32*(3), 216–222. https://doi.org/10.1097/CCO.0000000000000617.

Mathew, R., Kongara, S., Beaudoin, B., Karp, C. M., Bray, K., Degenhardt, K., et al. (2007). Autophagy suppresses tumor progression by limiting chromosomal instability. *Genes & Development*, *21*(11), 1367–1381. https://doi.org/10.1101/gad.1545107.

Matsuya, Y., McMahon, S. J., Tsutsumi, K., Sasaki, K., Okuyama, G., Yoshii, Y., et al. (2018). Investigation of dose-rate effects and cell-cycle distribution under protracted exposure to ionizing radiation for various dose-rates. *Scientific Reports*, *8*(1), 8287. https://doi.org/10.1038/s41598-018-26556-5.

Mazumder, S., Gong, B., Chen, Q., Drazba, J. A., Buchsbaum, J. C., & Almasan, A. (2002). Proteolytic cleavage of cyclin E leads to inactivation of associated kinase activity and amplification of apoptosis in hematopoietic cells. *Molecular and Cellular Biology*, *22*(7), 2398–2409. https://doi.org/10.1128/mcb.22.7.2398-2409.2002.

Mazumder, S., Plesca, D., & Almasan, A. (2007). A jekyll and hyde role of cyclin E in the genotoxic stress response: Switching from cell cycle control to apoptosis regulation. *Cell Cycle*, *6*(12), 1437–1442. https://doi.org/10.4161/cc.6.12.4432.
Mazumder, S., Plesca, D., Kinter, M., & Almasan, A. (2007). Interaction of a cyclin E fragment with Ku70 regulates Bax-mediated apoptosis. *Molecular and Cellular Biology*, *27*(9), 3511–3520. MCB.01448-06 [pii]. https://doi.org/10.1128/MCB.01448-06.
Michaud, M., Martins, I., Sukkurwala, A. Q., Adjemian, S., Ma, Y., Pellegatti, P., et al. (2011). Autophagy-dependent anticancer immune responses induced by chemotherapeutic agents in mice. *Science*, *334*(6062), 1573–1577. https://doi.org/10.1126/science.1208347.
Mikula-Pietrasik, J., Niklas, A., Uruski, P., Tykarski, A., & Ksiazek, K. (2020). Mechanisms and significance of therapy-induced and spontaneous senescence of cancer cells. *Cellular and Molecular Life Sciences*, 77(2), 213–229. https://doi.org/10.1007/s00018-019-03261-8.
Mitchell, C. R., Folkard, M., & Joiner, M. C. (2002). Effects of exposure to low-dose-rate (60)co gamma rays on human tumor cells in vitro. *Radiation Research*, *158*(3), 311–318. https://doi.org/10.1667/0033-7587(2002)158[0311:eoetld]2.0.co;2.
Mithal, P., Allott, E., Gerber, L., Reid, J., Welbourn, W., Tikishvili, E., et al. (2014). PTEN loss in biopsy tissue predicts poor clinical outcomes in prostate cancer. *International Journal of Urology*, *21*(12), 1209–1214. https://doi.org/10.1111/iju.12571.
Mitra, S., Nguyen, L. N., Akter, M., Park, G., Choi, E. H., & Kaushik, N. K. (2019). Impact of ROS generated by chemical, physical, and plasma techniques on cancer attenuation. *Cancers (Basel)*, *11*(7). https://doi.org/10.3390/cancers11071030.
Mizushima, N. (2007). Autophagy: Process and function. *Genes & Development*, *21*(22), 2861–2873. https://doi.org/10.1101/gad.1599207.
Mizushima, N. (2010). The role of the Atg1/ULK1 complex in autophagy regulation. *Current Opinion in Cell Biology*, *22*(2), 132–139. S0955-0674(09)00233-6 [pii]. https://doi.org/10.1016/j.ceb.2009.12.004.
Moiseeva, O., Mallette, F. A., Mukhopadhyay, U. K., Moores, A., & Ferbeyre, G. (2006). DNA damage signaling and p53-dependent senescence after prolonged beta-interferon stimulation. *Molecular Biology of the Cell*, *17*(4), 1583–1592. https://doi.org/10.1091/mbc.e05-09-0858.
Moscat, J., & Diaz-Meco, M. T. (2009). p62 at the crossroads of autophagy, apoptosis, and cancer. *Cell*, *137*(6), 1001–1004. S0092-8674(09)00619-9 [pii]. https://doi.org/10.1016/j.cell.2009.05.023.
Nag, S. (2000). Brachytherapy for prostate cancer: Summary of American brachytherapy society recommendations. *Seminars in Urologic Oncology*, *18*(2), 133–136. Retrieved from http://www.ncbi.nlm.nih.gov/pubmed/10875454.
Nam, H. Y., Han, M. W., Chang, H. W., Lee, Y. S., Lee, M., Lee, H. J., et al. (2013). Radioresistant cancer cells can be conditioned to enter senescence by mTOR inhibition. *Cancer Research*, *73*(14), 4267–4277. https://doi.org/10.1158/0008-5472.CAN-12-3516.
Narita, M., Nunez, S., Heard, E., Narita, M., Lin, A. W., Hearn, S. A., et al. (2003). Rb-mediated heterochromatin formation and silencing of E2F target genes during cellular senescence. *Cell*, *113*(6), 703–716. https://doi.org/10.1016/s0092-8674(03)00401-x.
Narita, M., Young, A. R., Arakawa, S., Samarajiwa, S. A., Nakashima, T., Yoshida, S., et al. (2011). Spatial coupling of mTOR and autophagy augments secretory phenotypes. *Science*, *332*(6032), 966–970. https://doi.org/10.1126/science.1205407.
Ohanna, M., Giuliano, S., Bonet, C., Imbert, V., Hofman, V., Zangari, J., et al. (2011). Senescent cells develop a PARP-1 and nuclear factor-{kappa}B-associated secretome (PNAS). *Genes & Development*, *25*(12), 1245–1261. https://doi.org/10.1101/gad.625811.

Panda, P. K., Patra, S., Naik, P. P., Praharaj, P. P., Mukhopadhyay, S., Meher, B. R., et al. (2020). Deacetylation of LAMP1 drives lipophagy-dependent generation of free fatty acids by Abrus agglutinin to promote senescence in prostate cancer. *Journal of Cellular Physiology*, *235*(3), 2776–2791. https://doi.org/10.1002/jcp.29182.

Panier, S., & Boulton, S. J. (2014). Double-strand break repair: 53BP1 comes into focus. *Nature Reviews. Molecular Cell Biology*, *15*(1), 7–18. https://doi.org/10.1038/nrm3719.

Parisotto, M., Grelet, E., El Bizri, R., Dai, Y., Terzic, J., Eckert, D., et al. (2018). PTEN deletion in luminal cells of mature prostate induces replication stress and senescence in vivo. *The Journal of Experimental Medicine*, *215*(6), 1749–1763. https://doi.org/10.1084/jem.20171207.

Parisotto, M., Grelet, E., El Bizri, R., & Metzger, D. (2019). Senescence controls prostatic neoplasia driven by Pten loss. *Molecular & Cellular Oncology*, *6*(1), 1511205. https://doi.org/10.1080/23723556.2018.1511205.

Park, Y.-E., Hayashi, Y. K., Bonne, G., Arimura, T., Noguchi, S., Nonaka, I., et al. (2009). Autophagic degradation of nuclear components in mammalian cells. *Autophagy*, *5*(6), 795–804. https://doi.org/10.4161/auto.8901.

Park, J. K., Jung, H. Y., Park, S. H., Kang, S. Y., Yi, M. R., Um, H. D., et al. (2008). Combination of PTEN and gamma-ionizing radiation enhances cell death and G (2)/M arrest through regulation of AKT activity and p21 induction in non-small-cell lung cancer cells. *International Journal of Radiation Oncology, Biology, Physics*, *70*(5), 1552–1560. https://doi.org/10.1016/j.ijrobp.2007.11.069.

Patel, N. H., Sohal, S. S., Manjili, M. H., Harrell, J. C., & Gewirtz, D. A. (2020). The roles of autophagy and senescence in the tumor cell response to radiation. *Radiation Research*, *194*(2), 103–115. https://doi.org/10.1667/RADE-20-00009.

Patterson-Fortin, J., & D'Andrea, A. D. (2020). Exploiting the microhomology-mediated end-joining pathway in cancer therapy. *Cancer Research*. https://doi.org/10.1158/0008-5472.CAN-20-1672.

Piscitello, D., Varshney, D., Lilla, S., Vizioli, M. G., Reid, C., Gorbunova, V., et al. (2018). AKT overactivation can suppress DNA repair via p70S6 kinase-dependent downregulation of MRE11. *Oncogene*, *37*(4), 427–438. https://doi.org/10.1038/onc.2017.340.

Plesca, D., Crosby, M. E., Gupta, D., & Almasan, A. (2007). E2F4 function in G2: Maintaining G2-arrest to prevent mitotic entry with damaged DNA. *Cell Cycle*, *6*(10), 1147–1152. https://doi.org/10.4161/cc.6.10.4259.

Raderschall, E., Golub, E. I., & Haaf, T. (1999). Nuclear foci of mammalian recombination proteins are located at single-stranded DNA regions formed after DNA damage. *Proceedings of the National Academy of Sciences of the United States of America*, *96*(5), 1921–1926. https://doi.org/10.1073/pnas.96.5.1921.

Rao, S., Tortola, L., Perlot, T., Wirnsberger, G., Novatchkova, M., Nitsch, R., et al. (2014). A dual role for autophagy in a murine model of lung cancer. *Nature Communications*, *5*, 3056. https://doi.org/10.1038/ncomms4056.

Ritschka, B., Storer, M., Mas, A., Heinzmann, F., Ortells, M. C., Morton, J. P., et al. (2017). The senescence-associated secretory phenotype induces cellular plasticity and tissue regeneration. *Genes & Development*, *31*(2), 172–183. https://doi.org/10.1101/gad.290635.116.

Rogakou, E. P., Pilch, D. R., Orr, A. H., Ivanova, V. S., & Bonner, W. M. (1998). DNA double-stranded breaks induce histone H2AX phosphorylation on serine 139. *The Journal of Biological Chemistry*, *273*(10), 5858–5868. https://doi.org/10.1074/jbc.273.10.5858.

Rouleau, M., Patel, A., Hendzel, M. J., Kaufmann, S. H., & Poirier, G. G. (2010). PARP inhibition: PARP1 and beyond. *Nature Reviews. Cancer*, *10*(4), 293–301. nrc2812 [pii]. https://doi.org/10.1038/nrc2812.

Rubin, M. A., Maher, C. A., & Chinnaiyan, A. M. (2011). Common gene rearrangements in prostate cancer. *Journal of Clinical Oncology*, *29*(27), 3659–3668. JCO.2011.35.1916 [pii]. https://doi.org/10.1200/JCO.2011.35.1916.
Sai, X., Qin, C., Wu, Y., Zhao, Y., & Bian, T. (2020). Downregulation of PTEN mediates bleomycin-induced premature senescence in lung cancer cells by suppressing autophagy. *The Journal of International Medical Research*, *48*(5), 300060520923522. https://doi.org/10.1177/0300060520923522.
Saleh, T., Bloukh, S., Carpenter, V. J., Alwohoush, E., Bakeer, J., Darwish, S., et al. (2020). Therapy-induced senescence: An "old" friend becomes the enemy. *Cancers (Basel)*, *12*(4). https://doi.org/10.3390/cancers12040822.
Saleh, T., Tyutyunyk-Massey, L., & Gewirtz, D. A. (2019). Tumor cell escape from therapy-induced senescence as a model of disease recurrence after dormancy. *Cancer Research*, *79*(6), 1044–1046. https://doi.org/10.1158/0008-5472.CAN-18-3437.
Saleh, T., Tyutyunyk-Massey, L., Patel, N. H., Cudjoe, E. K., Jr., Alotaibi, M., & Gewirtz, D. A. (2020). Studies of non-protective autophagy provide evidence that recovery from therapy-induced senescence is independent of early autophagy. *International Journal of Molecular Sciences*, *21*(4). https://doi.org/10.3390/ijms21041427.
San Filippo, J., Sung, P., & Klein, H. (2008). Mechanism of eukaryotic homologous recombination. *Annual Review of Biochemistry*, 77, 229–257. https://doi.org/10.1146/annurev.biochem.77.061306.125255.
Schiewer, M. J., Goodwin, J. F., Han, S., Brenner, J. C., Augello, M. A., Dean, J. L., et al. (2012). Dual roles of PARP-1 promote cancer growth and progression. *Cancer Discovery*, *2*, 1134–1149. https://doi.org/10.1158/2159-8290.CD-12-0120.
Schmid, T. E., Zlobinskaya, O., & Multhoff, G. (2012). Differences in phosphorylated histone H2AX foci formation and removal of cells exposed to low and high linear energy transfer radiation. *Current Genomics*, *13*(6), 418–425. https://doi.org/10.2174/138920212802510501.
Schwarze, S. R., Fu, V. X., Desotelle, J. A., Kenowski, M. L., & Jarrard, D. F. (2005). The identification of senescence-specific genes during the induction of senescence in prostate cancer cells. *Neoplasia*, 7(9), 816–823. https://doi.org/10.1593/neo.05250.
Seluanov, A., Danek, J., Hause, N., & Gorbunova, V. (2007). Changes in the level and distribution of Ku proteins during cellular senescence. *DNA Repair (Amst)*, *6*(12), 1740–1748. S1568-7864(07)00216-9 [pii]. https://doi.org/10.1016/j.dnarep.2007.06.010.
Sharma, A., & Almasan, A. (2020). USP14 regulates DNA damage response and is a target for radiosensitization in non-small cell lung cancer. *International Journal of Molecular Sciences*, *21*(17). https://doi.org/10.3390/ijms21176383.
Sharma, A., Alswillah, T., Kapoor, I., Debjani, P., Willard, B., Summers, M. K., et al. (2020). USP14 is a deubiquitinase for Ku70 and critical determinant of non-homologous end joining repair in autophagy and PTEN-deficient cells. *Nucleic Acids Research*, *48*(2), 736–747. https://doi.org/10.1093/nar/gkz1103.
Sharma, A., Alswillah, T., Singh, K., Chatterjee, P., Willard, B., Venere, M., et al. (2018). USP14 regulates DNA damage repair by targeting RNF168-dependent ubiquitination. *Autophagy*, *14*(11), 1976–1990. https://doi.org/10.1080/15548627.2018.1496877.
Sharma, A., Singh, K., Mazumder, S., Hill, B. T., Kalaycio, M., & Almasan, A. (2013). BECN1 and BIM interactions with MCL-1 determine fludarabine resistance in leukemic B cells. *Cell Death & Disease*, *4*, e628. https://doi.org/10.1038/cddis.2013.155.
Shimi, T., Butin-Israeli, V., Adam, S. A., Hamanaka, R. B., Goldman, A. E., Lucas, C. A., et al. (2011). The role of nuclear Lamin B1 in cell proliferation and senescence. *Genes & Development*, *25*(24), 2579–2593. https://doi.org/10.1101/gad.179515.111.
Singh, K., Matsuyama, S., Drazba, J. A., & Almasan, A. (2012). Autophagy-dependent senescence in response to DNA damage and chronic apoptotic stress. *Autophagy*, *8*(2), 236–251. https://doi.org/10.4161/auto.8.2.18600.

Smith, H. L., Southgate, H., Tweddle, D. A., & Curtin, N. J. (2020). DNA damage checkpoint kinases in cancer. *Expert Reviews in Molecular Medicine*, *22*, e2. https://doi.org/10.1017/erm.2020.3.

Sokolova, A. O., Yu, E. Y., & Cheng, H. H. (2020). Honing in on PARPi response in prostate cancer: From HR pathway to gene-by-gene granularity. *Clinical Cancer Research*, *26*(11), 2439–2440. https://doi.org/10.1158/1078-0432.CCR-20-0707.

Stevens, D. L., Bradley, S., Goodhead, D. T., & Hill, M. A. (2014). The influence of dose rate on the induction of chromosome aberrations and gene mutation after exposure of plateau phase V79-4 cells with high-LET alpha particles. *Radiation Research*, *182*(3), 331–337. https://doi.org/10.1667/RR13746.1.

Takahashi, A., Loo, T. M., Okada, R., Kamachi, F., Watanabe, Y., Wakita, M., et al. (2018). Downregulation of cytoplasmic DNases is implicated in cytoplasmic DNA accumulation and SASP in senescent cells. *Nature Communications*, *9*(1), 1249. https://doi.org/10.1038/s41467-018-03555-8.

Tanida, I., Sou, Y. S., Minematsu-Ikeguchi, N., Ueno, T., & Kominami, E. (2006). Atg8L/Apg8L is the fourth mammalian modifier of mammalian Atg8 conjugation mediated by human Atg4B, Atg7 and Atg3. *FEBS Journal*, *273*(11), 2553–2562. EJB5260 [pii]. https://doi.org/10.1111/j.1742-4658.2006.05260.x.

Tasdemir, E., Chiara Maiuri, M., Morselli, E., Criollo, A., D'Amelio, M., Djavaheri-Mergny, M., et al. (2008). A dual role of p53 in the control of autophagy. *Autophagy*, *4*(6), 810–814. https://doi.org/10.4161/auto.6486.

Taylor, B. S., Schultz, N., Hieronymus, H., Gopalan, A., Xiao, Y., Carver, B. S., et al. (2010). Integrative genomic profiling of human prostate cancer. *Cancer Cell*, *18*(1), 11–22. https://doi.org/10.1016/j.ccr.2010.05.026.

Trotman, L. C., Alimonti, A., Scaglioni, P. P., Koutcher, J. A., Cordon-Cardo, C., & Pandolfi, P. P. (2006). Identification of a tumour suppressor network opposing nuclear Akt function. *Nature*, *441*(7092), 523–527. https://doi.org/10.1038/nature04809.

Ulyanenko, S., Pustovalova, M., Koryakin, S., Beketov, E., Lychagin, A., Ulyanenko, L., et al. (2019). Formation of gammaH2AX and pATM foci in human mesenchymal stem cells exposed to low dose-rate gamma-radiation. *International Journal of Molecular Sciences*, *20*(11). https://doi.org/10.3390/ijms20112645.

Varga, T., & Aplan, P. D. (2005). Chromosomal aberrations induced by double strand DNA breaks. *DNA Repair (Amst)*, *4*(9), 1038–1046. https://doi.org/10.1016/j.dnarep.2005.05.004.

Wan, D., Jiang, W., & Hao, J. (2020). Research advances in how the cGAS-STING pathway controls the cellular inflammatory response. *Frontiers in Immunology*, *11*, 615. https://doi.org/10.3389/fimmu.2020.00615.

Wang, Y., Boerma, M., & Zhou, D. (2016). Ionizing radiation-induced endothelial cell senescence and cardiovascular diseases. *Radiation Research*, *186*(2), 153–161. https://doi.org/10.1667/RR14445.1.

Wang, Y. U., Chen, Y. N., Zhang, W., Yang, Y. U., Bai, W. K., Shen, E., et al. (2016). Upregulation of ULK1 expression in PC-3 cells following tumor protein P53 transfection by sonoporation. *Oncology Letters*, *11*(1), 699–704. https://doi.org/10.3892/ol.2015.3946.

Wang, M., Kern, A. M., Hulskotter, M., Greninger, P., Singh, A., Pan, Y., et al. (2014). EGFR-mediated chromatin condensation protects KRAS-mutant cancer cells against ionizing radiation. *Cancer Research*, *74*(10), 2825–2834. https://doi.org/10.1158/0008-5472.CAN-13-3157.

Wang, M., Morsbach, F., Sander, D., Gheorghiu, L., Nanda, A., Benes, C., et al. (2011). EGF receptor inhibition radiosensitizes NSCLC cells by inducing senescence in cells sustaining DNA double-strand breaks. *Cancer Research*, *71*(19), 6261–6269. https://doi.org/10.1158/0008-5472.CAN-11-0213.

Wang, Y., Zhang, N., Zhang, L., Li, R., Fu, W., Ma, K., et al. (2016). Autophagy regulates chromatin ubiquitination in DNA damage response through elimination of SQSTM1/p62. *Molecular Cell, 63*(1), 34–48. https://doi.org/10.1016/j.molcel.2016.05.027.

Wang, N., Zhu, C., Xu, Y., Qian, W., & Zheng, M. (2018). Negative regulation of PTEN by MicroRNA-221 and its association with drug resistance and cellular senescence in lung cancer cells. *BioMed Research International, 2018*, 7908950. https://doi.org/10.1155/2018/7908950.

Wiley, C. D., Velarde, M. C., Lecot, P., Liu, S., Sarnoski, E. A., Freund, A., et al. (2016). Mitochondrial dysfunction induces senescence with a distinct secretory phenotype. *Cell Metabolism, 23*(2), 303–314. https://doi.org/10.1016/j.cmet.2015.11.011.

Xue, W., Zender, L., Miething, C., Dickins, R. A., Hernando, E., Krizhanovsky, V., et al. (2007). Senescence and tumour clearance is triggered by p53 restoration in murine liver carcinomas. *Nature, 445*, 656–660. https://doi.org/10.1038/nature05529.

Xu, J., Patel, N. H., Saleh, T., Cudjoe, E. K., Jr., Alotaibi, M., Wu, Y., et al. (2018). Differential radiation sensitivity in p53 wild-type and p53-deficient tumor cells associated with senescence but not apoptosis or (nonprotective) autophagy. *Radiation Research, 190*(5), 538–557. https://doi.org/10.1667/RR15099.1.

Yamazaki, T., Kirchmair, A., Sato, A., Buque, A., Rybstein, M., Petroni, G., et al. (2020). Mitochondrial DNA drives abscopal responses to radiation that are inhibited by autophagy. *Nature Immunology, 21*(10), 1160–1171. https://doi.org/10.1038/s41590-020-0751-0.

Yang, H., Wang, H., Ren, J., Chen, Q., & Chen, Z. J. (2017). cGAS is essential for cellular senescence. *Proceedings of the National Academy of Sciences of the United States of America, 114*(23), E4612–E4620. https://doi.org/10.1073/pnas.1705499114.

Yu, Q., Katlinskaya, Y. V., Carbone, C. J., Zhao, B., Katlinski, K. V., Zheng, H., et al. (2015). DNA-damage-induced type I interferon promotes senescence and inhibits stem cell function. *Cell Reports, 11*(5), 785–797. https://doi.org/10.1016/j.celrep.2015.03.069.

Zaremba, T., Thomas, H. D., Cole, M., Coulthard, S. A., Plummer, E. R., & Curtin, N. J. (2011). Poly(ADP-ribose) polymerase-1 (PARP-1) pharmacogenetics, activity and expression analysis in cancer patients and healthy volunteers. *The Biochemical Journal, 436*(3), 671–679. https://doi.org/10.1042/BJ20101723.

Zhang, Y., Dong, Y., Melkus, M. W., Yin, S., Tang, S. N., Jiang, P., et al. (2018). Role of P53-senescence induction in suppression of LNCaP prostate cancer growth by cardiotonic compound bufalin. *Molecular Cancer Therapeutics, 17*(11), 2341–2352. https://doi.org/10.1158/1535-7163.MCT-17-1296.

CHAPTER SEVEN

Cellular senescence and tumor promotion: Role of the Unfolded Protein Response

Olivier Pluquet* and Corinne Abbadie
Univ Lille, CNRS, Inserm, CHU Lille, UMR9020-U1277-CANTHER-Cancer Heterogeneity Plasticity and Resistance to Therapies, Lille, France
*Corresponding author: e-mail address: olivier.pluquet@ibl.cnrs.fr

Contents

Abstract

Senescence is a cellular state which can be viewed as a stress response phenotype implicated in various physiological and pathological processes, including cancer. Therefore, it is of fundamental importance to understand why and how a cell acquires and maintains a senescent phenotype. Direct evidence has pointed to the homeostasis of the endoplasmic reticulum whose control appears strikingly affected during senescence. The endoplasmic reticulum is one of the sensing organelles that transduce signals between different pathways in order to adapt a functional proteome upon intrinsic or extrinsic

Advances in Cancer Research, Volume 150
ISSN 0065-230X
https://doi.org/10.1016/bs.acr.2021.01.001

challenges. One of these signaling pathways is the Unfolded Protein Response (UPR), which has been shown to be activated during senescence. Its exact contribution to senescence onset, maintenance, and escape, however, is still poorly understood. In this article, we review the mechanisms through which the UPR contributes to the appearance and maintenance of characteristic senescent features. We also discuss whether the perturbation of the endoplasmic reticulum proteostasis or accumulation of misfolded proteins could be possible causes of senescence, and—as a consequence—to what extent the UPR components could be considered as therapeutic targets allowing for the elimination of senescent cells or altering their secretome to prevent neoplastic transformation.

Abbreviations

DDR	DNA damage response
ECM	extracellular matrix
ER	endoplasmic reticulum
HMEC	human mammary epithelial cell
LBS	lipid bilayer stress
MMP	matrix metalloproteinase
NHDF	normal human dermal fibroblast
NHEK	normal human epidermal keratinocyte
OIS	oncogene-induced senescence
PSNE	post-senescence neoplastic escape
RIDD	regulated-IRE1 dependent decay
ROS	reactive oxygen species
SA-β-Gal	senescence-associated-β-galactosidase
SAHF	senescence-associated heterochromatin foci
SASP	senescence-associated secretory phenotype
SIPS	stress-induced premature senescence
TIS	therapy-induced senescence
UPR	unfolded protein response
UV	ultra violet

1. Introduction

Senescence is a complex cellular state which can be viewed as a stress response phenotype. However, the mechanisms that contribute to senescence remain unclear. Still, it is now clearly established that advanced age is correlated with an accumulation of senescent cells in tissues and organs (Childs, Durik, Baker, & van Deursen, 2015; Herbig, Ferreira, Condel, Carey, & Sedivy, 2006). Using a transgenic mouse model expressing a suicide gene specifically in senescent cells and small pharmacological molecules specifically targeting senescent cells, called senolytics, it was demonstrated that senescence plays a causal role in aging and associated pathologies (Baar et al., 2017; Baker et al., 2011, 2016).

The senescent phenotype is characterized by a stable proliferation arrest mediated by the $p53/p21^{WAF1}$ and/or $p16^{INK4}/Rb$ pathways, accompanied by various molecular alterations. Those include epigenetic changes, increased oxidative stress, accumulation of unrepaired DNA damage, changes in metabolism, as well as an increase in autophagic activity and in the activity of the Senescence-Associated-β-Galactosidase (SA-β-Gal) (reviewed in (Abbadie, Pluquet, & Pourtier, 2017; Hernandez-Segura, Nehme, & Demaria, 2018)). Senescent cells also undergo major morphological changes, such as cell enlargement and cytoskeleton-mediated shape changes (Hwang, Yoon, & Kang, 2009; Nishio, Inoue, Qiao, Kondo, & Mimura, 2001; Rodier & Campisi, 2011; Wang & Gundersen, 1984). Remarkably, senescent cells synthesize and secrete large amounts of a variety of proteins constituting the so-called senescence-associated secretory phenotype (SASP). The SASP is enriched in pro-inflammatory cytokines, growth factors, and remodeling enzymes of the extracellular matrix (Abbadie et al., 2017; Acosta et al., 2008; Coppé et al., 2008; Dörr et al., 2013; Malaquin, Martinez, & Rodier, 2016; Nacarelli, Liu, & Zhang, 2017).

Senescence can be induced by various stimuli, including telomere shortening, oxidative stress, exposure to UV or ionizing radiation, chemotherapeutic drugs, mitochondrial dysfunction, and activation of oncogenes, such as Ras or NF-kB (reviewed in (Hernandez-Segura et al., 2018; Kuilman, Michaloglou, Mooi, & Peeper, 2010; McHugh & Gil, 2018; Rodier & Campisi, 2011)). Depending on these different inducers, senescence is referred to as replicative senescence (when it is induced by telomere shortening), stress-induced premature senescence (SIPS; induced by oxidative stress), oncogene-induced senescence (OIS; induced by the activation of an oncogene), or therapy-induced senescence (TIS; induced by an anti-cancer treatment). Because all of these inducers are damaging agents that affect DNA or other cells components, the senescence program can be viewed as a stress response phenotype, with a potential variability in the intensity and/or expression of characteristic senescent features, depending on the nature and intensity of the upstream inducer.

Senescence is well known to act as an anti-tumor barrier. The foundation to understand the interconnections between senescence and neoplastic transformation has been laid mainly by the group of Judith Campisi whose studies led to two dogmas. Based on high stability of the cell cycle arrest, they proposed that senescence is a cell-autonomous tumor suppressor mechanism (Campisi, 2001). Additionally, after characterizing the SASP and its pro-inflammatory nature, they suggested that senescence promotes tumorigenesis by paracrine pathways (Campisi, Andersen, Kapahi, & Melov, 2011). Indeed, senescent cells oversecrete such proteins as growth factors, remodeling enzymes of the extracellular matrix and inflammatory cytokines (Coppé, Desprez, Krtolica, & Campisi, 2010) which influence the tissue microenvironment and impact cellular differentiation and proliferation, in particular in cancer cells (Coppé et al., 2010). Removing senescent cells removes this pro-inflammatory environment and could delay cancer development (Velarde & Demaria, 2016).

One should bear in mind that most of the data on senescence were generated from fibroblast models, which are not the most relevant model of tumorigenesis, since the most prevalent human cancers (carcinomas) originate in epithelial cells. Of note, the incidence of the latter dramatically increases with age (de Magalhães, 2013). We and others have shown that senescence may also play a tumor promoting role through a mechanism which is intrinsic to epithelial cells. Normal Human Mammary Epithelial Cells (HMECs) and Normal Human Epidermal Keratinocytes (NHEKs) can spontaneously and systematically escape from the senescent state. Some rare senescent cells re-enter the cell cycle *via* a process we called Post-Senescence Neoplastic Emergence (PSNE) (Gosselin et al., 2009), with characteristics similar to those observed during the early stages of tumor initiation (Gosselin, Martien, et al., 2009; Malaquin et al., 2013; Martin et al., 2014; Romanov et al., 2001). More importantly, PSNE cells xenografted in *nude* mice developed into disseminated skin lesions, such as hyperplasia and small non-melanoma skin carcinoma, evincing their tumorigenic potential (Malaquin et al., 2013; Martin et al., 2014). Post-senescence neoplastic emergence also occurs in a model of BRAF-induced senescence with depleted PTEN (Vredeveld et al., 2012). Importantly, cancer cells exposed to irinotecan were shown to undergo senescence but some of the cancer cells in therapy-induced senescence (TIS) escaped through MCL-1 activation and evolved into cancer cells which were more aggressive than the parental cells (Jonchère et al., 2015). Similarly, the Gorgoulis group demonstrated that in senescence triggered by p21, a subset of cells escaped senescence to re-enter the cell cycle and generate daughter cells with a more aggressive behavior (Galanos et al., 2016). Escape from TIS was demonstrated for a variety of cancer cell types (reviewed in (Pluquet, Abbadie, & Coqueret, 2019)). It is worth to note that escape from TIS is not restricted to *in vitro* cell culture models: it is also found *in vivo* after TIS cell transplantation (Saleh et al., 2019). Therefore, senescence, and especially TIS of cancer cells, cannot be viewed as a definitive arrest but rather as a transient adaptive mechanism making cells prone to switch to a neoplastic cell state.

Even though the molecular mechanisms underlying senescence have been extensively studied, the exact contribution of stress signaling pathways that drive senescence and interfere with the neoplastic transformation is still not fully understood. However, emerging evidence suggests that a key mechanism controlling cellular senescence, and in particular the SASP, could be the unfolded protein response (UPR) (Abbadie & Pluquet, 2020). The UPR is a stress response mechanism that is activated in reply to protein misfolding which dysregulates the ER proteostasis. The UPR is responsible for restoring the correct folding and processing of the proteins synthesized in the endoplasmic reticulum, including those destined for secretion (Pluquet, Pourtier, & Abbadie, 2015). In this review, we focus on the role of the ER stress and the UPR in senescence and on how this stress response can interfere with the neoplastic transformation. We underline the role of the UPR as a potential key controller of the senescence hallmarks. We also discuss the possibility of targeting the UPR components by senolytic treatments as an alternative way to reduce senescence and post-senescence neoplastic escape.

2. Control of endoplasmic reticulum proteostasis: Signaling and function

2.1 Endoplasmic reticulum stress

The endoplasmic reticulum (ER) is a large intracellular organelle made of a network of sheets and tubules. It extends around the chromatin, forming the nuclear envelope, and in the entire cell volume where it makes contacts with other organelles and with the plasma membrane. It constitutes more than a half of the total membrane surface of a cell. The ER is a dynamic structure, organized in functional subregions playing multiple roles. These include calcium storage, lipid synthesis, lipid droplet and autophagosome biogenesis, as well as the synthesis, folding, processing, and transport of luminal or transmembrane proteins of the ER itself, of the Golgi apparatus, of endosomes, lysosomes, plasma membrane proteins, and of secreted proteins (Baumann & Walz, 2001). In the ER homeostasis state, the incoming flow of proteins equals the outgoing flow (normal proteostasis). When the incoming and outgoing protein flows are no longer equal, or when misfolded or unfolded proteins or protein aggregates accumulate, the ER homeostasis is perturbed, a situation referred to as the ER stress (Almanza et al., 2019; Wang & Kaufman, 2016). Various intrinsic and extrinsic factors can cause the ER stress, including redox alterations, hypoxia, temperature changes, and somatic mutations which may lead to oncogene activation. The ER has the ability to sense the proteostasis imbalance and to induce appropriate stress-response pathways, among which the Unfolded Protein Response (UPR), in order to restore the ER homeostasis or to allow the cell to adapt to such new conditions.

2.2 The unfolded protein response

The UPR is a stress response program composed of three distinct subpathways (arms) induced by three sensor proteins: PERK, ATF6α, and IRE1α (Hetz, Chevet, & Oakes, 2015) (Fig. 1). The UPR activation aims at reducing the ER protein load by attenuating the general translational activity and retranslocating the altered proteins to the cytosol for degradation. It also increases the ER folding capacity by activating transcription factors that transactivate genes encoding the ER chaperones and by expanding the ER volume. Finally, it reduces the protein flux into the ER by selectively releasing mRNAs encoding ER-targeted proteins from the ER membrane (Reid, Chen, Tay, Shenolikar, & Nicchitta, 2014) and by degrading the remaining ER-associated mRNAs (Maurel, Chevet, Tavernier, & Gerlo, 2014).

The three UPR arms have been thoroughly described in several recent reviews (Almanza et al., 2019; Karagöz, Aragón, & Acosta-Alvear, 2019; Nam & Jeon, 2019; Pluquet et al., 2015). Briefly, the UPR involves at least three ER resident transmembrane proteins: the PERK kinase, the IRE1α endoribonuclease, and ATF6α. The PERK kinase reduces the overall synthesis of proteins through the phosphorylation of the eIF2α translation initiation factor. eIF2α phosphorylation indirectly induces the translation of mRNAs containing short upstream open reading frames (uORFs), including the one encoding the ATF4 transcription factor which activates the transcription of downstream genes which also participate in the cell response to the ER stress. IRE1α, once activated, induces the unconventional splicing of the XBP1 mRNA, allowing the translation of an active transcription factor whose main targets are the genes involved in the quality control of the ER proteins. In addition, other mRNAs are degraded by IRE1α through a mechanism called regulated IRE1α-dependent decay (RIDD), to prevent the ER overload (Carew, Nelson, Liang, Smith, & Milcarek, 2018; Maurel et al., 2014) (Fig. 1). ATF6α is activated in

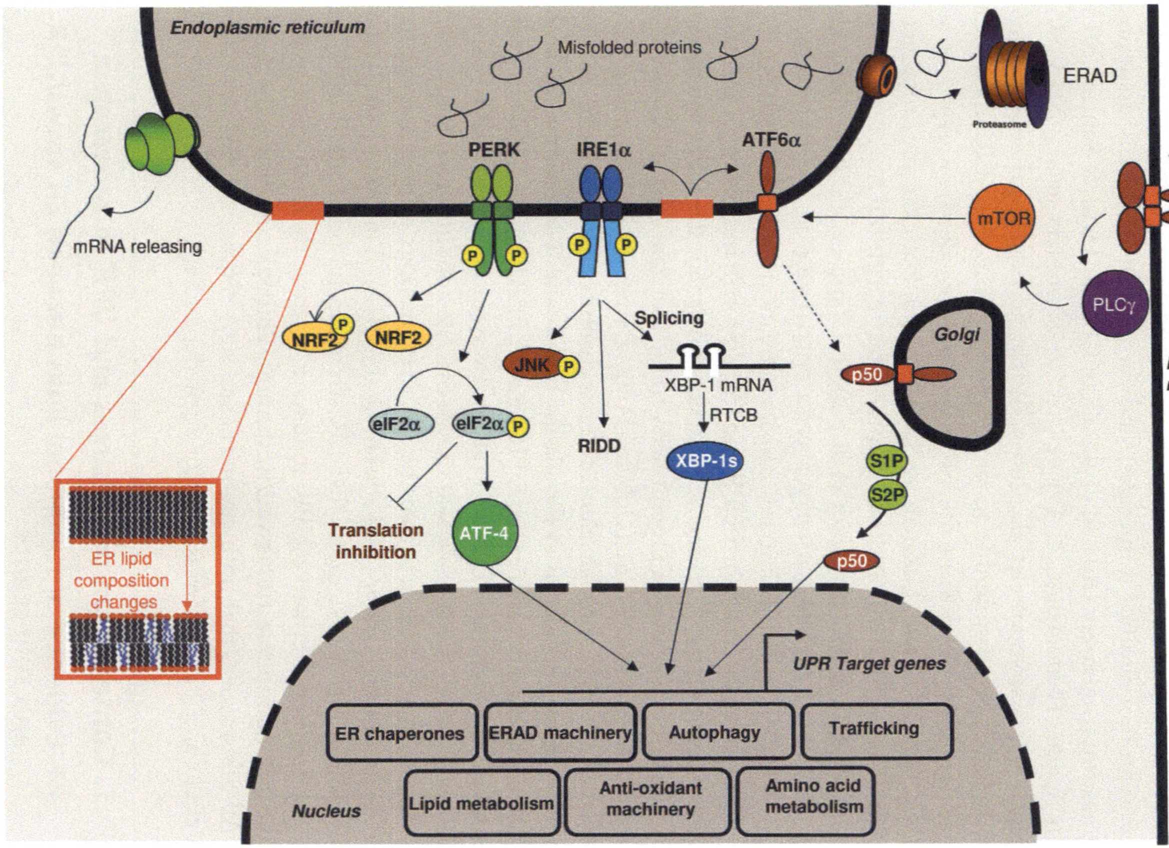

Fig. 1 Activation of the three arms of the Unfolded Protein Response (UPR). The three arms: PERK, IRE1α, and ATF6α, initiate the activation of signaling cascades which increase protein folding, prevent ER overload, and control various cellular processes. In essence, it is an adaptive mechanism that aims to restore the ER homeostasis and promote cell survival.

response to protein misfolding—like PERK and IRE1α—but also in response to alterations in the redox state (Higa et al., 2014; Higa & Chevet, 2012). Once activated, ATF6α migrates from the ER to the Golgi apparatus membrane where it is cleaved to give rise to a soluble active transcription factor targeting genes encoding chaperones, folding enzymes (such as protein disulfide isomerases), and components of the ER-associated degradation machinery (Adachi et al., 2008; Chevet, Hetz, & Samali, 2015; Shen, Chen, Hendershot, & Prywes, 2002; Shoulders et al., 2013; Wu et al., 2007; Yamamoto et al., 2007). When adaptation of cells through the UPR is not possible due to prolonged or unresolved ER stress, new signals are transmitted from the ER to induce cell death (Karagöz et al., 2019; Szegezdi, Logue, Gorman, & Samali, 2006). For example, it has been shown that IRE1α can induce apoptosis by recruiting ASK1 and JNK (Nishitoh et al., 2002; Szegezdi et al., 2006; Urano et al., 2000). In addition, the PERK/ATF4 pathway plays a crucial role in the induction of CHOP, which, in turn, initiates caspase activation (Nishitoh, 2012).

While the main established UPR inducer is protein misfolding or increased protein folding demand, emerging evidence shows that the UPR can also be activated in the absence of protein stress ((Hu, Dougan, McGehee, Love, & Ploegh, 2009), reviewed in (Cubillos-Ruiz, Bettigole, & Glimcher, 2017; Hetz et al., 2015)), through alternative mechanisms. One of them involves ligand-receptor interactions at the plasma membrane. For example, secreted VEGF was shown to induce two UPR sensors: PERK and ATF6α, in endothelial cells through the PLCγ/mTOR pathway in the absence of ER stress (Karali et al., 2014). Besides, it was shown that the conditioned medium from ER-stressed cancer cells activated the ER stress markers in macrophages in a TLR4-dependent manner (Mahadevan et al., 2011). This last result is especially interesting as it suggests that ER stress can be transmitted from cell to cell. Another alternative UPR inducer is altering the lipid composition of the ER membrane (reviewed in (Fun & Thibault, 2019)).

All this suggests that both canonical and non-canonical UPR activation can occur and may be stress- and/or aged-dependent.

3. Contribution of ER stress and UPR to senescence

3.1 Evidence of functional links between ER stress/UPR and senescence

Senescent cells *in vitro* are characterized by an irregular flat shape and enlarged size. Morphological changes are even observed at the organelle level, in particular in the ER which expands in correlation with the cell size expansion (Druelle et al., 2016; Wang & Gundersen, 1984). The activation of senescence by oncogenic HRas or BRaf in human melanocytes was shown to give rise to similar morphological changes, with, however, additional extensive vacuolization (Denoyelle et al., 2006). The ER ultrastructure alterations were also reported in a model of adriamycin-induced senescence in lymphoma cells (Dörr et al., 2013). Drugs known to induce the ER membrane reorganization, such as Troglitazone and nordihydroguaiaretic acid (NDGA) (Varadarajan et al., 2012), were shown to accelerate the onset of senescence in WI38 fibroblasts (Troglitazone) and in cancer cells (NDGA) (Cui et al., 2008; Gan et al., 2008). Moreover, numerous reports indicated that the three UPR sensors and their effectors are involved in senescence, even though not always together, the involvement of the particular players depending on the senescence inducer and the cell type (see Table 1 for details). In particular, we demonstrated that the three UPR arms were activated with different kinetics in replicative senescent dermal fibroblasts, while the overall level of the UPR activation was rather mild and progressive, making the senescent fibroblasts still able to respond to an acute ER stressor (Cormenier et al., 2017; Druelle et al., 2016). Among the three UPR sensors, ATF6α appeared to be the main actor responsible for inducing some senescence hallmarks (Cormenier et al., 2017; Druelle et al., 2016). Moreover, ATF6α was also shown to be the main UPR sensor involved in senescence in keratinocytes undergoing stress-induced senescence, but in that case in conjunction with PERK (Drullion et al., 2018). ATF6α was also activated in mammary epithelial cells undergoing senescence induced by the expression of the HRasV12 oncogene, by exposure to UV, or by NBR1 abrogation (Kim, Kim, et al., 2019). In contrast, the results of Blazanin and colleagues have pointed out to the involvement of IRE1a in keratinocytes undergoing HRasV12-induced senescence. Interestingly, they underlined that IRE1α RIDD and XBP1 splicing activities seemed to play opposite roles in these cells. They found that the degradation of Id1 mediated by the IRE1α RIDD activity promoted HRas-induced senescence, whereas XBP1 splicing favored proliferation (Blazanin et al., 2017). Besides, it was shown that when the expression of PTEN and SKP2 was lost, the phosphorylation of eIF2a and the activation of ATF4, the two main effectors of the PERK arm of the UPR, induced senescence (Lin et al., 2010).

There is also *in vivo* evidence supporting the claim that the UPR activation is associated with senescence. One report showed that spitz

Table 1 UPR sensors and their effectors whose expression was shown to be altered during senescence.

Senescence program	Cell type	Expression of UPR sensors and/or their effectors	References
Replicative senescence			
	NHDFs	Increased Herpud1, Orp150, Ero1lβ, Chop, Erdj4 mRNA levels Increased CANX, PDI, ATF4, XBP1s, P-eIF2α, protein levels	Druelle et al. (2016)
	WI38	Increased Bip, Xbp1s mRNA levels Increased BIP, ERO1Lα protein levels, increased P-eIF2α levels, increased ATF6(p50) expression	Matos, Gouveia, and Almeida (2014)
	Rat embryo fibroblasts	Increased α-glucosidase II	Benvenuti et al. (2002)
	Human Dermal Fibroblasts	Decreased Calreticulin, BIP/GRP78, PDI protein Levels	Boraldi, Annovi, Tiozzo, Sommer, and Quaglino (2010)
	HUVECs	Increased Calreticulin, PDIA1 protein levels, decreased GRP94 protein levels	Klooster et al. (2009)
	HUVECs and BJ fibroblasts	Increased Chop/Gadd153 mRNA levels	Shelton, Chang, Whittier, Choi, and Funk (1999)
	Human Dermal Fibroblasts	Decreased CANX protein levels	Choi and Kim (2004)
	Human fibroblasts	Decreased HSP47 protein levels	Nizard et al. (2004)
	Primary Kidney Macrophages	Decreased Bip/Grp78 mRNA levels	Barreda, Hanington, Walsh, Wong, and Belosevic (2004)

Oncogene-induced senescence			
HRAS	Keratinocytes	IRE1α-mediated RIDD activation Reduced PERK/ATF4 activity Reduced XBP1s activity	Blazanin et al. (2017)
HRAS	Melanocytes	Increased Xbp1s mRNA levels. Increased BIP/GRP78, CHOP, ATF4, P-PERK protein levels. Decreased calreticulin protein levels	Denoyelle et al. (2006)
HRAS	Primary murine keratinocytes	Increased Atf4 mRNA levels Increased CNX, ATF6α, and P-PERK protein levels	Zhu, Ferry, et al. (2014)
HRAS	IMR90 cells	Increased Atf4, Xbp1s, Bip, Chop mRNA levels	Chen et al. (2015)
HRAS	human diploid embryonic lung fibroblasts)	Increased CHOP, ATF4 protein levels	Dörr et al. (2013)
HRAS	MCF7	Increased BIP, P-eiF2α, cleaved ATF6α protein levels	Kim, Kim, et al. (2019)
Stress-induced premature senescence			
CuSO4	WI38	Increased Bip, Xbp1s mRNA levels Increased BIP, ERO1Lα protein levels, increased P-eIF2α levels, increased ATF6(p50) expression	Matos et al. (2014)
H2O2		Decreased Bip, Xbp1s, Chop mRNA levels Increased ERO1Lα protein levels, increased P-eIF2α levels	

Continued

Table 1 UPR sensors and their effectors whose expression was shown to be altered during senescence.—cont'd

Senescence program	Cell type	Expression of UPR sensors and/or their effectors	References
Endogenous ROS	NHEKs	Increased Bip, Grp94, Herpud1, Gadd34, chop, Atf4, Hrd1, Edem1, Xbp1s mRNA levels	Drullion et al. (2018)
AGE, Thapsigargin	Proximal tubular epithelial cells	Increased BIP/GRP78 protein levels	Liu et al. (2014)
Tetrasubstituted Naphthalene Diimide Analogs	MIA, PaCa-2 pancreatic cells	Increased Chop mRNA levels	Micco et al. (2013)
Low glucose treatment	Normal human skin fibroblasts	Increased ORP150 protein levels	Krętowski, Stypułkowska, and Cechowska-Pasko (2013)
Metformin	MDA-MB-468	Increased Chop mRNA levels	Williams, Singleton, Llopis, and Skripnikova (2013)
Oroxin A	MDA-MB-231 and MCF7	Increased Atf4, Bip mRNA levels	He et al. (2016)
AGE	Renal tubular epithelial cells (RTECs)	Increased Atf4 mRNA levels Increased ATF4, Bip protein levels	Liu et al. (2015)
Tunicamycin, Thapsigargin		Increased ATF4 protein levels	
DTT	NHDFs	Increased Chop, Bip, Herpud1, Ero1lb mRNA levels	Cormenier et al. (2017)

E235	HT1080	Increased Atf4, Xbp1s mRNA levels Increased ATF4, P-eIF2α protein levels	Sayers et al. (2013)
UV	MCF7	Increased BIP, P-eiF2α, cleaved ATF6α protein levels	Kim, Kim, et al. (2019)
TIS			
Adriamycin	Lymphoma cells	Increased ATF4, CHOP protein levels	Dörr et al. (2013)
X-Rays	Pulmonary artery endothelial cells	Increased Bip/Grp78, Chop, and Gadd34 mRNA transient phosphorylation of eIF2α	Panganiban, Mungunsukh, and Day (2013)
Genetically-induced senescence			
SKIP2$^{-/-}$ mouse	MEFs	Increased ATF4 activity	Lin et al. (2010)
Silencing of NBR1 (siRNA)	MCF7	Increased BIP, P-eIF2α, XBP1s, cleaved ATF6α protein levels	Kim, Kim, et al. (2019)
Mitochondrial dysfunction-induced senescence	OPA1$^{-/-}$ muscle cells	Increased XBP1 splicing and PERK/ATF4 activation	Tezze et al. (2017)

naevi-containing melanocytes expressing mutant HRasV12 displayed increased expression of BiP/GRP78, an ER chaperone mainly induced by the ATF6a arm of the UPR, and of the PDI protein disulfide isomerase which is involved in the control of disulfide bond formation (Denoyelle et al., 2006). By using a mouse model mimicking the development of retinal dysfunction through constant endothelial activation by transmembrane TNF-α, Lenin and colleagues showed exacerbated age-induced visual deficits resulting from the ER stress in correlation with premature senescence (Lenin, Nagy, Gentry, & Gangaraju, 2018). Xbp1 splicing by IRE1α and the activation of the ATF4 transcription factor by PERK were also found in muscle cells undergoing mitochondrial dysfunction-induced senescence (Tezze et al., 2017). Apart from that, IRE1α endoribonuclease activity was shown to play a key role in inducing a senescent phenotype and the SASP in ischemic retinal cells (Oubaha et al., 2016).

The fact that senescence is associated with a certain degree of the UPR activation raises a question whether UPR activation alone could be sufficient to drive cell senescence. An argument speaking in favor of this hypothesis is our finding that ER stressors, such as 00 and dithiothreitol, induce senescence hallmarks, *e.g.,* SA-β-Gal activity and DNA damage, in proliferating fibroblasts (Druelle et al., 2016). The same results were obtained by others in mouse primary proximal tubular epithelial cells (Liu et al., 2014). Besides, it has been shown that the activation of ATF4 by pharmacological compound E235 increased the expression of senescence markers in cancer cells (Sayers et al., 2013).

Some reports, however, suggested that the UPR activation may protect from senescence. Zhu and colleagues reported an increased number of SA-β-Gal-positive cells and a reduced $p21^{WAF1}$ expression during oncogene-induced senescence (OIS) in primary murine keratinocytes silenced for XBP1 (Zhu, Ferry, et al., 2014). In the same line, deleting ATF4 was shown to trigger senescence in mouse embryonic fibroblasts (MEFs) transformed by HRasV12 and the SV40 large T antigen (Horiguchi et al., 2012). Finally, inhibiting PERK—but not ATF6a and IRE1a—induced cellular senescence in MCF7 and HT1080 cells (Kim, Jang, Choe, Sohn, & Kim, 2012), and silencing eIF2α was shown to accelerate senescence in HT1090 cells treated with doxorubicin and in xenografted tumors in which senescent cells were quantified by SA-β-Gal staining (Rajesh et al., 2013).

Interestingly, there are also some data showing that the ER stress may be a consequence of senescence, following the increased protein synthesis demand related to the SASP production which induces proteotoxic stress impairing the ER function and resulting in the UPR activation (Chen et al., 2015; Dörr et al., 2013) (see also Tables 1 and 2, and Fig. 2). Conversely, the senescent status could decrease the ability of cells to activate the UPR. For instance, it has been shown that the recruitment of p53 to synoviolin, an ER transmembrane E3 ubiquitin ligase promotes IRE1α ubiquitination and its subsequent degradation by the proteasome (Namba et al., 2015). This suggested that once p53-mediated senescence is established, the sensitivity of the senescent cell to the ER stress could decline.

Further investigations are needed to determine the cause-effect relationship between senescence and the UPR as well as the kinetics and specificity of the different UPR arms' activation in different types of senescence. However, emerging data already indicate different mechanisms through which the UPR contributes to the development of characteristic features (senescence hallmarks) in senescent cells.

3.2 UPR impacts most of the senescence hallmarks

3.2.1 UPR could contribute to controlling the SASP

One of the hallmarks of senescent programs is the senescence-associated secretory phenotype (SASP). Since the ER is the site of synthesis, folding and maturation of the secreted proteins, one can assume that the ER stress and the UPR activation could be involved in establishing the secretome changes associated with senescence. The SASP is an important driver of chronic inflammation acting *via* autocrine/paracrine pathways. It is mainly controlled by NF-kB (Ferrand et al., 2015), mTOR (Herranz et al., 2015; Laberge et al., 2015), p38MAPK (Freund, Patil, & Campisi, 2011), and Notch signaling (Hoare et al., 2016). The pro-inflammatory molecules secreted by senescent cells include cytokines and growth factors, such as IL6 and IL8 which reinforce the senescent phenotype (Acosta et al., 2008; Kuilman et al., 2008), IL1β and TGFβ which cooperatively induce senescence through ROS and persistent DDR (Hubackova, Krejcikova, Bartek, & Hodny, 2012), as well as IL1α which plays a role in controlling the IL6/8 network (Orjalo, Bhaumik, Gengler, Scott, & Campisi, 2009). Many connections between the ER stress and inflammation have been described (Dandekar, Mendez, & Zhang, 2015; Schmitz, Shaban, Albert, Gökçen, & Kracht, 2018). Interestingly, all UPR sensors have been shown to be involved in the production of cytokines and chemokines, including IL1β, IL6, and IL8, in cancer cells (Auf et al., 2010; Dejeans et al., 2012; Garg et al., 2012; Reverendo, Mendes, Argüello, Gatti, & Pierre, 2019).

Table 2 Targeting UPR effectors impacting senescence hallmarks in functional studies.

Senescence program	Cell type	Type of targeting of UPR effectors	Senescence phenotype	References
Replicative senescence				
	NHDFs	siATF6α	Decreased number of SA-β-Gal positive cells Reversing cell enlargement Decreased Dcr2, IL6 mRNA levels	Druelle et al. (2016)
		siIRE1α	Decreased number of SA-β-Gal positive cells	
Oncogene-induced senescence				
HRAS	Melanocytes	siATF4	Decreased number of SA-β-Gal positive cells	Denoyelle et al. (2006)
		siXBP1, Dominant Negative IRE1α	Decreased number of SA-β-Gal positive cells	
		Dominant Negative ATF6α	Decreased number of SA-β-Gal positive cells	
HRAS + SV40 Large T antigen	Primary embryo fibroblasts	$ATF4^{-/-}$	Increased $p16^{INK4}$ and $p19^{ARF}$ protein levels	Horiguchi et al. (2012)
HRAS	Primary murine keratinocytes	siATF4	Increased number of SA-β-Gal positive cells Decreased p21 protein levels	Zhu, Ferry, et al. (2014)
HRAS	MCF7	siATF6α	Decreased number of SA-β-Gal positive cells	Kim, Kim, et al. (2019)

Stress-induced premature senescence				
CuSO4	WI38	PERK inhibition: GSK2606414 IRE1α inhibition: 4u8c	Decreased number of SA-β-Gal positive cells	Matos et al. (2014)
H_2O_2	WI38	PERK inhibition: GSK2606414	Decreased number of SA-β-Gal positive cells	
DTT	NHDFs	siATF6α	Decreased number of SA-β-Gal positive cells Decrease of cell enlargement	Cormenier et al. (2017)
		siATF6α	Decreased number of SA-β-Gal positive cells Decreased Dcr2, IL6 mRNA levels	Drullion et al. (2018)
		siPERK	Decreased number of SA-β-Gal positive cells Decreased IL6 mRNA levels	
AGE	Renal tubular epithelial cells (RTECs)	4-PBA	Decreased number of SA-β-Gal and SAHF positive cells	Liu, Yang, et al. (2015)
		siATF4	Decreased number of SA-β-Gal and SAHF positive cells	
E235	HT1080	shATF4	Decreased gH2AX protein levels	Sayers et al. (2013)
UV	MCF7	siATF6α siPERK	Decreased number of SA-β-Gal positive cells	Kim, Kim, et al. (2019)

Continued

Table 2 Targeting UPR effectors impacting senescence hallmarks in functional studies.—cont'd

Senescence program	Cell type	Type of targeting of UPR effectors	Senescence phenotype	References
Genetically- induced senescence				
Silencing of NBR1 (siRNA)	MCF7	siATF6α	Decreased number of SA-β-Gal positive cells	Kim, Kim, et al. (2019)
siPERK	HT1080	siPERK	Increased number of SA-β-Gal positive cells Increased p21 protein levels	Kim, Jang, et al. (2012)
siCalreticulin	Human endometrial stromal cells (ESCs)	siCALR	Increased number of SA-β-Gal positive cells	Kusama et al. (2014)
sieIF2α	Doxorubicin-induced senescence in HT1080 cells and tumor xenograft assays		Increased number of SA-β-Gal positive cells	Rajesh et al. (2013)
$PERK^{-/-}$ $eIF2\alpha^{A/A}$	Primary MEFs		Increased number of SA-β-Gal positive cells	Rajesh et al. (2013)

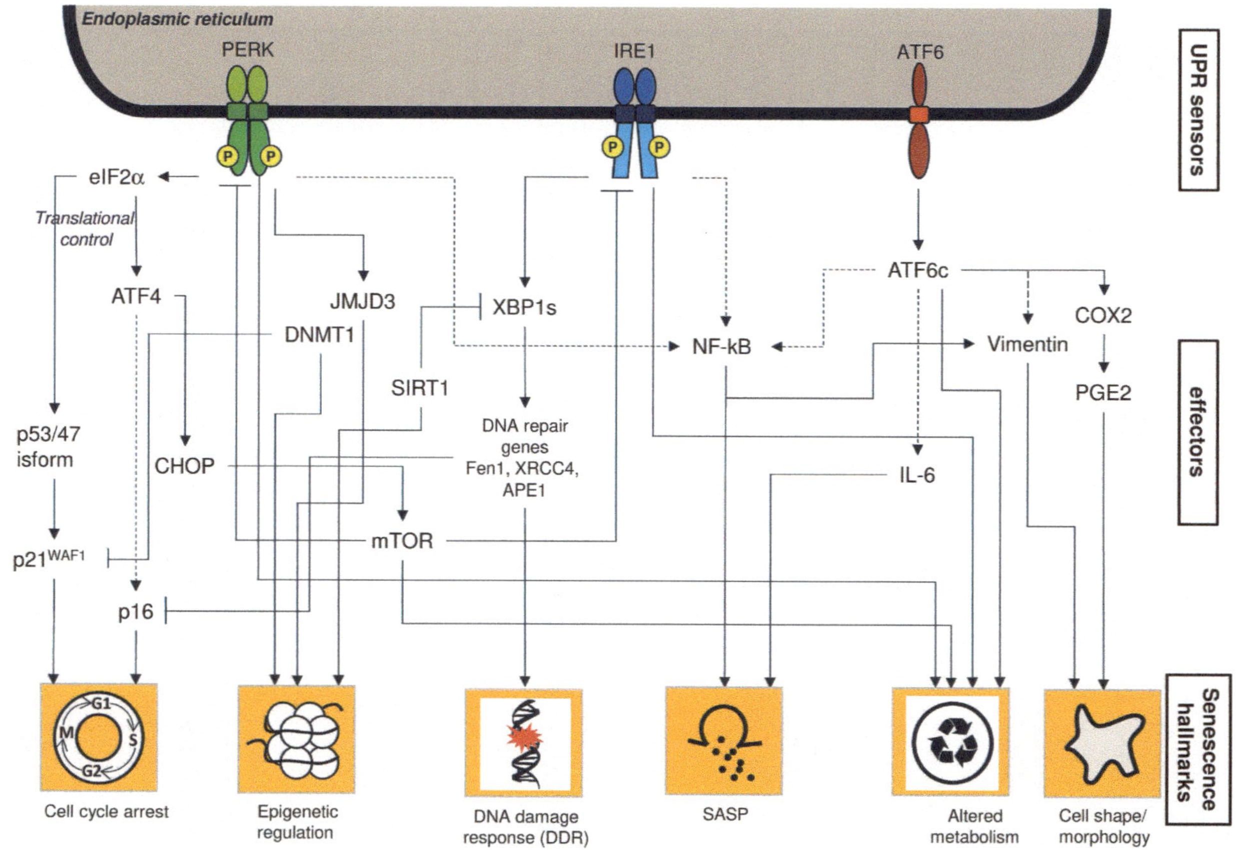

Fig. 2 The Unfolded Protein Response (UPR) network associated with senescence hallmarks. The UPR activates downstream effectors which mediate the appearance of senescence hallmarks. Dotted lines indicate pathways with no direct interaction. Arrows indicate positive regulation and dot-end—inhibition. JMJD3: Jumonji domain-containing 3 histone lysine demethylase; mTOR: mechanistic target of rapamycin; DDR: DNA Damage Response pathway; SASP: Senescence-Associated Secretory Phenotype.

Conversely, the release of cytokines in the microenvironment results in the activation of the UPR arms (Brozzi et al., 2015; Kharroubi et al., 2004; Xue et al., 2005).

In addition to growth factors and cytokines, the SASP comprises extracellular matrix (ECM) components and enzymes remodeling the ECM, including matrix metalloproteinases (MMPs). These molecules have been reported to be upregulated by the UPR in both normal and cancer cells. For example, XBP1 controls the expression of MMP1 and MMP3 in oral squamous cell carcinoma (Sun et al., 2018), whereas ATF4 induces MMP2 and MMP7 expression in human esophageal squamous cell carcinoma cells (Zhu, Chen, et al., 2014).

Although one could assume that the ER stress and the UPR contribute to the secretome of senescent cells, as they do for the secretome of cancer cells, the data directly linking the ER stress with the SASP are scarce. One study showed that knocking down ATF6α in senescent fibroblasts decreased IL6 mRNA levels (Druelle et al., 2016). Another one demonstrated that Hras-mediated senescence induced the UPR and also mediated the expression of macroH2A1 (involved in the formation of senescence-associated heterochromatin foci (SAHF)), which, in turn, induced the expression of various SASP-associated genes, including cytokines, MMPs, and growth factors. In this model, macroH2A1 was regulated by a positive feedback loop mediated by the SASP components. The ER overloaded with SASP components activates the UPR which restricts the recruitment of macroH2A1 through a ROS/DNA damage-activated ATM-dependent mechanism (Chen et al., 2015). Clearly, the exact role of the UPR effectors will have to be further addressed in the context of SASP.

3.2.2 UPR contributes to alterations in the cell morphology associated with senescence

The UPR was shown to contribute to morphological changes of senescent cells. This is likely to occur by changes in the expression of vimentin, a cytoskeleton filament intermediate. Indeed, the vimentin protein levels were shown to be increased in senescent fibroblasts and seemed to induce morphological changes giving rise to the characteristic cell shape associated with senescence (Nishio et al., 2001). In the same line, data from our lab suggests that ATF6α-dependent morphological changes observed during senescence are linked to remodeling of the vimentin cytoskeleton (Druelle et al., 2016). However, it remains unclear if increased vimentin levels are due to increased vimentin synthesis or due to its decreased degradation.

The fact that the vimentin promoter can be activated by NF-kB, a transcription factor playing a key role in senescence, speaks in favor of the first hypothesis. Besides, the SPARC protein, involved in the extracellular matrix synthesis but also in the control of cell size and cell spreading (Salvatierra et al., 2015), was shown to be upregulated in human fibroblasts undergoing SIPS induced by *tert*-butylhydroperoxide (*t*-BHP) (Dumont et al., 2000) or UVB (Chainiaux, Magalhaes, Eliaers, Remacle, & Toussaint, 2002). Interestingly, SPARC protein expression correlated with the expression of vimentin (López-Moncada, Torres, Castellón, & Contreras, 2019; Shi et al., 2016), and SPARC mRNA levels were found to be regulated by IRE1α in U87 cancer cells (Dejeans et al., 2012). In addition, the FLNA protein, an actin crosslinker involved in the regulation of the cytoskeleton remodeling, plays a role in mitochondrial fission-associated myocardial senescence by increasing aberrant interactions between mitochondria and actin cytoskeleton (Nishimura et al., 2018). Interestingly, IRE1α and PERK were demonstrated to bind to FLNA, and deleting IRE1α or PERK impaired the actin cytoskeleton dynamics (Urra et al., 2018; van Vliet et al., 2017). Finally, LOXL2, a protein known to catalyze the first step in the formation of crosslinks in collagens and elastin, is expressed during cellular senescence (Pascal et al., 2005; Pollheimer et al., 2018) and has been shown to be involved in the architecture of the actin cytoskeleton (Zhan et al., 2017). It has been shown that LOXL2 accumulating in the ER interacts with GRP78 and activates the IRE1α/XBP1 UPR arm (Cuevas et al., 2017), while nuclear LOXL2 is associated with the upregulation of vimentin (Hong & Yu, 2019; Moon et al., 2013). Given the cell shape changes occurring during senescence, these data seem to support the hypothesis that the UPR effectors may play a role in the remodeling and reorganization of actin and intermediate filament cytoskeletons during senescence.

3.2.3 UPR contributes to the senescence-associated cell cycle arrest

Senescence is a cellular state characterized by a stable cell cycle arrest. The senescence-associated cell cycle arrest is mainly controlled by the p53/$p21^{WAF1}$ and/or $p16^{INK4}$/Rb pathways (Campisi, 2001). The stability of the cell cycle arrest results from the accumulation of unrepaired DNA damage. It is well described that in the context of replicative senescence or oncogene-induced senescence, the unrepaired DNA damage include shortened telomeres, DNA double-strand breaks and DNA replication stress, all of which trigger a persistent DNA Damage Response (DDR). DDR induces the recruitment and binding of 53BP1 and γH2AX to the DNA damage site to promote the activation of a signaling cascade involving the ATM/ATR and CHKs kinases, followed by a sustained activation of the p53/p21(TP53/CDKN1A) pathway (Abbadie et al., 2017).

In oxidative stress-induced senescence, the main type of DNA damage are single-strand breaks (SSBs). They remain unrepaired because of a downregulated expression of *PARP1*, the first-line enzyme of the SSB repair pathway, which is also a consequence of the oxidative stress. The accumulation of unresolved SSB repair foci becomes the initiating event of a signaling cascade which involves the p38MAPK, leads to the upregulation of p16, to the hypophosphorylation of the Rb tumor suppressor and to cell cycle arrest (Nassour et al., 2016).

It has been known for decades that the ER stress can impair cell cycle progression (Brewer & Diehl, 2000; Brewer, Hendershot, Sherr, & Diehl, 1999; Malzer et al., 2010; Raven et al., 2008). Yet, the ER stress and the UPR were shown not to activate p53 but rather enhance its degradation (Pluquet, Qu, Baltzis, & Koromilas, 2005). Therefore, the question is whether the UPR could induce senescence by activating these cell cycle arrest pathway components independently of their canonical inducers.

We showed that silencing ATF6α, IRE1α or PERK in senescent fibroblasts failed to alleviate the senescent cell cycle arrest (Druelle et al., 2016). In contrast, silencing PERK and ATF6α—but not IRE1α—in senescent keratinocytes (NHEKs) restored their proliferation (Drullion et al., 2018) to nearly normal levels. This difference could be explained by the fact that senescence in fibroblasts is telomere-dependent, whereas in NHEKs it depends on oxidative stress (Abbadie et al., 2017; Nassour et al., 2016).

HRASV12 is an oncogene which induces cell hyperproliferation within 24 h following its activation. This hyperproliferation leads to replicative stress which then induces a delayed cell cycle arrest relevant of senescence within the 5–10 following days. The activation of the UPR pathway is involved in both of these steps. Indeed, knocking down IRE1α reduced proliferation during the first phase but blocked the induction of senescence and sustained proliferation during the second phase. These effects were shown to be independent of *Xbp1* splicing since XBP1 knockdown blocked proliferation and accelerated senescence (Blazanin et al., 2017).

The XBP1s transcription factor may play a role in senescence when downregulated. Indeed, XBP1 has been suggested to control a cluster of DNA repair genes (Acosta-Alvear et al., 2007), including Fen1 (an endonuclease), XRCC4 (involved in the NHEJ repair), APE1/Ref1 (involved in the BER pathway and redox regulation). Interestingly, the inhibition of Fen1 led to senescence induction in colorectal cancer cells through the p53/p21^{WAF1} pathway (Jaiswal et al., 2015). Overexpressing the XRCC4-Lig4 complex that ligates the broken ends during NHEJ reduced p16^{INK4A} protein expression and suppressed senescence in human fibroblasts that underwent X-ray treatment (Li et al., 2016). The downregulation of APE1/Ref1 is observed in a H_2O_2-induced senescence model of the human bone marrow derived mesenchymal stem cells (hBMSCs). Its overexpression has been shown to rescue cells from SIPS by decreasing ROS production (Heo et al., 2009). In addition, silencing APE1/Ref1 resulted in premature senescence in normal human BJ fibroblasts and a conditional knockout of APE1/Ref1 in 6-weeks old mice increased p16^{INK4A}, p21^{WAF1}, and γH2AX protein expression as well as the SA-β-Gal activity in colon and skin (Li et al., 2018). It has been shown that ATF4 is involved in DNA repair induced by arsenite stress (Fung, Liu, & Demple, 2007). Recently, a bioinformatic study predicted that ATF4 may play a key role in increasing the DNA repair capacity of senescent cells (Zhao et al., 2019).

Overall, these data support the hypothesis that the UPR mediates senescence by controlling DDR and DNA repair capacity.

3.2.4 UPR could contribute to chromatin remodeling associated with senescence

Senescent cells undergo alterations of chromatin and a redistribution of epigenetic marks. Remarkably, the inhibition of DNA methylases or histone acetylases can induce senescence (Petrova, Velichko, Razin, & Kantidze, 2016), demonstrating the key role played by epigenetic mechanisms and associated chromatin modifications. Interestingly, several studies have shown that the ER stress can impact chromatin remodeling by regulating chromatin posttranslational modifications, such as methylation and acetylation, which, in turn, impact the transcription of the ER target genes (reviewed in (Dicks, Gutierrez, Michalak, Bordignon, & Agellon, 2015)). In the same line, several proteins involved in the remodeling of chromatin have been shown to interact, either directly or indirectly, with different components of the UPR.

For example, the expression of the GRP78 ER chaperone is mainly controlled by ATF6α. Its transcriptional activation involves the recruitment of coactivators, including the PRMT1 arginine methyltransferase (Baumeister et al., 2005). Interestingly, we did not find increased GRP78 expression during replicative senescence of normal human dermal fibroblasts (NHDFs) (Druelle et al., 2016). This is in agreement with the fact that PRMT1 protein expression is reduced in WI38 fibroblasts undergoing replicative senescence or SIPS induced by a H_2O_2 treatment (Lim, Lee, Lee, Oh, & Kim, 2008). Moreover, silencing PRMT1 in breast cancer cells or in neuroblastoma SK-N-SH cells was shown to induce cellular senescence (Gao et al., 2016; Lee et al., 2019).

Furthermore, histone acetyltransferase p300/PCAF acts as a coactivator of XBP-1s (Lew et al., 2011) and ATF6α transcriptional activity (Baumeister et al., 2005). p300/PCAF was recently shown to drive the senescent phenotype (Sen et al., 2019), and the functional role of ATF6α in the onset of senescent programs was underlined in several studies (Druelle et al., 2016; Kim, Kim, et al., 2019).

The UPR is also an important regulator of methyltransferases, and altered expression levels of the latter have been shown in different senescence types. The gene encoding Jumonji domain-containing 3 histone lysine demethylase (JMJD3) that specifically demethylates di- or tri-methylated lysine 27 of histone H3 (H3K27me2 or H3K27me3) is an ATF4 target gene (Shan, Fu, Balasubramanian, Anthony, & Kilberg, 2012). The expression of JMJD3 is upregulated during OIS in human fibroblasts (Agger et al., 2009). Lower expression levels of SUV39H1, a methyltransferase for lysine 9 on histone H3, during senescence are associated with reduced H3K9 trimethylation in WI38 cells (Sidler et al., 2014) and with the ER stress activation mediated by the upregulation of ATF3 and CHOP in NSCLC cells (Liu, Guo, Liu, & Su, 2015). Finally, the DNMT1 methyltransferase is downregulated during senescence and it is involved in $p16^{INK4A}$ promoter regulation (So, Jung, Lee, Kim, & Kang, 2011). Inhibiting DNMT1 was shown to induce a senescent phenotype in human umbilical cord blood-derived multipotent stem cells (So et al., 2011) and in 2BS cells (human embryonic lung fibroblasts) (Pan et al., 2013). Interestingly, inducers of the ER stress, such as valproic acid and Tunicamycin, reduce the DNMT1 protein levels in human lens epithelial cells (Palsamy, Bidasee, & Shinohara, 2014).

Another key chromatin modifier involved in cellular senescence and therapy-induced senescence is MacroH2A1 (Chen et al., 2015; Zhang et al., 2005), a variant of histone H2A (Borghesan et al., 2016). MacroH2A1.1 interacts with PARP-1 and the CREB-binding protein (CBP) to indirectly mediate acetylation of H2BK120, which, in turn, positively regulates the expression of the SASP genes upon OIS (Ruiz & Gamble, 2018; Yu et al., 2017). Chen and colleagues have established that removing macroH2A1 from SASP gene promoters during OIS is dependent on the ER stress activation. In their study, OIS induced ER stress, which led to the production of ROS and persistent DNA damage response through the activation of ATM which bound to macroH2A1 on the SASP genes promoters, altering their functions (Chen et al., 2015).

In addition, the levels of SIRT1, a modulator of chromatin function, were reported to decrease in senescent mouse embryonic fibroblasts (Sasaki, Maier, Bartke, & Scrable, 2006). The inhibition of SIRT1 was also shown to promote premature senescence in endothelial cells (Ota et al., 2007). In contrast, its overexpression suppressed the senescent phenotype in primary porcine aortic endothelial cells (Zu et al., 2010). Interestingly, SIRT1 inhibited the XBP1s transcriptional activity by its deacetylation (Wang, Chen, & Ouyang, 2011).

Taken together, all these data suggest a functional link between the UPR and epigenetic regulators, which could be associated with the onset of senescence.

3.2.5 Crosstalk between UPR and autophagy during senescence

Macro-autophagy, here simply called autophagy, is an evolutionarily conserved lysosome-mediated process allowing the cell to degrade some altered macromolecules or organelles. It occurs through dynamic rearrangements of cellular membranes that enclose cytoplasmic constituents to form a double membrane vesicle called autophagosome. This process is controlled by a set of ATG genes (Kroemer, Mariño, & Levine, 2010). The origin of the double membrane is still debated though and it is actually likely that is does not come from a single source. Depending on the cell and stress type, fragments of endoplasmic reticulum as well as of mitochondrial and plasma membranes may be incorporated in the autophagic membrane (Rubinsztein, Shpilka, & Elazar, 2012).

Autophagy could either help cells to survive or induce cell death, depending on the duration and the degree of stress. A majority of studies demonstrated that autophagy plays a critical role during the establishment of senescence triggered by various types of stress (Deruy et al., 2010; Gosselin et al., 2009; Ivanov et al., 2013; Singh, Matsuyama, Drazba, & Almasan, 2012; Young et al., 2009; Young, Narita, & Narita, 2011). However, one study showed the contrary: that its inhibition favored senescence (García-Prat et al., 2016). Moreover, some evidence indicate that autophagy and senescence, the two clearly interconnected, are also linked to ER homeostasis (Smith & Wilkinson, 2017; Yan, Ni, Song, Ding, & Huang, 2015).

Overall, most of the data lead to the hypothesis that the senescence-associated ER stress contributes to increase the autophagic activity. For example, inhibiting the PERK and IRE1α arms of the UPR by siRNA significantly decreased the autophagy-mediated senescence induced by exposure to the penta-1,2,3,4,6-O-galloyl-β-D-glucose (PGG) chemopreventive polyphenolic compound (Dong et al., 2014). The mechanisms by which the ER stress induces autophagy are not fully characterized but it has been shown that several transcription factors of the UPR can partly regulate the expression of ATG genes (Ding et al., 2007) at different steps of the process. For example, the MAP1LC3B/ATG8 and ATG1/ULK1 genes are direct transcriptional targets of the PERK/ATF4 UPR arm, and the ATG5 gene is among those of the PERK/CHOP one (Pike et al., 2013; Rouschop et al., 2010). Silencing ATF4 in TIS glioblastoma cells significantly decreased the levels of MAP1LC3-II which is indispensable to the autophagic activity (Wang et al., 2018). The role of the increased autophagic activity during senescence is not yet clear. The group of Clemens Schmitt proposed that its function could be to relieve the proteotoxic stress induced by the increased concentrations of proteins composing the SASP in the ER (Dörr et al., 2013).

Even though the mechanisms interconnecting the UPR, autophagy, and senescence need further investigations, it seems likely that the UPR may be required for controlling autophagy-mediated senescence. It is worth noting that the connections between the UPR and chaperone-mediated autophagy (CMA) or micro-autophagy during senescence have not yet been explored.

3.2.6 UPR controls senescence-associated β-galactosidase activity

Increased β-Galactosidase (β-Gal) activity is one of the characteristics of senescent cells, and the lysosomal senescence-associated β-Gal activity (SA-β-Gal) is often used as a marker of increased lysosomal content (Dimri et al., 1995). The inhibitors of PERK and IRE1 were shown to decrease SA-β-Gal activity in $CuSO_4$-induced SIPS (Matos et al., 2014). In the same line, silencing ATF6α was shown to revert several senescent features in normal fibroblasts, including SA-β-Gal activity (Druelle et al., 2016). The hypothesis that ATF6α plays an important role in regulating the SA-β-Gal activity is also supported by the results of studies where senescence was induced by various stress conditions, including NBR1 abrogation, H-Ras^{V12} activation and UV irradiation (Denoyelle et al., 2006; Kim, Kim, et al., 2019). Taking all this into consideration, the SA-β-Gal activity may reflect the increase in lysosome and autophagic vacuole content, which may be controlled by the UPR.

3.2.7 *UPR is connected to the senescent signaling of mTOR*

The Mechanistic Target of Rapamycin (mTOR) protein controls numerous cellular processes, including autophagy, and plays an important role in promoting the secretory phenotype of senescent cells (Herranz et al., 2015; Laberge et al., 2015; Narita et al., 2011). Several studies showed a cross-talk between mTOR and the UPR. Indeed, mTOR can attenuate IRE1α activity by controlling its autophosphorylation state (Sanchez-Alvarez, Del Pozo, & Bakal, 2017). mTOR can also decrease PERK activity by downregulating the eIF2α phosphorylation in cells in which the Tuberous Sclerosis Complex (TSC) is lost (Krishnamoorthy et al., 2018). Inhibiting mTOR by rapamycin suppresses the activation of PERK and IRE1 (Bachar et al., 2009; Pfaffenbach et al., 2010). Conversely, prolonged mTORC1 activity upregulates the expression of ATF4 and CHOP (Panda et al., 2018). To date, only one study demonstrated the existence of a signaling cascade through which the ER stress regulates the mTOR activity, leading to senescence. Wang and colleagues showed that flavokawain B-induced senescence in glioblastoma cells depends on autophagy through the ATF4-CHOP-TRIB3-AKT-mTOR-RPS6KB1 signaling pathway (Wang et al., 2018). Further studies are needed to better understand the role of the UPR in metabolic dysfunctions associated with senescence.

3.2.8 *UPR could contribute to lipid accumulation that triggers senescence*

Many studies considered the senescent phenotype as a DNA damage response. However, emerging evidence indicates that also deregulation of lipid metabolism or treatment with specific lipids induce senescence (Flor, Wolfgeher, Wu, & Kron, 2017). A signature of altered lipid homeostasis is found in replicative and therapy-induced senescence (Flor et al., 2017; Lizardo, Lin, Gokcumen, & Atilla-Gokcumen, 2017). Numerous studies highlighted that changes in the ER/cellular lipid compositions or perturbations of lipid metabolism can activate the three arms of the UPR (Cunha et al., 2008; Fu et al., 2011; Volmer, van der Ploeg, & Ron, 2013); reviewed in (Fun & Thibault, 2019). Importantly, the disruption of the ER lipid homeostasis (called Lipid Bilayer Stress, LBS) is sensed by PERK and IRE1, while ATF6α activation is more sensitive to increasing levels of specific sphingolipids (Fun & Thibault, 2019). Moreover, this UPR activation was shown to be independent of misfolded protein accumulation (Koh, Wang, Beaudoin-Chabot, & Thibault, 2018; Tam et al., 2018; Volmer et al., 2013). Indeed, the absence of IRE1 and PERK luminal domain was shown to prevent their activation upon proteotoxic stress but not upon LBS. In the latter case, the subsequent decrease in the membrane fluidity probably favors oligomerization of their respective transmembrane domains necessary to trigger the UPR downstream effectors (Volmer et al., 2013). Consistently, the ATF6α transmembrane domain was shown to be highly sensitive to increased levels of specific sphingolipids (dihydrosphingosine, DHS, and dihydroceramide, DHC) but not to other ceramides, since mutations within the ATF6α transmembrane domain abolished its activation upon increased levels of DHS and DHC, but not upon proteotoxic stress (Tam et al., 2018). Interestingly, one study showed that DHC and triacylglycerols (components of lipid droplets), accumulate during replicative senescence (Lizardo et al., 2017), suggesting that ATF6α activation may result from increased DHC levels.

Moreover, several studies using ER stress inducers pointed out to a link between the UPR, senescence, and lipid metabolism. DTT was shown to induce an increase in ceramide levels (Epstein et al., 2012), and the injection of tunicamycin in mice resulted in the upregulation of sphingolipid genes in hepatocytes (Lee et al., 2015). We previously demonstrated that DTT, tunicamycin, and thapsigargin induced premature senescence in NHDFs (Druelle et al., 2016). Lizardo and colleagues showed that numerous genes involved in unsaturated fatty acid biosynthesis as well as in sphingolipid and glycerolipid metabolism are upregulated during replicative senescence (Lizardo et al., 2017). In the same line, we found that DTT-mediated premature senescence was dependent on the overproduction of prostaglandin E_2 (PGE_2) in wildtype fibroblasts through the ATF6α/COX2 pathway (Cormenier et al., 2017). Moreover, the CD36 protein, encoded by one of lipid metabolism genes, is involved in PGE2 release upon Thapsigargin induction in CHO cells (Kuda et al., 2011). Upregulating CD36 by ox-LDL activates the ER stress, and silencing CD36 decreases the ox-LDL-induced nuclear translocation of ATF6α as well as the IRE1α/XBP1 functions in macrophages (Yao et al., 2014). Ox-LDL is known to promote senescence in various types of cells, including endothelial progenitor cells (Imanishi, Hano, Sawamura, & Nishio, 2004; Rosso et al., 2006), chondrocytes (Zushi et al., 2009), and hematopoietic stem cells (Zhang et al., 2013). In addition, CD36 overexpression in human fibroblasts has recently been shown to induce a premature senescent-like phenotype by inducing the accumulation of phosphatidylcholine remodeling membranes (Saitou et al., 2018).

Overall, these data indicate that lipids and the UPR may be inter connected and the interplay between them may cause cellular senescence.

4. Is UPR activated following accumulation of misfolded proteins in the ER during senescence?

The accumulation of misfolded proteins leads to proteotoxic stress which disrupts normal cellular metabolism, giving rise to the accumulation of protein aggregates (Crum et al., 2015). It also induces cell enlargement and stimulates the expression of inflammatory cytokines (Haines, Juhasz, & Tosaki, 2013). All this may lead to the emergence of the senescent cellular phenotypes. This raises the question whether proteotoxic stress occurring during senescence due to oxidized misfolded protein, SASP protein overload in the ER lumen, and/or alternative mechanisms may activate the UPR. Indeed, an increasing demand for protein synthesis, maturation, and secretion necessary for the production of the SASP (which overloads the ER) could induce the UPR. Such a situation was demonstrated in a model of TIS applied to lymphoma cells (Dörr et al., 2013).

4.1 Arguments in favor of the UPR being activated as a consequence of misfolded protein accumulation

N-glycosylation inhibitor tunicamycin, which induces the ER stress by disrupting protein processing and folding, has been shown to induce premature senescence in normal fibroblasts (Druelle et al., 2016). Similar results were found in RTK-dependent tumor cells upon inhibiting OST, a protein complex catalyzing the N-glycosylation of the nascent polypeptide entering the ER (Lopez-Sambrooks et al., 2016). Also other pharmacological inducers of the ER stress known to induce the accumulation of misfolded proteins in the ER lumen, such as thapsigargin and dithiothreitol, can cause premature senescence. This premature senescence is strongly impaired in fibroblasts deficient for ATF6α (Cormenier et al., 2017; Druelle et al., 2016), suggesting that ATF6α is a key player in the senescence onset. Moreover, a recent study showed that the activation of wildtype ATF6α alone—but not of a dominant negative ATF6α mutant—was sufficient to induce premature senescence (Kim, Kim, et al., 2019), suggesting that ATF6α transcriptional activity is required for senescence onset. In the same line, our own studies demonstrated that specific pharmacological activation of ATF6 by compound 147 (Plate et al., 2016) increased the number of SA-β-Gal-positive (*i.e.*, senescent) cells (Pluquet et al., unpublished data).

The increased level of oxidative stress was shown to trigger the onset of senescence (Kuilman et al., 2010). While overactivity of the ER could be a source of senescence-associated oxidative stress, this oxidative stress may target ER chaperones and foldases, impairing the ER folding capacity. Indeed, carbonylation (an oxidative damage) of calreticulin and ERp29 was reported in senescent WI-38 human embryonic fibroblasts (Ahmed, Rogowska-Wrzesinska, Roepstorff, Bulteau, & Friguet, 2010; Baraibar, Liu, Ahmed, & Friguet, 2012). Besides, a significant increase in ROS production (upon treatment by oroxin A) induced the expression of ATF4 and GRP78, and increased the phosphorylation of p38MAPK, resulting in the ER stress-mediated senescence (He et al., 2016).

The UPR activation depends also on the activity of NOX (Kim, Kim, et al., 2019) and may result in ROS production (Cullinan & Diehl, 2006). Silencing NOX4 was shown to suppress senescence triggered by HRas V12 (Kodama et al., 2013). Moreover, oxidized lipoproteins promoted ROS-dependent senescence in endothelial progenitor cells (Imanishi et al., 2004) and in vascular endothelial cells (Riahi et al., 2015). Besides, lysosomal LDL oxidation was shown to trigger senescence in human cells associated with secretion of inflammatory cytokines (Ahmad & Leake, 2019).

Increased levels of oxidized proteins observed in senescent cells may also be related to the activity of the Hsc70-interacting protein (CHIP). Recently, CHIP was shown to play a role in the regulation of autophagy (Guo et al., 2015; Maan & Pati, 2018) and to modulate oxidized protein levels in senescent cells (Sisoula & Gonos, 2011). CHIP$^{-/-}$ MEFs had increased levels of oxidized proteins compared to wildtype cells (Sisoula & Gonos, 2011) and reached the senescence plateau faster (Min et al., 2008). Min and colleagues showed that CHIP$^{-/-}$ mice displayed more SA-β-Gal staining in kidneys at three and 12 months compared to their wildtype counterparts (Min et al., 2008). It has been also demonstrated that overexpression of CHIP prevents CHOP expression and eIF2a phosphorylation upon the ER stress (Cabral Miranda et al., 2014).

Finally, ischemic retinal cells engage the IRE1α UPR arm to induce the senescence state in which the cells adopt the senescence-associated secretory phenotype (SASP) (Oubaha et al., 2016). Overall, this suggests that the overload of (oxidized) proteins may result in the accumulation of misfolded and/or aggregated proteins, which, in turn, induces the UPR and may accelerate the senescence onset.

4.2 Arguments in favor of the UPR being activated by alternative pathways and not by accumulation of misfolded protein in the ER

Even though the main function of the UPR is to cope with protein misfolding or with an increased protein folding demand occurring in various physiological and pathological conditions, some studies have shown that the UPR may be activated by alternative pathways ((Hu et al., 2009), reviewed in (Hetz et al., 2015)). This may involve a cell-to-cell transmission of the ER stress through secreted factors that elicit a pro-inflammatory response in the microenvironment, such as VEGF (Karali et al., 2014), or through the stimulation of Toll-like receptor 4 (TLR) (Martinon, Chen, Lee, & Glimcher, 2010). Consistently with this statement, it has been described that ER-stressed cancer cells secrete factors which upregulate the UPR markers in macrophages (Mahadevan et al., 2011). In human dermal fibroblasts (HDFs), the expression of TLR4 pathway genes was higher in senescent cells than in those in the exponential growth phase (Seo, Park, Oh, & Shin, 2018). Cytokine secretion through TLR-4 signaling in senescent fibroblasts was shown to be dependent on secreted HMGB1 (Davalos et al., 2013), and the latter can induce the ER stress (He et al., 2018). In addition, while VEGF was shown to be actively secreted in different models of senescence (Coppé, Kauser, Campisi, & Beauséjour, 2006; Oubaha et al., 2016; Regina et al., 2016), there is no evidence for an autocrine/paracrine role of VEGF as a factor that induces or reinforces senescence, in contrast to IL1α/β, IL6, IL8, TGFβ. Although the UPR transcription factors ATF4 and XBP1s can drive the production of inflammatory cytokines (*e.g.*, IL1β, IL-6, and IL8) and of VEGF in various cell types (Auf et al., 2010; Iwasaki et al., 2014; Toosi, Orlow, & Manga, 2012), it remains unclear whether it occurs during senescence. We can speculate that senescent cells may secrete unknown factors that activate the UPR, which, in turn, may drive the production of cytokines involved in inducing and/or enhancing senescence.

Structural changes in the ER membrane can activate the UPR independently of the accumulation of misfolded proteins. One of the best examples are membrane lipid changes (Fun & Thibault, 2019). ATF6α—but not PERK and IRE1α—is preferentially activated by specific sphingolipids, such as dihydrosphingosine (DHS) and dihydroceramide (DHC), to transactivate lipid biosynthetic and metabolic genes rather than the ER chaperone genes (Tam et al., 2018). The ER activation in *C. elegans* leads to the restructuring of the ER morphology in intestinal cells through a lipophagy process that depletes lipid droplets. Interestingly, the same report concluded that increased lipophagy was independent of chaperone function (Daniele et al., 2020). In addition, inhibiting lipid mediator sphingosine kinase 1 (SPHK1) accelerated cellular senescence in adipose-derived stromal cells. This senescence was reversed upon treatment with a combination of a ceramide synthesis inhibitor and a sphingosine 1 phosphate (Kim, Lee, et al., 2019). Finally, primary bone-derived mesenchymal stem cells (BMSCs) entered senescence following a treatment with serum from aged people (containing high levels of ceramide), unlike those which were treated with serum from young people (low levels of ceramide) (Khayrullin et al., 2019).

These data clearly illustrate that the UPR activation may occur in the absence of misfolded proteins and could potentially contribute to the establishment and/or maintenance of senescent features.

5. Role of the UPR in neoplastic transformation by senescence evasion

We demonstrated that NHEKs escaping from senescence downregulate the UPR genes expressed in senescent cells, suggesting that senescent cells that re-enter the cell cycle are able to successfully resolve the ER stress (Drullion et al., 2018). Therefore, it seems that the UPR activation prevents senescent NHEKs from switching to pre-malignant transformation. Specifically, we showed that knocking down PERK and to a lesser extent ATF6α—but not IRE1a—increased senescence escape in these cells. This indicates that the PERK arm of the UPR could act as an oncosuppressive barrier. This is consistent with the results of a study by others showing that the deletion of CHOP (a component of the PERK UPR arm) in a mouse model of K-rasV12-driven lung cancer increases early tumor incidence (Huber et al., 2013). We also showed that treating senescent NHEKs with ER stress inducers, such as thapsigargin, decreased post-senescence evasion (Drullion et al., 2018), probably due to an increase of cell death in senescent NHEKs by excess of autophagy. Indeed, we had previously demonstrated that the survival and the fate of senescent NHEKs were controlled by autophagy levels: high levels of autophagy led to senescent cell death, whereas moderate levels favored neoplastic escape (Abbadie et al., 2017; Deruy et al., 2014). The level and/or the duration of the PERK and ATF6α activation at NHEK senescence may contribute to determining the level of autophagic activity in these cells. Therefore, the ability of senescent NHEK to re-enter the cell cycle probably partly results from their ability to prevent autophagic cell death by resolving some of its upstream inducer stress, including the ER stress.

Furthermore, we had previously demonstrated that the neoplastic escape from senescence involves the presence of unrepaired DNA single-strand breaks (Nassour et al., 2016). It is noteworthy that some studies linked the UPR with the DNA damage repair pathway and genome stability (Dufey, Urra, & Hetz, 2015) (see Section 3.2). It was suggested that XBP1 controls a cluster of DNA repair genes (Acosta-Alvear et al., 2007). In addition, the ER stress can impact chromatin remodeling by regulating methylation and acetylation, which may, in turn, reduce the transcription of the ER target genes (Dicks et al., 2015). These data indicate that the ER stress and the UPR are closely involved in determining the cell fate, in particular in senescence escape. Further investigations are needed to better understand the role of the UPR in this process.

6. Concluding remarks and perspectives

Both *in vitro* and *in vivo* data described here above clearly support the hypothesis that the UPR and senescence are closely linked. The UPR seems to contribute to establishing the senescence hallmarks *via* various pathways. However, the upstream UPR activator in the context of senescence remains unknown. Still, we may claim that it probably relies on oxidative stress, proteotoxic stress, lipid stress, autophagy, or a blend of them. Clarifying these issues is necessary to understand the origin(s) of the UPR activation observed in senescent cells. The UPR activation during senescence seems to be more complex than a simple dependency of the accumulation of misfolded proteins. Pinning down the exact connections between the UPR and the various forms of autophagy occurring during senescence, such as CMA, lipophagy, ER phagy, and macroautophagy, would certainly bring important insights into this issue. Another point that deserves further investigations is the dynamic activation of the UPR in senescence, its level of activation, as well as the possible existence of positive and negative feedback loops. Further experiments are also needed to better characterize how the UPR contributes to the SASP in a context of oxidative, proteotoxic, and lipid stresses, as well as in that of increased autophagy. This question is highly relevant because the senescence-associated secretory phenotype (SASP) plays an important physiological and pathological role in aging.

The effects of aging are correlated with an increase in the number of senescent cells in tissues and organs (Herbig et al., 2006), which increases the incidence of many age-related conditions (Baar et al., 2017; Baker et al., 2011, 2016). The hope to overcome these effects is in the senolytic drugs which were shown to alleviate some age-related symptoms and extend the healthy lifespan (Zhu et al., 2015). Modulating the UPR to limit senescence may in fact represent a potential therapeutic tool whenever senescence is involved. It is tempting to speculate that senescent cells require an optimal UPR duration/intensity for their long-term survival and that some specific pharmacological compounds—unknown to date—may act as senolytics by increasing the UPR activity and inducing senescent cell death. This would have the advantage of preventing senescence escape, and as a consequence maybe also—in a longer run—neoplastic transformation. Indeed, numerous studies on different senescence types and using different models showed that a sustained UPR activation leads to cell death (reviewed in (Merksamer & Papa, 2010; Walter & Ron, 2011)). Alternatively, preventing the UPR adaptation, in particular inhibiting ATF6α in replicative senescence or IRE1α in Hras-driven senescence, by using existing compounds, such as Ceapin A7 (ATF6α inhibitor), Kira6 or STF083010 (two IRE1 inhibitors), could be beneficial by reducing or suppressing deleterious aspects of senescent cells, in particular the SASP. If this turns out to be true, the UPR inhibitors should work as senomorphics able to at least partially suppress some senescent characteristics. Therefore, by slowing down the onset of senescence, they could reduce the post-senescence neoplastic escape.

On the whole, the unfolded protein response (UPR) appears to play multiple roles in both proliferating and senescent cells. It may also contribute to tumorigenesis. Understanding the mechanisms through which the UPR regulates cellular processes, and in particular cellular senescence, may pave the way for developing new therapeutics for the treatment of some age-related diseases and also cancer.

Acknowledgments

This work was funded by grants from the French *Centre National de la Recherche Scientifique (CNRS)*, the University of Lille (France), and the French *Contrat Plan Etat Région (CPER)* to O.P.

Declaration of interests

The authors declare no conflict of interest.

References

Abbadie, C., & Pluquet, O. (2020). Unfolded protein response (UPR) controls major senescence hallmarks. *Trends in Biochemical Sciences*, *45*(5), 371–374. https://doi.org/10.1016/j.tibs.2020.02.005.

Abbadie, C., Pluquet, O., & Pourtier, A. (2017). Epithelial cell senescence: An adaptive response to pre-carcinogenic stresses? *Cellular and Molecular Life Sciences: CMLS*, *74*, 4471–4509. https://doi.org/10.1007/s00018-017-2587-9.

Acosta, J. C., O'Loghlen, A., Banito, A., Guijarro, M. V., Augert, A., Raguz, S., et al. (2008). Chemokine signaling via the CXCR2 receptor reinforces senescence. *Cell*, *133*(6), 1006–1018. https://doi.org/10.1016/j.cell.2008.03.038.

Acosta-Alvear, D., Zhou, Y., Blais, A., Tsikitis, M., Lents, N. H., Arias, C., et al. (2007). XBP1 controls diverse cell type- and condition-specific transcriptional regulatory networks. *Molecular Cell*, *27*(1), 53–66. https://doi.org/10.1016/j.molcel.2007.06.011.

Adachi, Y., Yamamoto, K., Okada, T., Yoshida, H., Harada, A., & Mori, K. (2008). ATF6 is a transcription factor specializing in the regulation of quality control proteins in the endoplasmic reticulum. *Cell Structure and Function*, *33*(1), 75–89.

Agger, K., Cloos, P. A. C., Rudkjaer, L., Williams, K., Andersen, G., Christensen, J., et al. (2009). The H3K27me3 demethylase JMJD3 contributes to the activation of the INK4A-ARF locus in response to oncogene- and stress-induced senescence. *Genes & Development*, *23*(10), 1171–1176. https://doi.org/10.1101/gad.510809.

Ahmad, F., & Leake, D. S. (2019). Lysosomal oxidation of LDL alters lysosomal pH, induces senescence, and increases secretion of pro-inflammatory cytokines in human macrophages. *Journal of Lipid Research*, *60*(1), 98–110. https://doi.org/10.1194/jlr.M088245.

Ahmed, E. K., Rogowska-Wrzesinska, A., Roepstorff, P., Bulteau, A.-L., & Friguet, B. (2010). Protein modification and replicative senescence of WI-38 human embryonic fibroblasts. *Aging Cell*, *9*(2), 252–272. https://doi.org/10.1111/j.1474-9726.'2010.00555.x.

Almanza, A., Carlesso, A., Chintha, C., Creedican, S., Doultsinos, D., Leuzzi, B., et al. (2019). Endoplasmic reticulum stress signalling—From basic mechanisms to clinical applications. *The FEBS Journal*, *286*(2), 241–278. https://doi.org/10.1111/febs.14608.

Auf, G., Jabouille, A., Guérit, S., Pineau, R., Delugin, M., Bouchecareilh, M., et al. (2010). Inositol-requiring enzyme 1alpha is a key regulator of angiogenesis and invasion in malignant glioma. *Proceedings of the National Academy of Sciences of the United States of America*, *107*(35), 15553–15558. https://doi.org/10.1073/pnas.0914072107.

Baar, M. P., Brandt, R. M. C., Putavet, D. A., Klein, J. D. D., Derks, K. W. J., Bourgeois, B. R. M., et al. (2017). Targeted apoptosis of senescent cells restores tissue homeostasis in response to chemotoxicity and aging. *Cell*, *169*(1), 132–147.e16. https://doi.org/10.1016/j.cell.2017.02.031.

Bachar, E., Ariav, Y., Ketzinel-Gilad, M., Cerasi, E., Kaiser, N., & Leibowitz, G. (2009). Glucose amplifies fatty acid-induced endoplasmic reticulum stress in pancreatic beta-cells via activation of mTORC1. *PLoS One*, *4*(3), e4954. https://doi.org/10.1371/journal.pone.0004954.

Baker, D. J., Childs, B. G., Durik, M., Wijers, M. E., Sieben, C. J., Zhong, J., et al. (2016). Naturally occurring p16(Ink4a)-positive cells shorten healthy lifespan. *Nature*, *530*(7589), 184–189. https://doi.org/10.1038/nature16932.

Baker, D. J., Wijshake, T., Tchkonia, T., LeBrasseur, N. K., Childs, B. G., van de Sluis, B., et al. (2011). Clearance of p16Ink4a-positive senescent cells delays ageing-associated disorders. *Nature*, *479*(7372), 232–236. https://doi.org/10.1038/nature10600.

Baraibar, M. A., Liu, L., Ahmed, E. K., & Friguet, B. (2012). Protein oxidative damage at the crossroads of cellular senescence, aging, and age-related diseases. *Oxidative Medicine and Cellular Longevity*, *2012*, 919832. https://doi.org/10.1155/2012/919832.

Barreda, D. R., Hanington, P. C., Walsh, C. K., Wong, P., & Belosevic, M. (2004). Differentially expressed genes that encode potential markers of goldfish macrophage development in vitro. *Developmental and Comparative Immunology, 28*(7–8), 727–746. https://doi.org/10.1016/j.dci.2003.11.005.
Baumann, O., & Walz, B. (2001). Endoplasmic reticulum of animal cells and its organization into structural and functional domains. *International Review of Cytology, 205*, 149–214.
Baumeister, P., Luo, S., Skarnes, W. C., Sui, G., Seto, E., Shi, Y., et al. (2005). Endoplasmic reticulum stress induction of the Grp78/BiP promoter : Activating mechanisms mediated by YY1 and its interactive chromatin modifiers. *Molecular and Cellular Biology, 25*(11), 4529–4540. https://doi.org/10.1128/MCB.25.11.4529-4540.2005.
Benvenuti, S., Cramer, R., Quinn, C. C., Bruce, J., Zvelebil, M., Corless, S., et al. (2002). Differential proteome analysis of replicative senescence in rat embryo fibroblasts. *Molecular & Cellular Proteomics: MCP, 1*(4), 280–292.
Blazanin, N., Son, J., Craig-Lucas, A. B., John, C. L., Breech, K. J., Podolsky, M. A., et al. (2017). ER stress and distinct outputs of the IRE1α RNase control proliferation and senescence in response to oncogenic Ras. *Proceedings of the National Academy of Sciences of the United States of America, 114*(37), 9900–9905. https://doi.org/10.1073/pnas.1701757114.
Boraldi, F., Annovi, G., Tiozzo, R., Sommer, P., & Quaglino, D. (2010). Comparison of ex vivo and in vitro human fibroblast ageing models. *Mechanisms of Ageing and Development, 131*(10), 625–635. https://doi.org/10.1016/j.mad.2010.08.008.
Borghesan, M., Fusilli, C., Rappa, F., Panebianco, C., Rizzo, G., Oben, J. A., et al. (2016). DNA hypomethylation and histone variant macroH2A1 synergistically attenuate chemotherapy-induced senescence to promote hepatocellular carcinoma progression. *Cancer Research, 76*(3), 594–606. https://doi.org/10.1158/0008-5472.CAN-15-1336.
Brewer, J. W., & Diehl, J. A. (2000). PERK mediates cell-cycle exit during the mammalian unfolded protein response. *Proceedings of the National Academy of Sciences of the United States of America, 97*(23), 12625–12630. https://doi.org/10.1073/pnas.220247197.
Brewer, J. W., Hendershot, L. M., Sherr, C. J., & Diehl, J. A. (1999). Mammalian unfolded protein response inhibits cyclin D1 translation and cell-cycle progression. *Proceedings of the National Academy of Sciences of the United States of America, 96*(15), 8505–8510. https://doi.org/10.1073/pnas.96.15.8505.
Brozzi, F., Nardelli, T. R., Lopes, M., Millard, I., Barthson, J., Igoillo-Esteve, M., et al. (2015). Cytokines induce endoplasmic reticulum stress in human, rat and mouse beta cells via different mechanisms. *Diabetologia, 58*(10), 2307–2316. https://doi.org/10.1007/s00125-015-3669-6.
Cabral Miranda, F., Adão-Novaes, J., Hauswirth, W. W., Linden, R., Petrs-Silva, H., & Chiarini, L. B. (2014). CHIP, a carboxy terminus HSP-70 interacting protein, prevents cell death induced by endoplasmic reticulum stress in the central nervous system. *Frontiers in Cellular Neuroscience, 8*, 438. https://doi.org/10.3389/fncel.2014.00438.
Campisi, J. (2001). Cellular senescence as a tumor-suppressor mechanism. *Trends in Cell Biology, 11*(11), S27–S31.
Campisi, J., Andersen, J. K., Kapahi, P., & Melov, S. (2011). Cellular senescence: A link between cancer and age-related degenerative disease? *Seminars in Cancer Biology, 21*(6), 354–359. https://doi.org/10.1016/j.semcancer.2011.09.001.
Carew, N. T., Nelson, A. M., Liang, Z., Smith, S. M., & Milcarek, C. (2018). Linking endoplasmic reticular stress and alternative splicing. *International Journal of Molecular Sciences, 19*(12), 3919. https://doi.org/10.3390/ijms19123919.

Chainiaux, F., Magalhaes, J.-P., Eliaers, F., Remacle, J., & Toussaint, O. (2002). UVB-induced premature senescence of human diploid skin fibroblasts. *The International Journal of Biochemistry & Cell Biology*, *34*(11), 1331–1339. https://doi.org/10.1016/s1357-2725(02)00022-5.

Chen, H., Ruiz, P. D., McKimpson, W. M., Novikov, L., Kitsis, R. N., & Gamble, M. J. (2015). MacroH2A1 and ATM play opposing roles in paracrine senescence and the senescence-associated secretory phenotype. *Molecular Cell*, *59*(5), 719–731. https://doi.org/10.1016/j.molcel.2015.07.011.

Chevet, E., Hetz, C., & Samali, A. (2015). Endoplasmic reticulum stress-activated cell reprogramming in oncogenesis. *Cancer Discovery*, *5*(6), 586–597. https://doi.org/10.1158/2159-8290.CD-14-1490.

Childs, B. G., Durik, M., Baker, D. J., & van Deursen, J. M. (2015). Cellular senescence in aging and age-related disease: From mechanisms to therapy. *Nature Medicine*, *21*(12), 1424–1435. https://doi.org/10.1038/nm.4000.

Choi, B.-H., & Kim, J.-S. (2004). Age-related decline in expression of calnexin. *Experimental & Molecular Medicine*, *36*(5), 499–503. https://doi.org/10.1038/emm.2004.63.

Coppé, J.-P., Desprez, P.-Y., Krtolica, A., & Campisi, J. (2010). The senescence-associated secretory phenotype: The dark side of tumor suppression. *Annual Review of Pathology*, *5*, 99–118. https://doi.org/10.1146/annurev-pathol-121808-102144.

Coppé, J.-P., Kauser, K., Campisi, J., & Beauséjour, C. M. (2006). Secretion of vascular endothelial growth factor by primary human fibroblasts at senescence. *The Journal of Biological Chemistry*, *281*(40), 29568–29574. https://doi.org/10.1074/jbc.M603307200.

Coppé, J.-P., Patil, C. K., Rodier, F., Sun, Y., Muñoz, D. P., Goldstein, J., et al. (2008). Senescence-associated secretory phenotypes reveal cell-nonautonomous functions of oncogenic RAS and the p53 tumor suppressor. *PLoS Biology*, *6*(12), 2853–2868. https://doi.org/10.1371/journal.pbio.0060301.

Cormenier, J., Martin, N., Deslé, J., Salazar-Cardozo, C., Pourtier, A., Abbadie, C., et al. (2017). The ATF6α arm of the unfolded protein response mediates replicative senescence in human fibroblasts through a COX2/prostaglandin E2 intracrine pathway. *Mechanisms of Ageing and Development*, *170*, 82–91. https://doi.org/10.1016/j.mad.2017.08.003.

Crum, T. S., Gleixner, A. M., Posimo, J. M., Mason, D. M., Broeren, M. T., Heinemann, S. D., et al. (2015). Heat shock protein responses to aging and proteotoxicity in the olfactory bulb. *Journal of Neurochemistry*, *133*(6), 780–794. https://doi.org/10.1111/jnc.13041.

Cubillos-Ruiz, J. R., Bettigole, S. E., & Glimcher, L. H. (2017). Tumorigenic and immunosuppressive effects of endoplasmic reticulum stress in cancer. *Cell*, *168*(4), 692–706. https://doi.org/10.1016/j.cell.2016.12.004.

Cuevas, E. P., Eraso, P., Mazón, M. J., Santos, V., Moreno-Bueno, G., Cano, A., et al. (2017). LOXL2 drives epithelial-mesenchymal transition via activation of IRE1-XBP1 signalling pathway. *Scientific Reports*, 7, 44988. https://doi.org/10.1038/srep44988.

Cui, Y., Lu, C., Liu, L., Sun, D., Yao, N., Tan, S., et al. (2008). Reactivation of methylation-silenced tumor suppressor gene p16INK4a by nordihydroguaiaretic acid and its implication in G1 cell cycle arrest. *Life Sciences*, *82*(5–6), 247–255. https://doi.org/10.1016/j.lfs.2007.11.013.

Cullinan, S. B., & Diehl, J. A. (2006). Coordination of ER and oxidative stress signaling: The PERK/Nrf2 signaling pathway. *The International Journal of Biochemistry & Cell Biology*, *38*(3), 317–332. https://doi.org/10.1016/j.biocel.2005.09.018.

Cunha, D. A., Hekerman, P., Ladrière, L., Bazarra-Castro, A., Ortis, F., Wakeham, M. C., et al. (2008). Initiation and execution of lipotoxic ER stress in pancreatic beta-cells. *Journal of Cell Science*, *121*(Pt 14), 2308–2318. https://doi.org/10.1242/jcs.026062.

Dandekar, A., Mendez, R., & Zhang, K. (2015). Cross talk between ER stress, oxidative stress, and inflammation in health and disease. *Methods in Molecular Biology (Clifton, N.J.)*, *1292*, 205–214. https://doi.org/10.1007/978-1-4939-2522-3_15.

Daniele, J. R., Higuchi-Sanabria, R., Durieux, J., Monshietehadi, S., Ramachandran, V., Tronnes, S. U., et al. (2020). UPRER promotes lipophagy independent of chaperones to extend life span. *Science Advances*, *6*(1), eaaz1441. https://doi.org/10.1126/sciadv.aaz1441.

Davalos, A. R., Kawahara, M., Malhotra, G. K., Schaum, N., Huang, J., Ved, U., et al. (2013). P53-dependent release of Alarmin HMGB1 is a central mediator of senescent phenotypes. *The Journal of Cell Biology*, *201*(4), 613–629. https://doi.org/10.1083/jcb.201206006.

de Magalhães, J. P. (2013). How ageing processes influence cancer. *Nature Reviews. Cancer*, *13*(5), 357–365. https://doi.org/10.1038/nrc3497.

Dejeans, N., Pluquet, O., Lhomond, S., Grise, F., Bouchecareilh, M., Juin, A., et al. (2012). Autocrine control of glioma cells adhesion and migration through IRE1α-mediated cleavage of SPARC mRNA. *Journal of Cell Science*, *125*(Pt 18), 4278–4287. https://doi.org/10.1242/jcs.099291.

Denoyelle, C., Abou-Rjaily, G., Bezrookove, V., Verhaegen, M., Johnson, T. M., Fullen, D. R., et al. (2006). Anti-oncogenic role of the endoplasmic reticulum differentially activated by mutations in the MAPK pathway. *Nature Cell Biology*, *8*(10), 1053–1063. https://doi.org/10.1038/ncb1471.

Deruy, E., Gosselin, K., Vercamer, C., Martien, S., Bouali, F., Slomianny, C., et al. (2010). MnSOD upregulation induces autophagic programmed cell death in senescent keratinocytes. *PLoS One*, *5*(9), e12712. https://doi.org/10.1371/journal.pone.0012712.

Deruy, E., Nassour, J., Martin, N., Vercamer, C., Malaquin, N., Bertout, J., et al. (2014). Level of macroautophagy drives senescent keratinocytes into cell death or neoplastic evasion. *Cell Death & Disease*, *5*, e1577. https://doi.org/10.1038/cddis.2014.533.

Dicks, N., Gutierrez, K., Michalak, M., Bordignon, V., & Agellon, L. B. (2015). Endoplasmic reticulum stress, genome damage, and cancer. *Frontiers in Oncology*, *5*, 11. https://doi.org/10.3389/fonc.2015.00011.

Dimri, G. P., Lee, X., Basile, G., Acosta, M., Scott, G., Roskelley, C., et al. (1995). A biomarker that identifies senescent human cells in culture and in aging skin in vivo. *Proceedings of the National Academy of Sciences of the United States of America*, *92*(20), 9363–9367.

Ding, W.-X., Ni, H.-M., Gao, W., Hou, Y.-F., Melan, M. A., Chen, X., et al. (2007). Differential effects of endoplasmic reticulum stress-induced autophagy on cell survival. *The Journal of Biological Chemistry*, *282*(7), 4702–4710. https://doi.org/10.1074/jbc.M609267200.

Dong, Y., Yin, S., Jiang, C., Luo, X., Guo, X., Zhao, C., et al. (2014). Involvement of autophagy induction in penta-1,2,3,4,6-O-galloyl-β-D-glucose-induced senescence-like growth arrest in human cancer cells. *Autophagy*, *10*(2), 296–310. https://doi.org/10.4161/auto.27210.

Dörr, J. R., Yu, Y., Milanovic, M., Beuster, G., Zasada, C., Däbritz, J. H. M., et al. (2013). Synthetic lethal metabolic targeting of cellular senescence in cancer therapy. *Nature*, *501*(7467), 421–425. https://doi.org/10.1038/nature12437.

Druelle, C., Drullion, C., Deslé, J., Martin, N., Saas, L., Cormenier, J., et al. (2016). ATF6α regulates morphological changes associated with senescence in human fibroblasts. *Oncotarget*, *7*, 67699–67715. https://doi.org/10.18632/oncotarget.11505.

Drullion, C., Marot, G., Martin, N., Desle, J., Saas, L., Salazar-Cardozo, C., et al. (2018). Pre-malignant transformation by senescence evasion is prevented by the PERK and ATF6alpha branches of the Unfolded Protein Response. *Cancer Letters*, *438*, 187–196. https://doi.org/10.1016/j.canlet.2018.09.008.

Dufey, E., Urra, H., & Hetz, C. (2015). ER proteostasis addiction in cancer biology: Novel concepts. *Seminars in Cancer Biology*, *33*, 40–47. https://doi.org/10.1016/j.semcancer.2015.04.003.

Dumont, P., Burton, M., Chen, Q. M., Gonos, E. S., Frippiat, C., Mazarati, J. B., et al. (2000). Induction of replicative senescence biomarkers by sublethal oxidative stresses in normal human fibroblast. *Free Radical Biology & Medicine*, *28*(3), 361–373.

Epstein, S., Kirkpatrick, C. L., Castillon, G. A., Muñiz, M., Riezman, I., David, F. P. A., et al. (2012). Activation of the unfolded protein response pathway causes ceramide accumulation in yeast and INS-1E insulinoma cells. *Journal of Lipid Research*, *53*(3), 412–420. https://doi.org/10.1194/jlr.M022186.

Ferrand, M., Kirsh, O., Griveau, A., Vindrieux, D., Martin, N., Defossez, P.-A., et al. (2015). Screening of a kinase library reveals novel pro-senescence kinases and their common NF-κB-dependent transcriptional program. *Aging*, 7(11), 986–1003. https://doi.org/10.18632/aging.100845.

Flor, A. C., Wolfgeher, D., Wu, D., & Kron, S. J. (2017). A signature of enhanced lipid metabolism, lipid peroxidation and aldehyde stress in therapy-induced senescence. *Cell Death Discovery*, *3*, 17075. https://doi.org/10.1038/cddiscovery.2017.75.

Freund, A., Patil, C. K., & Campisi, J. (2011). P38MAPK is a novel DNA damage response-independent regulator of the senescence-associated secretory phenotype. *The EMBO Journal*, *30*(8), 1536–1548. https://doi.org/10.1038/emboj.2011.69.

Fu, S., Yang, L., Li, P., Hofmann, O., Dicker, L., Hide, W., et al. (2011). Aberrant lipid metabolism disrupts calcium homeostasis causing liver endoplasmic reticulum stress in obesity. *Nature*, *473*(7348), 528–531. https://doi.org/10.1038/nature09968.

Fun, X. H., & Thibault, G. (2019). Lipid bilayer stress and proteotoxic stress-induced unfolded protein response deploy divergent transcriptional and non-transcriptional programmes. *Biochimica Et Biophysica Acta. Molecular and Cell Biology of Lipids.*, *1865*, 158449. https://doi.org/10.1016/j.bbalip.2019.04.009.

Fung, H., Liu, P., & Demple, B. (2007). ATF4-dependent oxidative induction of the DNA repair enzyme Ape1 counteracts arsenite cytotoxicity and suppresses arsenite-mediated mutagenesis. *Molecular and Cellular Biology*, *27*(24), 8834–8847. https://doi.org/10.1128/MCB.00974-07.

Galanos, P., Vougas, K., Walter, D., Polyzos, A., Maya-Mendoza, A., Haagensen, E. J., et al. (2016). Chronic p53-independent p21 expression causes genomic instability by deregulating replication licensing. *Nature Cell Biology*, *18*(7), 777–789. https://doi.org/10.1038/ncb3378.

Gan, Q., Huang, J., Zhou, R., Niu, J., Zhu, X., Wang, J., et al. (2008). PPAR{gamma} accelerates cellular senescence by inducing p16INK4{alpha} expression in human diploid fibroblasts. *Journal of Cell Science*, *121*(Pt 13), 2235–2245. https://doi.org/10.1242/jcs.026633.

Gao, Y., Zhao, Y., Zhang, J., Lu, Y., Liu, X., Geng, P., et al. (2016). The dual function of PRMT1 in modulating epithelial-mesenchymal transition and cellular senescence in breast cancer cells through regulation of ZEB1. *Scientific Reports*, *6*, 19874. https://doi.org/10.1038/srep19874.

García-Prat, L., Martínez-Vicente, M., Perdiguero, E., Ortet, L., Rodríguez-Ubreva, J., Rebollo, E., et al. (2016). Autophagy maintains stemness by preventing senescence. *Nature*, *529*(7584), 37–42. https://doi.org/10.1038/nature16187.

Garg, A. D., Kaczmarek, A., Krysko, O., Vandenabeele, P., Krysko, D. V., & Agostinis, P. (2012). ER stress-induced inflammation: Does it aid or impede disease progression? *Trends in Molecular Medicine*, *18*(10), 589–598. https://doi.org/10.1016/j.molmed.2012.06.010.

Gosselin, K., Deruy, E., Martien, S., Vercamer, C., Bouali, F., Dujardin, T., et al. (2009). Senescent keratinocytes die by autophagic programmed cell death. *The American Journal of Pathology*, *174*(2), 423–435. https://doi.org/10.2353/ajpath.2009.080332.

Gosselin, K., Martien, S., Pourtier, A., Vercamer, C., Ostoich, P., Morat, L., et al. (2009). Senescence-associated oxidative DNA damage promotes the generation of neoplastic cells. *Cancer Research*, *69*(20), 7917–7925. https://doi.org/10.1158/0008-5472.CAN-08-2510.

Guo, D., Ying, Z., Wang, H., Chen, D., Gao, F., Ren, H., et al. (2015). Regulation of autophagic flux by CHIP. *Neuroscience Bulletin*, *31*(4), 469–479. https://doi.org/10.1007/s12264-015-1543-7.

Haines, D. D., Juhasz, B., & Tosaki, A. (2013). Management of multicellular senescence and oxidative stress. *Journal of Cellular and Molecular Medicine*, *17*(8), 936–957. https://doi.org/10.1111/jcmm.12074.

He, J., Du, L., Bao, M., Zhang, B., Qian, H., Zhou, Q., et al. (2016). Oroxin A inhibits breast cancer cell growth by inducing robust endoplasmic reticulum stress and senescence. *Anti-Cancer Drugs*, *27*(3), 204–215. https://doi.org/10.1097/CAD.0000000000000318.

He, Q., Fu, Y., Ding, X., Li, D., Wang, Z., Tian, D., et al. (2018). High-mobility group box 1 induces endoplasmic reticulum stress and activates hepatic stellate cells. *Laboratory Investigation; a Journal of Technical Methods and Pathology*, *98*(9), 1200–1210. https://doi.org/10.1038/s41374-018-0085-9.

Heo, J.-Y., Jing, K., Song, K.-S., Seo, K.-S., Park, J.-H., Kim, J.-S., et al. (2009). Downregulation of APE1/Ref-1 is involved in the senescence of mesenchymal stem cells. *Stem Cells (Dayton, Ohio)*, *27*(6), 1455–1462. https://doi.org/10.1002/stem.54.

Herbig, U., Ferreira, M., Condel, L., Carey, D., & Sedivy, J. M. (2006). Cellular senescence in aging primates. *Science (New York, N.Y.)*, *311*(5765), 1257. https://doi.org/10.1126/science.1122446.

Hernandez-Segura, A., Nehme, J., & Demaria, M. (2018). Hallmarks of cellular senescence. *Trends in Cell Biology*, *28*(6), 436–453. https://doi.org/10.1016/j.tcb.2018.02.001.

Herranz, N., Gallage, S., Mellone, M., Wuestefeld, T., Klotz, S., Hanley, C. J., et al. (2015). MTOR regulates MAPKAPK2 translation to control the senescence-associated secretory phenotype. *Nature Cell Biology*, *17*(9), 1205–1217. https://doi.org/10.1038/ncb3225.

Hetz, C., Chevet, E., & Oakes, S. A. (2015). Proteostasis control by the unfolded protein response. *Nature Cell Biology*, *17*(7), 829–838. https://doi.org/10.1038/ncb3184.

Higa, A., & Chevet, E. (2012). Redox signaling loops in the unfolded protein response. *Cellular Signalling*, *24*(8), 1548–1555. https://doi.org/10.1016/j.cellsig.2012.03.011.

Higa, A., Taouji, S., Lhomond, S., Jensen, D., Fernandez-Zapico, M. E., Simpson, J. C., et al. (2014). Endoplasmic reticulum stress-activated transcription factor ATF6α requires the disulfide isomerase PDIA5 to modulate chemoresistance. *Molecular and Cellular Biology*, *34*(10), 1839–1849. https://doi.org/10.1128/MCB.01484-13.

Hoare, M., Ito, Y., Kang, T.-W., Weekes, M. P., Matheson, N. J., Patten, D. A., et al. (2016). NOTCH1 mediates a switch between two distinct secretomes during senescence. *Nature Cell Biology*, *18*(9), 979–992. https://doi.org/10.1038/ncb3397.

Hong, X., & Yu, J.-J. (2019). Silencing of lysyl oxidase-like 2 inhibits the migration, invasion and epithelial-to-mesenchymal transition of renal cell carcinoma cells through the Src/FAK signaling pathway. *International Journal of Oncology*, *54*(5), 1676–1690. https://doi.org/10.3892/ijo.2019.4726.

Horiguchi, M., Koyanagi, S., Okamoto, A., Suzuki, S. O., Matsunaga, N., & Ohdo, S. (2012). Stress-regulated transcription factor ATF4 promotes neoplastic transformation by suppressing expression of the INK4a/ARF cell senescence factors. *Cancer Research*, *72*(2), 395–401. https://doi.org/10.1158/0008-5472.CAN-11-1891.

Hu, C.-C. A., Dougan, S. K., McGehee, A. M., Love, J. C., & Ploegh, H. L. (2009). XBP-1 regulates signal transduction, transcription factors and bone marrow colonization in B cells. *The EMBO Journal*, *28*(11), 1624–1636. https://doi.org/10.1038/emboj.2009.117.

Hubackova, S., Krejcikova, K., Bartek, J., & Hodny, Z. (2012). IL1- and TGFβ-Nox4 signaling, oxidative stress and DNA damage response are shared features of replicative, oncogene-induced, and drug-induced paracrine 'bystander senescence'. *Aging*, *4*(12), 932–951. https://doi.org/10.18632/aging.100520.

Huber, A.-L., Lebeau, J., Guillaumot, P., Pétrilli, V., Malek, M., Chilloux, J., et al. (2013). P58(IPK)-mediated attenuation of the proapoptotic PERK-CHOP pathway allows malignant progression upon low glucose. *Molecular Cell*, *49*(6), 1049–1059. https://doi.org/10.1016/j.molcel.2013.01.009.

Hwang, E. S., Yoon, G., & Kang, H. T. (2009). A comparative analysis of the cell biology of senescence and aging. *Cellular and Molecular Life Sciences: CMLS*, *66*(15), 2503–2524. https://doi.org/10.1007/s00018-009-0034-2.

Imanishi, T., Hano, T., Sawamura, T., & Nishio, I. (2004). Oxidized low-density lipoprotein induces endothelial progenitor cell senescence, leading to cellular dysfunction. *Clinical and Experimental Pharmacology & Physiology*, *31*(7), 407–413. https://doi.org/10.1111/j.1440-1681.2004.04022.x.

Ivanov, A., Pawlikowski, J., Manoharan, I., van Tuyn, J., Nelson, D. M., Rai, T. S., et al. (2013). Lysosome-mediated processing of chromatin in senescence. *The Journal of Cell Biology*, *202*(1), 129–143. https://doi.org/10.1083/jcb.201212110.

Iwasaki, Y., Suganami, T., Hachiya, R., Shirakawa, I., Kim-Saijo, M., Tanaka, M., et al. (2014). Activating transcription factor 4 links metabolic stress to interleukin-6 expression in macrophages. *Diabetes*, *63*(1), 152–161. https://doi.org/10.2337/db13-0757.

Jaiswal, A. S., Panda, H., Law, B. K., Sharma, J., Jani, J., Hromas, R., et al. (2015). NSC666715 and its analogs inhibit strand-displacement activity of DNA polymerase β and potentiate temozolomide-induced DNA damage, senescence and apoptosis in colorectal cancer cells. *PLoS One*, *10*(5), e0123808. https://doi.org/10.1371/journal.pone.0123808.

Jonchère, B., Vétillard, A., Toutain, B., Lam, D., Bernard, A. C., Henry, C., et al. (2015). Irinotecan treatment and senescence failure promote the emergence of more transformed and invasive cells that depend on anti-apoptotic Mcl-1. *Oncotarget*, *6*(1), 409–426. https://doi.org/10.18632/oncotarget.2774.

Karagöz, G. E., Aragón, T., & Acosta-Alvear, D. (2019). Recent advances in signal integration mechanisms in the unfolded protein response. *F1000Research*, *8*, 1840. https://doi.org/10.12688/f1000research.19848.1.

Karali, E., Bellou, S., Stellas, D., Klinakis, A., Murphy, C., & Fotsis, T. (2014). VEGF Signals through ATF6 and PERK to promote endothelial cell survival and angiogenesis in the absence of ER stress. *Molecular Cell*, *54*(4), 559–572. https://doi.org/10.1016/j.molcel.2014.03.022.

Kharroubi, I., Ladrière, L., Cardozo, A. K., Dogusan, Z., Cnop, M., & Eizirik, D. L. (2004). Free fatty acids and cytokines induce pancreatic beta-cell apoptosis by different mechanisms: Role of nuclear factor-kappaB and endoplasmic reticulum stress. *Endocrinology*, *145*(11), 5087–5096. https://doi.org/10.1210/en.2004-0478.

Khayrullin, A., Krishnan, P., Martinez-Nater, L., Mendhe, B., Fulzele, S., Liu, Y., et al. (2019). Very long-chain C24:1 ceramide is increased in serum extracellular vesicles with aging and can induce senescence in bone-derived mesenchymal stem cells. *Cell*, *8*(1), 37. https://doi.org/10.3390/cells8010037.

Kim, H. D., Jang, C.-Y., Choe, J. M., Sohn, J., & Kim, J. (2012). Phenylbutyric acid induces the cellular senescence through an Akt/p21(WAF1) signaling pathway. *Biochemical and Biophysical Research Communications*, *422*(2), 213–218. https://doi.org/10.1016/j.bbrc.2012.04.086.

Kim, H. S., Kim, Y., Lim, M. J., Park, Y.-G., Park, S. I., & Sohn, J. (2019). The p38-activated ER stress-ATF6α axis mediates cellular senescence. *FASEB Journal : Official Publication of the Federation of American Societies for Experimental Biology*, *33*(2), 2422–2434. https://doi.org/10.1096/fj.201800836R.

Kim, M. K., Lee, W., Yoon, G.-H., Chang, E.-J., Choi, S.-C., & Kim, S. W. (2019). Links between accelerated replicative cellular senescence and down-regulation of SPHK1 transcription. *BMB Reports*, *52*(3), 220–225.

Klooster, R., Eman, M. R., le Duc, Q., Verheesen, P., Verrips, C. T., Roovers, R. C., et al. (2009). Selection and characterization of KDEL-specific VHH antibody fragments and their application in the study of ER resident protein expression. *Journal of Immunological Methods*, *342*(1–2), 1–12. https://doi.org/10.1016/j.jim.2008.10.009.

Kodama, R., Kato, M., Furuta, S., Ueno, S., Zhang, Y., Matsuno, K., et al. (2013). ROS-generating oxidases Nox1 and Nox4 contribute to oncogenic Ras-induced premature senescence. *Genes to Cells: Devoted to Molecular & Cellular Mechanisms*, *18*(1), 32–41. https://doi.org/10.1111/gtc.12015.

Koh, J. H., Wang, L., Beaudoin-Chabot, C., & Thibault, G. (2018). Lipid bilayer stress-activated IRE-1 modulates autophagy during endoplasmic reticulum stress. *Journal of Cell Science*, *131*(22), 1–12. https://doi.org/10.1242/jcs.217992.

Krętowski, R., Stypułkowska, A., & Cechowska-Pasko, M. (2013). Low-glucose medium induces ORP150 expression and exerts inhibitory effect on apoptosis and senescence of human breast MCF7 cells. *Acta Biochimica Polonica*, *60*(2), 167–173.

Krishnamoorthy, J., Tenkerian, C., Gupta, J., Ghaddar, N., Wang, S., Darini, C., et al. (2018). Downregulation of PERK activity and eIF2α serine 51 phosphorylation by mTOR complex 1 elicits pro-oxidant and pro-death effects in tuberous sclerosis-deficient cells. *Cell Death & Disease*, *9*(3), 254. https://doi.org/10.1038/s41419-018-0326-2.

Kroemer, G., Mariño, G., & Levine, B. (2010). Autophagy and the integrated stress response. *Molecular Cell*, *40*(2), 280–293. https://doi.org/10.1016/j.molcel.2010.09.023.

Kuda, O., Jenkins, C. M., Skinner, J. R., Moon, S. H., Su, X., Gross, R. W., et al. (2011). CD36 protein is involved in store-operated calcium flux, phospholipase A2 activation, and production of prostaglandin E2. *The Journal of Biological Chemistry*, *286*(20), 17785–17795. https://doi.org/10.1074/jbc.M111.232975.

Kuilman, T., Michaloglou, C., Mooi, W. J., & Peeper, D. S. (2010). The essence of senescence. *Genes & Development*, *24*(22), 2463–2479. https://doi.org/10.1101/gad.1971610.

Kuilman, T., Michaloglou, C., Vredeveld, L. C. W., Douma, S., van Doorn, R., Desmet, C. J., et al. (2008). Oncogene-induced senescence relayed by an interleukin-dependent inflammatory network. *Cell*, *133*(6), 1019–1031. https://doi.org/10.1016/j.cell.2008.03.039.

Kusama, K., Yoshie, M., Tamura, K., Nakayama, T., Nishi, H., Isaka, K., et al. (2014). The role of exchange protein directly activated by cyclic AMP 2-mediated calreticulin expression in the decidualization of human endometrial stromal cells. *Endocrinology*, *155*(1), 240–248. https://doi.org/10.1210/en.2013-1478.

Laberge, R.-M., Sun, Y., Orjalo, A. V., Patil, C. K., Freund, A., Zhou, L., et al. (2015). MTOR regulates the pro-tumorigenic senescence-associated secretory phenotype by promoting IL1A translation. *Nature Cell Biology*, *17*(8), 1049–1061. https://doi.org/10.1038/ncb3195.

Lee, Y.-J., Chang, W.-W., Chang, C.-P., Liu, T.-Y., Chuang, C.-Y., Qian, K., et al. (2019). Downregulation of PRMT1 promotes the senescence and migration of a non-MYCN amplified neuroblastoma SK-N-SH cells. *Scientific Reports*, *9*(1), 1771. https://doi.org/10.1038/s41598-018-38394-6.

Lee, S.-Y., Hong, I.-K., Kim, B.-R., Shim, S.-M., Sung Lee, J., Lee, H.-Y., et al. (2015). Activation of sphingosine kinase 2 by endoplasmic reticulum stress ameliorates hepatic steatosis and insulin resistance in mice. *Hepatology (Baltimore, Md.)*, *62*(1), 135–146. https://doi.org/10.1002/hep.27804.

Lenin, R., Nagy, P. G., Gentry, J., & Gangaraju, R. (2018). Featured article: Deterioration of visual function mediated by senescence-associated endoplasmic reticulum stress in

inflammatory tie2-TNF mice. *Experimental Biology and Medicine (Maywood, N.J.), 243*(12), 976–984. https://doi.org/10.1177/1535370218794915.
Lew, Q. J., Chu, K. L., Lee, J., Koh, P. L., Rajasegaran, V., Teo, J. Y., et al. (2011). PCAF interacts with XBP-1S and mediates XBP-1S-dependent transcription. *Nucleic Acids Research, 39*(2), 429–439. https://doi.org/10.1093/nar/gkq785.
Li, Z., Scott, M. J., Fan, E. K., Li, Y., Liu, J., Xiao, G., et al. (2016). Tissue damage negatively regulates LPS-induced macrophage necroptosis. *Cell Death and Differentiation, 23*(9), 1428–1447. https://doi.org/10.1038/cdd.2016.21.
Li, M., Yang, X., Lu, X., Dai, N., Zhang, S., Cheng, Y., et al. (2018). APE1 deficiency promotes cellular senescence and premature aging features. *Nucleic Acids Research, 46*(11), 5664–5677. https://doi.org/10.1093/nar/gky326.
Lim, Y., Lee, E., Lee, J., Oh, S., & Kim, S. (2008). Down-regulation of asymmetric arginine methylation during replicative and H2O2-induced premature senescence in WI-38 human diploid fibroblasts. *Journal of Biochemistry, 144*(4), 523–529. https://doi.org/10.1093/jb/mvn097.
Lin, H.-K., Chen, Z., Wang, G., Nardella, C., Lee, S.-W., Chan, C.-H., et al. (2010). Skp2 targeting suppresses tumorigenesis by Arf-p53-independent cellular senescence. *Nature, 464*(7287), 374–379. https://doi.org/10.1038/nature08815.
Liu, X., Guo, S., Liu, X., & Su, L. (2015). Chaetocin induces endoplasmic reticulum stress response and leads to death receptor 5-dependent apoptosis in human non-small cell lung cancer cells. *Apoptosis: An International Journal on Programmed Cell Death, 20*(11), 1499–1507. https://doi.org/10.1007/s10495-015-1167-4.
Liu, J., Huang, K., Cai, G.-Y., Chen, X.-M., Yang, J.-R., Lin, L.-R., et al. (2014). Receptor for advanced glycation end-products promotes premature senescence of proximal tubular epithelial cells via activation of endoplasmic reticulum stress-dependent p21 signaling. *Cellular Signalling, 26*(1), 110–121. https://doi.org/10.1016/j.cellsig.2013.10.002.
Liu, J., Yang, J.-R., Chen, X.-M., Cai, G.-Y., Lin, L.-R., & He, Y.-N. (2015). Impact of ER stress-regulated ATF4/p16 signaling on the premature senescence of renal tubular epithelial cells in diabetic nephropathy. *American Journal of Physiology. Cell Physiology, 308*(8), C621–C630. https://doi.org/10.1152/ajpcell.00096.2014.
Lizardo, D. Y., Lin, Y.-L., Gokcumen, O., & Atilla-Gokcumen, G. E. (2017). Regulation of lipids is central to replicative senescence. *Molecular BioSystems, 13*(3), 498–509. https://doi.org/10.1039/c6mb00842a.
López-Moncada, F., Torres, M. J., Castellón, E. A., & Contreras, H. R. (2019). Secreted protein acidic and rich in cysteine (SPARC) induces epithelial-mesenchymal transition, enhancing migration and invasion, and is associated with high Gleason score in prostate cancer. *Asian Journal of Andrology, 21*(6), 557–564. https://doi.org/10.4103/aja.aja_23_19.
Lopez-Sambrooks, C., Shrimal, S., Khodier, C., Flaherty, D. P., Rinis, N., Charest, J. C., et al. (2016). Oligosaccharyltransferase inhibition induces senescence in RTK-driven tumor cells. *Nature Chemical Biology, 12*(12), 1023–1030. https://doi.org/10.1038/nchembio.2194.
Maan, M., & Pati, U. (2018). CHIP promotes autophagy-mediated degradation of aggregating mutant p53 in hypoxic conditions. *The FEBS Journal, 285*(17), 3197–3214. https://doi.org/10.1111/febs.14602.
Mahadevan, N. R., Rodvold, J., Sepulveda, H., Rossi, S., Drew, A. F., & Zanetti, M. (2011). Transmission of endoplasmic reticulum stress and pro-inflammation from tumor cells to myeloid cells. *Proceedings of the National Academy of Sciences of the United States of America, 108*(16), 6561–6566. https://doi.org/10.1073/pnas.1008942108.
Malaquin, N., Martinez, A., & Rodier, F. (2016). Keeping the senescence secretome under control : Molecular reins on the senescence-associated secretory phenotype. *Experimental Gerontology, 82*, 39–49. https://doi.org/10.1016/j.exger.2016.05.010.

Malaquin, N., Vercamer, C., Bouali, F., Martien, S., Deruy, E., Wernert, N., et al. (2013). Senescent fibroblasts enhance early skin carcinogenic events via a paracrine MMP-PAR-1 axis. *PLoS One*, *8*(5), e63607. https://doi.org/10.1371/journal.pone.0063607.

Malzer, E., Daly, M.-L., Moloney, A., Sendall, T. J., Thomas, S. E., Ryder, E., et al. (2010). Impaired tissue growth is mediated by checkpoint kinase 1 (CHK1) in the integrated stress response. *Journal of Cell Science*, *123*(Pt 17), 2892–2900. https://doi.org/10.1242/jcs.070078.

Martin, N., Salazar-Cardozo, C., Vercamer, C., Ott, L., Marot, G., Slijepcevic, P., et al. (2014). Identification of a gene signature of a pre-transformation process by senescence evasion in normal human epidermal keratinocytes. *Molecular Cancer*, *13*, 151. https://doi.org/10.1186/1476-4598-13-151.

Martinon, F., Chen, X., Lee, A.-H., & Glimcher, L. H. (2010). TLR activation of the transcription factor XBP1 regulates innate immune responses in macrophages. *Nature Immunology*, *11*(5), 411–418. https://doi.org/10.1038/ni.1857.

Matos, L., Gouveia, A. M., & Almeida, H. (2014). ER stress response in human cellular models of senescence. *The Journals of Gerontology. Series A, Biological Sciences and Medical Sciences*, *70*, 924–935. https://doi.org/10.1093/gerona/glu129.

Maurel, M., Chevet, E., Tavernier, J., & Gerlo, S. (2014). Getting RIDD of RNA: IRE1 in cell fate regulation. *Trends in Biochemical Sciences*, *39*(5), 245–254. https://doi.org/10.1016/j.tibs.2014.02.008.

McHugh, D., & Gil, J. (2018). Senescence and aging: Causes, consequences, and therapeutic avenues. *The Journal of Cell Biology*, *217*(1), 65–77. https://doi.org/10.1083/jcb.201708092.

Merksamer, P. I., & Papa, F. R. (2010). The UPR and cell fate at a glance. *Journal of Cell Science*, *123*(Pt 7), 1003–1006. https://doi.org/10.1242/jcs.035832.

Micco, M., Collie, G. W., Dale, A. G., Ohnmacht, S. A., Pazitna, I., Gunaratnam, M., et al. (2013). Structure-based design and evaluation of naphthalene diimide G-quadruplex ligands as telomere targeting agents in pancreatic cancer cells. *Journal of Medicinal Chemistry*, *56*(7), 2959–2974. https://doi.org/10.1021/jm301899y.

Min, J.-N., Whaley, R. A., Sharpless, N. E., Lockyer, P., Portbury, A. L., & Patterson, C. (2008). CHIP deficiency decreases longevity, with accelerated aging phenotypes accompanied by altered protein quality control. *Molecular and Cellular Biology*, *28*(12), 4018–4025. https://doi.org/10.1128/MCB.00296-08.

Moon, H.-J., Finney, J., Xu, L., Moore, D., Welch, D. R., & Mure, M. (2013). MCF-7 cells expressing nuclear associated lysyl oxidase-like 2 (LOXL2) exhibit an epithelial-to-mesenchymal transition (EMT) phenotype and are highly invasive in vitro. *The Journal of Biological Chemistry*, *288*(42), 30000–30008. https://doi.org/10.1074/jbc.C113.502310.

Nacarelli, T., Liu, P., & Zhang, R. (2017). Epigenetic basis of cellular senescence and its implications in aging. *Genes*, *8*(12), 343. https://doi.org/10.3390/genes8120343.

Nam, S. M., & Jeon, Y. J. (2019). Proteostasis in the endoplasmic reticulum: Road to cure. *Cancers*, *11*(11), 1793. https://doi.org/10.3390/cancers11111793.

Namba, T., Chu, K., Kodama, R., Byun, S., Yoon, K. W., Hiraki, M., et al. (2015). Loss of p53 enhances the function of the endoplasmic reticulum through activation of the IRE1α/XBP1 pathway. *Oncotarget*, *6*(24), 19990–20001. https://doi.org/10.18632/oncotarget.4598.

Narita, M., Young, A. R. J., Arakawa, S., Samarajiwa, S. A., Nakashima, T., Yoshida, S., et al. (2011). Spatial coupling of mTOR and autophagy augments secretory phenotypes. *Science (New York, N.Y.)*, *332*(6032), 966–970. https://doi.org/10.1126/science.1205407.

Nassour, J., Martien, S., Martin, N., Deruy, E., Tomellini, E., Malaquin, N., et al. (2016). Defective DNA single-strand break repair is responsible for senescence and neoplastic escape of epithelial cells. *Nature Communications*, 7, 10399. https://doi.org/10.1038/ncomms10399.

Nishimura, A., Shimauchi, T., Tanaka, T., Shimoda, K., Toyama, T., Kitajima, N., et al. (2018). Hypoxia-induced interaction of filamin with Drp1 causes mitochondrial hyperfission-associated myocardial senescence. *Science Signaling*, *11*(556), eaat5185. https://doi.org/10.1126/scisignal.aat5185.

Nishio, K., Inoue, A., Qiao, S., Kondo, H., & Mimura, A. (2001). Senescence and cytoskeleton: Overproduction of vimentin induces senescent-like morphology in human fibroblasts. *Histochemistry and Cell Biology*, *116*(4), 321–327. https://doi.org/10.1007/s004180100325.

Nishitoh, H. (2012). CHOP is a multifunctional transcription factor in the ER stress response. *Journal of Biochemistry*, *151*(3), 217–219. https://doi.org/10.1093/jb/mvr143.

Nishitoh, H., Matsuzawa, A., Tobiume, K., Saegusa, K., Takeda, K., Inoue, K., et al. (2002). ASK1 is essential for endoplasmic reticulum stress-induced neuronal cell death triggered by expanded polyglutamine repeats. *Genes & Development*, *16*(11), 1345–1355. https://doi.org/10.1101/gad.992302.

Nizard, C., Noblesse, E., Boisdé, C., Moreau, M., Faussat, A.-M., Schnebert, S., et al. (2004). Heat shock protein 47 expression in aged normal human fibroblasts: Modulation by Salix alba extract. *Annals of the New York Academy of Sciences*, *1019*, 223–227. https://doi.org/10.1196/annals.1297.037.

Orjalo, A. V., Bhaumik, D., Gengler, B. K., Scott, G. K., & Campisi, J. (2009). Cell surface-bound IL-1alpha is an upstream regulator of the senescence-associated IL-6/IL-8 cytokine network. *Proceedings of the National Academy of Sciences of the United States of America*, *106*(40), 17031–17036. https://doi.org/10.1073/pnas.0905299106.

Ota, H., Akishita, M., Eto, M., Iijima, K., Kaneki, M., & Ouchi, Y. (2007). Sirt1 modulates premature senescence-like phenotype in human endothelial cells. *Journal of Molecular and Cellular Cardiology*, *43*(5), 571–579. https://doi.org/10.1016/j.yjmcc.2007.08.008.

Oubaha, M., Miloudi, K., Dejda, A., Guber, V., Mawambo, G., Germain, M.-A., et al. (2016). Senescence-associated secretory phenotype contributes to pathological angiogenesis in retinopathy. *Science Translational Medicine*, *8*(362), 362ra144. https://doi.org/10.1126/scitranslmed.aaf9440.

Palsamy, P., Bidasee, K. R., & Shinohara, T. (2014). Valproic acid suppresses Nrf2/Keap1 dependent antioxidant protection through induction of endoplasmic reticulum stress and Keap1 promoter DNA demethylation in human lens epithelial cells. *Experimental Eye Research*, *121*, 26–34. https://doi.org/10.1016/j.exer.2014.01.021. In this issue.

Pan, K., Chen, Y., Roth, M., Wang, W., Wang, S., Yee, A. S., et al. (2013). HBP1-mediated transcriptional regulation of DNA methyltransferase 1 and its impact on cell senescence. *Molecular and Cellular Biology*, *33*(5), 887–903. https://doi.org/10.1128/MCB.00637-12.

Panda, D. K., Bai, X., Sabbagh, Y., Zhang, Y., Zaun, H.-C., Karellis, A., et al. (2018). Defective interplay between mTORC1 activity and endoplasmic reticulum stress-unfolded protein response in uremic vascular calcification. *American Journal of Physiology. Renal Physiology*, *314*(6), F1046–F1061. https://doi.org/10.1152/ajprenal.00350.2017.

Panganiban, R. A. M., Mungunsukh, O., & Day, R. M. (2013). X-irradiation induces ER stress, apoptosis, and senescence in pulmonary artery endothelial cells. *International Journal of Radiation Biology*, *89*(8), 656–667. https://doi.org/10.3109/09553002.2012.711502.

Pascal, T., Debacq-Chainiaux, F., Chrétien, A., Bastin, C., Dabée, A.-F., Bertholet, V., et al. (2005). Comparison of replicative senescence and stress-induced premature senescence combining differential display and low-density DNA arrays. *FEBS Letters*, *579*(17), 3651–3659. https://doi.org/10.1016/j.febslet.2005.05.056.

Petrova, N. V., Velichko, A. K., Razin, S. V., & Kantidze, O. L. (2016). Small molecule compounds that induce cellular senescence. *Aging Cell*, *15*(6), 999–1017. https://doi.org/10.1111/acel.12518.

Pfaffenbach, K. T., Nivala, A. M., Reese, L., Ellis, F., Wang, D., Wei, Y., et al. (2010). Rapamycin inhibits postprandial-mediated X-box-binding protein-1 splicing in rat liver. *The Journal of Nutrition, 140*(5), 879–884. https://doi.org/10.3945/jn.109.119883.

Pike, L. R. G., Singleton, D. C., Buffa, F., Abramczyk, O., Phadwal, K., Li, J.-L., et al. (2013). Transcriptional up-regulation of ULK1 by ATF4 contributes to cancer cell survival. *The Biochemical Journal, 449*(2), 389–400. https://doi.org/10.1042/BJ20120972.

Plate, L., Cooley, C. B., Chen, J. J., Paxman, R. J., Gallagher, C. M., Madoux, F., et al. (2016). Small molecule proteostasis regulators that reprogram the ER to reduce extracellular protein aggregation. *eLife, 5*, e15550. https://doi.org/10.7554/eLife.15550.

Pluquet, O., Abbadie, C., & Coqueret, O. (2019). Connecting cancer relapse with senescence. *Cancer Letters, 463*, 50–58. https://doi.org/10.1016/j.canlet.2019.08.004.

Pluquet, O., Pourtier, A., & Abbadie, C. (2015). The unfolded protein response and cellular senescence. A review in the theme : Cellular mechanisms of endoplasmic reticulum stress signaling in health and disease. *American Journal of Physiology. Cell Physiology, 308*(6), C415–C425. https://doi.org/10.1152/ajpcell.00334.2014.

Pluquet, O., Qu, L.-K., Baltzis, D., & Koromilas, A. E. (2005). Endoplasmic reticulum stress accelerates p53 degradation by the cooperative actions of Hdm2 and glycogen synthase kinase 3beta. *Molecular and Cellular Biology, 25*(21), 9392–9405. https://doi.org/10.1128/MCB.25.21.9392-9405.2005.

Pollheimer, M. J., Racedo, S., Mikels-Vigdal, A., Marshall, D., Bowlus, C., Lackner, C., et al. (2018). Lysyl oxidase-like protein 2 (LOXL2) modulates barrier function in cholangiocytes in cholestasis. *Journal of Hepatology, 69*(2), 368–377. https://doi.org/10.1016/j.jhep.2018.04.009.

Rajesh, K., Papadakis, A. I., Kazimierczak, U., Peidis, P., Wang, S., Ferbeyre, G., et al. (2013). EIF2α phosphorylation bypasses premature senescence caused by oxidative stress and pro-oxidant antitumor therapies. *Aging, 5*(12), 884–901.

Raven, J. F., Baltzis, D., Wang, S., Mounir, Z., Papadakis, A. I., Gao, H. Q., et al. (2008). PKR and PKR-like endoplasmic reticulum kinase induce the proteasome-dependent degradation of cyclin D1 via a mechanism requiring eukaryotic initiation factor 2alpha phosphorylation. *The Journal of Biological Chemistry, 283*(6), 3097–3108. https://doi.org/10.1074/jbc.M709677200.

Regina, C., Panatta, E., Candi, E., Melino, G., Amelio, I., Balistreri, C. R., et al. (2016). Vascular ageing and endothelial cell senescence: Molecular mechanisms of physiology and diseases. *Mechanisms of Ageing and Development, 159*, 14–21. https://doi.org/10.1016/j.mad.2016.05.003.

Reid, D. W., Chen, Q., Tay, A. S.-L., Shenolikar, S., & Nicchitta, C. V. (2014). The unfolded protein response triggers selective mRNA release from the endoplasmic reticulum. *Cell, 158*(6), 1362–1374. https://doi.org/10.1016/j.cell.2014.08.012.

Reverendo, M., Mendes, A., Argüello, R. J., Gatti, E., & Pierre, P. (2019). At the crossway of ER-stress and proinflammatory responses. *The FEBS Journal, 286*(2), 297–310. https://doi.org/10.1111/febs.14391.

Riahi, Y., Kaiser, N., Cohen, G., Abd-Elrahman, I., Blum, G., Shapira, O. M., et al. (2015). Foam cell-derived 4-hydroxynonenal induces endothelial cell senescence in a TXNIP-dependent manner. *Journal of Cellular and Molecular Medicine, 19*(8), 1887–1899. https://doi.org/10.1111/jcmm.12561.

Rodier, F., & Campisi, J. (2011). Four faces of cellular senescence. *The Journal of Cell Biology, 192*(4), 547–556. https://doi.org/10.1083/jcb.201009094.

Romanov, S. R., Kozakiewicz, B. K., Holst, C. R., Stampfer, M. R., Haupt, L. M., & Tlsty, T. D. (2001). Normal human mammary epithelial cells spontaneously escape senescence and acquire genomic changes. *Nature, 409*(6820), 633–637. https://doi.org/10.1038/35054579.

Rosso, A., Balsamo, A., Gambino, R., Dentelli, P., Falcioni, R., Cassader, M., et al. (2006). P53 Mediates the accelerated onset of senescence of endothelial progenitor cells in diabetes. *The Journal of Biological Chemistry*, *281*(7), 4339–4347. https://doi.org/10.1074/jbc.M509293200.

Rouschop, K. M. A., van den Beucken, T., Dubois, L., Niessen, H., Bussink, J., Savelkouls, K., et al. (2010). The unfolded protein response protects human tumor cells during hypoxia through regulation of the autophagy genes MAP1LC3B and ATG5. *The Journal of Clinical Investigation*, *120*(1), 127–141. https://doi.org/10.1172/JCI40027.

Rubinsztein, D. C., Shpilka, T., & Elazar, Z. (2012). Mechanisms of autophagosome biogenesis. *Current Biology: CB*, *22*(1), R29–R34. https://doi.org/10.1016/j.cub.2011.11.034.

Ruiz, P. D., & Gamble, M. J. (2018). MacroH2A1 chromatin specification requires its docking domain and acetylation of H2B lysine 20. *Nature Communications*, *9*(1), 5143. https://doi.org/10.1038/s41467-018-07189-8.

Saitou, M., Lizardo, D. Y., Taskent, R. O., Millner, A., Gokcumen, O., & Atilla-Gokcumen, G. E. (2018). An evolutionary transcriptomics approach links CD36 to membrane remodeling in replicative senescence. *Molecular Omics*, *14*(4), 237–246. https://doi.org/10.1039/c8mo00099a.

Saleh, T., Tyutyunyk-Massey, L., Murray, G. F., Alotaibi, M. R., Kawale, A. S., Elsayed, Z., et al. (2019). Tumor cell escape from therapy-induced senescence. *Biochemical Pharmacology*, *162*, 202–212. https://doi.org/10.1016/j.bcp.2018.12.013.

Salvatierra, E., Alvarez, M. J., Leishman, C. C., Rivas Baquero, E., Lutzky, V. P., Chuluyan, H. E., et al. (2015). SPARC controls melanoma cell plasticity through Rac1. *PLoS One*, *10*(8), e0134714. https://doi.org/10.1371/journal.pone.0134714.

Sanchez-Alvarez, M., Del Pozo, M. A., & Bakal, C. (2017). AKT-mTOR signaling modulates the dynamics of IRE1 RNAse activity by regulating ER-mitochondria contacts. *Scientific Reports*, 7(1), 16497. https://doi.org/10.1038/s41598-017-16662-1.

Sasaki, T., Maier, B., Bartke, A., & Scrable, H. (2006). Progressive loss of SIRT1 with cell cycle withdrawal. *Aging Cell*, *5*(5), 413–422. https://doi.org/10.1111/j.1474-9726.2006.00235.x.

Sayers, C. M., Papandreou, I., Guttmann, D. M., Maas, N. L., Diehl, J. A., Witze, E. S., et al. (2013). Identification and characterization of a potent activator of p53-independent cellular senescence via a small-molecule screen for modifiers of the integrated stress response. *Molecular Pharmacology*, *83*(3), 594–604. https://doi.org/10.1124/mol.112.081810.

Schmitz, M. L., Shaban, M. S., Albert, B. V., Gökçen, A., & Kracht, M. (2018). The crosstalk of endoplasmic reticulum (ER) stress pathways with NF-κB: Complex mechanisms relevant for cancer, inflammation and infection. *Biomedicine*, *6*(2), 58. https://doi.org/10.3390/biomedicines6020058.

Sen, P., Lan, Y., Li, C. Y., Sidoli, S., Donahue, G., Dou, Z., et al. (2019). Histone acetyltransferase p300 induces de novo super-enhancers to drive cellular senescence. *Molecular Cell*, *73*(4), 684–698.e8. https://doi.org/10.1016/j.molcel.2019.01.021.

Seo, S.-W., Park, S.-K., Oh, S.-J., & Shin, O. S. (2018). TLR4-mediated activation of the ERK pathway following UVA irradiation contributes to increased cytokine and MMP expression in senescent human dermal fibroblasts. *PLoS One*, *13*(8), e0202323. https://doi.org/10.1371/journal.pone.0202323.

Shan, J., Fu, L., Balasubramanian, M. N., Anthony, T., & Kilberg, M. S. (2012). ATF4-dependent regulation of the JMJD3 gene during amino acid deprivation can be rescued in Atf4-deficient cells by inhibition of deacetylation. *The Journal of Biological Chemistry*, *287*(43), 36393–36403. https://doi.org/10.1074/jbc.M112.399600.

Shelton, D. N., Chang, E., Whittier, P. S., Choi, D., & Funk, W. D. (1999). Microarray analysis of replicative senescence. *Current Biology: CB*, *9*(17), 939–945.

Shen, J., Chen, X., Hendershot, L., & Prywes, R. (2002). ER stress regulation of ATF6 localization by dissociation of BiP/GRP78 binding and unmasking of Golgi localization signals. *Developmental Cell*, *3*(1), 99–111. https://doi.org/10.1016/s1534-5807(02)00203-4.

Shi, D., Jiang, K., Fu, Y., Fang, R., Liu, X. I., & Chen, J. (2016). Overexpression of SPARC correlates with poor prognosis in patients with cervical carcinoma and regulates cancer cell epithelial-mesenchymal transition. *Oncology Letters*, *11*(5), 3251–3258. https://doi.org/10.3892/ol.2016.4399.

Shoulders, M. D., Ryno, L. M., Genereux, J. C., Moresco, J. J., Tu, P. G., Wu, C., et al. (2013). Stress-independent activation of XBP1s and/or ATF6 reveals three functionally diverse ER proteostasis environments. *Cell Reports*, *3*(4), 1279–1292. https://doi.org/10.1016/j.celrep.2013.03.024.

Sidler, C., Woycicki, R., Li, D., Wang, B., Kovalchuk, I., & Kovalchuk, O. (2014). A role for SUV39H1-mediated H3K9 trimethylation in the control of genome stability and senescence in WI38 human diploid lung fibroblasts. *Aging*, *6*(7), 545–563. https://doi.org/10.18632/aging.100678.

Singh, K., Matsuyama, S., Drazba, J. A., & Almasan, A. (2012). Autophagy-dependent senescence in response to DNA damage and chronic apoptotic stress. *Autophagy*, *8*(2), 236–251. https://doi.org/10.4161/auto.8.2.18600.

Sisoula, C., & Gonos, E. S. (2011). CHIP E3 ligase regulates mammalian senescence by modulating the levels of oxidized proteins. *Mechanisms of Ageing and Development*, *132*(5), 269–272. https://doi.org/10.1016/j.mad.2011.04.003.

Smith, M., & Wilkinson, S. (2017). ER homeostasis and autophagy. *Essays in Biochemistry*, *61*(6), 625–635. https://doi.org/10.1042/EBC20170092.

So, A.-Y., Jung, J.-W., Lee, S., Kim, H.-S., & Kang, K.-S. (2011). DNA methyltransferase controls stem cell aging by regulating BMI1 and EZH2 through microRNAs. *PLoS One*, *6*(5), e19503. https://doi.org/10.1371/journal.pone.0019503.

Sun, Y., Jiang, F., Pan, Y., Chen, X., Chen, J., Wang, Y., et al. (2018). XBP1 promotes tumor invasion and is associated with poor prognosis in oral squamous cell carcinoma. *Oncology Reports*, *40*(2), 988–998. https://doi.org/10.3892/or.2018.6498.

Szegezdi, E., Logue, S. E., Gorman, A. M., & Samali, A. (2006). Mediators of endoplasmic reticulum stress-induced apoptosis. *EMBO Reports*, *7*(9), 880–885. https://doi.org/10.1038/sj.embor.7400779.

Tam, A. B., Roberts, L. S., Chandra, V., Rivera, I. G., Nomura, D. K., Forbes, D. J., et al. (2018). The UPR activator ATF6 responds to proteotoxic and lipotoxic stress by distinct mechanisms. *Developmental Cell*, *46*(3), 327–343.e7. https://doi.org/10.1016/j.devcel.2018.04.023.

Tezze, C., Romanello, V., Desbats, M. A., Fadini, G. P., Albiero, M., Favaro, G., et al. (2017). Age-associated loss of OPA1 in muscle impacts muscle mass, metabolic homeostasis, systemic inflammation, and epithelial senescence. *Cell Metabolism*, *25*(6), 1374–1389.e6. https://doi.org/10.1016/j.cmet.2017.04.021.

Toosi, S., Orlow, S. J., & Manga, P. (2012). Vitiligo-inducing phenols activate the unfolded protein response in melanocytes resulting in upregulation of IL6 and IL8. *The Journal of Investigative Dermatology*, *132*(11), 2601–2609. https://doi.org/10.1038/jid.2012.181.

Urano, F., Wang, X., Bertolotti, A., Zhang, Y., Chung, P., Harding, H. P., et al. (2000). Coupling of stress in the ER to activation of JNK protein kinases by transmembrane protein kinase IRE1. *Science (New York, N.Y.)*, *287*(5453), 664–666.

Urra, H., Henriquez, D. R., Cánovas, J., Villarroel-Campos, D., Carreras-Sureda, A., Pulgar, E., et al. (2018). IRE1α governs cytoskeleton remodelling and cell migration through a direct interaction with filamin A. *Nature Cell Biology*, *20*(8), 942–953. https://doi.org/10.1038/s41556-018-0141-0.

van Vliet, A. R., Giordano, F., Gerlo, S., Segura, I., Van Eygen, S., Molenberghs, G., et al. (2017). The ER stress sensor PERK coordinates ER-plasma membrane contact site

formation through interaction with filamin-A and F-actin remodeling. *Molecular Cell, 65*(5), 885–899.e6. https://doi.org/10.1016/j.molcel.2017.01.020.
Varadarajan, S., Bampton, E. T. W., Smalley, J. L., Tanaka, K., Caves, R. E., Butterworth, M., et al. (2012). A novel cellular stress response characterised by a rapid reorganisation of membranes of the endoplasmic reticulum. *Cell Death and Differentiation, 19*(12), 1896–1907. https://doi.org/10.1038/cdd.2012.108.
Velarde, M. C., & Demaria, M. (2016). Targeting senescent cells : Possible implications for delaying skin aging: A mini-review. *Gerontology, 62*(5), 513–518. https://doi.org/10.1159/000444877.
Volmer, R., van der Ploeg, K., & Ron, D. (2013). Membrane lipid saturation activates endoplasmic reticulum unfolded protein response transducers through their transmembrane domains. *Proceedings of the National Academy of Sciences of the United States of America, 110*(12), 4628–4633. https://doi.org/10.1073/pnas.1217611110.
Vredeveld, L. C. W., Possik, P. A., Smit, M. A., Meissl, K., Michaloglou, C., Horlings, H. M., et al. (2012). Abrogation of BRAFV600E-induced senescence by PI3K pathway activation contributes to melanomagenesis. *Genes & Development, 26*(10), 1055–1069. https://doi.org/10.1101/gad.187252.112.
Walter, P., & Ron, D. (2011). The unfolded protein response: From stress pathway to homeostatic regulation. *Science (New York, N.Y.), 334*(6059), 1081–1086. https://doi.org/10.1126/science.1209038.
Wang, F.-M., Chen, Y.-J., & Ouyang, H.-J. (2011). Regulation of unfolded protein response modulator XBP1s by acetylation and deacetylation. *The Biochemical Journal, 433*(1), 245–252. https://doi.org/10.1042/BJ20101293.
Wang, E., & Gundersen, D. (1984). Increased organization of cytoskeleton accompanying the aging of human fibroblasts in vitro. *Experimental Cell Research, 154*(1), 191–202.
Wang, M., & Kaufman, R. J. (2016). Protein misfolding in the endoplasmic reticulum as a conduit to human disease. *Nature, 529*(7586), 326–335. https://doi.org/10.1038/nature17041.
Wang, J., Qi, Q., Zhou, W., Feng, Z., Huang, B., Chen, A., et al. (2018). Inhibition of glioma growth by flavokawain B is mediated through endoplasmic reticulum stress induced autophagy. *Autophagy, 14*(11), 2007–2022. https://doi.org/10.1080/15548627.2018.1501133.
Williams, C. C., Singleton, B. A., Llopis, S. D., & Skripnikova, E. V. (2013). Metformin induces a senescence-associated gene signature in breast cancer cells. *Journal of Health Care for the Poor and Underserved, 24*(1 Suppl), 93–103. https://doi.org/10.1353/hpu.2013.0044.
Wu, J., Rutkowski, D. T., Dubois, M., Swathirajan, J., Saunders, T., Wang, J., et al. (2007). ATF6alpha optimizes long-term endoplasmic reticulum function to protect cells from chronic stress. *Developmental Cell, 13*(3), 351–364. https://doi.org/10.1016/j.devcel.2007.07.005.
Xue, X., Piao, J.-H., Nakajima, A., Sakon-Komazawa, S., Kojima, Y., Mori, K., et al. (2005). Tumor necrosis factor alpha (TNFalpha) induces the unfolded protein response (UPR) in a reactive oxygen species (ROS)-dependent fashion, and the UPR counteracts ROS accumulation by TNFalpha. *The Journal of Biological Chemistry, 280*(40), 33917–33925. https://doi.org/10.1074/jbc.M505818200.
Yamamoto, K., Sato, T., Matsui, T., Sato, M., Okada, T., Yoshida, H., et al. (2007). Transcriptional induction of mammalian ER quality control proteins is mediated by single or combined action of ATF6alpha and XBP1. *Developmental Cell, 13*(3), 365–376. https://doi.org/10.1016/j.devcel.2007.07.018.
Yan, M.-M., Ni, J.-D., Song, D., Ding, M., & Huang, J. (2015). Interplay between unfolded protein response and autophagy promotes tumor drug resistance. *Oncology Letters, 10*(4), 1959–1969. https://doi.org/10.3892/ol.2015.3508.

Yao, S., Miao, C., Tian, H., Sang, H., Yang, N., Jiao, P., et al. (2014). Endoplasmic reticulum stress promotes macrophage-derived foam cell formation by up-regulating cluster of differentiation 36 (CD36) expression. *The Journal of Biological Chemistry*, *289*(7), 4032–4042. https://doi.org/10.1074/jbc.M113.524512.

Young, A. R. J., Narita, M., Ferreira, M., Kirschner, K., Sadaie, M., Darot, J. F. J., et al. (2009). Autophagy mediates the mitotic senescence transition. *Genes & Development*, *23*(7), 798–803. https://doi.org/10.1101/gad.519709.

Young, A. R. J., Narita, M., & Narita, M. (2011). Spatio-temporal association between mTOR and autophagy during cellular senescence. *Autophagy*, 7(11), 1387–1388. https://doi.org/10.4161/auto.7.11.17348.

Yu, S., Wang, X., Geng, P., Tang, X., Xiang, L., Lu, X., et al. (2017). Melatonin regulates PARP1 to control the senescence-associated secretory phenotype (SASP) in human fetal lung fibroblast cells. *Journal of Pineal Research*, *63*(1), e12405. https://doi.org/10.1111/jpi.12405.

Zhan, X.-H., Jiao, J.-W., Zhang, H.-F., Li, C.-Q., Zhao, J.-M., Liao, L., et al. (2017). A three-gene signature from protein-protein interaction network of LOXL2- and actin-related proteins for esophageal squamous cell carcinoma prognosis. *Cancer Medicine*, *6*(7), 1707–1719. https://doi.org/10.1002/cam4.1096.

Zhang, R., Poustovoitov, M. V., Ye, X., Santos, H. A., Chen, W., Daganzo, S. M., et al. (2005). Formation of MacroH2A-containing senescence-associated heterochromatin foci and senescence driven by ASF1a and HIRA. *Developmental Cell*, *8*(1), 19–30. https://doi.org/10.1016/j.devcel.2004.10.019.

Zhang, X.-P., Zhang, G.-H., Wang, Y.-Y., Liu, J., Wei, Q., Xu, C.-Y., et al. (2013). Oxidized low-density lipoprotein induces hematopoietic stem cell senescence. *Cell Biology International*, *37*(9), 940–948. https://doi.org/10.1002/cbin.10121.

Zhao, Z., Dong, Q., Liu, X., Wei, L., Liu, L., Li, Y., et al. (2019). Dynamic transcriptome profiling in DNA damage-induced cellular senescence and transient cell-cycle arrest. *Genomics*, *112*, 1309–1317. https://doi.org/10.1016/j.ygeno.2019.07.020.

Zhu, H., Chen, X., Chen, B., Chen, B., Song, W., Sun, D., et al. (2014). Activating transcription factor 4 promotes esophageal squamous cell carcinoma invasion and metastasis in mice and is associated with poor prognosis in human patients. *PLoS One*, *9*(7), e103882. https://doi.org/10.1371/journal.pone.0103882.

Zhu, B., Ferry, C. H., Markell, L. K., Blazanin, N., Glick, A. B., Gonzalez, F. J., et al. (2014). The nuclear receptor peroxisome proliferator-activated receptor-β/δ (PPARβ/δ) promotes oncogene-induced cellular senescence through repression of endoplasmic reticulum stress. *The Journal of Biological Chemistry*, *289*(29), 20102–20119. https://doi.org/10.1074/jbc.M114.551069.

Zhu, Y., Tchkonia, T., Pirtskhalava, T., Gower, A. C., Ding, H., Giorgadze, N., et al. (2015). The Achilles' heel of senescent cells: From transcriptome to senolytic drugs. *Aging Cell*, *14*(4), 644–658. https://doi.org/10.1111/acel.12344.

Zu, Y., Liu, L., Lee, M. Y. K., Xu, C., Liang, Y., Man, R. Y., et al. (2010). SIRT1 promotes proliferation and prevents senescence through targeting LKB1 in primary porcine aortic endothelial cells. *Circulation Research*, *106*(8), 1384–1393. https://doi.org/10.1161/CIRCRESAHA.109.215483.

Zushi, S., Akagi, M., Kishimoto, H., Teramura, T., Sawamura, T., & Hamanishi, C. (2009). Induction of bovine articular chondrocyte senescence with oxidized low-density lipoprotein through lectin-like oxidize d low-density lipoprotein receptor 1. *Arthritis and Rheumatism*, *60*(10), 3007–3016. https://doi.org/10.1002/art.24816.

CHAPTER EIGHT

mTOR as a senescence manipulation target: A forked road

Sarah Saoudaoui[a,b,†], Monique Bernard[a,b,†], Guillaume B. Cardin[a,b,†], Nicolas Malaquin[a,b,†], Apostolos Christopoulos[a,b,c], and Francis Rodier[a,b,d,*]

[a]Centre de recherche du Centre hospitalier de l'Université de Montréal (CRCHUM), Montreal, QC, Canada
[b]Institut du cancer de Montréal, Montreal, QC, Canada
[c]Otolaryngology-Head and Neck Surgery Service, Centre Hospitalier de l'Université de Montréal (CHUM), Montreal, QC, Canada
[d]Université de Montréal, Département de radiologie, radio-oncologie et médicine nucléaire, Montreal, QC, Canada
*Corresponding author: e-mail addresses: rodierf@mac.com; francis.rodier@umontreal.ca

Contents

Abstract

Cellular senescence, cancer and aging are highly interconnected. Among many important molecular machines that lie at the intersection of this triad, the mechanistic (formerly mammalian) target of rapamycin (mTOR) is a central regulator of cell

† These authors contributed equally.

Advances in Cancer Research, Volume 150
ISSN 0065-230X
https://doi.org/10.1016/bs.acr.2021.02.002

metabolism, proliferation, and survival. The mTOR signaling cascade is essential to maintain cellular homeostasis in normal biological processes or in response to stress, and its dysregulation is implicated in the progression of many disorders, including age-associated diseases. Accordingly, the pharmacological implications of mTOR inhibition using rapamycin or others rapalogs span the treatment of various human diseases from immune disorders to cancer. Importantly, rapamycin is one of the only known pan-species drugs that can extend lifespan. The molecular and cellular mechanisms explaining the phenotypic consequences of mTOR are vast and heavily studied. In this review, we will focus on the potential role of mTOR in the context of cellular senescence, a tumor suppressor mechanism and a pillar of aging. We will explore the link between senescence, autophagy and mTOR and discuss the opportunities to exploit senescence-associated mTOR functions to manipulate senescence phenotypes in age-associated diseases and cancer treatment.

1. mTOR and aging

1.1 Hallmarks of aging

Aging is defined by a progressive deterioration of physiological functions accompanied by increased vulnerability to develop age-associated diseases, including cancer and cardiovascular, respiratory, metabolic, neural or muscular disorders (Lopez-Otin, Blasco, Partridge, Serrano, & Kroemer, 2013). Cellular and molecular hallmarks of aging include telomere shortening, persistent DNA damage, genomic instability, epigenetic alterations, loss of proteostasis, deregulation of nutrient sensing, mitochondrial dysfunction and the presence of senescent cells (Hernandez-Segura, Nehme, & Demaria, 2018; Lopez-Otin et al., 2013). Apart from other molecular hallmarks, which can slowly accumulate in most cells of an aging organism, senescent cells represent a binary state; activation of the senescence program in a non-senescent cell rapidly confers highly specialized functions to the newly created senescent cell. Interestingly, most of the molecular hallmarks of aging are also found in senescent cells suggesting that these cells represent a particularly relevant target to modulate age-associated diseases (Hernandez-Segura et al., 2018; Rodier & Campisi, 2011). Senescent cells have been shown to accumulate with age in tissues of rodents, primates and humans (Dimri et al., 1995; Herbig & Sedivy, 2006; van Deursen, 2014; Wang et al., 2009). In rodent models, it is clear that senescent cells contribute in part to the deterioration of tissue homeostasis associated with age-associated diseases and aging. Indeed, the clearance of senescent cells using either genetic or pharmacologic approaches can restore organ functionality for a plethora of age-associated diseases and can extend healthspan or lifespan (Baker et al., 2016; Bussian et al., 2018; Chang et al., 2016; Jeon et al., 2017). Senescent cells are very active microenvironmentally via the production of a pro-inflammatory secretome termed the SASP (senescence-associated secretory phenotype), which has been linked to chronic, sterile age-associated inflammation (inflammaging) (Acosta et al., 2008; Basisty et al., 2020; Coppe et al., 2008; Franceschi & Campisi, 2014; Malaquin, Martinez, & Rodier, 2016). Paradoxically, while senescent cells contribute to age-associated tissue dysfunctions, these cells are essential for normal embryonic patterning, placental biology, tissue repair and tumor suppression (Demaria et al., 2014; Munoz-Espin et al., 2013; Rodier & Campisi, 2011; Storer et al., 2013; van Deursen, 2014). These beneficial functions of senescence are confirmed using rodent models in which the elimination of senescence as a whole or the constant elimination of resident senescent cells leads to tissue dysfunction and cancer (Baker et al., 2011; Demaria et al., 2014; Jeon et al., 2017; Rodier & Campisi, 2011). This suggests that to achieve a beneficial impact in the context of aging, the manipulation of senescence phenotypes such as SASP or the elimination of senescent cells (senolysis) must be precisely targeted and timed. In this review we will explore whether the mammalian target of rapamycin (mTOR) pathway represents a viable pharmacologic strategy to manipulate the phenotypes of cell senescence.

1.2 Stress, metabolic sensors, and mTOR in aging

The search for longevity, rejuvenation or even immortality has been an obsession since the beginning of the history of humankind. Despite the long conceptual history of aging research, the molecular equivalent has only emerged recently through the discovery of the genetic basis of longevity in the nematode *Caenorhabditis elegans* (Klass, 1983). For example, single mutations in the *age-1* or *daf-2* genes were sufficient to extend the lifespan of this microscopic worm, providing hints of molecular targets that regulate aging (Friedman & Johnson, 1988; Kenyon, Chang, Gensch, Rudner, & Tabtiang, 1993). *age-1* and *daf-2* are components of the evolutionary conserved insulin/IGF-1/FOXO signaling cascade, and mutations in this pathway can extend the lifespan of mammals or are linked to longevity in humans (Kenyon, 2011). The insulin/IGF-1/FOXO pathway is key to regulating energy and metabolism, and many other gene mutations involved in food intake (*eat* mutants) result in the partial starvation of nematodes, causing prolonged longevity (Avery, 1993; Lakowski & Hekimi, 1998). These observations laid the foundation for the concept that caloric restriction can slow the progression of aging across species, including nematodes, rodents and primates (Fontana, Nehme, & Demaria, 2018; Leonov et al., 2017). One of the mechanisms connecting caloric restriction with the hallmarks of aging involves mitochondrial oxidative stress and the generation of free radicals at the mitochondrial inner membrane, which are found at lower levels in animals with a restricted diet (Sohal & Weindruch, 1996). Still, metabolic balance remains a key element associated with caloric restriction and mTOR is a key regulator of metabolism. Not surprisingly, numerous reports have documented that the inhibition of mTOR signaling enhances longevity in different model organisms, including *Saccharomyces cerevisae*, *C. elegans*, *Drosophila* or *Mus musculus* (Harrison et al., 2009; Kaeberlein et al., 2005; Kapahi et al., 2004; Papadopoli et al., 2019; Vellai et al., 2003; Weichhart, 2018). Exactly how mTOR extends the longevity phenotype via alterations of cellular phenotypes is covered in other reviews and remains under investigation (Papadopoli et al., 2019). Interestingly, links between mTOR and cellular senescence suggest that part of the life extension provided via mTOR inhibition involves alterations of cellular senescence biology.

1.3 mTOR signaling

mTOR is a serine/threonine protein kinase of the phosphoinositide 3-kinase (PI3K)-related kinase family with a central role in the regulation of cellular metabolism. mTOR is a component of two distinct complexes: mTORC1 and mTORC2 (Fig. 1). The two complexes share common proteins in addition to mTOR; the mTORC1 complex is composed specifically of the regulatory-associated protein of mammalian target of *rapamycin* (RAPTOR) and the proline-rich Akt substrate 40kDa (PRAS40) (Kim et al., 2002, 2003), while the mTORC2 complex specifically includes the rapamycin-insensitive companion of mTOR (RICTOR), mammalian stress-activated map kinase-interacting protein 1 (mSin1) and protein observed with Rictor 1 and 2 (Protor 1/2) (Frias et al., 2006; Sarbassov et al., 2004). mTORC1 was originally described as the target of rapamycin, while mTORC2 was defined as insensitive to rapamycin (hence the name RICTOR). However, subsequent studies have demonstrated that mTORC2 can be inhibited by rapamycin under certain conditions and in certain cell types (Lamming et al., 2012; Sarbassov et al., 2006). Prolonged exposure to

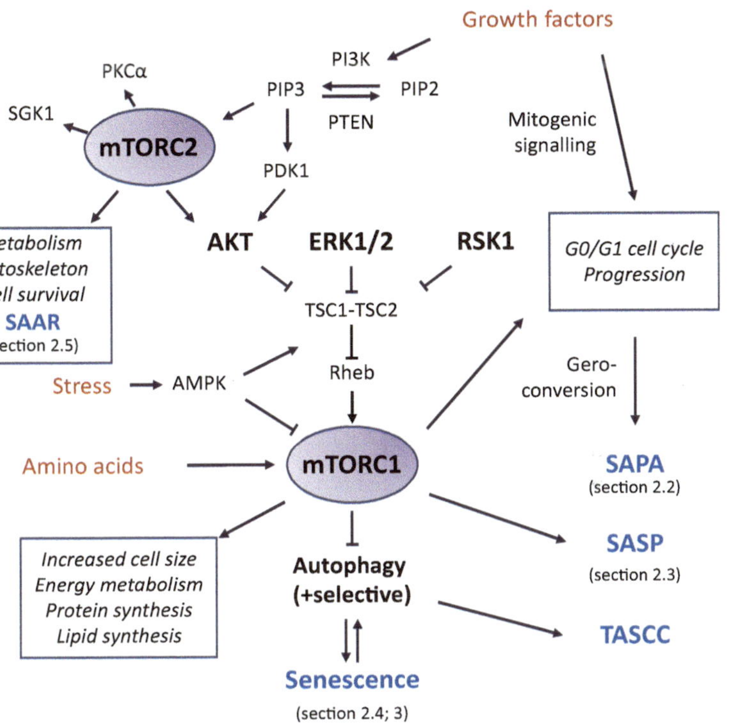

Fig. 1 The PI3K/AKT/mTOR signaling pathway and its association with senescence phenotypes. The principal signaling nodes and outputs of the mTORC1 and mTORC2 complexes in response to growth factors are indicated. Several aspects of the senescence program are regulated by mTOR as described in more detail in highlighted sections of the text.

rapamycin prevents the formation of new mTORC2 complexes, thereby inhibiting its activity (Liu & Sabatini, 2020; Sarbassov et al., 2006). Overall, the two mTOR complexes are involved in many biological activities of the cells with specificities for each complex.

1.3.1 Role and regulation of mTORC1

The mTORC1 complex is central for metabolism. It contributes to *cell proliferation* (cell cycle progression) in the presence of pro-mitotic signals and controls *cell growth* (in size) by regulating protein synthesis, lipid synthesis, energy metabolism and *autophagy*. mTORC1 responds to various signals: growth factors, energy status (glucose, AMP/ATP ratio), amino acids, oxygen and stress. Hence, mTORC1 is considered a sensor and integrator of the cellular environment in order to regulate homeostasis. mTORC1 promotes cell growth when conditions are favorable or conversely, promotes catabolism when conditions are unfavorable (Laplante & Sabatini, 2012; Liu & Sabatini, 2020). In addition to cell growth, mTORC1 stimulates early G1 cell cycle progression in the presence of pro-mitotic signals via its ability to regulate the S6K1 and 4E-BP1/eukaryotic translation initiation factor 4E (eIF4E) pathways to promote cyclin D1 accumulation (Fingar et al., 2004; Wee & Wang, 2017).

mTORC1 is regulated by the tuberous sclerosis 1 and 2 complex (TSC1-TSC2 or hamartin-tuberin), which itself negatively regulates the activity of Ras homolog enriched in brain (Rheb), a GTPase necessary to activate the mTOR kinase (Huang & Manning, 2008). Several kinases like AKT, ERK1/2, and RSK1 contribute to inhibit TSC1-TSC2 to *activate* mTOR. Alternatively, activating phosphorylation of TSC1-TSC2 *inhibits* mTORC1. For example, AMPK is a kinase that is activated by energy imbalances when the AMP to ATP ratio increases or by hypoxic stress and can activate TSC2 via phosphorylation. AMPK can also phosphorylate Raptor, which is inhibitory for mTORC1. These mechanisms are involved in the regulation of *autophagy* by energy status and hypoxia (Hardie, 2011). The response to *DNA damage* can inhibit mTORC1 by increasing the expression of TSC2 and PTEN (phosphatase and tensin homolog) (Laplante & Sabatini, 2012). Amino acids can also regulate mTORC1 activity, independent of TSC1-TSC2. In the presence of amino acids, mTORC1 binds to a complex called the "Ragulator," which enables mTORC1 localization to the lysosomal membrane and promotes its activation (Sancak et al., 2010).

1.3.2 Role and regulation of mTORC2

There are three identified targets for mTORC2: AKT, SGK1 (serum and glucocorticoid-induced protein kinase 1) and PKC-α (protein kinase C-α). These kinases regulate cell survival, metabolism and cytoskeleton organization. The PI3KI pathway is central for the control of mTOR by growth factors and hormones. After binding of the growth factor to its receptor, the second messenger phosphatidylinositol (3, 4, 5) P3 (PIP3) is formed with the activation of class I PI3K. The PI3K can be inhibited by the phosphatase PTEN. The second messenger PIP3 promotes the recruitment of PDK1 to the plasma membrane, which in turn, recruits AKT and phosphorylates it at thr308. PIP3 production is also directly responsible for mTORC2 activation and mTORC2 is responsible for the phosphorylation of AKT at ser473. Both phosphorylations are necessary for full activation of AKT (Manning & Toker, 2017). Also, the binding of mTORC2 to ribosomes is necessary for its activation and is dependent on the activity

could lead to an irreversible senescence arrest (Blagosklonny, 2014). Geroconversion requires opposing signals: activity of growth-promoting, mitogen-/nutrient-sensing pathways confronted with CDKi activity that maintains cell cycle arrest. In this context, it is proposed that mTOR inhibition using rapamycin can prevent geroconversion by preventing futile cell growth (Blagosklonny, 2018; Chitikova et al., 2014). Comparably, the prevention of cellular senescence in cells nearing replicative senescence is achieved using AZD8055, an ATP mimetic pan-mTORC inhibitor. Within the cell population, a loss of SABGAL and re-acquisition of fibroblastic spindle morphology was observed, suggesting the prevention of a transition to senescence (Walters, Deneka-Hannemann, & Cox, 2016).

2.3 mTOR and the senescence secretome

Much like the SAPA, the SASP is critical for the biological activities of senescent cells and presents itself with beneficial and deleterious consequences. For example, some SASP factors can reinforce the SAPA in neighboring cells or can stimulate the immune system, contributing to the elimination of senescent cells (Acosta et al., 2013; Kale, Sharma, Stolzing, Desprez, & Campisi, 2020; Prata, Ovsyannikova, Tchkonia, & Kirkland, 2018). Alternatively, the SASP in persistent senescent cells can create a chronic, pro-inflammatory microenvironment proposed to contribute to inflammaging, and is thus associated with chronic inflammatory disorders or cancer cell stimulation (Franceschi & Campisi, 2014; He & Sharpless, 2017; Ovadya & Krizhanovsky, 2014). Once senescence is initiated, the regulation of SASP factors expression involves many interlaced signaling pathways, including the DDR (Coppe et al., 2008; Kang et al., 2015; Malaquin et al., 2020; Malaquin, Carrier-Leclerc, Dessureault, & Rodier, 2015; Ohanna et al., 2011; Pazolli et al., 2012; Rodier et al., 2009, 2011), the p38-MAPK stress response (Alspach et al., 2014; Freund et al., 2011), the cytoplasmic chromatin sensing cyclic GMP-AMP synthase (cGAS) and stimulator of interferon genes (STING) (Dou et al., 2017; Gluck et al., 2017; Yang, Wang, Ren, Chen, & Chen, 2017), and the inflammasome (Acosta et al., 2013). Overall, these pathways contribute to the activation of NF-κB, the major transcription factor regulating the expression of SASP factors (Acosta et al., 2008; Alspach et al., 2014; Chien et al., 2011; Coppe et al., 2008; Freund et al., 2011; Ohanna et al., 2011; Rovillain et al., 2011).

Given its central homeostatic functions, several studies have demonstrated that mTOR interfaces with the establishment of SASP in different contexts of senescence (Bent, Gilbert, & Hemann, 2016; Herranz et al., 2015;

Laberge et al., 2015; Zhang et al., 2018). In a simple summary, the genetic depletion of mTOR or Raptor proteins or the pharmacological inhibition of mTOR using chemical inhibitors (Rapamycin, Everolimus, Torin1) abolishes most of the SASP. Under a more encompassing vision, mTOR could be involved at different levels to regulate SASP. First, mTOR has been shown to directly impact NF-κB activity. Indeed, mTOR and Raptor can control NF-κB DNA-binding activity and RelA/p65 phosphorylation in pre-senescent prostatic cancer cells but also in senescent cancer cells obtained from treatment with chemotherapeutics agents or irradiation (Dan et al., 2008; Zhang et al., 2018). In this context, activation of PI3K-AKT by p38-MAPK signaling promotes the activation of mTOR and its association with the IKK complex, inhibiting its negative activity against NF-κB. Alternatively, mTOR can control the expression of SASP factors at the post-translational level by enhancing mRNA stability (Herranz et al., 2015). More specifically, in senescent cells, the ability of the RNA-binding protein ZFP36 ring finger protein-like 1 (ZFP36L1) to degrade the transcripts of numerous SASP factors is inhibited by the kinase MAPKAPK2, which is also positively modulated at the translational level by mTOR activity (Herranz et al., 2015). Similarly, mTOR has been shown to promote the translation of IL-1α, which is itself one of the first factors involved in the autocrine amplification loop regulating the levels of many other SASP factors (Laberge et al., 2015). Thus, in general, mTOR positively promotes the SASP via NF-κB activity stimulation and enhanced translation of SASP factors.

2.4 mTOR and SA autophagy

Crosstalk between autophagy and senescence has been reported numerous times, but the relationship between the two pathways appears to be context specific: many studies find pro-senescence functions for autophagy, while others find the opposite (Gewirtz, 2013; Kwon, Kim, Jeoung, Kim, & Kang, 2017).

2.4.1 Autophagy is key for the senescence program

The induction of autophagy during cellular senescence was first documented in the context of oncogenic Ras-induced OIS in human diploid fibroblasts (Young et al., 2009). A strong confirmation of this link came with the knockdown of the autophagy genes *ATG5* or *ATG7*, which caused a delay in the expression of the senescence hallmarks SABGAL and SASP, and revealed that autophagy effectively contributes to the OIS mitotic-senescence transition. By definition, the activity of pro-mitogenic MAPK and PI3K pathways is upregulated at the beginning of OIS, consistent with a period of rapid proliferation referred to as the mitotic burst. Following several days and many cell divisions, this highly proliferative state is replaced by SAPA, and the downregulation of the PI3K pathway is explained by the existence of PI3K negative feedback loops (Courtois-Cox et al., 2006; Julien, Carriere, Moreau, & Roux, 2010). Simultaneous with the decreasing PI3K signaling, decreasing mTORC1 and mTORC2 activities induce autophagy and contribute to cell cycle exit (Young et al., 2009). Interestingly, autophagy activation with the concomitant negative feedback in the mTOR pathways is also observed during DNA damage-induced senescence, suggesting that the OIS mitotic burst is not a prerequisite for the autophagy-senescence connection (Young et al., 2009). It has been established that senescence phenotypes, particularly the SASP, develop over a period of multiple days and logically, autophagy as a general process could play a role in specific parts of senescence maturation. For example, during OIS, mTOR was found to colocalize with autophagic markers (ULK1, P62 and LAMP2) in a special compartment where autophagy can occur, known as the TOR-autophagy spatial coupling compartment (TASCC) (Narita et al., 2011). mTOR recruitment to the TASCC was amino acid-dependent and TASCC formation was key for the synthesis of SASP factors such as IL-6 and IL-8 (Narita et al., 2011). This discovery implied that mTOR activity can be regulated in function of its localization within the cell, reflecting nutriment availability. The authors suggested that the TASCC allowed for protein turnover to favor the synthesis of new proteins required for the phenotypic changes linked to senescence.

While senescence-autophagy studies do not always probe mTOR, its role as a master regulator of autophagy nevertheless suggests underlying links. Examples of additional key senescence regulators linked to autophagy include proteins such as laminB1, GATA4 and SIRT1. Decreased expression of laminB1 is recognized as a hallmark of senescence in conjunction with alterations in nuclear lamina biology (Castro-Obregon, 2020; Freund, Laberge, Demaria, & Campisi, 2012). During OIS, laminB1 is degraded via autophagy, and the key autophagosome component LC3 is found to interact with laminB1 at the nuclear lamina at sites of nuclear membrane "blebbing." The blebs are proposed to contain chromatin fragments that are also degraded by autophagy in the cytoplasm (Dou et al., 2015; Dou, Ivanov, Adams, & Berger, 2016). SA cytosolic chromatin fragments processed by autophagosomes are connected to the regulation of the SASP via cGAS-STING innate immune sensing and interferon signaling (Dou et al., 2017; Gluck et al., 2017; Yang et al., 2017). The transcription factor GATA4 provides a different example of SA selective autophagy: the selective autophagy that constitutively degrades GATA4 is shut down by the ATM/ATR senescence DDR, leading to GATA4 stabilization, which then contributes to NF-κB activation (Kang et al., 2015).

Noteworthy in the context of this review is SIRT1, a member of the sirtuins family, which harbors strong links to age-associated processes (Grabowska, Sikora, & Bielak-Zmijewska, 2017; Pyo, Yun, Yoon, Choi, & Lee, 2020). SIRT1 is a protein essential for NAD^+ lysine deacetylation; one principal role for SIRT1 is epigenetic regulation through histones deacetylation (Lagunas-Rangel, 2019). SIRT1 regulates ROS levels and inflammation by deacetylation of key transcription factors (FOXO, p65 subunit of NFkB) (Salminen, Kaarniranta, & Kauppinen, 2013). SIRT1 also regulates the autophagy resulting from serum starvation by deacetylation of the autophagy protein LC3 in the nucleus (Huang et al., 2015; Lee et al., 2008). Recently, SIRT1 levels were shown to be regulated by autophagy in fibroblasts during OIS, replicative senescence and therapy-induced senescence. SIRT1 levels were shown to decrease in cellular senescence during which the protein is transferred from the nucleus to the cytoplasm where it is degraded by autophagy. This is dependent upon SIRT1 binding to a special motif in LC3, the LC3-interacting region (LIR), again suggesting a form of selective autophagy (Xu et al., 2020). Based on these results, the authors proposed that blocking lysosomal degradation could enhance SIRT1 activity during senescence (Tan & Finkel, 2020; Xu et al., 2020).

The link between senescence and autophagy is robust and occurs in varied cell types, including cancer cells. For example, the overexpression of the CDKi $p16^{INK4A}$ or $p21^{WAF1/CIP1}$ induces senescence and autophagy in human fibroblasts and MDA-MB-231 breast cancer cells (Capparelli et al., 2012). The downregulation of ATG5 in normal melanocytes facilitates the bypass of BRAF-mediated OIS, suggesting a pro-senescence role for autophagy in this context (Liu et al., 2013). BRAF-melanocytes undergo aberrant, but limited, proliferation and enter into senescence that is reminiscent of the benign tumor cells that form melanocytic nevus, exhibiting normal levels of ATG5 and autophagy. However, if ATG5 and autophagy are downregulated, melanocytes undergo malignant transformation and evolve into melanoma. Autophagy may play a protective role for cancer development and in fact, the authors found that both ATG5 and LC3 levels were decreased in patients with melanomas compared to those with benign nevi (Liu et al., 2013; Liu, He, & Simon, 2014). Thus, autophagy and OIS are together tumor suppressive.

It should also be noted that the study of interactions between senescence and autophagy can be complicated by the heterogeneity of the cell populations under study, particularly when using biochemical extraction techniques on bulk populations. For example, autophagy induced in human diploid fibroblasts via serum starvation leads to senescence hallmarks such as SABGAL, p21 and $p16^{INK4a}$ combined with myofibroblastic hallmarks. However, when examining single cells in this model, sustained autophagy leads to mTORC2 activation followed by senescence or myofibroblast differentiation in non-overlapping cells (Bernard et al., 2020). In another study that induced senescence in a glioma cell line via genotoxic stress, single cell senescence was observed by morphometric parameters and autophagy by GFP-LC3 puncta formation. The results obtained at the bulk population level of cells revealed that on average, autophagy preceded senescence. However, in single cell analysis, this observation was not confirmed and autophagy was at times decreasing before, during, or after the acquisition of the senescence phenotype. Overall, the inhibition of autophagy diminished senescence and promoted apoptosis (Filippi-Chiela, Bueno e Silva, Thome, & Lenz, 2015).

2.4.2 Lack of autophagy induces senescence

While autophagy is recognized as a multifaceted process in the context of senescence, it remains to be seen whether autophagy can be easily manipulated in a uniformly beneficial manner. It was shown that pharmacological or genetic inhibition of autophagy in breast tumor cell treated with adriamycin or camptothecin delayed senescence but not abrogated it (Goehe et al., 2012). It was also shown in colon carcinoma cells treated by irradiation that the induction of autophagy and senescence took place in parallel; however, the inhibition of autophagy failed to suppress senescence, indicating that the two processes were dissociable (Alotaibi et al., 2016). In some cases, cancer cells can recover from an apparent senescence-like state and can resume proliferation several days after the end of treatment. In this context autophagy inhibition did not alter either proliferation arrest or proliferative recovery suggesting that autophagy did not significantly contribute to senescence itself or to the survival of senescent cells (Saleh et al., 2020; Sharma et al., 2014). To add further complexity, serious autophagy defects may promote senescence rather than inhibit it. Mitochondria, particularly defective mitochondria, are primary sources of intracellular ROS, which generate further deleterious intracellular damage such as organelle, DNA, or protein alterations leading to impaired cell functions and senescence or organismal aging (Baraibar, Liu, Ahmed, & Friguet, 2012; Ogrodnik, Salmonowicz, & Gladyshev, 2019). Autophagy has the capacity to remove those stressors and it remains logical to assume that in this context, autophagy acts as an anti-senescence mechanism (Kongara & Karantza, 2012; Mathew et al., 2009). For example, prolonged autophagy inhibition using small hairpin RNA against key autophagy genes induces premature senescence in a p53-dependent manner in primary human fibroblasts due to increased damaged mitochondria and ROS (Kang, Lee, Kim, Choi, & Park, 2011). Similarly, defective autophagy compromises the ability of cells to adapt to metabolic stress, which may lead to insufficient ATP generation and accumulation of damaged mitochondria with increased ROS levels (Jin & White, 2007). These experiments suggest that cellular senescence can be a consequence of defective autophagy. In normal human foreskin epidermal melanocytes, silencing of the autophagy gene *ATG7* and exposure to ROS (H_2O_2 or UVB) caused elevated levels of SABGAL-positive cells and increased levels of the CDKi $p16^{INK4A}$, again suggesting a protective effect of autophagy against premature senescence induction (Qiao et al., 2020).

2.5 mTOR and SA survival signals

Resistance to apoptosis in senescent cells (SAAR) is a phenotype related to inherent stresses found in senescent cells. Briefly, senescent cells develop an addiction to anti-apoptotic Bcl2-family members such as Bcl2 and Bcl-XL, which prevent spontaneous apoptosis of these cells (Childs, Baker, Kirkland, Campisi, & van Deursen, 2014; Wang, 1995). This vulnerability has been exploited to describe a new class of drugs, senolytic agents, which preferentially promote apoptosis in senescent cells when compared to their pre-senescent counterparts (Chang et al., 2016; Fleury et al., 2019; Kirkland, Tchkonia, Zhu, Niedernhofer, & Robbins, 2017; Lafontaine et al., 2021; Malaquin, Vancayseele, et al., 2020; Saleh et al., 2020; Wang et al., 2016; Zhu et al., 2015, 2017). For example, BH3-only mimetics such as Navitoclax or ABT-737 target anti-apoptotic Bcl2 proteins, thus causing Bak/Bax-dependent apoptosis in situations when Mcl-1 is absent or also inhibited (Chang et al., 2016; Saleh, Carpenter, et al., 2020; van Delft et al., 2006). We have recently demonstrated a novel one-two punch strategy to treat cancer where senescence induction represents the first hit and senolysis as the second hit that converts senescence to death (Fleury et al., 2019; Lafontaine et al., 2021; Malaquin, Vancayseele, et al., 2020). Exactly why senescent cells become addicted to anti-apoptotic Bcl2 proteins remain unclear but may relate to SA metabolic changes. For example, mTOR is an important Mcl-1 regulator, acting as an inducer of Mcl-1 translation (Hsieh et al., 2010; Mills et al., 2008). Alternatively, mTORC2 can stabilize the Mcl-1 protein, preventing its degradation mediated by GSK-3 and SCF-FBXW7 (Koo, Yue, Deng, Khuri, & Sun, 2015). This pro-survival regulation is key given that both *MCL-1* mRNA and protein have a very short half-life (Nijhawan et al., 2003; Yang, Buchan, Townsend, & Craig, 1996). This suggests that mTOR could be an important protein for SA survival signals following BH3 mimetic treatments. Although the specific contributions of senescence and cell death were not evaluated, mTOR inhibition has been shown to enhance BH3 mimetic therapies for acute myeloid leukemia (Rahmani et al., 2015; van Delft et al., 2006), non-small cell lung carcinoma (Kim, Moretti, Mitchell, Jung, & Lu, 2009), rhabdomyosarcoma (Preuss, Hugle, Reimann, Schlecht, & Fulda, 2013), acute lymphocytic leukemia (Iacovelli et al., 2015), glioblastoma (Venkatesan et al., 2016), triple-negative breast cancer (Li, Liu, Chang, Zou, & Xing, 2018), estrogen receptor α breast cancer (Williams et al., 2019) and renal cell cancer (Nayman et al., 2019). In at least one case, the synergy between mTOR inhibition and BH3 mimetics is dependent on p53 status. Co-treatment induced senescence and a SASP, activating the AKT/ERK–mTORC1–4EBP1–Mcl-1/surviving signaling axis in p53 wild-type cells but not in p53 defective cancer cells, thereby promoting their cell death. The synergy was restored in p53 wild-type cells following NF-kB inhibition or treatment with neutralizing antibodies against SASP components (Li et al., 2017). Still, while inhibition of mTOR often sensitizes many cells to BH3 mimetics, this is not always the case.

3. Exploring mTOR manipulation during senescence

A first step to deciphering the impact of mTOR inhibition on the senescence phenotypes important for age-associated diseases and cancer therapy is to summarize the senescence phenotypes regulated by mTOR. Briefly, mTOR impacts three key senescence phenotypes, the SASP, SAPA and SAAR (summarized in Fig. 1). *mTOR positively promotes the*

SASP via NF-κB activity stimulation and enhanced translation of SASP factors. In addition, *mTOR regulates the SASP* via autophagy inhibition by preventing TASCC (negative regulation) and through the promotion of cytoplasmic chromatin fragments accumulation (positive regulation), which triggers cGAS-STING and NF-κB. From a SASP point of view, mTOR appears to have a relatively positive impact. From a SAPA point of view, in combination with other pro-mitogenic signals, *mTOR can paradoxically promote geroconversion* toward senescence via early G0/G1 cell cycle promotion. Given that cells will be pushed to enter S-phase in the presence of unrepaired DNA lesions, this geroconversion also promotes stresses that result in a more stable SAPA and may indirectly promote the creation of additional DNA lesions that stimulate the SASP. From a SAAR point of view, *mTOR promotes cell survival* via increased stabilization of the anti-apoptotic Bcl2 family member Mcl-1. Finally, mTOR regulates autophagy, and selective autophagy has varied effects on senescence phenotypes, yielding an unclear status for mTOR influence at this level. For example, autophagy is required for SA loss of laminB1 and SIRT1, but selective autophagy must also be blocked for the stabilization of the key senescence transcription factor GATA4. Overall, with the exception of its impact on autophagy, normal basal mTOR activity appears to positively influence SASP, geroconversion, SAPA and SAAR. Given this understanding, we can envision two scenarios where mTOR manipulation has an impact on cell senescence: (1) when mTOR inhibition is used on cells nearing senescence or during senescence activation, or (2) when mTOR inhibition is used on established senescent cells.

In the first scenario, the data suggest that mTOR inhibition can prevent or reduce entry into senescence, instead promoting quiescence or slow cell cycling and perhaps allowing more time to repair DNA, thereby preserving the pool of non-senescent cells. Indirectly, this strategy also prevents the activation of the SASP and may prevent low levels of chronic inflammation (inflammaging). However, this could simultaneously allow the increased proliferative survival of oncogenically damaged cells, allowing the perpetuation of genomic instability and potentially increasing long-term cancer risk. Given the similarities between the effects of mTOR inhibition and caloric restriction (Papadopoli et al., 2019; Weichhart, 2018), it is interesting to note that both mTOR inhibition and caloric restriction mimetics in culture can extend the lifespan of normal cell cultures (Antonioli, Torres, Ferretti, Piccinato, & Sertie, 2019; Kaeberlein et al., 2005). While inhibiting or delaying the entry into senescence of normal cells may appear useful and could explain some of the beneficial effects of mTOR inhibition on healthspan and longevity, it is not immediately obvious that this strategy will be appropriate or useful during cancer therapy. Whether senescence is beneficial or detrimental in the context of cancer therapy remains mostly controversial. However, inhibiting senescence induction using rapamycin in specific contexts in which it is detrimental could be an interesting cancer treatment strategy.

In the second scenario involving inhibition of mTOR in established senescent cells, the data suggests that the inhibition of SAAR via decreased Mcl-1 levels could trigger apoptosis in a manner similar to senolytic drugs. However, this has not been observed as multiple groups have shown that senescent cells treated with rapamycin display inhibition of the SASP and none have reported apoptosis or elimination of senescent cells (Bent et al., 2016; Herranz et al., 2015; Laberge et al., 2015; Zhang et al., 2018). This suggests that mTOR inhibition does not impair senescent cell survival but affects the microenvironment created by senescent cells. This would also be consistent with observations that SAAR depends mostly on Bcl2 and Bcl-XL, not on Mcl-1 (Chang et al., 2016; Wang, 1995; Zhu et al., 2015, 2017). In a recent study, the short-term treatment of elderly mice with rapamycin restored oral cavity health, including regeneration of periodontal bone and attenuation of bone inflammation (An et al., 2020). Interestingly, markers of SASP were decreased in the regenerated tissue, suggesting the clearance or inactivation of senescent cells (An et al., 2020). Similarly, other groups have reported the diminishment of senescence markers in the lungs or artery of rapamycin-treated old mice (Hinojosa et al., 2012; Lesniewski et al., 2017). Finally, human sun-exposed skin treated with topical rapamycin also displayed diminished markers of senescence (Chung et al., 2019). These interesting observations suggest several possibilities: rapamycin is indeed senolytic in vivo for established senescent cells, the inhibition of mTOR in senescent cells can reverse some of their phenotypes in addition to inhibiting the SASP, or mTOR inhibition can trigger indirect senolysis via enhanced elimination of resident senescent cells by the immune system. Although not yet tested in vivo in the context of cancer therapy-induced senescence, inhibiting mTOR after the induction of senescence in tumors could yield interesting results, particularly in immune-competent models.

Overall, using mTOR inhibition to manipulate senescence remains a complex problem. Despite this, there is little doubt that if used appropriately, in time and space, mTOR inhibition can yield beneficial phenotypes via the manipulation of senescence. Other compounds explored for their potential in lifespan extension may also manipulate senescence, further supporting the link between mTOR and senescence. For example, natural products with anti-oxidative and anti-inflammatory properties such as fisetin, curcuma, or piperlongumine have anti-age potential and show beneficial impact on age-related diseases (Beevers, Zhou, & Huang, 2013; Bielak-Zmijewska et al., 2019; Yousefzadeh et al., 2018). Not surprisingly, these molecules can attenuate phenotypes of senescent cells, notably the SASP, and can sometimes serve as senolytics (Wang et al., 2016; Yousefzadeh et al., 2018). *In fine*, more experiments will be required to clearly answer these possibilities, particularly a more systematic evaluation of senescence hallmarks in tissues in which senescence is possibly manipulated using mTOR or other life-extending compounds.

Authors contributions

All authors read and approved the final manuscript. S.S., M.B., G.B.C. and N.M wrote the manuscript and interpreted the relevant literature. A.C. supervision, revised critically the manuscript, provided financial support, technical support, expertise, conceptual advice and obtained funding for the study. F.R. supervision, wrote the manuscript, interpreted the relevant literature, provided financial support, technical support, expertise, conceptual advice and obtained funding for the study.

Acknowledgments

We thank members of the Rodier and Christopoulos laboratories for valuable comments and discussions as well as Jacqueline Chung for manuscript editing. F.R. and A.C. are researchers of the CRCHUM and Institut du cancer de Montréal (ICM), which receive support from the Fonds de recherche du Québec—Santé (FRQS). This work was supported by the ICM and by grants to F.R. from the Oncopole (265877), the Canadian Institute for Health Research (173341) and the Terry Fox Research Institute and to A.C. from the Cancer Research Society (25198) and from the Dr. Azar-Angelil Research Chair in Head and Neck Oncology. F.R. is supported by a Fonds de Recherche Québec Santé Senior career award (281706). S.S. is supported by a Canderel student fellowship from the ICM and by a "La vie en rose" master fellowship from the CRCHUM.

Declaration of interests

The authors declare no competing interests.

References

Acosta, J. C., Banito, A., Wuestefeld, T., Georgilis, A., Janich, P., Morton, J. P., et al. (2013). A complex secretory program orchestrated by the inflammasome controls paracrine senescence. *Nature Cell Biology*, *15*(8), 978–990. https://doi.org/10.1038/ncb2784.

Acosta, J. C., O'Loghlen, A., Banito, A., Guijarro, M. V., Augert, A., Raguz, S., et al. (2008). Chemokine signaling via the CXCR2 receptor reinforces senescence. *Cell*, *133*(6), 1006–1018. https://doi.org/10.1016/j.cell.2008.03.038.

Alotaibi, M., Sharma, K., Saleh, T., Povirk, L. F., Hendrickson, E. A., & Gewirtz, D. A. (2016). Radiosensitization by PARP inhibition in DNA repair proficient and deficient tumor cells: Proliferative recovery in senescent cells. *Radiation Research*, *185*(3), 229–245. https://doi.org/10.1667/RR14202.1.

Alspach, E., Flanagan, K. C., Luo, X., Ruhland, M. K., Huang, H., Pazolli, E., et al. (2014). p38MAPK plays a crucial role in stromal-mediated tumorigenesis. *Cancer Discovery*, *4*(6), 716–729. https://doi.org/10.1158/2159-8290.CD-13-0743.

An, J. Y., Kerns, K. A., Ouellette, A., Robinson, L., Morris, H. D., Kaczorowski, C., et al. (2020). Rapamycin rejuvenates oral health in aging mice. *eLife*, *9*. https://doi.org/10.7554/eLife.54318.

Antonioli, E., Torres, N., Ferretti, M., Piccinato, C. A., & Sertie, A. L. (2019). Individual response to mTOR inhibition in delaying replicative senescence of mesenchymal stromal cells. *PLoS One*, *14*(1). https://doi.org/10.1371/journal.pone.0204784, e0204784.

Avery, L. (1993). The genetics of feeding in *Caenorhabditis elegans*. *Genetics*, *133*(4), 897–917.

Baker, D. J., Childs, B. G., Durik, M., Wijers, M. E., Sieben, C. J., Zhong, J., et al. (2016). Naturally occurring p16(Ink4a)-positive cells shorten healthy lifespan. *Nature*, *530*(7589), 184–189. https://doi.org/10.1038/nature16932.

Baker, D. J., Wijshake, T., Tchkonia, T., LeBrasseur, N. K., Childs, B. G., van de Sluis, B., et al. (2011). Clearance of p16Ink4a-positive senescent cells delays ageing-associated disorders. *Nature*, *479*(7372), 232–236. https://doi.org/10.1038/nature10600.

Baraibar, M. A., Liu, L., Ahmed, E. K., & Friguet, B. (2012). Protein oxidative damage at the crossroads of cellular senescence, aging, and age-related diseases. *Oxidative Medicine and Cellular Longevity*, *2012*, 919832. https://doi.org/10.1155/2012/919832.

Basisty, N., Kale, A., Jeon, O. H., Kuehnemann, C., Payne, T., Rao, C., et al. (2020). A proteomic atlas of senescence-associated secretomes for aging biomarker development. *PLoS Biology*, *18*(1), e3000599. https://doi.org/10.1371/journal.pbio.3000599.

Beausejour, C. M., Krtolica, A., Galimi, F., Narita, M., Lowe, S. W., Yaswen, P., et al. (2003). Reversal of human cellular senescence: Roles of the p53 and p16 pathways. *The EMBO Journal*, *22*(16), 4212–4222. https://doi.org/10.1093/emboj/cdg417.

Beevers, C. S., Zhou, H., & Huang, S. (2013). Hitting the golden TORget: Curcumin's effects on mTOR signaling. *Anti-Cancer Agents in Medicinal Chemistry*, *13*(7), 988–994. https://doi.org/10.2174/18715206113130070004.

Bent, E. H., Gilbert, L. A., & Hemann, M. T. (2016). A senescence secretory switch mediated by PI3K/AKT/mTOR activation controls chemoprotective endothelial secretory responses. *Genes & Development*, *30*(16), 1811–1821. https://doi.org/10.1101/gad.284851.116.

Bernard, M., Yang, B., Migneault, F., Turgeon, J., Dieude, M., Olivier, M. A., et al. (2020). Autophagy drives fibroblast senescence through MTORC2 regulation. *Autophagy*, *16*(11), 2004–2016. https://doi.org/10.1080/15548627.2020.1713640.

Bielak-Zmijewska, A., Grabowska, W., Ciolko, A., Bojko, A., Mosieniak, G., Bijoch, L., et al. (2019). The role of curcumin in the modulation of ageing. *International Journal of Molecular Sciences*, *20*(5). https://doi.org/10.3390/ijms20051239.

Blagosklonny, M. V. (2014). Geroconversion: Irreversible step to cellular senescence. *Cell Cycle*, *13*(23), 3628–3635. https://doi.org/10.4161/15384101.2014.985507.

Blagosklonny, M. V. (2018). Rapamycin, proliferation and geroconversion to senescence. *Cell Cycle*, *17*(24), 2655–2665. https://doi.org/10.1080/15384101.2018.1554781.

Bonner, W. M., Redon, C. E., Dickey, J. S., Nakamura, A. J., Sedelnikova, O. A., Solier, S., et al. (2008). GammaH2AX and cancer. *Nature Reviews. Cancer*, *8*(12), 957–967. https://doi.org/10.1038/nrc2523.

Bussian, T. J., Aziz, A., Meyer, C. F., Swenson, B. L., van Deursen, J. M., & Baker, D. J. (2018). Clearance of senescent glial cells prevents tau-dependent pathology and cognitive decline. *Nature*, *562*(7728), 578–582. https://doi.org/10.1038/s41586-018-0543-y.

Campisi, J., & d'Adda di Fagagna, F. (2007). Cellular senescence: When bad things happen to good cells. *Nature Reviews. Molecular Cell Biology*, *8*(9), 729–740. https://doi.org/10.1038/nrm2233.

Capparelli, C., Chiavarina, B., Whitaker-Menezes, D., Pestell, T. G., Pestell, R. G., Hulit, J., et al. (2012). CDK inhibitors (p16/p19/p21) induce senescence and autophagy in cancer-associated fibroblasts, "fueling" tumor growth via paracrine interactions, without an increase in neo-angiogenesis. *Cell Cycle*, *11*(19), 3599–3610. https://doi.org/10.4161/cc.21884.

Castro-Obregon, S. (2020). Lamin B receptor: Role on chromatin structure, cellular senescence and possibly aging. *The Biochemical Journal*, *477*(14), 2715–2720. https://doi.org/10.1042/BCJ20200165.

Chang, B. D., Broude, E. V., Dokmanovic, M., Zhu, H., Ruth, A., Xuan, Y., et al. (1999). A senescence-like phenotype distinguishes tumor cells that undergo terminal proliferation arrest after exposure to anticancer agents. *Cancer Research*, *59*(15), 3761–3767.

Chang, J., Wang, Y., Shao, L., Laberge, R. M., Demaria, M., Campisi, J., et al. (2016). Clearance of senescent cells by ABT263 rejuvenates aged hematopoietic stem cells in mice. *Nature Medicine*, *22*(1), 78–83. https://doi.org/10.1038/nm.4010.

Chatterjee, N., & Walker, G. C. (2017). Mechanisms of DNA damage, repair, and mutagenesis. *Environmental and Molecular Mutagenesis*, *58*(5), 235–263. https://doi.org/10.1002/em.22087.

Chien, Y., Scuoppo, C., Wang, X., Fang, X., Balgley, B., Bolden, J. E., et al. (2011). Control of the senescence-associated secretory phenotype by NF-kappaB promotes senescence and enhances chemosensitivity. *Genes & Development*, *25*(20), 2125–2136. https://doi.org/10.1101/gad.17276711.

Childs, B. G., Baker, D. J., Kirkland, J. L., Campisi, J., & van Deursen, J. M. (2014). Senescence and apoptosis: Dueling or complementary cell fates? *EMBO Reports*, *15*(11), 1139–1153. https://doi.org/10.15252/embr.201439245.

Chitikova, Z. V., Gordeev, S. A., Bykova, T. V., Zubova, S. G., Pospelov, V. A., & Pospelova, T. V. (2014). Sustained activation of DNA damage response in irradiated apoptosis-resistant cells induces reversible senescence associated with mTOR downregulation and expression of stem cell markers. *Cell Cycle*, *13*(9), 1424–1439. https://doi.org/10.4161/cc.28402.

Chung, C. L., Lawrence, I., Hoffman, M., Elgindi, D., Nadhan, K., Potnis, M., et al. (2019). Topical rapamycin reduces markers of senescence and aging in human skin: An exploratory, prospective, randomized trial. *Geroscience*, *41*(6), 861–869. https://doi.org/10.1007/s11357-019-00113-y.

Coppe, J. P., Patil, C. K., Rodier, F., Sun, Y., Munoz, D. P., Goldstein, J., et al. (2008). Senescence-associated secretory phenotypes reveal cell-nonautonomous functions of oncogenic RAS and the p53 tumor suppressor. *PLoS Biology*, *6*(12), 2853–2868. https://doi.org/10.1371/journal.pbio.0060301.

Courtois-Cox, S., Genther Williams, S. M., Reczek, E. E., Johnson, B. W., McGillicuddy, L. T., Johannessen, C. M., et al. (2006). A negative feedback signaling network underlies oncogene-induced senescence. *Cancer Cell*, *10*(6), 459–472. https://doi.org/10.1016/j.ccr.2006.10.003.

d'Adda di Fagagna, F. (2008). Living on a break: Cellular senescence as a DNA-damage response. *Nature Reviews. Cancer*, *8*(7), 512–522. https://doi.org/10.1038/nrc2440.

d'Adda di Fagagna, F., Reaper, P. M., Clay-Farrace, L., Fiegler, H., Carr, P., Von Zglinicki, T., et al. (2003). A DNA damage checkpoint response in telomere-initiated senescence. *Nature*, *426*(6963), 194–198. https://doi.org/10.1038/nature02118.

Dan, H. C., Cooper, M. J., Cogswell, P. C., Duncan, J. A., Ting, J. P., & Baldwin, A. S. (2008). Akt-dependent regulation of NF-{kappa}B is controlled by mTOR and Raptor in association with IKK. *Genes & Development, 22*(11), 1490–1500. https://doi.org/10.1101/gad.1662308.

Demaria, M., Ohtani, N., Youssef, S. A., Rodier, F., Toussaint, W., Mitchell, J. R., et al. (2014). An essential role for senescent cells in optimal wound healing through secretion of PDGF-AA. *Developmental Cell, 31*(6), 722–733. https://doi.org/10.1016/j.devcel.2014.11.012.

Dimri, G. P., Lee, X., Basile, G., Acosta, M., Scott, G., Roskelley, C., et al. (1995). A biomarker that identifies senescent human cells in culture and in aging skin in vivo. *Proceedings of the National Academy of Sciences of the United States of America, 92*(20), 9363–9367.

Dou, Z., Ghosh, K., Vizioli, M. G., Zhu, J., Sen, P., Wangensteen, K. J., et al. (2017). Cytoplasmic chromatin triggers inflammation in senescence and cancer. *Nature, 550*(7676), 402–406. https://doi.org/10.1038/nature24050.

Dou, Z., Ivanov, A., Adams, P. D., & Berger, S. L. (2016). Mammalian autophagy degrades nuclear constituents in response to tumorigenic stress. *Autophagy, 12*(8), 1416–1417. https://doi.org/10.1080/15548627.2015.1127465.

Dou, Z., Xu, C., Donahue, G., Shimi, T., Pan, J. A., Zhu, J., et al. (2015). Autophagy mediates degradation of nuclear lamina. *Nature, 527*(7576), 105–109. https://doi.org/10.1038/nature15548.

Filippi-Chiela, E. C., Bueno e Silva, M. M., Thome, M. P., & Lenz, G. (2015). Single-cell analysis challenges the connection between autophagy and senescence induced by DNA damage. *Autophagy, 11*(7), 1099–1113. https://doi.org/10.1080/15548627.2015.1009795.

Fingar, D. C., Richardson, C. J., Tee, A. R., Cheatham, L., Tsou, C., & Blenis, J. (2004). mTOR controls cell cycle progression through its cell growth effectors S6K1 and 4E-BP1/eukaryotic translation initiation factor 4E. *Molecular and Cellular Biology, 24*(1), 200–216. https://doi.org/10.1128/mcb.24.1.200-216.2004.

Fleury, H., Malaquin, N., Tu, V., Gilbert, S., Martinez, A., Olivier, M. A., et al. (2019). Exploiting interconnected synthetic lethal interactions between PARP inhibition and cancer cell reversible senescence. *Nature Communications, 10*(1), 2556. https://doi.org/10.1038/s41467-019-10460-1.

Fontana, L., Nehme, J., & Demaria, M. (2018). Caloric restriction and cellular senescence. *Mechanisms of Ageing and Development, 176*, 19–23. https://doi.org/10.1016/j.mad.2018.10.005.

Franceschi, C., & Campisi, J. (2014). Chronic inflammation (inflammaging) and its potential contribution to age-associated diseases. *The Journals of Gerontology. Series A, Biological Sciences and Medical Sciences, 69*(Suppl. 1), S4–S9. https://doi.org/10.1093/gerona/glu057.

Freund, A., Laberge, R. M., Demaria, M., & Campisi, J. (2012). Lamin B1 loss is a senescence-associated biomarker. *Molecular Biology of the Cell, 23*(11), 2066–2075. https://doi.org/10.1091/mbc.E11-10-0884.

Freund, A., Patil, C. K., & Campisi, J. (2011). p38MAPK is a novel DNA damage response-independent regulator of the senescence-associated secretory phenotype. *The EMBO Journal, 30*(8), 1536–1548. https://doi.org/10.1038/emboj.2011.69.

Frias, M. A., Thoreen, C. C., Jaffe, J. D., Schroder, W., Sculley, T., Carr, S. A., et al. (2006). mSin1 is necessary for Akt/PKB phosphorylation, and its isoforms define three distinct mTORC2s. *Current Biology, 16*(18), 1865–1870. https://doi.org/10.1016/j.cub.2006.08.001.

Friedman, D. B., & Johnson, T. E. (1988). A mutation in the age-1 gene in *Caenorhabditis elegans* lengthens life and reduces hermaphrodite fertility. *Genetics, 118*(1), 75–86.

Gewirtz, D. A. (2013). Autophagy and senescence: A partnership in search of definition. *Autophagy*, *9*(5), 808–812. https://doi.org/10.4161/auto.23922.

Gluck, S., Guey, B., Gulen, M. F., Wolter, K., Kang, T. W., Schmacke, N. A., et al. (2017). Innate immune sensing of cytosolic chromatin fragments through cGAS promotes senescence. *Nature Cell Biology*, *19*(9), 1061–1070. https://doi.org/10.1038/ncb3586.

Goehe, R. W., Di, X., Sharma, K., Bristol, M. L., Henderson, S. C., Valerie, K., et al. (2012). The autophagy-senescence connection in chemotherapy: Must tumor cells (self) eat before they sleep? *The Journal of Pharmacology and Experimental Therapeutics*, *343*(3), 763–778. https://doi.org/10.1124/jpet.112.197590.

Gonzalez, L. C., Ghadaouia, S., Martinez, A., & Rodier, F. (2015). Premature aging/senescence in cancer cells facing therapy: Good or bad? *Biogerontology*. https://doi.org/10.1007/s10522-015-9593-9.

Grabowska, W., Sikora, E., & Bielak-Zmijewska, A. (2017). Sirtuins, a promising target in slowing down the ageing process. *Biogerontology*, *18*(4), 447–476. https://doi.org/10.1007/s10522-017-9685-9.

Hardie, D. G. (2011). AMPK and autophagy get connected. *The EMBO Journal*, *30*(4), 634–635. https://doi.org/10.1038/emboj.2011.12.

Harding, S. M., Benci, J. L., Irianto, J., Discher, D. E., Minn, A. J., & Greenberg, R. A. (2017). Mitotic progression following DNA damage enables pattern recognition within micronuclei. *Nature*, *548*(7668), 466–470. https://doi.org/10.1038/nature23470.

Harrison, D. E., Strong, R., Sharp, Z. D., Nelson, J. F., Astle, C. M., Flurkey, K., et al. (2009). Rapamycin fed late in life extends lifespan in genetically heterogeneous mice. *Nature*, *460*(7253), 392–395. https://doi.org/10.1038/nature08221.

Hayflick, L. (1965). The limited in vitro lifetime of human diploid cell strains. *Experimental Cell Research*, *37*, 614–636. https://doi.org/10.1016/0014-4827(65)90211-9.

He, S., & Sharpless, N. E. (2017). Senescence in health and disease. *Cell*, *169*(6), 1000–1011. https://doi.org/10.1016/j.cell.2017.05.015.

Herbig, U., Ferreira, M., Condel, L., Carey, D., & Sedivy, J. M. (2006). Cellular senescence in aging primates. *Science*, *311*(5765), 1257. https://doi.org/10.1126/science.1122446.

Herbig, U., & Sedivy, J. M. (2006). Regulation of growth arrest in senescence: Telomere damage is not the end of the story. *Mechanisms of Ageing and Development*, *127*(1), 16–24. https://doi.org/10.1016/j.mad.2005.09.002.

Hernandez-Segura, A., Nehme, J., & Demaria, M. (2018). Hallmarks of cellular senescence. *Trends in Cell Biology*, *28*(6), 436–453. https://doi.org/10.1016/j.tcb.2018.02.001.

Herranz, N., Gallage, S., Mellone, M., Wuestefeld, T., Klotz, S., Hanley, C. J., et al. (2015). mTOR regulates MAPKAPK2 translation to control the senescence-associated secretory phenotype. *Nature Cell Biology*, *17*(9), 1205–1217. https://doi.org/10.1038/ncb3225.

Hinojosa, C. A., Mgbemena, V., Van Roekel, S., Austad, S. N., Miller, R. A., Bose, S., et al. (2012). Enteric-delivered rapamycin enhances resistance of aged mice to pneumococcal pneumonia through reduced cellular senescence. *Experimental Gerontology*, *47*(12), 958–965. https://doi.org/10.1016/j.exger.2012.08.013.

Hsieh, A. C., Costa, M., Zollo, O., Davis, C., Feldman, M. E., Testa, J. R., et al. (2010). Genetic dissection of the oncogenic mTOR pathway reveals druggable addiction to translational control via 4EBP-eIF4E. *Cancer Cell*, *17*(3), 249–261. https://doi.org/10.1016/j.ccr.2010.01.021.

Huang, J., & Manning, B. D. (2008). The TSC1-TSC2 complex: A molecular switchboard controlling cell growth. *The Biochemical Journal*, *412*(2), 179–190. https://doi.org/10.1042/BJ20080281.

Huang, R., Xu, Y., Wan, W., Shou, X., Qian, J., You, Z., et al. (2015). Deacetylation of nuclear LC3 drives autophagy initiation under starvation. *Molecular Cell*, *57*(3), 456–466. https://doi.org/10.1016/j.molcel.2014.12.013.

Iacovelli, S., Ricciardi, M. R., Allegretti, M., Mirabilii, S., Licchetta, R., Bergamo, P., et al. (2015). Co-targeting of Bcl-2 and mTOR pathway triggers synergistic apoptosis in BH3 mimetics resistant acute lymphoblastic leukemia. *Oncotarget*, *6*(31), 32089–32103. https://doi.org/10.18632/oncotarget.5156.

Jackson, S. P., & Bartek, J. (2009). The DNA-damage response in human biology and disease. *Nature*, *461*(7267), 1071–1078. https://doi.org/10.1038/nature08467.

Jeon, O. H., Kim, C., Laberge, R. M., Demaria, M., Rathod, S., Vasserot, A. P., et al. (2017). Local clearance of senescent cells attenuates the development of post-traumatic osteoarthritis and creates a pro-regenerative environment. *Nature Medicine*, *23*(6), 775–781. https://doi.org/10.1038/nm.4324.

Jin, S., & White, E. (2007). Role of autophagy in cancer: Management of metabolic stress. *Autophagy*, *3*(1), 28–31. https://doi.org/10.4161/auto.3269.

Julien, L. A., Carriere, A., Moreau, J., & Roux, P. P. (2010). mTORC1-activated S6K1 phosphorylates Rictor on threonine 1135 and regulates mTORC2 signaling. *Molecular and Cellular Biology*, *30*(4), 908–921. https://doi.org/10.1128/MCB.00601-09.

Kaeberlein, M., Powers, R. W., 3rd, Steffen, K. K., Westman, E. A., Hu, D., Dang, N., et al. (2005). Regulation of yeast replicative life span by TOR and Sch9 in response to nutrients. *Science*, *310*(5751), 1193–1196. https://doi.org/10.1126/science.1115535.

Kale, A., Sharma, A., Stolzing, A., Desprez, P. Y., & Campisi, J. (2020). Role of immune cells in the removal of deleterious senescent cells. *Immunity & Ageing*, *17*, 16. https://doi.org/10.1186/s12979-020-00187-9.

Kang, H. T., Lee, K. B., Kim, S. Y., Choi, H. R., & Park, S. C. (2011). Autophagy impairment induces premature senescence in primary human fibroblasts. *PLoS One*, *6*(8), e23367. https://doi.org/10.1371/journal.pone.0023367.

Kang, C., Xu, Q., Martin, T. D., Li, M. Z., Demaria, M., Aron, L., et al. (2015). The DNA damage response induces inflammation and senescence by inhibiting autophagy of GATA4. *Science*, *349*(6255). https://doi.org/10.1126/science.aaa5612, aaa5612.

Kapahi, P., Zid, B. M., Harper, T., Koslover, D., Sapin, V., & Benzer, S. (2004). Regulation of lifespan in Drosophila by modulation of genes in the TOR signaling pathway. *Current Biology*, *14*(10), 885–890. https://doi.org/10.1016/j.cub.2004.03.059.

Kenyon, C. (2011). The first long-lived mutants: Discovery of the insulin/IGF-1 pathway for ageing. *Philosophical Transactions of the Royal Society of London. Series B, Biological Sciences*, *366*(1561), 9–16. https://doi.org/10.1098/rstb.2010.0276.

Kenyon, C., Chang, J., Gensch, E., Rudner, A., & Tabtiang, R. (1993). A *C. elegans* mutant that lives twice as long as wild type. *Nature*, *366*(6454), 461–464. https://doi.org/10.1038/366461a0.

Kim, K. W., Moretti, L., Mitchell, L. R., Jung, D. K., & Lu, B. (2009). Combined Bcl-2/mammalian target of rapamycin inhibition leads to enhanced radiosensitization via induction of apoptosis and autophagy in non-small cell lung tumor xenograft model. *Clinical Cancer Research*, *15*(19), 6096–6105. https://doi.org/10.1158/1078-0432.CCR-09-0589.

Kim, D. H., Sarbassov, D. D., Ali, S. M., King, J. E., Latek, R. R., Erdjument-Bromage, H., et al. (2002). mTOR interacts with raptor to form a nutrient-sensitive complex that signals to the cell growth machinery. *Cell*, *110*(2), 163–175. https://doi.org/10.1016/s0092-8674(02)00808-5.

Kim, D. H., Sarbassov, D. D., Ali, S. M., Latek, R. R., Guntur, K. V., Erdjument-Bromage, H., et al. (2003). Gbeta L, a positive regulator of the rapamycin-sensitive pathway required for the nutrient-sensitive interaction between raptor and mTOR. *Molecular Cell*, *11*(4), 895–904. https://doi.org/10.1016/s1097-2765(03)00114-x.

Kirkland, J. L., Tchkonia, T., Zhu, Y., Niedernhofer, L. J., & Robbins, P. D. (2017). The clinical potential of senolytic drugs. *Journal of the American Geriatrics Society*. https://doi.org/10.1111/jgs.14969.

Klass, M. R. (1983). A method for the isolation of longevity mutants in the nematode *Caenorhabditis elegans* and initial results. *Mechanisms of Ageing and Development*, *22*(3-4), 279–286. https://doi.org/10.1016/0047-6374(83)90082-9.

Kongara, S., & Karantza, V. (2012). The interplay between autophagy and ROS in tumorigenesis. *Frontiers in Oncology*, *2*, 171. https://doi.org/10.3389/fonc.2012.00171.

Koo, J., Yue, P., Deng, X., Khuri, F. R., & Sun, S. Y. (2015). mTOR complex 2 stabilizes Mcl-1 protein by suppressing its glycogen synthase kinase 3-dependent and SCF-FBXW7-mediated degradation. *Molecular and Cellular Biology*, *35*(13), 2344–2355. https://doi.org/10.1128/MCB.01525-14.

Kwon, Y., Kim, J. W., Jeoung, J. A., Kim, M. S., & Kang, C. (2017). Autophagy is pro-senescence when seen in close-up, but anti-senescence in long-shot. *Molecules and Cells*, *40*(9), 607–612. https://doi.org/10.14348/molcells.2017.0151.

Laberge, R. M., Sun, Y., Orjalo, A. V., Patil, C. K., Freund, A., Zhou, L., et al. (2015). MTOR regulates the pro-tumorigenic senescence-associated secretory phenotype by promoting IL1A translation. *Nature Cell Biology*, *17*(8), 1049–1061. https://doi.org/10.1038/ncb3195.

Lafontaine, J., Cardin, G. B., Malaquin, N., Boisvert, J. S., Rodier, F., & Wong, P. (2021). Senolytic targeting of Bcl-2 anti-apoptotic family increases cell death in irradiated sarcoma cells. *Cancers (Basel)*, *13*(3), 386. https://doi.org/10.3390/cancers13030386.

Lagunas-Rangel, F. A. (2019). Current role of mammalian sirtuins in DNA repair. *DNA Repair (Amst)*, *80*, 85–92. https://doi.org/10.1016/j.dnarep.2019.06.009.

Lakowski, B., & Hekimi, S. (1998). The genetics of caloric restriction in *Caenorhabditis elegans*. *Proceedings of the National Academy of Sciences of the United States of America*, *95*(22), 13091–13096. https://doi.org/10.1073/pnas.95.22.13091.

Lamming, D. W., Ye, L., Katajisto, P., Goncalves, M. D., Saitoh, M., Stevens, D. M., et al. (2012). Rapamycin-induced insulin resistance is mediated by mTORC2 loss and uncoupled from longevity. *Science*, *335*(6076), 1638–1643. https://doi.org/10.1126/science.1215135.

Laplante, M., & Sabatini, D. M. (2012). mTOR signaling in growth control and disease. *Cell*, *149*(2), 274–293. https://doi.org/10.1016/j.cell.2012.03.017.

Lee, I. H., Cao, L., Mostoslavsky, R., Lombard, D. B., Liu, J., Bruns, N. E., et al. (2008). A role for the NAD-dependent deacetylase Sirt1 in the regulation of autophagy. *Proceedings of the National Academy of Sciences of the United States of America*, *105*(9), 3374–3379. https://doi.org/10.1073/pnas.0712145105.

Leonov, A., Feldman, R., Piano, A., Arlia-Ciommo, A., Lutchman, V., Ahmadi, M., et al. (2017). Caloric restriction extends yeast chronological lifespan via a mechanism linking cellular aging to cell cycle regulation, maintenance of a quiescent state, entry into a non-quiescent state and survival in the non-quiescent state. *Oncotarget*, *8*(41), 69328–69350. https://doi.org/10.18632/oncotarget.20614.

Lesniewski, L. A., Seals, D. R., Walker, A. E., Henson, G. D., Blimline, M. W., Trott, D. W., et al. (2017). Dietary rapamycin supplementation reverses age-related vascular dysfunction and oxidative stress, while modulating nutrient-sensing, cell cycle, and senescence pathways. *Aging Cell*, *16*(1), 17–26. https://doi.org/10.1111/acel.12524.

Li, X., Li, B., Ni, Z., Zhou, P., Wang, B., He, J., et al. (2017). Metformin synergizes with BCL-XL/BCL-2 inhibitor ABT-263 to induce apoptosis specifically in p53-defective cancer cells. *Molecular Cancer Therapeutics*, *16*(9), 1806–1818. https://doi.org/10.1158/1535-7163.MCT-16-0763.

Li, H., Liu, L., Chang, H., Zou, Z., & Xing, D. (2018). Downregulation of MCL-1 and upregulation of PUMA using mTOR inhibitors enhance antitumor efficacy of BH3 mimetics in triple-negative breast cancer. *Cell Death & Disease*, *9*(2), 137. https://doi.org/10.1038/s41419-017-0169-2.

Liu, H., He, Z., & Simon, H. U. (2014). Autophagy suppresses melanoma tumorigenesis by inducing senescence. *Autophagy, 10*(2), 372–373. https://doi.org/10.4161/auto.27163.

Liu, H., He, Z., von Rutte, T., Yousefi, S., Hunger, R. E., & Simon, H. U. (2013). Down-regulation of autophagy-related protein 5 (ATG5) contributes to the pathogenesis of early-stage cutaneous melanoma. *Science Translational Medicine, 5*(202). https://doi.org/10.1126/scitranslmed.3005864, 202ra123.

Liu, G. Y., & Sabatini, D. M. (2020). mTOR at the nexus of nutrition, growth, ageing and disease. *Nature Reviews. Molecular Cell Biology, 21*(4), 183–203. https://doi.org/10.1038/s41580-019-0199-y.

Lopez-Otin, C., Blasco, M. A., Partridge, L., Serrano, M., & Kroemer, G. (2013). The hallmarks of aging. *Cell, 153*(6), 1194–1217. https://doi.org/10.1016/j.cell.2013.05.039.

Lowe, S. W., Cepero, E., & Evan, G. (2004). Intrinsic tumour suppression. *Nature, 432*(7015), 307–315. https://doi.org/10.1038/nature03098.

Malaquin, N., Carrier-Leclerc, A., Dessureault, M., & Rodier, F. (2015). DDR-mediated crosstalk between DNA-damaged cells and their microenvironment. *Frontiers in Genetics, 6*, 94. https://doi.org/10.3389/fgene.2015.00094.

Malaquin, N., Martinez, A., & Rodier, F. (2016). Keeping the senescence secretome under control: Molecular reins on the senescence-associated secretory phenotype. *Experimental Gerontology, 82*, 39–49. https://doi.org/10.1016/j.exger.2016.05.010.

Malaquin, N., Olivier, M. A., Martinez, A., Nadeau, S., Sawchyn, C., Coppe, J. P., et al. (2020). Non-canonical ATM/MRN activities temporally define the senescence secretory program. *EMBO Reports*, e50718. https://doi.org/10.15252/embr.202050718.

Malaquin, N., Vancayseele, A., Gilbert, S., Antenor-Habazac, L., Olivier, M. A., Ait Ali Brahem, Z., et al. (2020). DNA damage- but not enzalutamide-induced senescence in prostate cancer promotes senolytic Bcl-xL inhibitor sensitivity. *Cell, 9*(7), 1593. https://doi.org/10.3390/cells9071593.

Manning, B. D., & Toker, A. (2017). AKT/PKB signaling: Navigating the network. *Cell, 169*(3), 381–405. https://doi.org/10.1016/j.cell.2017.04.001.

Mathew, R., Karp, C. M., Beaudoin, B., Vuong, N., Chen, G., Chen, H. Y., et al. (2009). Autophagy suppresses tumorigenesis through elimination of p62. *Cell, 137*(6), 1062–1075. https://doi.org/10.1016/j.cell.2009.03.048.

Mijit, M., Caracciolo, V., Melillo, A., Amicarelli, F., & Giordano, A. (2020). Role of p53 in the regulation of cellular senescence. *Biomolecules, 10*(3). https://doi.org/10.3390/biom10030420.

Mills, J. R., Hippo, Y., Robert, F., Chen, S. M., Malina, A., Lin, C. J., et al. (2008). mTORC1 promotes survival through translational control of Mcl-1. *Proceedings of the National Academy of Sciences of the United States of America, 105*(31), 10853–10858. https://doi.org/10.1073/pnas.0804821105.

Munoz-Espin, D., Canamero, M., Maraver, A., Gomez-Lopez, G., Contreras, J., Murillo-Cuesta, S., et al. (2013). Programmed cell senescence during mammalian embryonic development. *Cell, 155*(5), 1104–1118. https://doi.org/10.1016/j.cell.2013.10.019.

Munoz-Espin, D., & Serrano, M. (2014). Cellular senescence: From physiology to pathology. *Nature Reviews. Molecular Cell Biology, 15*(7), 482–496. https://doi.org/10.1038/nrm3823.

Narita, M., Nunez, S., Heard, E., Narita, M., Lin, A. W., Hearn, S. A., et al. (2003). Rb-mediated heterochromatin formation and silencing of E2F target genes during cellular senescence. *Cell, 113*(6), 703–716.

Narita, M., Young, A. R., Arakawa, S., Samarajiwa, S. A., Nakashima, T., Yoshida, S., et al. (2011). Spatial coupling of mTOR and autophagy augments secretory phenotypes. *Science, 332*(6032), 966–970. https://doi.org/10.1126/science.1205407.

Nayman, A. H., Siginc, H., Zemheri, E., Yencilek, F., Yildirim, A., & Telci, D. (2019). Dual-inhibition of mTOR and Bcl-2 enhances the anti-tumor effect of Everolimus against renal cell carcinoma in vitro and in vivo. *Journal of Cancer, 10*(6), 1466–1478. https://doi.org/10.7150/jca.29192.

Nijhawan, D., Fang, M., Traer, E., Zhong, Q., Gao, W., Du, F., et al. (2003). Elimination of Mcl-1 is required for the initiation of apoptosis following ultraviolet irradiation. *Genes & Development, 17*(12), 1475–1486. https://doi.org/10.1101/gad.1093903.

Ogrodnik, M., Salmonowicz, H., & Gladyshev, V. N. (2019). Integrating cellular senescence with the concept of damage accumulation in aging: Relevance for clearance of senescent cells. *Aging Cell, 18*(1), e12841. https://doi.org/10.1111/acel.12841.

Ohanna, M., Giuliano, S., Bonet, C., Imbert, V., Hofman, V., Zangari, J., et al. (2011). Senescent cells develop a PARP-1 and nuclear factor-{kappa}B-associated secretome (PNAS). *Genes & Development, 25*(12), 1245–1261. https://doi.org/10.1101/gad.625811.

Ovadya, Y., & Krizhanovsky, V. (2014). Senescent cells: SASPected drivers of age-related pathologies. *Biogerontology, 15*(6), 627–642. https://doi.org/10.1007/s10522-014-9529-9.

Papadopoli, D., Boulay, K., Kazak, L., Pollak, M., Mallette, F., Topisirovic, I., et al. (2019). mTOR as a central regulator of lifespan and aging. *F1000Res, 8*, 998. https://doi.org/10.12688/f1000research.17196.1. F1000 Faculty Rev.

Pazolli, E., Alspach, E., Milczarek, A., Prior, J., Piwnica-Worms, D., & Stewart, S. A. (2012). Chromatin remodeling underlies the senescence-associated secretory phenotype of tumor stromal fibroblasts that supports cancer progression. *Cancer Research, 72*(9), 2251–2261. https://doi.org/10.1158/0008-5472.CAN-11-3386.

Prata, L., Ovsyannikova, I. G., Tchkonia, T., & Kirkland, J. L. (2018). Senescent cell clearance by the immune system: Emerging therapeutic opportunities. *Seminars in Immunology, 40*, 101275. https://doi.org/10.1016/j.smim.2019.04.003.

Preuss, E., Hugle, M., Reimann, R., Schlecht, M., & Fulda, S. (2013). Pan-mammalian target of rapamycin (mTOR) inhibitor AZD8055 primes rhabdomyosarcoma cells for ABT-737-induced apoptosis by down-regulating Mcl-1 protein. *The Journal of Biological Chemistry, 288*(49), 35287–35296. https://doi.org/10.1074/jbc.M113.495986.

Pyo, I. S., Yun, S., Yoon, Y. E., Choi, J. W., & Lee, S. J. (2020). Mechanisms of aging and the preventive effects of resveratrol on age-related diseases. *Molecules, 25*(20). https://doi.org/10.3390/molecules25204649.

Qiao, Z., Xu, Z., Xiao, Q., Yang, Y., Ying, J., Xiang, L., et al. (2020). Dysfunction of ATG7-dependent autophagy dysregulates the antioxidant response and contributes to oxidative stress-induced biological impairments in human epidermal melanocytes. *Cell Death Discovery, 6*, 31. https://doi.org/10.1038/s41420-020-0266-3.

Rahmani, M., Aust, M. M., Hawkins, E., Parker, R. E., Ross, M., Kmieciak, M., et al. (2015). Co-administration of the mTORC1/TORC2 inhibitor INK128 and the Bcl-2/Bcl-xL antagonist ABT-737 kills human myeloid leukemia cells through Mcl-1 down-regulation and AKT inactivation. *Haematologica, 100*(12), 1553–1563. https://doi.org/10.3324/haematol.2015.130351.

Rodier, F., & Campisi, J. (2011). Four faces of cellular senescence. *The Journal of Cell Biology, 192*(4), 547–556. https://doi.org/10.1083/jcb.201009094.

Rodier, F., Coppe, J. P., Patil, C. K., Hoeijmakers, W. A., Munoz, D. P., Raza, S. R., et al. (2009). Persistent DNA damage signalling triggers senescence-associated inflammatory cytokine secretion. *Nature Cell Biology, 11*(8), 973–979. https://doi.org/10.1038/ncb1909.

Rodier, F., Munoz, D. P., Teachenor, R., Chu, V., Le, O., Bhaumik, D., et al. (2011). DNA-SCARS: Distinct nuclear structures that sustain damage-induced senescence growth arrest and inflammatory cytokine secretion. *Journal of Cell Science, 124*(Pt. 1), 68–81. https://doi.org/10.1242/jcs.071340.

Rogakou, E. P., Pilch, D. R., Orr, A. H., Ivanova, V. S., & Bonner, W. M. (1998). DNA double-stranded breaks induce histone H2AX phosphorylation on serine 139. *The Journal of Biological Chemistry*, *273*(10), 5858–5868.

Rovillain, E., Mansfield, L., Caetano, C., Alvarez-Fernandez, M., Caballero, O. L., Medema, R. H., et al. (2011). Activation of nuclear factor-kappa B signalling promotes cellular senescence. *Oncogene*, *30*(20), 2356–2366. https://doi.org/10.1038/onc.2010.611.

Saleh, T., Bloukh, S., Carpenter, V. J., Alwohoush, E., Bakeer, J., Darwish, S., et al. (2020). Therapy-induced senescence: An "Old" friend becomes the enemy. *Cancers (Basel)*, *12*(4). https://doi.org/10.3390/cancers12040822.

Saleh, T., Carpenter, V. J., Tyutyunyk-Massey, L., Murray, G., Leverson, J. D., Souers, A. J., et al. (2020). Clearance of therapy-induced senescent tumor cells by the senolytic ABT-263 via interference with BCL-XL -BAX interaction. *Molecular Oncology*, *14*(10), 2504–2519. https://doi.org/10.1002/1878-0261.12761.

Saleh, T., Tyutyunyk-Massey, L., Patel, N. H., Cudjoe, E. K., Jr., Alotaibi, M., & Gewirtz, D. A. (2020). Studies of non-protective autophagy provide evidence that recovery from therapy-induced senescence is independent of early autophagy. *International Journal of Molecular Sciences*, *21*(4). https://doi.org/10.3390/ijms21041427.

Salminen, A., Kaarniranta, K., & Kauppinen, A. (2013). Crosstalk between oxidative stress and SIRT1: Impact on the aging process. *International Journal of Molecular Sciences*, *14*(2), 3834–3859. https://doi.org/10.3390/ijms14023834.

Sancak, Y., Bar-Peled, L., Zoncu, R., Markhard, A. L., Nada, S., & Sabatini, D. M. (2010). Ragulator-Rag complex targets mTORC1 to the lysosomal surface and is necessary for its activation by amino acids. *Cell*, *141*(2), 290–303. https://doi.org/10.1016/j.cell.2010.02.024.

Sarbassov, D. D., Ali, S. M., Kim, D. H., Guertin, D. A., Latek, R. R., Erdjument-Bromage, H., et al. (2004). Rictor, a novel binding partner of mTOR, defines a rapamycin-insensitive and raptor-independent pathway that regulates the cytoskeleton. *Current Biology*, *14*(14), 1296–1302. https://doi.org/10.1016/j.cub.2004.06.054.

Sarbassov, D. D., Ali, S. M., Sengupta, S., Sheen, J. H., Hsu, P. P., Bagley, A. F., et al. (2006). Prolonged rapamycin treatment inhibits mTORC2 assembly and Akt/PKB. *Molecular Cell*, *22*(2), 159–168. https://doi.org/10.1016/j.molcel.2006.03.029.

Sharma, K., Goehe, R. W., Di, X., Hicks, M. A., 2nd, Torti, S. V., Torti, F. M., et al. (2014). A novel cytostatic form of autophagy in sensitization of non-small cell lung cancer cells to radiation by vitamin D and the vitamin D analog, EB 1089. *Autophagy*, *10*(12), 2346–2361. https://doi.org/10.4161/15548627.2014.993283.

Sohal, R. S., & Weindruch, R. (1996). Oxidative stress, caloric restriction, and aging. *Science*, *273*(5271), 59–63. https://doi.org/10.1126/science.273.5271.59.

Storer, M., Mas, A., Robert-Moreno, A., Pecoraro, M., Ortells, M. C., Di Giacomo, V., et al. (2013). Senescence is a developmental mechanism that contributes to embryonic growth and patterning. *Cell*, *155*(5), 1119–1130. https://doi.org/10.1016/j.cell.2013.10.041.

Tan, J. X., & Finkel, T. (2020). Autophagy goes nuclear. *Nature Cell Biology*, *22*(10), 1159–1161. https://doi.org/10.1038/s41556-020-00587-5.

van Delft, M. F., Wei, A. H., Mason, K. D., Vandenberg, C. J., Chen, L., Czabotar, P. E., et al. (2006). The BH3 mimetic ABT-737 targets selective Bcl-2 proteins and efficiently induces apoptosis via Bak/Bax if Mcl-1 is neutralized. *Cancer Cell*, *10*(5), 389–399. https://doi.org/10.1016/j.ccr.2006.08.027.

van Deursen, J. M. (2014). The role of senescent cells in ageing. *Nature*, *509*(7501), 439–446. https://doi.org/10.1038/nature13193.

Vellai, T., Takacs-Vellai, K., Zhang, Y., Kovacs, A. L., Orosz, L., & Muller, F. (2003). Genetics: Influence of TOR kinase on lifespan in *C. elegans*. *Nature*, *426*(6967), 620. https://doi.org/10.1038/426620a.

Venkatesan, S., Hoogstraat, M., Caljouw, E., Pierson, T., Spoor, J. K., Zeneyedpour, L., et al. (2016). TP53 mutated glioblastoma stem-like cell cultures are sensitive to dual mTORC1/2 inhibition while resistance in TP53 wild type cultures can be overcome by combined inhibition of mTORC1/2 and Bcl-2. *Oncotarget*, *36*(7), 58435–58444. https://doi.org/10.18632/oncotarget.11205.

Walters, H. E., Deneka-Hannemann, S., & Cox, L. S. (2016). Reversal of phenotypes of cellular senescence by pan-mTOR inhibition. *Aging (Albany NY)*, *8*(2), 231–244. https://doi.org/10.18632/aging.100872.

Wang, E. (1995). Senescent human fibroblasts resist programmed cell death, and failure to suppress bcl2 is involved. *Cancer Research*, *55*(11), 2284–2292.

Wang, Y., Chang, J., Liu, X., Zhang, X., Zhang, S., Zhang, X., et al. (2016). Discovery of piperlongumine as a potential novel lead for the development of senolytic agents. *Aging (Albany NY)*, *8*(11), 2915–2926. https://doi.org/10.18632/aging.101100.

Wang, C., Jurk, D., Maddick, M., Nelson, G., Martin-Ruiz, C., & von Zglinicki, T. (2009). DNA damage response and cellular senescence in tissues of aging mice. *Aging Cell*, *8*(3), 311–323. https://doi.org/10.1111/j.1474-9726.2009.00481.x.

Wee, P., & Wang, Z. (2017). Epidermal growth factor receptor cell proliferation signaling pathways. *Cancers (Basel)*, *9*(5), 52. https://doi.org/10.3390/cancers9050052.

Weichhart, T. (2018). mTOR as regulator of lifespan, aging, and cellular senescence: A mini-review. *Gerontology*, *64*(2), 127–134. https://doi.org/10.1159/000484629.

Williams, M. M., Elion, D. L., Rahman, B., Hicks, D. J., Sanchez, V., & Cook, R. S. (2019). Therapeutic inhibition of Mcl-1 blocks cell survival in estrogen receptor-positive breast cancers. *Oncotarget*, *10*(52), 5389–5402. https://doi.org/10.18632/oncotarget.27070.

Xu, Z., Han, X., Ou, D., Liu, T., Li, Z., Jiang, G., et al. (2020). Targeting PI3K/AKT/mTOR-mediated autophagy for tumor therapy. *Applied Microbiology and Biotechnology*, *104*(2), 575–587. https://doi.org/10.1007/s00253-019-10257-8.

Yang, T., Buchan, H. L., Townsend, K. J., & Craig, R. W. (1996). MCL-1, a member of the BLC-2 family, is induced rapidly in response to signals for cell differentiation or death, but not to signals for cell proliferation. *Journal of Cellular Physiology*, *166*(3), 523–536. https://doi.org/10.1002/(SICI)1097-4652(199603)166:3<523::AID-JCP7>3.0.CO;2-R.

Yang, H., Wang, H., Ren, J., Chen, Q., & Chen, Z. J. (2017). cGAS is essential for cellular senescence. *Proceedings of the National Academy of Sciences of the United States of America*, *114*(23), E4612–E4620. https://doi.org/10.1073/pnas.1705499114.

Young, A. R., Narita, M., Ferreira, M., Kirschner, K., Sadaie, M., Darot, J. F., et al. (2009). Autophagy mediates the mitotic senescence transition. *Genes & Development*, *23*(7), 798–803. https://doi.org/10.1101/gad.519709.

Yousefzadeh, M. J., Zhu, Y., McGowan, S. J., Angelini, L., Fuhrmann-Stroissnigg, H., Xu, M., et al. (2018). Fisetin is a senotherapeutic that extends health and lifespan. *eBioMedicine*, *36*, 18–28. https://doi.org/10.1016/j.ebiom.2018.09.015.

Zhang, B., Fu, D., Xu, Q., Cong, X., Wu, C., Zhong, X., et al. (2018). The senescence-associated secretory phenotype is potentiated by feedforward regulatory mechanisms involving Zscan4 and TAK1. *Nature Communications*, *9*(1), 1723. https://doi.org/10.1038/s41467-018-04010-4.

Zhu, Y., Doornebal, E. J., Pirtskhalava, T., Giorgadze, N., Wentworth, M., Fuhrmann-Stroissnigg, H., et al. (2017). New agents that target senescent cells: The flavone, fisetin, and the BCL-XL inhibitors, A1331852 and A1155463. *Aging (Albany NY)*, *9*(3), 955–963. https://doi.org/10.18632/aging.101202.

Zhu, Y., Tchkonia, T., Fuhrmann-Stroissnigg, H., Dai, H. M., Ling, Y. Y., Stout, M. B., et al. (2015). Identification of a novel senolytic agent, navitoclax, targeting the Bcl-2 family of anti-apoptotic factors. *Aging Cell, 15*(3), 428–435. https://doi.org/10.1111/acel.12445.

Zhu, Y., Tchkonia, T., Pirtskhalava, T., Gower, A. C., Ding, H., Giorgadze, N., et al. (2015). The Achilles' heel of senescent cells: From transcriptome to senolytic drugs. *Aging Cell, 14*(4), 644–658. https://doi.org/10.1111/acel.12344.

Zinzalla, V., Stracka, D., Oppliger, W., & Hall, M. N. (2011). Activation of mTORC2 by association with the ribosome. *Cell, 144*(5), 757–768. https://doi.org/10.1016/j.cell.2011.02.014.

Kolibri 360 GmbH
Ferdinand-Jühlke-Straße 7
99095 Erfurt, Deutschland
produktsicherheit@kolibri360.de